Photosynthetic Protein Complexes

Edited by
Petra Fromme

Related Titles

Collings, A. F., Critchley, C. (eds.)

Artificial Photosynthesis

From Basic Biology to Industrial Application

2005

ISBN: 978-3-527-31090-6

Kahl, G., Meksem, K. (eds.)

The Handbook of Plant Functional Genomics

Concepts and Protocols

2008

ISBN: 978-3-527-31885-8

Meksem, K., Kahl, G. (eds.)

The Handbook of Plant Genome Mapping

Genetic and Physical Mapping

2005

ISBN: 978-3-527-31116-3

Roberts, K. (ed.)

Handbook of Plant Science

2 Volume Set

2007

ISBN: 978-0-470-05723-0

Buchner, J., Kiefhaber, T. (eds.)

Protein Folding Handbook

2005

ISBN: 978-3-527-30784-5

Tamm, L. K. (ed.)

Protein-Lipid Interactions

From Membrane Domains to Cellular Networks

2005

ISBN: 978-3-527-31151-4

Photosynthetic Protein Complexes

A Structural Approach

Edited by
Petra Fromme

WILEY-VCH Verlag GmbH & Co. KGaA

The Editor

Prof. Dr. Petra Fromme
Department of Chemistry and Biochemistry
Arizona State University
PO Box 871604
Tempe, Arizona 85287-1604
USA

Library of Congress Card No.: applied for

British Library Cataloguing-in-Publication Data
A catalogue record for this book is available from the British Library.

Bibliographic information published by the Deutsche Nationalbibliothek
Die Deutsche Nationalbibliothek lists this publication in the Deutsche Nationalbibliografie; detailed bibliographic data are available on the Internet at <http://dnb.d-nb.de>.

Composition SNP Best-set Typesetter Ltd., Hong Kong
Printing betz-druck GmbH, Darmstadt
Bookbinding Litges & Dopf GmbH, Heppenheim

Printed in the Federal Republic of Germany
Printed on acid-free paper

ISBN: 978-3-527-31730-1

Dedication

Horst Tobias Witt
1922–2007

This book is dedicated to Horst Tobias Witt, whose heart beat for Photosynthesis, until his last breath. He was one of the leading figures of Photosynthesis and he is greatly missed by all of us.

Horst Tobias Witt devoted his entire scientific career to unraveling the secrets of Photosynthesis. Early on in his scientific career, he was fascinated by energy conversion, and by the ability of plants to split water. He once even told me the story of how it had very nearly barred him from getting his Ph.D.! While he was

Photo of Horst Tobias Witt published in *Photosynthesis Research*, **96**, 5–8. With kind permission from Springer.

Photosynthetic Protein Complexes: A Structural Approach. Edited by P. Fromme

ISBN: 978-3-527-31730-1

working in the physics department at the Georg-August-University of Göttingen, he secretly grew algae in his drawer to study the process of photosynthesis and water splitting. One weekend, there was a leak in the system – which led to a spill of green algae solution flowing out under the doorway of the lab. This was discovered by his Ph.D. advisor Professor Dr. Robert Pohl. He confronted Horst with two options: either concentrate his efforts on his assigned Ph.D. topic, in solid state physics or change his focus to Photosynthesis – but with the requirement of solving the mechanism of water splitting in his Ph.D. thesis. Witt, however, was very smart – and, knowing that the latter would be a lifetime project, decided to graduate as soon as possible by placing his effort on the given topic. He would then focus the rest of his life on the goal of unraveling the secrets of photosynthesis.

After finishing his Dissertation in the field of solid state physics, he joined the Max Planck Institute of Physical Chemistry in Göttingen, Germany, in 1950. There, he worked with Manfred Eigen and Theodor Förster and started to study the kinetics of the photoreactions in Photosynthesis using flash photometry to identify the major redox cofactors of the electron transport chain by their spectral properties. He moved to the University of Marburg in 1955, where he and his coworkers were key players in the discovery of two separate light reactions – a major breakthrough in the understanding of Photosynthesis – which were discovered at the same time and independently by three groups: Witt's group, the group of Bessel Kok and the group of Lou Dysens.

In 1962, he accepted the position of Director at the Max Volmer Institute, at the Technical University Berlin, where he changed its image and research focus from a physical-chemical one to one centered on the field of Biophysical Chemistry. This research institute subsequently became one of the major research institutions in the field of photosynthesis.

He accepted the position in Berlin one year after the wall was built and the transition from Marburg to Berlin was difficult for his wife Dr. Ingrid Witt and their three children, Roland, Carola and Ingrid. His family had to make many sacrifices over the years to his devotion to Photosynthesis, but this did not hinder Dr. Ingrid Witt in making the major discovery, along-side her husband, of the first crystals of Photosystem I in the late 1980s. Over the years, Witt had many offers to join other universities in Germany and around the globe, including offers to become Director of the Max Planck Institute, but he turned them all down. He did not wanted to leave the very productive research environment in Berlin, where he stayed and was active in research until his heart stopped beating on the 14th of May 2007.

The discoveries of Horst Tobias Witt and his coworkers are too numerous to list them all in this short dedication, but I want to highlight at least a few examples. It is to Witt and his coworkers that we can attribute identification of the reaction center pigment in Photosystem II as a chlorophyll a, with an absorption maximum of 680nm (P680). It was also they who identified a phylloquinone (Q_A) as the stable electron acceptor in Photosystem II – and discovered the role of the plastoquinone pool and the electrochromic effect as a consequence of the electrochromic potential – thereby providing strong experimental evidence for the chemiosmotic hypothesis of ATP synthesis by Mitchell.

HT Witt was a strong personality, and always worked with young, enthusiastic and creative people. Many of them became major key players in the field of Photosynthesis in their own right. Gernot Renger, Ulrich Siggel, Wolfgang Junge, Berd Rumberg, Wolfgang Haehnel, Peter Gräber, Eberhard Schlodder, Klaus Brettel, Matthias Rögner and Jan Dekker, just to name a few, have worked at the Max Volmer Institute and collaborated with HT Witt. They have all made major discoveries and are leading experts in the field.

By the beginning of the 1980's, most of the cofactors of the electron transport chain had been discovered; many of them by through the efforts of HT Witt and his group of collaborators. However, interpretation of the spectroscopic results was difficult without structural information on the spatial arrangement of the proteins and cofactors of the electron transport chain. HT Witt was very excited when the structure of the first membrane protein, the purple bacterial reaction center, was discovered in 1985 by the pioneering work of Hartmut Michel and Johann Deisenhofer, who crystallized the protein complex and received the Nobel award for their work together with Robert Huber. Now, Witt's dream became to crystallize both Photosystem I and II – a task which many people considered impossible, taking the much greater complexity and instability of the Photosystems into account. He first tried to crystallize Photosystem II, as this protein was his "heartblood". But his wife, Dr. Ingrid Witt, who worked with him on the crystallization project, convinced him to change gears and try isolating and crystallizing Photosystem I. They acquired the first crystals of Photosystem I in 1988. When I joined the group in 1990, I had the great pleasure of working with Ingrid Witt for three month, before she finally retired and I continued her work. The projects on structure determination of Photosystem I and II were a collaboration of our group at the Max Volmer Institute at the TU-Berlin and the group of Norbert Krauß and Wolfram Saenger at the FU-Berlin. In 1993, the first crystal structure of Photosystem I was determined at a resolution of 6 Å – and the atomic structure was finally solved at 2.5 Å in 2001. This is still the largest membrane protein that has ever been crystallized, consisting of 36 individual proteins and 381 cofactors.

At the end of the 1990, Athina Zouni joined our group as post-Doctoral fellow, to work with us on the crystallization of Photosystem II. I still remember that we packed a print-out of the diffraction pattern of the first PSII crystals in a gift box, which we gave HT Witt as a gift for his birthday on March 1, 1998. He was very excited. Taking all the experience with Photosystem I crystallization into account, it took only three years to improve the crystals and solve the first structure of water oxidizing complex of Photosystem II at a resolution of 3.8 Å, in 2001. For the first time, the location and shape of the water-oxidizing Mn cluster was discovered – and Witt's dream of so many years finally came true at the age of 79. He was now able to see, for the first time, the site of water splitting. All further structures that have been published at improved resolution are based on the same crystals from the thermophilic cyanobacterium, *TS. elongatus*, that had been discovered by HT Witt and his coworkers.

HT Witt was an elected member of the Berlin-Brandenburgische Akademie der Wissenschaften, Deutsche Akademie der Naturforscher Leopoldina Halle, Akademie der Wissenschaften zu Göttingen and Österreichischen Akademie der Wissenschaften. He has received numerous scientific awards and honors for his work including the Otto-Warburg Medal, the Peter-Mitchell Medal, the Feldberg-Prize and the Charles-F. Ketterling Prize. In 2001 he became honorary doctorate (Dr. h.c.) of the University of Göttingen and at the 4th of December 2006 he received one of the most prestigious honors of Germany: the "Bundesverdienstkreuz 1. Klasse" (Federal Cross of Merit 1st class).

HT Witt will be always remembered as a legend, and his life shows that keeping dreams and curiosity alive will allow scientists to finally unravel the secrets of one the great mysteries of Nature: Photosynthesis.

We will all keep HT Witt in our best memories. He is greatly missed by his colleagues, friends and his family.

Contents

Photosynthetic Protein Complexes: A Structural Approach. Edited by P. Fromme

ISBN: 978-3-527-31730-1

Preface

Photosynthesis is the most important biological processes on earth. It converts light energy from the sun into chemical energy, and provides a food source for all higher life on earth. All fossil fuels have been produced by the photosynthetic process. Oxygenic Photosynthesis changed the atmosphere from anoxic to oxygen-rich 2.5 billion years ago, by using water as the electron donor for the photosynthetic process. All of the oxygen in the atmosphere, which is essential for all respiratory processes, is produced by this route. The appearance and rise of abundant atmospheric oxygen has also resulted in huge changes in the geology of our planet and allowed formation of the ozone layer, which protects life on the surface of the earth from highly damaging UV radiation.

Interest in Photosynthesis goes far beyond the academic, since understanding of the structures and molecular details of the processes has huge implications for the future of mankind. Discovery of the molecular mechanisms of Photosynthesis holds the clue for solving the energy crisis, forming the basis for development of new routes towards biological energy sources.

Nature has been developing and optimizing Photosynthesis for the past 2.5 billion years. Light is captured by huge antenna systems and transferred to the photosynthetic reaction centers, which are large, nanoscale, biosolar energy converters consisting of more than 100 000 atoms each. The electrons for these events are extracted from water, which is split into oxygen and protons. Nature uses a fundamental electrical concept for the primary energy conversion process. First, the membrane is "charged", like a battery, during the event of electron and proton transfer. Then, the energy is stored in the form of chemical bonds, in the high-energy molecule ATP, as well as in the form of reduced hydrogen, as NADPH. These molecules are later used in the "dark" reactions of Photosynthesis, to build up carbohydrates and all other biomolecules in the biosphere. The primary processes in Photosynthesis drive all higher life on our planet Earth. Once we are able to understand how nature has accomplished this remarkable task, we will be better-equipped to secure the energy needs of humans through the conversion and utilization of solar energy.

The major structures of the photosynthetic complexes have only been revealed relatively recently. This is the first book to describe the structure and function of all major photosynthetic complexes on the basis of high-resolution structures. This

Photosynthetic Protein Complexes: A Structural Approach. Edited by P. Fromme

ISBN: 978-3-527-31730-1

book is also unique in that all 15 chapters are written by experts in the field, who are key players in the discovery of the structure and function of the protein complexes of Photosynthesis. The structures and functions of all of the major protein complexes that catalyze the primary events in Photosynthesis, from light capturing to electron transfer and ATP and NADPH production, are described in this book.

This book is an essential tool for comprehensive understanding of Photosynthesis, and is aimed at a very broad audience. Readers from high-school level to engineers working on bioenergy conversion, as well as experts in Photosynthesis, will enjoy reading it, with the beautiful and fascinating structures of the protein complexes shown in full color, and all color figures directly included in the text. Another very important feature is that it is designed as a teaching tool. It is accompanied by a website, at www.wiley-vch.de/publish/en/books/ISBN978-3-527-31730-1, where all figures from the book are freely accessible and can be downloaded without any password protection. The figures can be directly used for lectures and teaching in the classroom. The website is constantly updated with new animations and figures. In addition, abstracts of all the chapters are freely accessible, and individual chapters can be downloaded, using a pay-per-view option, from the publisher's website at www.interscience.wiley.com.

I want to thank all of the authors who have contributed to this book. They are very busy researchers from all over the world, on the verge of making new discoveries every day, and I am very happy that they so kindly agreed to devote so much of their busy time to write the chapters. These authors have brought to life a dream of publishing this unique and exciting book about one of the major discoveries in science – the unraveling of the secrets of Photosynthesis, which were invented by Nature 2.5 Billion years ago.

I am sure that you, as a reader, will love this book and find it a powerful tool for research and teaching.

Read it and enjoy!!

Tempe, July 2008

Petra Fromme

List of Contributors

Noam Adir
Technion – Israel Institute of Technology
Schulich Faculty of Chemistry
Technion City
Haifa 32000
Israel

James P. Allen
Arizona State University
Department of Chemistry and Biochemistry
PO Box 871604
Tempe, AZ 85287-1604
USA

Alexey Amunts
Tel Aviv University
Department Biochemistry
Sherman Building
Tel Aviv 69978
Israel

Danas Baniulis
Purdue University
Department of Biological Sciences
915 West State St.
West Lafayette, IN 47907
USA

Robert E. Blankenship
Washington University
Departments of Chemistry
Laboratory Sciences 401B
St. Louis, MO 63130-4899
USA

Egbert J. Boekema
University of Groningen
Groningen Biomolecular Sciences and Biotechnology Institute
Department of Biophysical Chemistry
Nijenborgh 4
9747 A.G. Groningen
The Netherlands

Bettina Böttcher
European Molecular Biology Laboratory
Meyerhofstraße 1
69126 Heidelberg
Germany

R. David Britt
University of California
Chemistry Department
One Shields Avenue
Davis, CA 95616
USA

Photosynthetic Protein Complexes: A Structural Approach. Edited by P. Fromme

ISBN: 978-3-527-31730-1

Wenrui Chang
Chinese Academy of Sciences
National Laboratory of Biomacromolecules
Institute of Biophysics
15 Datun Road
Chaoyang District
Beijing 100101
People's Republic of China

Richard J. Cogdell
University of Glasgow
Institute of Biomedical and Life Sciences
Division of Biochemistry and Molecular Biology
Glasgow G12 8QQ
Scotland
UK

William A. Cramer
Purdue University
Department of Biological Sciences
915 West State St.
West Lafayette, IN 47907
USA

Miguel A. De la Rosa
Universidad de Sevilla & CSIC
Instituto de Bioquímica Vegetal y Fotosíntesis
Américo Vespucio 49
41092 Sevilla
Spain

Jan P. Dekker
V.U. University Amsterdam
Faculty of Sciences
Division of Physics and Astronomy
De Boelelaan 1081
1081 H.V. Amsterdam
The Netherlands

Antonio Díaz-Quintana
Universidad de Sevilla & CSIC
Instituto de Bioquímica Vegetal y Fotosíntesis
Américo Vespucio 49
41092 Sevilla
Spain

Omri Drory
Tel Aviv University
Department of Biochemistry
Sherman Building
Tel Aviv 69978
Israel

Petra Fromme
Arizona State University
Department of Chemistry and Biochemistry
Main Campus Room PSC-307
PO Box 871604
Tempe, AZ 85287-1604
USA

Raimund Fromme
Arizona State University
Department of Chemistry and Biochemistry
PO Box 871604
Tempe, AZ 85287-1604
USA

Ritsuko Fujii
University of Glasgow
Institute of Biomedical and Life Sciences
Division of Biochemistry and Molecular Biology
Glasgow G12 8QQ
Scotland
UK

Mads Gabrielsen
University of Glasgow
Institute of Biomedical and Life Sciences
Division of Biochemistry and Molecular Biology
Glasgow G12 8QQ
Scotland
UK

Alastair T. Gardiner
University of Glasgow
Institute of Biomedical and Life Sciences
Division of Biochemistry and Molecular Biology
Glasgow G12 8QQ
Scotland
UK

Peter Gräber
Albert Ludwiqs Universität Freiburg
Institut für Physikalische Chemie
Albertstraße 23a
79104 Freiburg
Germany

Hideki Hashimoto
University of Glasgow
Institute of Biomedical and Life Sciences
Division of Biochemistry and Molecular Biology
Glasgow G12 8QQ
Scotland
UK

Michael P. Hendrich
Carnegie Mellon University
Department of Chemistry
Pittsburgh, PA 15213
USA

Takahiro Henmi
Osaka City University
The Graduate School of Science
Sugimoto 3-3-138, Sumiyoshi
Osaka 558-8585
Japan

Manuel Hervás
Universidad de Sevilla & CSIC
Instituto de Bioquímica Vegetal y Fotosíntesis
Américo Vespucio 49
41092 Sevilla
Spain

Martin F. Hohmann-Marriott
National Institutes of Health
National Institute of Biomedical Imaging and Bioengineering
9000 Rockville Pike
Bethesda, MD 20892
USA

Ingo Grotjohann
Arizona State University
Department of Chemistry and Biochemistry
PO Box 871604
Tempe, AZ 85287-1604
USA

Neil W. Isaacs
University of Glasgow
Institute of Biomedical and Life Sciences
Division of Biochemistry and Molecular Biology
Glasgow G12 8QQ
Scotland
UK

Nobuo Kamiya
Osaka City University
Graduate School of Science
Sugimoto 3-3-138
Sumiyoshi
Osaka 558-8585
Japan

Norbert Krauß
Queen Mary University of London
School of Biological and Chemical Sciences
Mile End Campus
London E1 4NS
UK

Zhenfeng Liu
California Institute of Technology
Howard Hughes Medical Institute
Pasadena, CA 91125
USA

Robert M. McCarrick
University of California
Chemistry Department
One Shields Avenue
Davis, CA 95616
USA

José A. Navarro
Universidad de Sevilla & CSIC
Instituto de Bioquímica Vegetal y Fotosíntesis
Américo Vespucio 49
41092 Sevilla
Spain

Nathan Nelson
Tel Aviv University
Department of Biochemistry
Sherman Building
Tel Aviv 69978
Israel

Aleksander W. Roszak
University of Glasgow
Institute of Biomedical and Life Sciences
Division of Biochemistry and Molecular Biology
Glasgow G12 8QQ
Scotland
UK

Jian-Ren Shen
Okayama University
The Graduate School of Natural Science and Technology
Division of Bio-sciences
Naka-Tsushima 3-1-1
Okayama 700-8530
Japan

June Southall
University of Glasgow
Institute of Biomedical and Life Sciences
Division of Biochemistry and Molecular Biology
Glasgow G12 8QQ
Scotland
UK

JoAnn C. Williams
Arizona State University
Department of Chemistry and Biochemistry
Tempe, AZ 85287-1604
USA

Eiki Yamashita
Osaka University
Institute of Protein Research
Toyonaka
Osaka
560-0043
Japan

Anna I. Zatsman
Carnegie Mellon University
Department of Chemistry
Pittsburgh, PA 15213
USA

Huamin Zhang
SSCI-Abtuit Inc.
West Lafayette, IN 47907
USA

Abbreviations

$\Delta\tilde{\mu}_{H}^{+}$	trans-membrane proton electrochemical potential gradient
accChl	accessory chlorophyll
AFM	atomic force microscopy
Ant	antheraxanthin
APC	allophycocyanin
BChl	bacteriochlorophyll
BIF	banded iron formation
C	carbon
Car	carotenoid
CCA	complimentary chromatic adaptation
Chl	chlorophyll
Chl_{D1} and Chl_{D2}	two accessory chlorophyll *a* molecules bound to D1 and D2 subunits, respectively
$ChlZ_{D1}$ and $ChlZ_{D2}$	two peripheral chlorophyll *a* molecules bound to D1 and D2 subunits, respectively
CL	cardiolipid
CP	chlorophyll binding protein
CP43	chlorophyll *a*-containing protein with apparent molecular mass of 43-kDa
CP47	chlorophyll *a*-containing protein with apparent molecular mass of 47-kDa
cyt	cytochrome
D1	reaction center subunits of PSII
D2	reaction center subunits of PSII
DBMIB	2,5-dibromo, 3-methyl, 6-isopropyl-benzoquinone
DGDG	digalactosyl diacylglycerol
ELIP	early light-induced protein
EM	electron microscopy
E_m	midpoint oxidation–reduction potential
EPR	electron paramagnetic resonance
ETC	electron transport (or transfer) chain
EXAFS	extended x-ray absorption fine structure
FAD	flavin adenine dinucleotide

Photosynthetic Protein Complexes: A Structural Approach. Edited by P. Fromme

ISBN: 978-3-527-31730-1

FAP	filamentous anoxygenic phototroph
Fd	ferredoxin
FeS-type	RCs that have FeS as final electron acceptors, also known as Type I RC
FMN	flavin mononucleotide
FMO	Fenna-Matthews-Olson protein
FNR	ferredoxin-$NADP^+$ reductase
GSB	green sulfur bacteria
H-, L-, M-	the 3 major integral polypeptide subunits of the *Rb. sphaeroides* purple bacterial reaction center
IsiA	iron-stress-induced protein A
k_2	second-order rate constant
K_A	equilibrium constant for complex association
k_{et}	electron transfer rate constant
K_R	equilibrium constant for complex reorganization
LHC-I	light-harvesting complex I in plants and algae
LHC-II	light-harvesting complex II in plants and algae
LH1	light-harvesting complex 1 in purple bacteria
LH2	light-harvesting complex 2 in purple bacteria
LMM	low-molecular mass
LPC	lysophosphatidylcholine
LP	linker protein
MGDG	monogalactosyldiacylglycerol
$NADP^+$	nicotinamide adenine dinucleotide phosphate
NASA	National Aeronautics and Space Administration
Neo	neoxanthin
NG	nonyl-+-D-glucoside
NMA	N-methyl asparagine
NMR	nuclear magnetic resonance
NPQ	non-photochemical quenching
NQNO	2-n-nonyl-4-hydroxyquinoline N-oxide
NRD	non-radiative dissipation
OEC	oxygen-evolving complex
PBP	phycobiliprotein
PBS	phycobilisome
Pc	plastocyanin
PC	phycocyanin
PCB	phycocyanobilin
P_{D1} and P_{D2}	PS II reaction center chlorophylls bound to D1 and D2 subunits, respectively
PDB	Protein Data Bank
PE	phycoerythrin
PEB	phycoerythrobilin
PEC	phycoerythrocyanin
PG	phosphatidyldiacylglycerol

pheo	pheophytin
PQ	plastoquinone
PQH_2	plastoquinol
pseudo-C2	pseudo-twofold
PS	photosystem (consisting of the RC fused or associated with the core antenna domain)
PSI	photosystem I
PSII	photosystem II
PUB	phycoeurobilin
PVB	phycoviolobilin
Q_A	tightly bound quinone in photosystem II and the purple bacterial reaction center
Q_B	mobile quinone in photosystem II and the purple bacterial reaction center
qE	high-energy quenching
qI	photoinhibition
QM/MM	quantum mechanical/molecular mechanical modeling
Q-type	RCs that have a mobile quinone as final electron acceptor, also known as Type II RC
RC	reaction center
rf	radio frequency
r.m.s.d.	root mean squared deviation
ROS	reactive oxygen species
SH3	Src homology 3
SQ	semiquinone
SQDG	sulfoquinovosyldiacylglycerol
suIV	subunit IV
TDS	tridecyl-stigmatellin
TMH	transmembrane helix
Tyr_Z	redox-active tyrosines, D1-Tyr161 in photosystem II
Tyr_D	redox-active tyrosines, D2-Tyr160 in photosystem II
UQ	ubiquinone
UQH_2	ubiquinol
UV	ultraviolet
VDE	violaxanthin de-epoxidase
Xanc	xanthophyll-cycle
XANES	x-ray absorption near edge structure
XAS	x-ray absorption spectroscopy
Zea	zeaxanthin

1
Overview of Photosynthesis

Petra Fromme and Ingo Grotjohann

1.1
Introduction to Photosynthesis: The Main Energy Source for Our Planet

Photosynthesis is the most important biological process on earth; it converts solar energy into chemical energy There are two types of photosynthesis: anoxygenic and oxygenic. Anoxygenic bacteria are the oldest photosynthetic organisms on earth. They contain a single photoreaction center and are not able to use water as an electron source for photosynthesis. Most of these bacteria are sensitive to oxygen and can be divided into two classes based on their photoreaction center. They contain either FeS clusters as terminal electron acceptors, which classify them as type-I reaction centers, or quinones, which lead to a classification as type-II reaction centers. Chapters 12 and 13 describe the non-oxygenic photoreaction centers/photosystems and present the implications for the evolution of photosynthesis.

Oxygenic photosynthesis was developed by the ancestors of cyanobacteria about 2.5 billion years ago. All photosynthetic organisms contain a photoreaction center that captures the light from the sun to drive a transmembrane charge separation, which builds up a membrane potential ($\Delta pH/\Delta\Psi$), thereby driving the synthesis of high energy products. All photosynthetic organisms contain a large membrane integral molecular motor–the ATP synthase–which uses the $\Delta pH/\Delta\Psi$ for the synthesis of ATP. Chapter 9 describes the function of this remarkable enzyme. While anoxygenic photosynthesis was limited by the supply of electron sources, oxygenic photosynthetic bacteria were able to use water as a universal electron donor. All oxygenic photosynthetic organisms contain two photosystems. Photosystem I contains a type-I reaction center with three 4Fe4S clusters as final electron acceptors. It catalyzes the light driven electron transfer from the luminal electron carriers plastocyanine or cytochrome c_6 to ferredoxin/flavodoxin at the stromal side of the membrane. Chapter 2 describes the structure and function of cyanobacterial Photosystem I. Photosystem II is unique as it is the only enzyme on earth which is able to split water into protons and molecular oxygen by the use of visible light. It contains a type-II reaction center which catalyzes the light-driven electron

Photosynthetic Protein Complexes: A Structural Approach. Edited by P. Fromme

ISBN: 978-3-527-31730-1

transport from water to a plastoquinone. The structure and function of Photosystem II is described in Chapters 3 and 4. Photosystems I and II work in series and are functionally coupled by the cytochrome b_6f complex, described in Chapter 7. The evolution of oxygen changed the atmosphere and also the metal available in the ocean. Chapter 5 discusses our present knowledge of the mechanism of the water splitting process. While Fe was abundant in the ocean three billion years ago, the formation of oxygen led to the formation of iron-oxide which sedimented, thereby depleting the oceans of iron, one of the most important metals for early life on earth. Even nowadays iron is the main limiting element for growth of biomass in the ocean. Photosynthetic organisms developed ways to adapt their photosynthetic apparatus as a response to this limited iron supply by the formation of new supercomplexes, which make the iron containing Photosystem I more efficient (see Chapter 6) and by adaptation of their soluble electron carriers, described in Chapters 8 and 15.

Endosymbiosis led to the development of green algae about 1.5 billion years ago; land plants settled on our planet about 1 billion years ago. The structure of Photosystem I from pea (described in Chapter 3), shows that the major parts of the photosynthetic process have been conserved over 1.5 billion years of evolution.

While all photosynthetic reaction centers are derived from a common ancestor, the peripheral antenna systems have developed independently in different organisms. The diversity of peripheral antenna systems is very likely a more recent evolutionary event, brought about by the competition for light among the different photosynthetic organisms. Purple bacteria have developed a membrane integral antenna system, consisting of light-harvesting complex 1 (LH1) and light-harvesting complex 2 (LH2). Both, LH1 and LH2 form rings of antenna proteins, with a large LH1 surrounding the reaction center. The smaller LH2 rings serve as a peripheral mobile antenna complex. Chapter 14 describes the structures of the purple bacterial antenna system.

In contrast, green non-sulfur bacteria (which contain a type-II reaction center) and green sulfur bacteria (which contain a type-I reaction center) developed a membrane peripheral organelle – the chlorosome – which serves as their antenna system. A chlorosome contains assemblies of more than 1000 bacteriochlorophyll molecules, which are surrounded by a lipid monolayer in which very few proteins are embedded. It is coupled to the reaction center by a "baseplate" and the FMO protein.

Green algae and higher plants contain membrane integral antenna complexes that serve as peripheral antenna systems for Photosystems I and II: the light-harvesting complexes I and II (LHC-I and LHC-II). Despite their similar name, they show no homologies in their structural organization to the LH1 and LH2 complexes of purple bacteria.

LHC-I serves exclusively as the peripheral antenna for Photosystem I; the structure of the supercomplex of PSI with four LHC-I proteins is described in Chapter 3. The function of LHC-II is even more complex: It can move from Photosystem II to Photosystem I to balance the light capturing capacity between the photosystems in the process of state transitions. Furthermore, LHC-II may play an important role in the dissipation of excess energy under high light intensities. The structure and function of LHC-II is described in Chapter 10; Chapter 6 presents indications of how LHC-II assembles with Photosystems I and II.

The majority of the peripheral antenna systems (including the LHC-I and -II antenna proteins from green algae and plants) contain similar pigments as the core antenna of the reaction center (i.e. (bacterio)chlorophylls and carotenoids), which allows them to increase the cross section for light capturing but leaves large spectral gaps in the coverage of light capturing.

Cyanobacteria and red algae contain the most highly developed antenna systems of all photosynthetic organisms, the phycobilisomes. The phycobilisomes are unique. They are able to absorb green light by open chain tetrapyroles that are covalently attached to the protein, thereby allowing the organisms to use the full spectrum of visible light for photosynthesis. The structure of the phycobilisomes is described in detail in Chapter 11.

1.2 The Protein Complexes of Oxygenic Photosynthesis

The major light reactions of photosynthesis are catalyzed by membrane proteins, a class of proteins for which structural information is scarce. Whereas more than 45 000 structures of soluble proteins have been determined, less than 200 different membrane protein structures have been unraveled so far . The present status of the structure analysis of photosynthetic proteins is fortuitous as structures of the majority of the large membrane proteins and the soluble protein complexes that are involved in the primary process of photosynthesis are known; all are described in this book. The primary reactions of photosynthesis are thereby the only membrane-bound process where at least partial structures of all major proteins and protein complexes have been determined. Figure 1.1 shows the structures of all proteins involved in the light reactions in oxygenic photosynthesis, based mainly on the structures of the protein complexes from cyanobacteria. The electron transfer chain of oxygenic photosynthesis is shown in Figure 1.2.

1.2.1 Photosystem II

The structure of Photosystem II (see Chapter 4) has only been determined from two different but closely related thermophilic cyanobacteria: *TS elongatus* [1, 13, 14] and *TS vulcanus* [15]. Photosystem II is a dimer in cyanobacteria, green algae and plants. One momomer of Photosystem II represents one functional unit. In cyanobacteria, it consists of 20 proteins, which bind 35 chlorophylls, 11 carotenoids, 14 lipids, 2 hemes, 1 non-heme iron, and the Mn_4Ca cluster that catalyzes water oxidation. The majority of the protein subunits and cofactors are membrane integral. The core of the complex, which binds all cofactors of the electron transport chain, is formed by the subunits D1 and D2, which show strong similarities

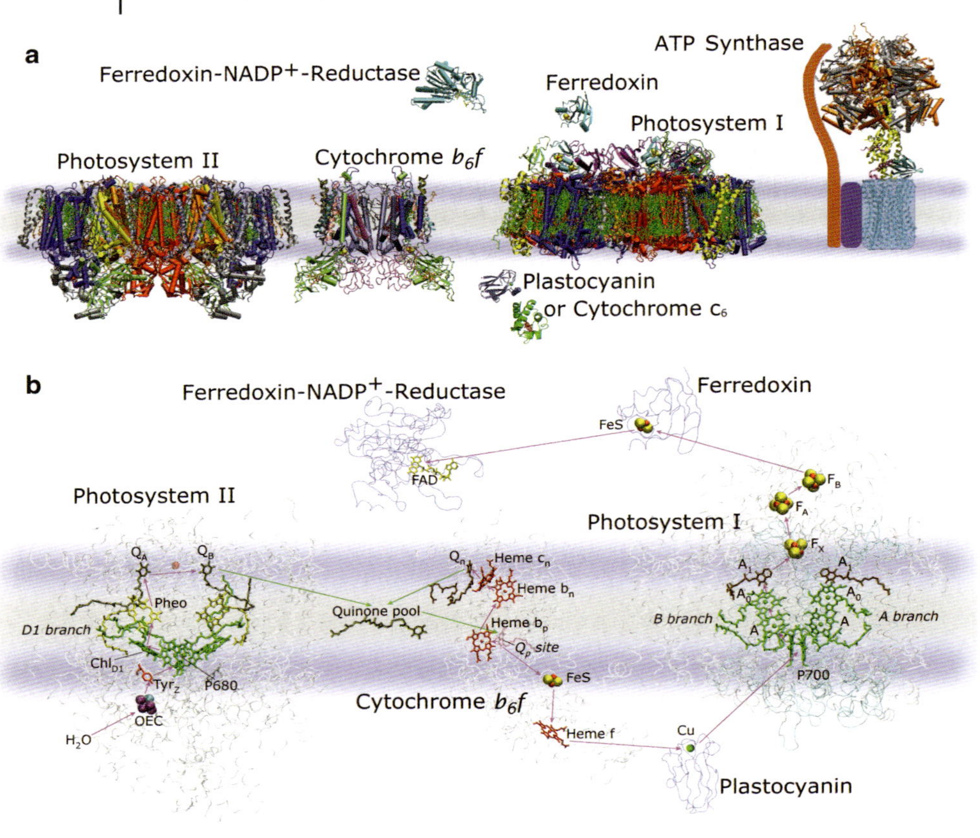

to the L and M subunits of the purple bacterial reaction center (see also Figure 1.3 and Chapter 12). The D1/D2 core is flanked by the antenna proteins CP43 and CP47, which bind the antenna chlorophylls. CP47 is located close to the dimerization domain while CP43 is located at the periphery of the PSII dimer. All small membrane intrinsic subunits are located peripheral to the PSII core subunits. Photosystem II is relatively flat on the stromal side but contains a hump that extends about 100 Å into the lumen. This luminal domain stabilizes the oxygen-evolving complex and consists of three membrane extrinsic subunits and large loops of the major core proteins of PSII: D1, D2, CP47, and CP43.

1.2.2 The Cytochrome b_6f Complex

In 2003, the structure of the cytochrome b_6f complex was solved both in the thermophilic cyanobacterium *Mastigocladus laminosus* [2, 18] and in the green algae *Chlamydomonas rheinhardtii* [19]. It is described in Chapter 7. The structures from both organisms are extremely similar which shows that the structure of the

Figure 1.1 The components of the light reactions of cyanobacterial oxygenic photosynthesis. **(a)** Structural models of the protein components of the electron transport chain and the ATP synthase in the order they appear in the electron transport chain. The proteins shown and the pdb files used are a dimer of Photosystem II from *Thermosynechococcus elongatus* (2AXT, [1], the dimer of cytochrome b_6f from *Mastigocladus laminosus* (1VF5, [2], the soluble protein plastocyanin from *Synechococcus sp. Pcc7942* (1BXV, [3], cytochrome c_6 from *Arthrospira maxima* (1F1F, [4], the trimeric complex of Photosystem I from *Thermosynechococcus elongatus* (1JB0, [5], the soluble protein ferredoxin from *Anabaena Pcc7119* (1CZP, [6], and the ferredoxin:$NADP^+$ reductase from *Anabaena Pcc7119* (1QUE, [7, 8]. The final protein, the ATP synthase, is not directly part of the electron transfer chain, but produces ATP from ADP and phosphate by using the electrochemical proton gradient generated during electron transfer. Whereas all other models have been derived from cyanobacterial sources, we used the model of the bovine mitochondrial F_1-ATPase (1H8E, [9] as placeholder for the membrane-extrinsic CF_1 part of the protein. The membrane-intrinsic F_0 part is even less well characterized, which is indicated by the cartoons. A glimpse at how the rotor ring composed of subunits c might look is given by the structure derived from a similar enzyme, the F-Type Na^+ ATPase from *Ilyobacter tartaricus* (1YCE, [10]). Color coding of cofactors: All chlorophylls are depicted in green, pheophytins in yellow, carotenoids in orange, hemes in red, lipids in brown, FeS clusters, S in yellow, Fe in red. Color coding of the individual protein subunits: Photosystem II: D1, yellow; D2 orange, CP47, red; CP 43, blue; PsbO, green; all other subunits are depicted in gray. Cytochrome b_6f complex: heme *b*, grey; Rieske FeS protein, deep purple; cytochrome *f* green, subunit IV, blue; PetG, light grey; PetM, turquoise; PetL, gold; and PetN, pink. Photosystem I: PsaA, blue; PsaB, red; PsaC, magenta; PsaD, cyan; PsaE, green; PsaF yellow; PsaL orange; all other small subunits are depicted in grey. ATP synthase: F1 subunits subunits α and subunits β, orange and grey; subunit γ, yellow; subunit ε, cyan. The ring of subunits III is represented by the structural model of the c-ring from *I. tartaricus* in a blue box. Subunit IV is depicted as a purple box. Subunits I and II that form the peripheral stalk are shown as orange cartoons. **(b)** The functional components of the electron transfer chain. Here, the same structural files as in panel A were used. The multimeric complexes of Photosystems I and II are represented by one monomer each, whereas the dimeric cytochrome b_6f is reduced to one functional half. The way of the electrons through the system can be seen by following the arrows. The magenta arrows represent direct electron transport, whereas the green arrows indicate the transport of an electron together with a proton, that is, in the form of a reduced plastoquinone (plastohydroquinone). Oxidized plastoquinones follow those arrows in the opposite direction. The blue molecule in the cytochrome b_6f complex labeled Q_p is a specific inhibitor sitting in the luminal plastoquinol binding site of the structure. For a detailed description of the processes in the electron transfer chain, see text. Photosystem II transport electrons from light-induced charge separation from the luminal (lower) side of the membrane to the stromal (upper) side, where the electrons, together with protons from the stroma, are used to reduce plastoquinone to plastoquinol. These plastoquinol exchanges with plastoquinones in the adjacent quinone pool. Photosystem II is re-reduced with electrons derived from water by the action of the oxygen-evolving complex (OEC). Plastoquinol binds to the luminal binding pocket of the cytochrome b_6f complex. One electron is cycling through the q-cycle which involves three hemes and the interaction with an (oxidized) plastoquinone in the stromal binding pocket, which can take up protons from the stroma. The other electron is given via several mediators to the soluble carrier plastocyanin, and the protons are released into the lumen. Cytochrome c_6 can replace plastocyanin in cyanobacteria. In the meanwhile, Photosystem I transports an electron by light-induced charge separation to the stromal side of the membrane, where it is used to reduce the soluble carrier ferredoxin. Ferredoxin can be replaced by flavodoxin under iron deficiency. The primary electron donor of Photosystem I is re-reduced from the luminal side by the reduced plastocyanin or cytochrome c_6. Ferredoxin delivers the electrons to the ferredoxin:$NADP^+$ reductase, where they are finally used to produce the reduction equivalent utilized in the fixation of carbon.

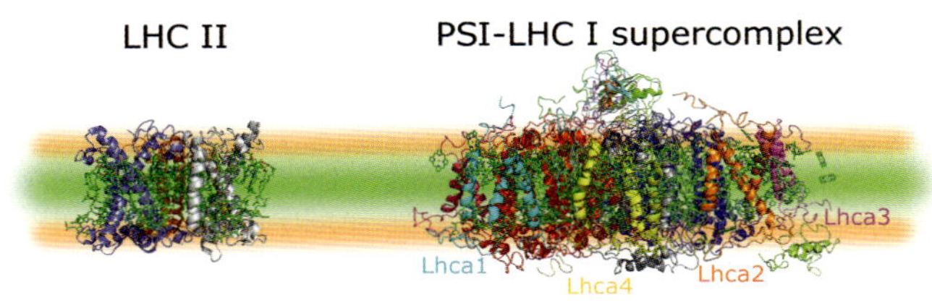

Figure 1.2 Structure of the light-harvesting complex II and the PSI–LHC-I supercomplex from higher plants. **Left**: structure of the trimeric LHC-II complex from spinach (1RWT, [11]). The individual subunits are colored blue, brown, and silver. The chlorophylls are shown in green, the carotenoids in orange. **Right**: structure of the plant PSI–LHC-I supercomplex from pea (2O01, [12]). The structure contains a monomeric photosystem that is connected to a half-moon belt of 4 LHC-I proteins. The view direction is along the membrane plane, from the side of the complex where the LHC-I belt is located. The LHC-I subunits are color-coded as follows: Lhca1, cyan; Lhca4, yellow; Lhca2, orange; and Lhca3 pink. The subunits of the reaction center core have the same color coding as shown in Figure 1.1 for the cyanobacterial PSI, except that all small membrane intrinsic subunits that are not plant specific are shown in grey. The plant specific subunits are color-coded as follows: PsaG, deep pink; PsaN, light green; PsaH, gold.

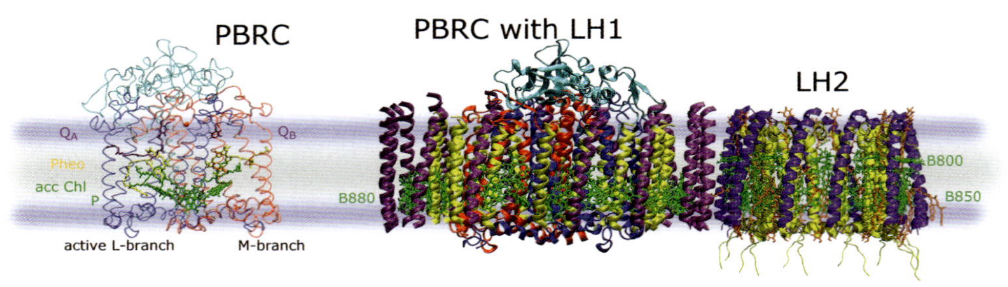

Figure 1.3 Structure of the purple bacterial reaction center with the peripheral Lh1 and Lh2 antenna rings. **Left**: Structure of the purple bacterial reaction center (PBRC). There are three proteins. The L subunit is depicted in blue, the M subunit in red, and the H subunit in cyan. Only the L-branch is involved in the first steps of the electron transfer. The cofactors of the electron transport chain consist of the special pair of bacteriochlorophylls (P) (green), the L-branch accessory chlorophylls (acc Chl) (green), the L-branch peophytin (Pheo) (yellow), and two quinones, Q_A and Q_B (purple). **Center**: Structure of the PbRC in complex with the LH1 ring from *Rps. palustris* (1PYH, [16]. The reaction center has the same color coding as on the left panel of the figure. The PbRC is surrounded by an open elipsoidal ring of 15 $\alpha\beta$ pairs of the LH1 proteins. The alpha chains are depicted in yellow and form the inner ring. The outer ring is formed by the β subunits, depicted in deep purple. The ring of LH1 proteins is broken by the helix W, which is depicted in dark yellow in the center of the picture. The bacteriochlorophyll molecules form a ring structure close to the luminal side of the membrane and are depicted as B880, as they have an absorption maximum at 880 nm. **Right**: Structure of the LH2 complex from *Rps. acidophila* (1NKZ, [17]). Nine pairs of the α and β subunits form a symmetric ring structure. The inner ring is formed by the α chain, depicted in yellow, while the outer ring consists of the β chain of the protein. Nine monomeric chlorophylls, which are oriented parallel to the membrane plane, form the B800 ring, while 18 chlorophylls, which are oriented perpendicular to the membrane plane and closely interact, form the B850 ring.

cytochrome b_6f complex has been conserved over 1.5 billion years of evolution. The cytochrome b_6f complex is a dimer. In contrast to Photosystems I and II, where the monomer is the functional unit, the dimerization of the b_6f complex is essential for the function, as there is a crossover of one essential subunit, the Rieske-iron sulfur protein which carries the 2Fe2S cluster between the two monomers. The monomer of the cytochrome b_6f complex consists of eight protein subunits. Three of them coordinate the cofactors of the electron transport chain. Cytochrome b_6 is the membrane integral core subunit of the complex and contains three heme groups (two b-type hemes and one covalently bound c-type heme) that form a major part of the electron transport chain. The Rieske-FeS-protein contains one transmembrane helix and a larger luminal extension that harbors a 2Fe2S cluster. The extrinsic subunit cytochrome f contains a c-type heme and mediates electron transfer from the cytochrome b_6f complex to plastocyanin. In addition, two quinone binding sites have been identified in the structure. A surprise was the presence of one chlorophyll and one β-carotene in the structures of the cytochrome b_6f complex. The function of the chlorophyll and the carotenoid is still under debate and is discussed in Chapter 7. The remaining five subunits of the complex are all membrane integral and stabilize the complex.

1.2.3
The Soluble Electron Transfer Proteins: Plastocyanin and Cytochrome c_6

The electron transport between the b_6f complex and Photosystem I is mediated by the soluble electron transfer proteins cytochrome c_6 and plastocyanin. The structure, function, and evolution of these electron transfer proteins are discussed in Chapter 8. Plastocyanin is the exclusive luminal electron carrier in plants, while most cyanobacteria contain genes for both cytochrome c_6 and plastocyanin, which are expressed as a function of the iron availability in the environment. Despite their functional similarity, the proteins show no structural homology. Plastocyanin consists mainly of beta-sheets and contains Cu as the redox active cofactor, while cytochrome c_6 is a typical c-type cytochrome which consists mainly of alpha helices and contains a c-type heme as the redox active cofactor.

As discussed in Chapter 8, there is strong evidence that cytochrome c_6 is the more ancient electron carrier and that plastocyanin was developed later in response to the limitation of iron and the increase of Mg availability in the ocean.

Structures of both plastocyanin and cytochrome have been determined from a wide variety of photosynthetic organisms ranging from thermophilic cyanobacteria to several higher plants (see [20–22, 48]) and references in Chapter 8).

1.2.4
Photosystem I

Photosystem I is the only photosystem, for which structures have been determined both from a thermophilic cyanobacterium [5, 23] and from a higher plant [12, 24].

In cyanobacteria, Photosystem I is a trimer with a molecular weight of 1 million Da; its structure is described in Chapter 2. It is so far the largest membrane protein for which a structure has been determined. One monomeric unit of Photosystem I consists of 12 proteins to which 127 cofactors are non-covalently bound. Photosystem I contains 96 chlorophylls, 22 carotenoids, 4 lipids, 3 4Fe4S clusters, 2 phylloquinones, and 1 Ca^{2+} ion. The core of Photosystem I is formed by the large subunits PsaA and PsaB which harbor all membrane intrinsic cofactors of the electron transport chain and also bind 79 of the 90 antenna chlorophylls. The C-terminal domain of PsaA/B surrounds the electron transfer chain like a fence and shows some structural but no sequence homologies to the core subunits D1/D2 of Photosystem II and the L and M subunits of the purple bacterial reaction center. The N-terminal antenna domain of PsaA/B shows homologies to the core antenna proteins CP43/47 of Photosystem II; this led to the suggestion that the large subunits of Photosystem I may have evolved by a gene fusion of an ancient reaction center protein with an antenna protein. The membrane intrinsic core of Photosystem I is surrounded by a field of seven small membrane intrinsic subunits that stabilize the antenna system and mediate trimerization of Photosystem I. Photosystem I is relatively flat on the luminal side, where the plastocyanin/cytochrome c_6 binding site is located at a 20 Å deep indentation of the complex (see also Chapter 8). However, Photosystem I extends the membrane by 90 Å at the stromal side where three extrinsic subunits form the docking site for the stromal electron carrier ferredoxin; under iron deficiency, flavodoxin replaces ferredoxin as the soluble electron carrier (see also Chapter 15). The smallest of the three extrinsic subunits, PsaC, carries the terminal FeS clusters that transfer the electron from Photosystem I to ferredoxin/flavodoxin.

The structure of plant Photosystem I from pea has been solved as a supercomplex with 4 LHC-I proteins bound to a monomeric Photosystem I. The structure of this supercomplex is shown in Figure 1.2 [12, 24] and is described in detail in Chapter 3. The structural comparison between plant and cyanobacterial Photosystem I revealed that the core structure of Photosystem I has been strongly conserved over 1.5 billion years of evolution. All cofactors of the electron transport chain as well as the large majority of the core antenna chlorophylls are identical in location and orientation in plants and cyanobacteria. The major differences can be found in the smaller membrane intrinsic subunits. While the acceptor side and the stromal hump of PSI are very similar in plants and cyanobacteria, the luminal docking site is more sophisticated in plants. PsaF is extended and mediates a tighter luminal docking of plastocyanin. Other major differences are found at the interface of PSI with the light-harvesting complexes and at the site of the trimerization domain in cyanobacteria.

1.2.5
Ferredoxin, Flavodoxin, and FNR

The electron is transferred from PSI to the Ferredoxin-NADP$^+$-reductase (FNR) by the soluble electron carrier ferredoxin. During iron deficiency, ferredoxin is replaced by flavodoxin. The structures of ferredoxin, flavodoxin, and FNR are described in detail in Chapter 15. Ferredoxin is a small soluble protein, consisting mainly of beta-sheets. It contains a 2Fe2S cluster as the redox active cofactor. Flavodoxin is slightly larger than ferredoxin; it consists mainly of α-helices and some β-sheets that run parallel to each other and contains a flavin as the redox active cofactor, which is exposed at the surface of the complex. Both ferredoxin and flavodoxin are one-electron carriers, so two subsequent docking events to the FNR protein must take place to allow reduction of $NADP^+ + H^+$ to NADPH by FNR. FNR has a molecular weight of 34 kDa; it contains FAD as the redox active cofactor [25]. The structure of FNR has also been solved as a complex with ferredoxin [6].

1.3 From Water to NADPH: Overview of the Electron Transfer Chain in Oxygenic Photosynthesis

We now briefly describe the path of the electrons from charge separation and water splitting in Photosystem II to the reduction of $NADP^+$ to NADPH by electrons provided by the light induced charge separation in Photosystem I.

The electrons for the electron transfer reactions are provided by Photosystem II in the process of water splitting. When the primary donor of Photosystem II, P680 is excited, leading to P680*, charge separation takes place and an electron is transferred across the membrane via the active D1-branch of the electron transport chain. With a redox potential of 1.1 V, the formed $P680^+$ is one of the strongest oxidants in nature. $P680^+$ extracts an electron from the oxygen-evolving complex, located at the luminal side of the membrane. A redox active tyrosine, Tyr_Z, serves as the intermediate electron carrier. The oxygen-evolving complex (OEC) contains a metal cluster consisting of four manganese atoms and one calcium ion. One electron is extracted from the Mn_4Ca cluster in each charge separation event, leading to the evolution of oxygen after four subsequent charge separations. During this process, four H^+ are released into the lumen and contribute to the establishment of the ΔpH across the membrane.

We want now to follow the electron on its way across the membrane to the acceptor side of Photosystem II. The acceptor side of this electron transfer chain consists of an accessory chlorophyll (Chl_{D1}), a pheophytin (Pheo), a tightly bound phylloquinone (Q_A) and the mobile phylloquinone (Q_B). After two subsequent charge separation events, Q_B is doubly reduced. Two protons access the Q_B binding pocket from the stromal side of the membrane; the reduced phylloquinol PQH_2 leaves Photosystem II through an exit channel into the membrane. It is subsequently replaced by a PQ from the plastoquinone pool.

The plastoquinol PQH_2 is hydrophobic and diffuses to the cytochrome b_6f complex, which functionally couples the light driven electron transfer events of Photosystems I and II. When PQH_2 binds to the cytochrome b_6f complex in a hydrophobic pocket close to the luminal side of the membrane, two protons are released into the lumen. The two electrons are diverted between two pathways. One electron is directly transferred to the 2Fe2S cluster that transfers the electron to cytochrome *f*, which is located at the luminal membrane extrinsic domain of the b_6f complex. Chapter 7, which discusses the structure of the cytochrome b_6f complex, refers to the lumen as the p (for positive) site of the membrane. The cytochrome *f* subunit contains a docking site for the soluble electron carriers plastocyanin or cytochrome c_6, which transfer the electron to Photosystem I.

The second electron goes through a cyclic path (the so called Q-cycle), which involves three hemes (heme b_p, heme b_n,and heme c_n) and a second quinone binding site. The two hemes b_n, heme c_n and the second quinone binding site are located closer to the stromal side (n-site) of the membrane. After the completion of the Q-cycle the second electron finally also reaches the FeS cluster and cytochrome *f* and leads to the reduction of a second molecule of plastocyanin. The Q-cycle serves two functions: it pumps additional protons across the membrane, thereby contributing to the establishment of a ΔpH, and it provides a "waiting loop" for the second electron, as the FeS cluster and cytochrome *f* can only transfer one electron at a time. In addition to its function in the linear electron transport chain, the cytochrome b_6f complex also plays a key role in the process of cyclic electron transfer, where the electrons cycle between Photosystem I and the b_6f complex. In this process, a ΔpH is established, which allows production of ATP. However, no NADPH is reduced in this process.

The electron transfer from the cytochrome b_6f complex to Photosystem I is a one-electron transfer mediated by either plastocyanin or cytochrome c_6 (see Chapter 8). Both proteins dock to cytochrome *f* in their oxidized state and dissociate from the cytochrome b_6f complex once they are reduced. They diffuse to Photosystem I where they dock at a luminal indention in close proximity to the oxidized $P700^+$. Once the electron has been transferred and used for re-reduction of $P700^+$ to P700, the oxidized plastocyanin or the cytochrome c_6 leave the binding site and move back to the cytochrome b_6f complex, ready to transfer the next electron. The binding mode of reduced plastocyanin and cytochrome c_6 varies between different organisms, from a weak diffusion controlled interaction in some cyanobacteria to the formation of a tight complex in higher plants.

Photosystem I catalyzes the light-driven electron transfer from plastocyanin/cytochrome c_6 at the luminal side of the membrane to ferredoxin/flavodoxin at the stromal side of the membrane. Once the excitation energy reaches the center of the complex and excites the primary electron donor P700, P700* is formed and charge separation takes place. An electron is ejected and transferred across the membrane by a chain of electron carriers consisting of two chlorophylls (A and A0) a phylloquinone (A1) and three 4Fe4S clusters F_X, F_A, and F_B. There are two branches of the electron transfer chain in Photosystem I; experimental evidence shows that both branches can function in electron transfer, but show differences in rates of electron transfer. The involvement of two branches in the electron transfer differs from the type-II reaction centers (PSII and the purple bacterial

RCs), where only one branch of the electron transfer chain is active. The differences between the two branches in PSI are currently a topic of research; which of the two branches is the more important is still controversial. There is evidence that the answer may even differ between different species (see Chapter 2 for a detailed discussion of this controversial topic). The symmetry between the two branches of the electron transfer chain in PSI breaks when the electron leaves the membrane integral part of PSI and reaches the two terminal iron sulfur clusters F_A and F_B, which are bound to the small subunit PsaC.

The electron is finally transferred from the terminal FeS cluster F_B to ferredoxin or flavodoxin. These proteins dock to PSI at the stromal hump and transfer the electron to FNR for the reduction of $NADP^+$ to NADPH, which is the final product of the electron transfer reactions. The structure and function of ferredoxin, flavodoxin, and FNR are described in Chapter 15. As ferredoxin and flavodoxin are one-electron carriers and two electrons are needed for the reduction of $NADP^+$ and H^+ to NADPH, two reduced ferredoxin proteins must subsequently bind to FNR and two electrons must be transferred to the redox active cofactor FAD before the product NADH is formed and released. It should also be noted that the reduction of $NADP^+$ contributes to the formation of the ΔpH across the membrane, as it lowers the proton concentration in the stroma.

The electron transfer reactions lead to the formation of an electrochemical proton gradient which is used for the synthesis of ATP from ADP and Pi (see Chapter 9 and Section 1.4). In the dark reactions, the high energy products ATP and reduced hydrogen in the form of NADPH are used to fix CO_2 and build up glucose in the Calvin cycle. Some organisms have also established active CO_2 enrichment reactions.

1.4
Coupling of Electrochemical Potential to ATP Synthesis

The electron transfer reactions build up an electrochemical potential of protons across the membrane which is used by the ATP synthase for the synthesis of ATP. The ATP synthase is a molecular motor, driven by $\Delta pH/\Delta\psi$, and synthesizes 3 ATP per full 360 degree rotation of the rotor parts of the enzyme. The ATP synthase can be found in nearly all organisms from *E. coli* to humans. The enzyme consists of a large stromal head – named F_1 – which contains the nucleotide binding sites. The F_1 head is coupled to the membrane integral proton conducting part – named F_0 – by at least two stalks.

The F_1 part of the chloroplast enzyme consists of five subunits with the stoichiometry $3\alpha3\beta\gamma\delta\varepsilon$. The three β subunits contain the three catalytic nucleotide binding sites, while the three α subunits contain regulatory nucleotide binding sites. The central stalk, consisting of the γ and ε subunits, rotates in the middle of the $3\alpha3\beta$ hexamer clockwise in the direction of the proton driven ATP synthesis and counterclockwise in the direction of ATP hydrolysis. One full rotation leads to the synthesis or hydrolysis of three ATP. The rotation of the gamma subunit changes the binding affinity of the catalytic sites from loose to tight to open. The synthesis of ATP by the ATP synthase follows the binding change mechanism, which was proposed by P. Boyer in 1989 and for which he received the Nobel Prize in 1998 [26]. The δ subunit does not participate in the rotation but forms part of the second stalk that connects the stator parts of the F1 and F0 domain of the enzyme.

The structure of the head of the ATP synthase (F1 part) from a bovine heart was solved in 1994 by John Walker and his colleagues [27]; he received the Nobel Prize for his work. The structure of the 3α3β hexamer of the CF1 part of the chloroplast enzyme has been determined [28], but no structural information on other parts of the plant enzyme is available so far from X-ray structure analysis . The structural knowledge of the intact chloroplast enzyme is entirely based on electron microscopic studies as described in Chapter 9.

The F_0 part of the chloroplast ATP synthase consists of four different subunits, historically named by roman numbers I to IV in the order of their discovery. This often leads to confusion when the F_0 part of the chloroplast enzyme (CF_0) is compared to the F_0 parts of other ATP synthases, where the letters "a" to "c" are used to name the subunits of the F_0 part. The F_0 part is a molecular motor driven by an electrochemical proton gradient. Only parts of the proton translocating membrane integral part of the ATP synthase (the F_0 part) are known to date [10, 29]. None of these proteins has been isolated from photosynthetic organisms.

The proton translocating F_0 part contains a stator, consisting of subunits I (b), II (b′), and IV (a) and a rotor consisting of a ring of subunits III (c). The subunits I (b) and II (b′) contain only one transmembrane helix and a large stromal membrane extrinsic domain that is very flexible and forms the second stalk connecting the F_0 part of the stator to the 3α3β hexamer and the δ subunit. Subunit IV (a) very likely contains two half proton translocating channels, one that delivers the protons to a glutamic/aspartic acid located in the center of the membrane at the second transmembrane helix of subunit III (c), and a second that allows the release of the protons to the stromal side after a full rotation is completed. The protonation of the individual subunits III in the ring drives the rotation of the ring of subunits III (c-ring), which in turn drives the rotation of the central stalk in the F_0 part, finally leading to the synthesis of three ATP for a full rotation of the ring of subunits III (c-ring). It is remarkable that the stoichiometry of the c-ring rotor differs between different organisms. A stoichiometry of 14 subunits III (c) has been determined for the spinach enzyme [30], while green algae may contain 15 subunits III (c) in the ring [31]. The crystal structure of a large subcomplex of the intact yeast ATP synthase that contains the F_1 part as well as the ring of subunits c [29] revealed that the yeast enzyme contains 10 subunits c, while the crystal structure of the Na^+ translocating ATP synthase from the bacterium *I. Tartaricus* showed a stoichiometry of 11 subunits c [10]. This agrees with AFM studies on the *E. coli* enzyme which also suggest 11 subunits per c-ring. A difference in stoichiometry has severe functional consequences as it changes the H^+/ATP ratio from 3.3 for *E.coli* to 5.0 for green algae.

1.5 Anoxygenic Photosynthesis

Detailed structural information on the more ancient reaction centers of non-oxygenic bacteria is only available for type-II reactions centers from purple bacteria [32, 33] (see Chapter 12). None of the ancient type-I photosystems from heliobacteria or green-sulfur bacteria have been crystallized so far, but biochemical, biophysical, and molecular biological studies shed light on the structural composition and organization of the ancient type-I reaction centers and the evolution of all photosystems as described in Chapter 13.

The structure of the purple bacterial reaction center is shown in Figure 1.3. The left side shows the structure of the reaction center; the right shows the structure of the supercomplex of the reaction center with the LH1 ring, described in Chapter 14. In addition, the LH2 ring is shown; it serves as a mobile peripheral antenna in purple bacteria.

The reaction center of purple bacteria represents a type-II reaction center. Chapter 12 describes in detail its structure, function, and evolution. The successful crystallization of the purple bacteria reaction center by Michel and Deisenhofer in 1985 [34] opened up the new research field of membrane protein structure determination. The purple bacterial photoreaction center was the first membrane protein, to have been crystallized, and led to the Nobel Prize in 1988 for their work.

The reaction center consists of three subunits: L, M, and H. Some purple bacteria (as e.g. *R. viridis)* also contain a cytochrome, bound at the outer (luminal) side of the complex. The L and M subunits form the central core of the reaction center and bind all the cofactors. The H subunit contains only one transmembrane helix and a larger cytosolic (stromal) domain that stabilizes the reaction center. The electron transport chain of the purple bacterial reaction centers is very similar to the electron transport chain in Photosystem II. It consists of a special pair of bacteriochlorophylls, P, which form the primary electron donor. When P is excited to P*, an electron is transported along the L branch of the electron transport chain; this involves an accessory chlorophyll (accChl), a pheophytin (Pheo), a tightly bound quinone (Q_A), and the mobile quinone Q_B. After two charge separation events the doubly reduced Q_B^- takes up two protons from the cytosolic (stromal) side of the membrane and leaves the binding pocket to transfer the electron to the bc complex. The bc complex shows homologies to the cytochrome b_6f complex in plants and pumps protons across the membrane. P^+ is reduced by a soluble cytochrome which shuttles electrons between the RC and the cytochrome *bc* complex. Note that no NADPH is produced in this process, but the charge separation leads to the transmembrane proton motive force that drives ATP synthesis.

Structural information on the photosystems from heliobacteria and green sulfur bacteria, which contain a type-I reaction center, is not yet available as none of these photosystems have been crystallized so far. It should be noted that, in this book, the term photosystem is used for all type-I reaction centers to reflect that they contain a reaction center and a core antenna system; this defines them as a photosystem. A description of the present knowledge of the ancient type-I photosystems is given in Chapter 13, which also contains a detailed discussion of the potential structure and function of the most ancient reaction center, the so called ur-reaction center.

The anoxygenic type-I photosystems contain homodimeric reaction centers. The core is formed by a homodimer of one large subunit (subunit A), which resembles the core subunits PsaA/B of PSI and harbors all the membrane intrinsic cofactors of the electron transport chain and core antenna chlorophylls. The electron transport chain consists of six chlorophylls, one 4Fe4S cluster, F_X, and very likely two quinones. The properties of the electron transport chain are similar to the electron transport chain in PSI, except that the role of the quinones in electron transport is still under debate [35, 36]. Recent evidence may indicate that a quinone is present and involved in the electron transfer reactions but may be only weakly bound and therefore easily lost during purification.

The type-I reaction centers also contain a stromal subunit (subunit B) that contains two 4Fe4S clusters that resemble F_A and F_B in PSI. However, this subunit is larger than subunit PsaC in oxygenic photosynthesis and is very loosely bound to the core. In addition to these two core subunits, the Photosystem from green sulfur bacteria also contains a membrane bound cytochrome subunit (subunit C).

1.6 The Antenna Systems

1.6.1 The Core Antenna Systems of the Photosystems

The first step in the photosynthetic process is the capturing of the light and the transfer of the excitation energy from the antenna complexes to the core of the photosystems – the reaction center – where charge separation takes place. In addition to the peripheral antenna systems that are flexible and vary between organisms, Photosystem I and II, as well as the photosystems from heliobacteria and green sulfur bacteria, contain a core antenna system. The core antenna is an intrinsic part of the photosystem and shows homologies between Photosystems I and II. The core antenna protein domain flanks the reaction center core on both sides. The core antenna domain of Photosystem I is fused with the reaction center domain and contains 90 chlorophylls, while there are two separate antenna proteins in Photosystem II (CP47 and CP43), containing together approximately 35 chlorophylls.

The core antenna size is much smaller in the anoxygenic type-I photosystems than in PSI. The photosystem from heliobacteria may contain about 35 chlorophylls, thereby resembling the core antenna size of PSII, while the photosystem of green sulfur bacteria has an even smaller antenna size with only about 10 chlorophylls. The cause of the reduced antenna size may be the presence of the huge chlorosome antenna, which could lead to the reduction of the core antenna size in green sulfur bacteria.

The purple bacteria reaction centers are the only photoreaction centers that do not contain a core antenna system,

In addition to the core antenna system, all photosynthetic organisms have developed peripheral antenna systems to increase the cross section for light absorption. In contrast to the reaction centers, which have all evolved from a common ancestor, nature has developed different types of antenna complexes, which is why the structure of the antenna systems differs greatly among different photosynthetic species.

1.6.2
LHC-I and LHC-II, the Antenna Systems of Higher Plants

Higher plants contain membrane integral antenna proteins, LHC-I and LHC-II. Both contain large amounts of these antenna complexes but especially LHC-II, which is the second most abundant protein on earth. In contrast to the core antenna system of PSI and PSII, which contain only chlorophyll *a*, the LHC-I and LHC-II complexes also contain chlorophyll *b*, which allows plants to absorb more light from the visible spectrum, thereby slightly narrowing the "green gap" of light that can not be used for photosynthesis.

The structure of LHC-II has been determined at atomic resolution from spinach [11] and pea [37] and is described in Chapter 10. The LHC-II mainly serves as the antenna to PSII but can also be transferred to PSI in the process of state transitions. The structure of the LHC-II complex is shown in Figure 1.2 (left). In plants and cyanobacteria large supercomplexes of Photosystems I and II with their peripheral antenna proteins are formed in the membrane; they are described in detail in Chapter 6.

In the crystals and the native membrane, LHC-II is a trimer; each monomer consists of three transmembrane helices. The structure unraveled that each monomeric LHC-II contains eight Chl *a* and six Chl *b* molecules, two lutein molecules, as well as one neoxanthin and one violaxanthin. In addition, two lipid molecules have been identified as phosphatidyl glycerol and digalactosyl diacylglycerol. All cofactors, including chlorophylls, carotenoids, and lipids, surround the polypeptide scaffolds. Most parts of the complex are deeply embedded within the membrane; the complex protrudes out of the membrane only by about 5–10 Å.

A unique feature of LHC-II is that it not only serves as an antenna complex, but also plays an important role in the dissipation of excess energy under high light intensities. The structure revealed a xanthophyll-cycle carotenoid binding site which contains a mixture of violaxanthin and lutein, indicating that the structure may represent the quenched state of LHC-II.

The movement of the LHC-II complex from PSII to PSI is regulated by an N-terminal phosphorylation site [38], which allows plants to adjust the distribution of excitation energy between the two photosystems.

The major peripheral antenna complex of Photosystem I in plants is the light-harvesting complex I (LHC-I). The structure of the plant PSI–LHC-I complex has been determined [12, 24] and is described in Chapter 3.

The plant Photosystem I–LHC-I complex is shown in Figure 1.2 (right). The supercomplex consists of a PSI monomer which binds four LHC-I proteins (Lhca1 to Lhca4). The four LHC-I proteins form a belt, surrounding one side of the PSI monomer in a half-moon shape. Each LHC-I protein consists of three transmembrane helices, which bind up to 15 chlorophyll molecules. While the transmembrane helices and the general structure of the LHC-I monomers resemble the same general fold as the LHC-II monomers, they differ significantly in the loop regions and show only moderate to low sequence homologies to the LHC-II complex. The transmembrane helices I and III form the central core of each of the LHC-I monomers and show sequence homology between different LHC proteins. Helix II shows low sequence homology between the different LHC-I proteins. For this reason, it can be used as the "signature" helix of the individual LHC-I proteins. The strongest interactions between PSI and the LHC-I antenna proteins are found between the plant specific subunit PsaG and the Lhca1 protein. The Lhca1 protein interacts with Lhca4, which is followed along the belt by Lhca2 and Lhca3. The functional coupling between the PSI core and the LHC-I belt is mediated by a number of chlorophylls located at the interface between the PSI core and the LHC-I complexes.

1.6.3
The Phycobilisomes

Cyanobacteria contain large membrane attached but extrinsic antenna complexes, the phycobilisomes. An artistic representation of the phycobilisomes is shown in Figure 1.4. The structures of all major pigment-containing proteins of the phycobilisomes have been determined [20, 41, 42] and are described in Chapter 11. It should be noted that no reports have been published on the crystallization and structure determination of the intact phycobilisome complex including the linkers, nor has the site where the phycobilisomes are attached to the photosystems been identified. However, X-ray pictures of the pigmented proteins and the biochemical as well as the biophysical evidence in combination with electron microscopy have made it possible to assemble a picture of the structural organization of the phycobilisome antenna.

The phycobilisome is probably the most efficient antenna system in nature as it allows the organism to fill in the "green gap" and use the full spectrum of visible light for photosynthesis. Phycobilisomes achieve this goal by using the unique feature of the phycobilisome pigments, which absorb strongly in the region 550–660 nm, thereby complimenting the spectral region covered by chlorophyll *a* in the blue and the red regions. The phycobilisomes mainly serve as peripheral antenna for Photosystem II, but can also move to Photosystem I in a process of state transitions. These are very fast processes that take place within a few seconds, thereby balancing the light capturing capacity in cyanobacteria between the two photosystems.

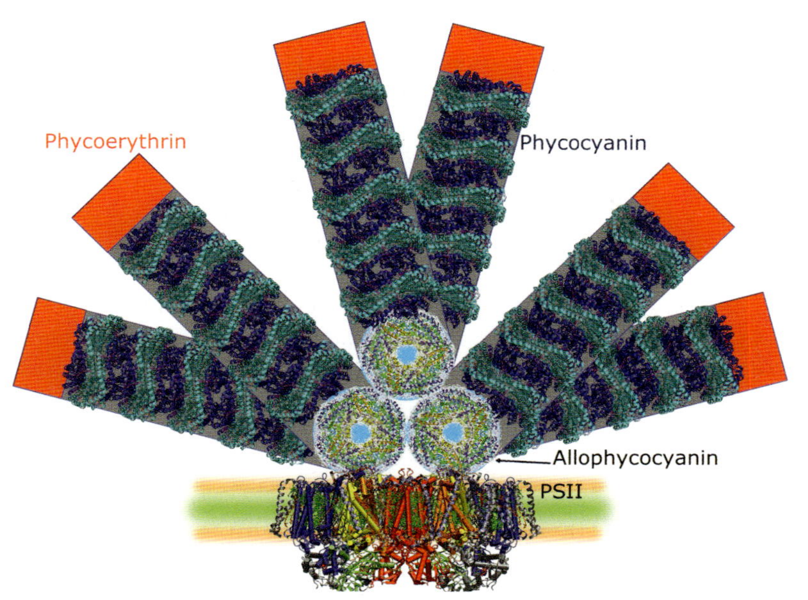

Figure 1.4 Artistic picture of the potential structural organization of the phycobilisome antenna. This picture shows the potential organization of the phycobilisome antenna system on top of the cyanobacterial Photosystem II structure. Three trimers of allophycocyanin from *Porphyra yezoensis* (1KN1, [39]) are shown at the core of the complex. The rods are depicted by the structure of phycocyanin from *Gracilaria chilensis* (2BV8, [40]). The alpha chains of phycocyanin are depicted in cyan, the beta chains in blue. Please note that this is a schematic picture, where important elements, such as the linker proteins are missing and the overall organization of the intact phycobilisome is only known at low resolution from electron microscopy. This picture is not an X-ray structure but combines high resolution X-ray structures of parts of the complex with an artist's rendition of the complex.

The structure of the entire phycobilisome can be described as a set of rod like stacks of disks that radiate from a central core of close-packed disks. The number of disks in the core and the height of the stacks is species dependent. There exists experimental evidence that allophycocyanin (APC) might be preferentially located in the core, while phycocyanin (PC) forms the inner part of the rods and phycoerythrin (PE) is located at the periphery. This arrangement leads to a funneling effect which allows energetically downhill excitation energy transfer from the highest energy pigments in PE at the periphery via PC to allophycocyanin in the central core of the phycobilisome. Figure 1.4 is an artistic overview of the proposed structural organization of the phycobilisome on top of the PSII dimer. All phycobiliproteins have the same general structural features. They consist of a heterodimer composed of two homologous subunits, α and β, which form an (α/β) heterodimer. Each subunit forms a compact globular structure that consists of six α-helices. The cofactors are covalently bound to the protein.

In addition to the core α/β structure, all phycobiliproteins contain two additional α-helices that extend out from the core and serve as the assembly interface of the monomer. Most of the phycobiliproteins form larger oligomeric assemblies. The most abundant are the (αβ)3 trimers, which form coin type structures that can come together into even larger assemblies, which can finally lead to the formation of the central core and the large rods, with the help of the linker proteins.

1.6.4
Chlorosomes

Anoxygenic photosynthetic organisms show a large variation of peripheral antenna. Both green sulfur bacteria and non-green sulfur bacteria contain chlorosomes. They can be regarded as a prokaryotic light-harvesting organelle and consist of huge assemblies of chlorophylls covered by a lipid monolayer. The chlorosomes may be functionally coupled to the reaction center by the FMO protein, which is soluble and contains seven chlorophylls. The FMO protein was the first pigment binding photosynthetic protein, for which the structure was determined [43]. Another soluble antenna protein, for which the structure has been determined, is the Peridinin-Chl-Protein, found in red algae [44].

1.6.5
LH1 and LH2

Purple bacteria contain two membrane integral antenna proteins – the LH1 and LH2 – which form ring structures. These structures are described in detail in Chapter 14.

The LH2 complex serves as a peripheral antenna for the purple bacterial reaction center. AFM images of native purple bacterial membranes have shown that the cells can increase the ratio of LH2 rings to RC cores in low light. The LH2 rings can form quasicrystalline lattices in the native membrane [45].

The structure of LH2 has been determined at atomic resolution [46] by X-ray crystallography (Figure 1.3 right shows the structures of the LH2 ring and the reaction center LH1 supercomplex). LH2 consists of two subunits, α and β, which contain a single transmembrane spanning α-helix each. The subunits form a ring, which is composed of eight or nine pairs of α,β–dimers. The stoichiometry of the ring is species dependent. The outer wall of the ring is formed by the α-helices of the β-proteins; the α-helices from the α-apoproteins form the inner wall of the ring. All of the pigments are arranged in two rings between these rings of α-helices.. The first ring of pigments, located closer to the cytoplasmic (stromal) side of the complex, consists of monomeric Bchl *a* molecules, which are oriented parallel to the plane of the membrane. The pigments absorb at 800 nm (B800 molecules). Each α,β-dimer binds one B800 Bchl *a*. The second ring consists of 16 or 18 tightly interacting bacteriochlorophyll molecules, which are oriented nearly at right angles to the membrane plane. Due to the stronger coupling between the pigments, these chlorophylls absorb light at a wavelength of 850 nm and are named B850 molecules. In addition to the chlorophylls, each α,β-dimer in the LH2 complex contains one carotenoid, which is essential for the assembly and stabilization of the LH2.

LH1 has been crystallized in the form of an elliptical open ring surrounding the reaction center [16, 47]. The structure is shown in Figure 1.3. The LH1 ring consists of 15 α,β-dimers that surround the RC. The ring is open, interrupted by a single transmembrane helix, which has been called Protein W and may be analogous to PufX. This ring opening is located opposite the Q_B binding site in the RC. This has led to the suggestion that the ring opening may be essential for the function as it would allow the reduced ubiquinol molecule to leave the complex to reach the cytochrome *bc* complex. Each α,β-dimer coordinates two bacteriochlorophyll molecules that strongly interact, leading to an absorption maximum of 880 nm (B880 molecules). The arrangement of the ring of pigments resembles the B850 ring of the LH2 complex, but the LH1 ring is interrupted. It should also be noted that the LH1 does not contain pigments that resemble the B800 ring in the LH2 complex.

The spectral distribution of the pigments is highly optimized and allows a very efficient excitation energy transfer from the periphery to the RC core. This involves the transfer of energy from the highest energy pigment B800, to the B850 ring in LH2, to the B880 ring in LH1, to the primary donor P in the core of the reaction center.

References

1 Loll, B., Kern, J., Saenger, W., Zouni, A. and Biesiadka, J. (2005) Towards complete cofactor arrangement in the 3.0 a resolution structure of photosystem II. *Nature*, **438**, 1040–4.

2 Kurisu, G., Zhang, H., Smith, J.L. and Cramer, W.A. (2003) Structure of the cytochrome b6f complex of oxygenic photosynthesis: tuning the cavity. *Science*, **302**, 1009–14.

3 Inoue, T., Sugawara, H., Hamanaka, S., Tsukui, H., Suzuki, E., Kohzuma, T. and Kai, Y. (1999) Crystal structure determinations of oxidized and reduced plastocyanin from the cyanobacterium Synechococcus sp. PCC 7942. *Biochemistry*, **38**, 6063–9.

4 Sawaya, M.R., Krogmann, D.W., Serag, A., Ho, K.K., Yeates, T.O. and Kerfeld, C.A. (2001) Structures of cytochrome c-549 and cytochrome c6 from the cyanobacterium Arthrospira maxima. *Biochemistry*, **40**, 9215–25.

5 Jordan, P., Fromme, P., Witt, H.T., Klukas, O., Saenger, W. and Krauss, N. (2001) Three-dimensional structure of cyanobacterial photosystem I at 2.5 A resolution. *Nature*, **411**, 909–17.

6 Morales, R., Charon, M.H., Hudry-Clergeon, G., Petillot, Y., Norager, S., Medina, M. and Frey, M. (1999) Refined X-ray structures of the oxidized, at 1.3 A, and reduced, at 1.17 A, [2Fe-2S] ferredoxin from the cyanobacterium Anabaena PCC7119 show redox-linked conformational changes. *Biochemistry*, **38**, 15764–73.

7 Serre, L., Vellieux, F., Medina, M., Gomez-Moreno, C., Fontecilla-Camps, J.C. and Frey, M. (1996) Crystal structures of a ferredoxin: NADP+ reductase and of a complex with NADP+. *Biochemical Society Transactions*, **24**, *10S*.

8 Serre, L., Vellieux, F.M., Medina, M., Gomez-Moreno, C., Fontecilla-Camps, J.C. and Frey, M. (1996) X-ray structure of the ferredoxin:NADP+ reductase from the

cyanobacterium Anabaena PCC 7119 at 1.8 A resolution, and crystallographic studies of NADP+ binding at 2.25 A resolution. *Journal of Molecular Biology*, **263**, 20–39.

9 Menz, R.I., Walker, J.E. and Leslie, A.G. (2001) Structure of bovine mitochondrial F(1)-ATPase with nucleotide bound to all three catalytic sites: implications for the mechanism of rotary catalysis. *Cell*, **106**, 331–41.

10 Meier, T., Polzer, P., Diederichs, K., Welte, W. and Dimroth, P. (2005) Structure of the rotor ring of F-Type Na+-ATPase from Ilyobacter tartaricus. *Science*, **308**, 659–62.

11 Liu, Z., Yan, H., Wang, K., Kuang, T., Zhang, J., Gui, L., An, X. and Chang, W. (2004) Crystal structure of spinach major light-harvesting complex at 2.72 A resolution. *Nature*, **428**, 287–92.

12 Amunts, A., Drory, O. and Nelson, N. (2007) The structure of a plant photosystem I supercomplex at 3.4 A resolution. *Nature*, **447**, 58–63.

13 Zouni, A., Witt, H.T., Kern, J., Fromme, P., Krauss, N., Saenger, W. and Orth, P. (2001) Crystal structure of photosystem II from Synechococcus elongatus at 3.8 A resolution. *Nature*, **409**, 739–43.

14 Ferreira, K.N., Iverson, T.M., Maghlaoui, K., Barber, J. and Iwata, S. (2004) Architecture of the photosynthetic oxygen-evolving center. *Science*, **303**, 1831–8.

15 Kamiya, N. and Shen, J.R. (2003) Crystal structure of oxygen-evolving photosystem II from Thermosynechococcus vulcanus at 3.7-A resolution. *Proceedings of the National Academy of Sciences of the United States of America*, **100**, 98–103.

16 Roszak, A.W., Howard, T.D., Southall, J., Gardiner, A.T., Law, C.J., Isaacs, N.W. and Cogdell, R.J. (2003) Crystal structure of the RC-LH1 core complex from Rhodopseudomonas palustris. *Science*, **302**, 1969–72.

17 Papiz, M.Z., Prince, S.M., Howard, T., Cogdell, R.J. and Isaacs, N.W. (2003) The structure and thermal motion of the B800-850 LH2 complex from Rps. acidophila at 2.0 A resolution and 100 K: new structural features and functionally relevant motions. *Journal of Molecular Biology*, **326**, 1523–38.

18 Zhang, H., Kurisu, G., Smith, J.L. and Cramer, W.A. (2003) A defined protein-detergent-lipid complex for crystallization of integral membrane proteins: the cytochrome b6f complex of oxygenic photosynthesis. *Proceedings of the National Academy of Sciences of the United States of America*, **100**, 5160–3.

19 Stroebel, D., Choquet, Y., Popot, J.L. and Picot, D. (2003) An atypical haem in the cytochrome b(6)f complex. *Nature*, **426**, 413–18.

20 Schirmer, T., Bode, W., Huber, R., Sidler, W. and Zuber, H. (1985) X-ray crystallographic structure of the light-harvesting biliprotein C-phycocyanin from the thermophilic cyanobacterium Mastigocladus laminosus and its resemblance to globin structures. *Journal of Molecular Biology*, **184**, 257–77.

21 Redinbo, M.R., Cascio, D., Choukair, M.K., Rice, D., Merchant, S. and Yeates, T.O. (1993) The 1.5-A crystal structure of plastocyanin from the green alga Chlamydomonas reinhardtii. *Biochemistry*, **32**, 105607.

22 Kerfeld, C.A., Anwar, H.P., Interrante, R., Merchant, S. and Yeates, T.O. (1995) The structure of chloroplast cytochrome c6 at 1.9 A resolution: evidence for functional oligomerization. *Journal of Molecular Biology*, **250**, 627–47.

23 Krauss, N., Schubert, W.D., Klukas, O., Fromme, P., Witt, H.T. and Saenger, W. (1996) Photosystem I at 4 Å resolution represents the first structural model of a joint photosynthetic reaction centre and core antenna system. *Nature Structural Biology*, **3**, 965–73.

24 Ben-Shem, A., Frolow, F. and Nelson, N. (2003) Crystal structure of plant photosystem I. *Nature*, **426**, 630–5.

25 Bruns, C.M. and Karplus, P.A. (1995) Refined crystal structure of spinach ferredoxin reductase at 1.7 A resolution: oxidized, reduced and 2′-phospho-5′-AMP bound states. *Journal of Molecular Biology*, **247**, 125–45.

26 Boyer, P.D. (1989) A perspective of the binding change mechanism for ATP synthesis. *The FASEB Journal*, **3**, 2164–78.

27 Abrahams, J.P., Leslie, A.G., Lutter, R. and Walker, J.E. (1994) Structure at 2.8 A resolution of F1-ATPase from bovine heart mitochondria. *Nature*, **370**, 621–8.

28 Groth, G. and Pohl, E. (2001) The structure of the chloroplast F1-ATPase at 3.2 A resolution. *The Journal of Biological Chemistry*, **276**, 1345–52.

29 Stock, D., Leslie, A.G. and Walker, J.E. (1999) Molecular architecture of the rotary motor in ATP synthase. *Science*, **286**, 1700–5.

30 Poetsch, A., Rexroth, S., Heberle, J., Link, T.A., Dencher, N.A. and Seelert, H. (2003) Characterisation of subunit III and its oligomer from spinach chloroplast ATP synthase. *Biochimica et Biophysica Acta*, **1618**, 59–66.

31 Pogoryelov, D., Yu, J., Meier, T., Vonck, J., Dimroth, P. and Muller, D.J. (2005) The c15 ring of the Spirulina platensis F-ATP synthase: F1/F0 symmetry mismatch is not obligatory. *EMBO Reports*, **6**, 1040–4.

32 Deisenhofer, J., Epp, O., Miki, K., Huber, R. and Michel, H. (1984) X-ray structure analysis of a membrane protein complex. Electron density map at 3 A resolution and a model of the chromophores of the photosynthetic reaction center from Rhodopseudomonas viridiz. *Journal of Molecular Biology*, **180**, 385–98.

33 Deisenhofer, J. and Michel, H. (1991) High-resolution structures of photosynthetic reaction centers. *Annual Review of Biophysics and Biophysical Chemistry*, **20**, 247–66.

34 Deisenhofer, J., Epp, O., Sinning, I. and Michel, H. (1995) Crystallographic refinement at 2.3 A resolution and refined model of the photosynthetic reaction centre from Rhodopseudomonas viridis. *Journal of Molecular Biology*, **246**, 429–57.

35 van der Est, A. (2001) Light-induced spin polarization in type-I photosynthetic reaction centres. *Biochimica et Biophysica Acta*, **1507**, 212–25.

36 Oh-oka, H. (2007) Type 1 reaction center of photosynthetic heliobacteria. *Photochemistry and Photobiology*, **83**, 177–86.

37 Standfuss, J., Terwisscha van Scheltinga, A.C., Lamborghini, M. and Kuhlbrandt, W. (2005) Mechanisms of photoprotection and nonphotochemical quenching in pea light-harvesting complex at 2.5 A resolution. *The EMBO Journal*, **24**, 919–28.

38 Nilsson, A., Stys, D., Drakenberg, T., Spangfort, M.D., Forsen, S. and Allen, J.F. (1997) Phosphorylation controls the three-dimensional structure of plant light-harvesting complex II. *The Journal of Biological Chemistry*, **272**, 18350–7.

39 Liu, J.Y., Jiang, T., Zhang, J.P. and Liang, D.C. (1999) Crystal structure of allophycocyanin from red algae Porphyra yezoensis at 2.2-A resolution. *The Journal of Biological Chemistry*, **274**, 16945–52.

40 Contreras-Martel, C., Matamala, A., Bruna, C., Poo-Caamano, G., Almonacid, D., Figueroa, M., Martinez-Oyanedel, J. and Bunster, M. (2007) The structure at 2 A resolution of Phycocyanin from Gracilaria chilensis and the energy transfer network in a PC-PC complex. *Biophysical Chemistry*, **125**, 388–96.

41 Ficner, R. and Huber, R. (1993) Refined crystal structure of phycoerythrin from Porphyridium cruentum at 0.23-nm resolution and localization of the gamma subunit. *European Journal of Biochemistry/FEBS*, **218**, 103–6.

42 Brejc, K., Ficner, R., Huber, R. and Steinbacher, S. (1995) Isolation, crystallization, crystal structure analysis and refinement of allophycocyanin from the cyanobacterium Spirulina platensis at 2.3 A resolution. *Journal of Molecular Biology*, **249**, 424–40.

43 Matthews, B.W., Fenna, R.E., Bolognesi, M.C., Schmid, M.F. and Olson, J.M. (1979) Structure of a bacteriochlorophyll a-protein from the green photosynthetic bacterium Prosthecochloris aestuarii. *Journal of Molecular Biology*, **131**, 259–85.

44 Hofmann, E., Wrench, P.M., Sharples, F.P., Hiller, R.G., Welte, W. and Diederichs, K. (1996) Structural basis of light-harvesting by carotenoids: peridinin-chlorophyll-protein from Amphidinium carterae. *Science*, **272**, 1788–91.

45 Sturgis, J.N. and Niederman, R.A. (2007) Atomic force microscopy reveals multiple patterns of antenna organization in purple bacteria: implications for energy transduction mechanisms and membrane modeling. *Photosynthesis Research*, **95**, 269–78.

46 Prince, S.M., Papiz, M.Z., Freer, A.A., McDermott, G., Hawthornthwaite-Lawless, A.M., Cogdell, R.J. and Isaacs, N.W. (1997) Apoprotein structure in the LH2 complex from Rhodopseudomonas acidophila strain 10050: modular assembly and protein pigment interactions. *Journal of Molecular Biology*, **268**, 412–23.

47 Law, C.J., Prince, S.M. and Cogdell, R.J. (1998) Crystallising the LH1-RC "core" complex of purple bacteria. *Biochemical Society Transactions*, **26**, *S160*.

48 Frazão, C., Soares, C.M., Carrondo, M.A., Pohl, E., Daurer, Z., Wilson, K.S., Hervás, M., Navarro, J.A., De la Rosa, M.A. and Sheldrick, G.M. (1995) Ab initio determination of the crystal structure of cytochrome c6 and comparison with plastocyanin. *Structure*, **3**, 1159–69.

2
Structure and Function of Cyanobacterial Photosystem I

Norbert Krauß

2.1
Introduction

In oxygenic photosynthesis, Photosystem I (PSI) is the protein-pigment complex that catalyzes the second step of light induced transmembrane charge separation in the series of electron transfer reactions that lead to the reduction of $NADP^+$ by water. PSI contains a large core antenna comprised of approximately 90 chlorophyll and about 22 carotenoid molecules that are excited by light. Excitation energy transfer within the core antenna leads to excitation of the primary donor P700, a pair of chlorophyll molecules located close to the lumenal side of PSI, which absorbs nearly 700 nm. P700 is part of a chain of cofactors, the electron transfer chain, which extends from the lumenal side to the stromal side of PSI and performs electron transfer through PSI across the photosynthetic membrane. Using spectroscopy, it has been shown that, in the first step of transmembrane charge separation, the strongly reducing excited state P700* donates an electron to the primary acceptor A_0, a chlorophyll *a* molecule, in about 1 ps (for a review of the kinetics of electron transfer reactions in PSI see Brettel & Leibl, 2001 [1]). To avoid re-reduction of the oxidized primary donor $P700^+$ by A_0^-, fast electron transfer from A_0^- has to occur in circa 30 ps to the secondary electron acceptor A_1, a phylloquinone molecule, from which the electron is sequentially transferred to the [4Fe4S] clusters F_X, F_A and F_B. Upon binding to the stromal side of PSI, the soluble protein ferredoxin, which has bound a redox-active [2Fe2S] cluster, accepts an electron from the reduced side of PSI, dissociates from PSI, and reduces $NADP^+$ in a reaction catalyzed by ferredoxin-$NADP^+$ reductase. In cyanobacteria and some algae, the flavin binding protein flavodoxin can replace ferredoxin under low iron conditions [2]. To complete the catalytic cycle in PSI, the oxidized primary donor $P700^+$ is re-reduced by soluble donor proteins, either plastocyanin or cytochrome c_6, which dock to PSI at the lumenal side. In plant chloroplasts, the copper protein plastocyanin is the only donor protein, whereas it is only the heme protein cytochrome c_6 in some cyanobacteria (see also Chapter 8). A number of algae and cyanobacteria

Photosynthetic Protein Complexes: A Structural Approach. Edited by P. Fromme

ISBN: 978-3-527-31730-1

can use both donor proteins depending on the availability of copper in the culture medium [3].

Cyanobacterial PSI consists of 11 or 12 protein subunits, 10 of which are homologous to subunits found in plant PSI. It can occur as homotrimers [4–8] and as monomers in photosynthetic membranes. Which of these two forms is dominant *in vivo* depends on the environmental conditions; the trimeric form is the more important one at low light intensities [9, 10].

Information on the three-dimensional structure of cyanobacterial PSI was obtained by electron microscopy and by X-ray crystallography. Electron microscopy provided the first pictures of PSI trimers at low resolution [11] and was used to elucidate the positions of individual protein subunits in the PSI complex [7, 12], and to determine the docking positions of ferredoxin [13] and flavodoxin [14]. In parallel, crystals of trimeric PSI from the thermophilic cyanobacterium *Thermosynechococcus elongatus* [15] were used to determine the structure of cyanobacterial PSI by X-ray crystallography. It took five years to determine the first crystal structure of PSI at a 6 Å resolution [16], and an additional eight years to improve the structure to a 2.5 Å resolution [17]. For a historical review of the various attempts to determine the structure of PSI see Fromme and Mathis, 2004 [18]. In the electron density map at 6 Å resolution, the positions of the three [4Fe4S] clusters were already visible and the models of the arrangements of α-helices and chlorophylls could be derived from it. Further improvement of the crystal quality [9] led to electron density maps at 4 Å resolution and were improved over several years, resulting in more complete models of the arrangements of α-helices and chlorophylls [19, 20] that could be used to model parts of the extra-membrane loop regions and the stromal subunits [21]. All of these structural models resulted from experiments carried out at room temperature (277 K). It was only when protocols to perform diffraction experiments at cryogenic temperatures were developed and the crystals further improved, that a significant progress in the structure analysis of PSI was possible, finally leading to the structure at 2.5 Å resolution [17]. This structure provides a picture at almost atomic detail and will be discussed in the present book chapter. The atomic coordinates of this complex are available from the Protein Data Bank under the accession code 1JB0. The crystal structure of plant PSI at 4.4 Å resolution [22] (see also Chapter 3) already showed that many structural features were conserved between PSI from cyanobacteria and plants. Many structural details in plant PSI, like amino acid side chain conformations and protein–cofactor interactions, were inferred from the structure of cyanobacterial PSI [23] and confirmed in an improved structure of plant PSI at 3.4 Å resolution [24] (see also Chapter 3).

The structure and function of cyanobacterial PSI are decribed in several review articles. The reader is referred to articles by Golbeck, 1994 [25], Chitnis, 1996 [26], Fromme and coworkers, 2001 [27], Grotjohann & Fromme, [28] and Fromme & Grotjohann [29].

2.2
Structural Overview

In the crystal structure of PSI from the thermophilic cyanobacterium *Thermosynechococcus elongatus* [17], the pigment–protein complex is present in its trimeric form, shown in Figure 2.1a. Each monomoric unit of the trimer has a mass of 356 kDa. It consists of: 12 protein subunits; a total of 128 cofactors divided into 96 chlorophylls, 22 carotenoids, 4 lipids, 2 phylloquinone molecules, 3 [4Fe4S] clusters, a Ca^{2+} ion located on an electron density map of 2.5 Å resolution [17]; and, 201 water molecules.

Viewed along the membrane normal, the PSI trimer has the shape of a clover leaf, with an approximate diameter of 220 Å; it is perfectly symmetrical in the crystals. This perfect symmetry is the result of the monomeric units being related by a crystallographic threefold (C_3) symmetry axis oriented perpendicular to the membrane and passing through the center of the trimer. The monomeric units are well separated at larger distances from the center; however, they show tight interactions in the central region surrounding the C_3 axis (Figure 2.1a and b). The intra-membrane space (the trimerization domain), defined by the phospholipid bilayer into which PSI is embedded, is filled by 32 transmembrane α-helices and an additional kinked α-helix belonging to a total of 9 different protein subunits (PsaA, PsaB, PsaF, PsaI, PsaJ, PsaK, PsaL, PsaM, and PsaX) (Figure 2.1c and d). Of these, the homologous subunits PsaA and PsaB are by far the largest and form the PSI core to which the cofactors of the electron transfer chain, from P700 to F_X, and the majority of antenna cofactors are bound. The PSI core is spatially surrounded by the seven smaller membrane-integral subunits, which can be divided into three groups according to their locations (Figure 2.1c). The first group consists of PsaI, PsaL, and PsaM located close the center of the PSI trimer (Figure 2.1b); of these, PsaL plays a central role in the formation of the trimeric complex. On the opposite side of the PSI core, PsaF, PsaJ, and PsaX, which form the second group of subunits, are exposed to the lipid bilayer of the membrane. In a position close to the long axis of the PSI core, and next to PsaA, subunit PsaK (which is identical with the third group of small membrane-integral subunits), was found. Since each of these protein subunits is involved, at least, in hydrophobic interactions with the antenna pigments, one of several possible roles of the small membrane-integral subunits is to stabilize the arrangement of the chlorophylls and carotenoids of the antenna system; this is one of the functional aspects of PSI clarified by the crystal structure analysis [17]. The observed differences in the molecular architectures of cyanobacterial [17] and plant PSI [22] are associated with the presence or absence of specific small membrane-integral subunits in these complexes. Plant PSI does not contain PsaM or PsaX, but has three additional membrane-integral subunits – PsaG, PsaH, and PsaO – which are not present in cyanobacteria [22, 30, 31] (see also Chapter 3). Almost no direct interactions between the small membrane-integral subunits and the PsaA/PsaB heterodimer can be found in the intra-membrane space [32] where the interactions are

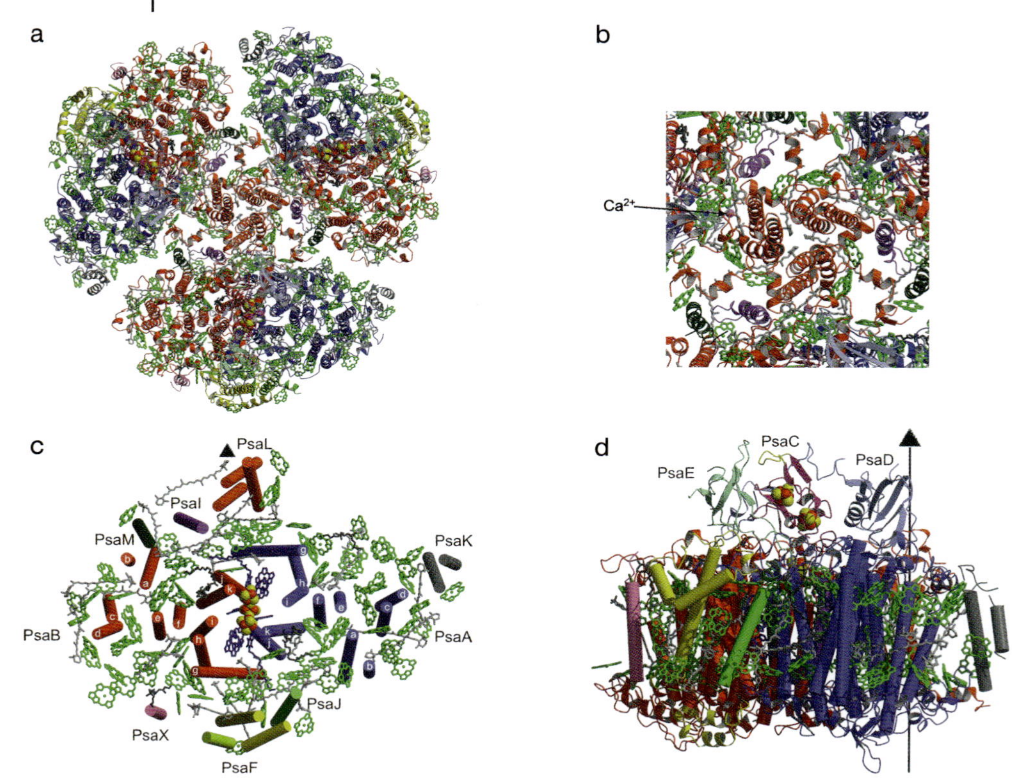

Figure 2.1 Overall structure of cyanobacterial PSI. **(a)** Schematic drawing of the PSI trimer, view direction perpendicular to the membrane plane from the stromal side. Protein subunits are shown in different colors, chlorophylls, carotenoids, phylloquinones and lipids are shown in stick representation; iron, sulfur and calcium are represented by red, yellow and violet spheres, respectively. Substituents to the chlorin systems of the chlorophylls have been omitted for clarity. **(b)** Close-up view of the center of the PSI trimer, showing that chlorophylls, carotenoids, and a putative Ca^{2+} ion which is marked by an arrow in one monomer are involved in interactions which stabilize the PSI trimer in addition to protein subunits, of which PsaL (in red) is by far the most important. **(c)** PSI monomer, viewed from the stromal side. Transmembrane α-helices are shown as cylinders and the stromal subunits PsaC, PsaD, and PsaE and the extra-membrane loop regions have been omitted for clarity. The transmembrane α-helices in PsaA and PsaB are labelled. The six chlorophylls and two phylloquinones of the electron transfer chain located in the center of the PSI monomer are shown in blue, 90 chlorophylls of the core antenna in green, 22 carotenoids in light grey, and 4 lipids in dark grey. The position of the crystallographic C_3 axis in the cemter of the PSI trimer is marked by a black triangle. **(d)** Side view, parallel to the membrane plane, of a PSI monomer. Extra-membrane loop regions and the stromal subunits are shown in ribbon representation. The loop which is inserted in PsaC with respect to bacterial ferredoxins is highlighted in yellow. The C_3 axis is indicated by a vertical line.

mediated by chlorophylls and carotenoids. Only in the extra-membrane space do direct contacts via hydrogen bonds, ionic [33], or hydrophobic interactions exist between these proteins.

The positions of the organic cofactors involved in electron and energy transfer – chlorophylls, carotenoids, and phylloquinones – are restricted to the intra-membrane space (Figure 2.1c and d). Only two of the three redox-active inorganic [4Fe4S] clusters which are bound to the small subunit PsaC (named F_A and F_B), are located in the extra-membrane space at the stromal side of the membrane. Together with the polypeptides PsaD and PsaE, PsaC forms a compact structure, the stromal ridge of PSI, which provides the docking site for the soluble proteins ferredoxin or flavodoxin at the acceptor side of PSI. None of these three subunits contains a transmembrane α-helix. In contrast to the pronounced hump on the stromal side, the lumenal side of PSI is more or less flat. It is noteworthy that, in this respect, the structure of Photosystem II (PSII) looks just the opposite: in this complex, the stromal side is almost flat, whereas the structure extends far into the lumen (see also Chapter 4).

2.3 The Protein Subunits

2.3.1 The Core Subunits PsaA and PsaB

The two subunits PsaA and PsaB have molecular masses of more than 80 kDa each and are homologous proteins [34] of 11 transmembrane α-helices each. As a consequence of this homology, PsaA and PsaB form a heterodimer of pseudo-twofold (pseudo-C_2) symmetry which makes up the core of PSI (Figure 2.1c). The pseudo-twofold symmetry axis is oriented perpendicular to the membrane and passes through the [4Fe4S] cluster F_X which structurally links PsaA and PsaB. The chlorophyll and phylloquinone cofactors of the electron transfer chain are arranged around this axis and surrounded by five transmembrane α-helices each from PsaA and PsaB. These transmembrane α-helices, which range from the seventh to the eleventh transmembrane helix (named *g* to *k* in alphabetical order) in the respective amino acid sequences, and the connecting loop regions were used to define the C-terminal domains of PsaA and PsaB and named “reaction center domains” [20]. This term reflects some similarity (including the same number of transmembrane α-helices, of the structure formed by the C-terminal domains of PsaA and PsaB) to the arrangement of subunits L and M of the photosynthetic reaction center from purple bacteria [35] which bind the cofactors of the electron transfer chain of this type-II reaction center (see also Chapter 12). Accordingly, the N-terminal domains of the PSI core subunits contain six transmembrane α-helices each, named *a* to *f*. They coordinate a total of 54 chlorophyll molecules of the core antenna [17, 27]. The transmembrane α-helices show the typical arrangement of the trimers of dimers also found in the proteins CP43 (PsbC) and CP47 (PsbB) which bind the core antenna chlorophylls in PSII [36–40]; it seems reasonable to call them core antenna domains [20]. However, the crystal structure at 2.5 Å resolution [17] showed that such a functional differentiation between the N- and C-terminal domains of PsaA and PsaB would be an oversimplification because the C-terminal reaction center domains also coordinate a total of 25 antenna chlorophylls, in addition to the 6 chlorophylls which are part of the electron transfer chain.

Recent structure analyses of PSII [36–40] have confirmed the idea of a modular architecture of all photosystems and photosynthetic reaction centers [41–43] (see Chapter 13). All photosystems and photosynthetic reaction centers contain a core structure of pseudo-C_2 symmetry, with a central domain consisting of 2×5 transmembrane α-helices that bind the intramembrane cofactors of the electron transfer chain. In type-II reaction centers, such as those from purple bacteria or PSII, this central domain consisting of subunits L and M in purple bacteria, or D1 and D2 in PSII, binds no (purple bacteria) or only a few (two in PSII) antenna (bacterio)chlorophylls. While the reaction center of purple bacteria does not contain a core antenna, it is surrounded by the light-harvesting complex LH1 (see Chapter 14). PSI and PSII are photosystems where the reaction center is either covalently linked to two core antenna domains or non-covalently linked to two core antenna proteins, respectively. The N-terminal core antenna domains of PsaA and PsaB in PSI as well as the core antenna proteins CP43 and CP47 in PSII contain a conserved structural motif of six transmembrane α-helices arranged as trimers of dimers. The overall structure formed by the 2×5 plus 2×6 transmembrane α-helices still obeys the pseudo-C_2 symmetry. In photosystems containing type-I reaction centers, these core antenna domains are always present and, as in PSI, each is covalently linked at its C-terminus with a single reaction center domain. It seems very likely that the structures formed by reaction center and core antenna domains of all these photosystems are very similar to the PsaA/PsaB core of PSI, with a total of 22 transmembrane α-helices [41]. In PSI, the core structure is pseudo-symmetric because it is a heterodimer formed by two different but homologous subunits. However, it can be expected that the core of photosystems from anoxygenic photosynthetic bacteria as found in green sulfur bacteria and heliobacteria (which contain a type-I reaction center), is of nearly perfect symmetry because it consists of a homodimer formed by two copies of the same protein subunit [44, 45] (see also Chapter 14). That similar chlorophyll-binding modules can be used by photosynthetic organisms in different contexts was shown when the structures of PSI supercomplexes were analyzed; they are induced in cyanobacteria under iron-deficiency and contain PSI trimers surrounded by an outer ring consisting of 18 copies of the IsiA protein [46, 47] (see also Chapter 6). This peripheral antenna protein is homologous to the chlorophyll-binding subunit CP43 of PSII. Therefore, it can be predicted that it contains six transmembrane α-helices, corresponding to the antenna binding module of photosynthetic reaction centers [48, 49].

The pseudo-symmetric shape of the PsaA/PsaB heterodimer is caused by the sequence similarity between these subunits, which is stronger for the transmembrane α-helices than for the loop regions[1)] where the symmetry is generally less well defined. Some exceptions are the stromal loop regions *hi* connecting the eighth and the ninth helices in PsaA and PsaB, which contain a segment highly conserved between PsaA and PsaB, and also between the corresponding PSI subunits from all species. Within this segment, two cysteines from each of the two subunits coordinate the iron-sulfur cluster F_X mediating inter-subunit interactions, as already suggested when the first amino acid sequences of subunits PsaA and PsaB were published [34, 50] and finally confirmed by the crystal structure at 2.5 Å resolution [17, 27]. This cluster plays a central role in stabilizing the PSI core, but the loop regions *hi* which contain the cysteine ligands are also important for the attachment of the stromal subunits, especially PsaC [51]. In addition to *hi*, the loop regions *fg* and *jk* in PsaA and PsaB are also involved in the attachment of the subunits of the stromal ridge via hydrogen bonds or salt bridges [51].

Looking at the structure of PSI, it is clear that part of the lumenal loop regions of PsaA and PsaB must be involved in the docking of the soluble electron donors to PSI, which can be plastocyanin or cytochrome c_6 in cyanobacteria (see Chapter 8). This is because the PSI core subunits constitute the surface region at the donor side of PSI that shields the cofactors of the electron transfer chain from the lumen. From this side, plastocyanin or cytochrome c_6 have to dock to PSI in a position close enough for efficient electron transfer to the oxidized primary donor $P700^+$ of PSI. This position is centered on the local pseudo-C_2 axis of PsaA/PsaB as already suggested on the basis of the PSI crystal structure at 6 Å resolution [16, 52], at an indentation of the lumenal surface with two surface exposed α-helices *ij* (2) of PsaA and PsaB at its base (Figure 2.2a) [27]. These helices constitute a hydrophobic patch at the surface which match the hydrophobic surface areas in plastocyanin or cytochrome c_6 [53, 54]. The importance of these helicies for the interaction between the soluble electron donors and PSI was shown by site directed mutagenesis [55, 56]. Within these helices, two tryptophans, W655 in PsaA and W631 in PsaB of PSI from *Thermosynechococcus elongatus*, (conserved between PsaA and PsaB in all known amino acid sequences, exposed to the lumen, and located next to the pseudo-C_2 axis of PsaA/PsaB) may be key residues in the recognition of plastocyanin/cytochrome c_6 by PSI, and may even be involved in electron transfer from the soluble proteins to $P700^+$ [56–58].

One of the most evident functions of PsaA and PsaB is to bind a total of 85 chlorophyll molecules, of which 79 belong to the core antenna and 6 are part of the electron transfer chain of PSI [17]. The PSI core subunits are, thus, mainly responsible for the higher density of antenna pigments in PSI compared to PSII [36–40]. In addition to the specific role of the C-terminal domains of PsaA/PsaB mentioned above, another interesting feature of the PSI core, first observed in the crystal structure [17], contributes to this unique chlorophyll-binding capacity; not only do the transmembrane α-helices coordinate chlorophylls, but also the loop regions bind a

1) Here, the term "loop regions" is used to describe those polypeptide segments which connect the transmembrane α-helices. In the structure of PSI they also contain α-helices and β-strands as secondary structure elements.

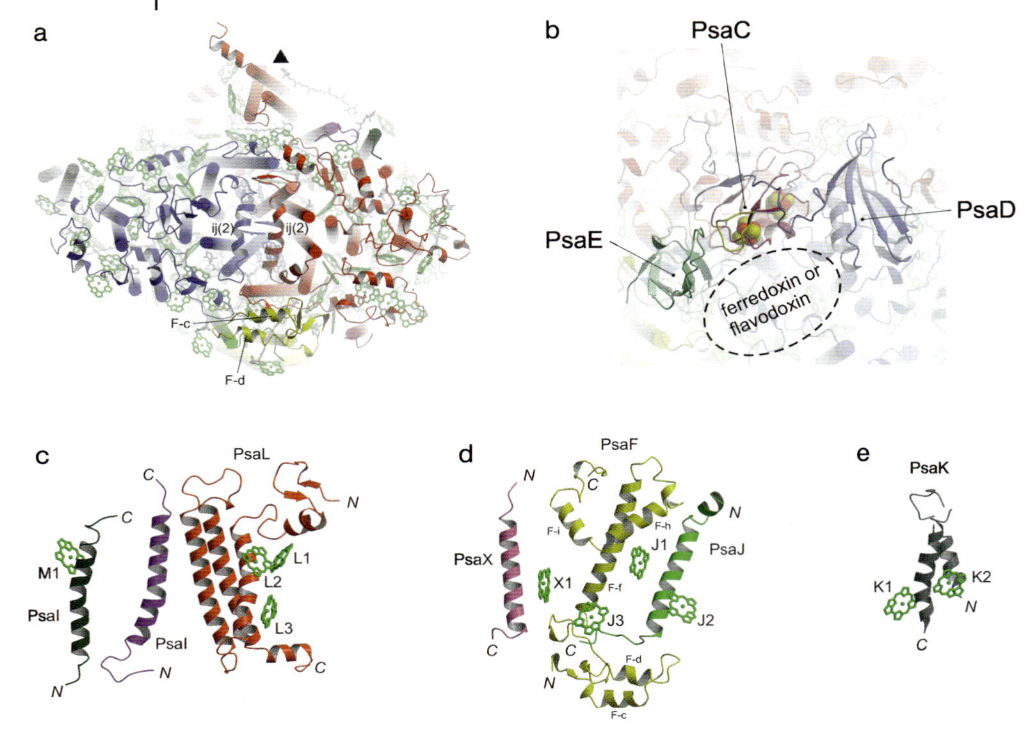

Figure 2.2 The protein subunits. The coloring scheme for the subunits is the same as in Figure 2.1. **(a)** Schematic view of the lumenal side of PSI. Surface exposed α-helices which are mentioned in the text are labelled. These are α-helices *ij*(2) of PsaA (blue) and PsaB (red) which are known to interact with plastocyanin or cytochrome c_6 and α-helices *F-c* and *F-d* of PsaF which are too far away from the putative binding site of these proteins to be important for their docking to PSI. **(b)** The stromal ridge of PSI, formed by subunits PsaC, PsaD, and PsaE. View direction perpendicular to the membrane plane. The putative docking site of the soluble electron acceptors ferredoxin or flavodoxin is marked by an ellipse; (c), (d), and (e). The small membrane-integral subunits are shown in ribbon representation. View direction parallel to the membrane plane. The chlorophyll molecules which are axially coordinated by these subunits are shown and labeled. The amino and carboxy termini of the polypeptide chains are indicated by the letters "N" and "C", respectively. **(c)** Subunits PsaL, PsaI and PsaM. **(d)** Subunits PsaF, PsaJ and PsaX. In PsaF, α-helices which are mentioned in the text are indicated. **(e)** Subunit PsaK.

total of 30 chlorophylls of the antenna, indicating a second function different from being involved in providing docking sites for soluble electron carrier proteins.

2.3.2 The Stromal Ridge: PsaC, PsaD and PsaE

The three subunits, PsaC, PsaD, and PsaE, that form the stromal ridge on the acceptor side of PSI (Figure 2.1d) can be removed from intact PSI with the use of chaotropic agents [59], already indicating that they are not anchored in the membrane. The central function of the stromal ridge is to dock ferredoxin or flavodoxin for reduction of these soluble proteins, but a role in cyclic electron flow around PSI is also discussed. A first topographic three-dimensional map of the arrangement of the stromal subunits was obtained by electron microscopy in 1997 [7]. A putative docking site of the soluble electron acceptors, indicated in Figure 2.2b, was already suggested on the basis of the PSI crystal structure at 6 Å resolution [16, 52] and seemed to fit well to the site identified by two-dimensional electron microscopy in a cross-linked complex of PSI with ferredoxin [13]. To find out where ferredoxin docks to PSI, a three-dimensional structure of the PSI-ferredoxin complex must be determined. A first step in this direction has been made by co-crystallizing PSI and ferredoxin [60], but to get a deeper insight into the interactions between these proteins the resolution has to be increased significantly. For more details see Chapter 15.

2.3.2.1 **Subunit PsaC**

This subunit is located in the center of the stromal ridge and binds the terminal acceptors of the electron transfer chain, the [4Fe4S] clusters F_A and F_B. The crystal structure confirmed that the overall fold of PsaC resembles 2[4Fe4S] bacterial ferredoxins; this had already been suggested based on the amino acid sequence which contains two conserved motifs CXXCXXCXXXCP like the bacterial proteins [25, 61]. The conserved cysteine residues are the ligands to the [4Fe4S] clusters in both PsaC [17] and the bacterial ferredoxins [62]. The repeat of the binding motifs gives rise to a weakly expressed pseudo-C_2 symmetry of the iron-sulfur cores of the structures, containing two single-turn α-helices separating the iron-sulfur clusters and two two-stranded antiparallel β-sheets. In each of the conserved motifs, the first three cysteine residues coordinate the same iron-sulfur cluster, but the fourth cysteine is positioned to coordinate the second cluster at the C-terminus of the single-turn α-helix. The most striking differences between the structures are due to an insertion of 10 amino acid residues in the region connecting the conserved cysteine containing motifs (Figures 2.1d and 2.2b) and extensions at the C-terminus by 14 residues and at the N-terminus by two residues in PsaC with respect to the ferredoxin from *Peptococcus aerogenes*. The insertion in PsaC folds into a loop which extends to the stromal surface of PSI and is involved in the docking of negatively charged ferredoxin or flavodoxin with the essential positively charged residue K34 in *Thermosynechococcus elongatus* [63, 64]. The C-terminal extension participates in hydrogen-bonding interactions with PsaB. The interaction of the C-terminal residue Y80 *(T. elongatus)* with the PSI core is an important symmetry-breaking element which helps to discriminate between the correct and the incorrect orientation when PsaC binds onto the pseudo-symmetric PsaA/PsaB heterodimer [51]. It is important to note that the pseudo-C_2 symmetry of PsaC does not apply to the amino acid residues involved in ionic interactions and hydrogen bonds with the PsaA/PsaB heterodimer. As a consequence, the pseudo-C_2 axes of PsaA/PsaB and of PsaC are oriented not in parallel with each other, but at an angle of 62°. This results in different distances from F_A to F_X (14.9 Å) and from F_B to F_X

(22.0 Å) [17], and is the geometric basis for the sequence $F_X \rightarrow F_A \rightarrow F_B$ in electron transfer (see below, 2.4.5). A comparison of the solution structure of PsaC, determined by NMR spectroscopy [65], with the structure of PsaC in the PSI complex [17], showed that conformational changes occur upon binding to PSI in the N- and C-terminal regions of PsaC, including a segment preceding the C-terminal extension [51, 66].

2.3.2.2 **Subunit PsaD**

Of the three subunits which form the stromal ridge, PsaD is closest to the C_3 axis of the PSI trimer [20, 21]. Its role in the docking of ferredoxin to PSI has been demonstrated [67–71]. In contrast to the other two stromal subunits, for which homology-based structural models existed in the case of PsaC and a solution structure was known in case of PsaE [72] before detailed structural information on these subunits became available from the crystal structure of PSI [17], the 2.5 Å resolution structure of PSI showed for the first time how PsaD is folded in the PSI complex (Figures 2.1d and 2.2b). This is because PsaD has no overall defined structure in solution [73], and folding probably takes place during the assembly into the PSI complex. The structure of PsaD in PSI can be described as consisting of three parts. The large N-terminal part contains a larger four-stranded antiparallel β-sheet neighbored by a smaller two-stranded antiparallel β-sheet, and an α-helix which in the amino acid sequence is inserted between the third and fourth strands of the larger β-sheet. The N-terminal part has a large number of contacts to PsaA/PsaB, to PsaC and even to PsaL. In the scenario of sequential binding of PsaC and PsaD to PSI compatible with the complex network of interactions between PsaA/PsaB, PsaC, and PsaD, the N-terminal part plays an essential role in stabilizing PsaC in its correct orientation on PsaA/PsaB [51]. A remarkable feature of the PsaD structure is the second part defined by residues Pro96 to Pro123 *(T. elongatus)*; it folds like a clamp over PsaC [27] and is involved in numerous hydrogen-bonding interactions with PsaC [51]. This is a part of the structure which adopts its conformation only when bound to PSI and which is critical for stabilizing the acceptor side of PSI [27]. For K106 in PsaD from *Synechocystis* sp. PCC 6803 (K104 in *T. elongatus*), a direct role in ferredoxin docking was shown by crosslinking with ferredoxin [69], and the crosslinked PSI-ferredoxin complex, still functional in electron transfer from PSI to ferredoxin, was characterized by electron microscopy, showing the ferredoxin docking site [13]. The short C-terminal part of PsaD forms several hydrogen bonds exclusively to PsaB.

2.3.2.3 **Subunit PsaE**

For this subunit (Figures 2.1d and 2.2b), which is located at the longest distance from the center of the PSI trimer within the stromal ridge, a solution structure was already known [72] before its structure could be modeled into an electron density map of PSI crystals. Surprisingly, the structure revealed similarity to Src homology 3 (SH3) domains found in eukaryotic proteins of signal transduction pathways [72]. Only after the resolution and quality of electron density maps determined by crystallographic methods had been improved over many years, could a first model of the polypeptide backbone of PsaE be fitted into an electron density map at 4 Å resolution [21] using the NMR structure of PsaE from *Synechococcus* sp. PCC 7002 [72]. This initial structural model was confirmed at higher resolution [17] and shows good agreement with the solution structure for those parts which fold into a five-strand anti-parallel β-sheet, forming a β-barrel [27]. The largest differences are found for the long CD-loop (connecting strands βC and βD, nomenclature as in Falzone and coworkers 1994 [72]) which adopts a twisted conformation and is described to be flexible in solution. Obviously, a change in its conformation upon binding to PSI enables the CD-loop to form hydrogen bonds with PsaA, PsaB and PsaC as found in the crystal structure [27, 51]. The CD-loop is the most prominent part of PsaE for the interaction with the PSI core. Compared to the extensive binding networks found between PsaC and PsaD, PsaE is only weakly integrated into the stromal ridge of PSI. This is consistent with the suggested scenario of a sequential assembly of the stromal subunits, where PsaE is the protein which binds last to PSI [51]. Like PsaD, which interacts with the small membrane-integral subunit PsaL, PsaE forms hydrogen bonds via its AB-loop with the membrane-integral subunit PsaF [27]. An interesting detail of the stromal ridge structure is the interaction between the BC-loop of PsaE and the inserted loop in PsaC, which contains residue K34 essential for ferredoxin docking. The pairwise stabilization of these two loop conformations could be critical for binding ferredoxin to PSI, especially as the loop in PsaE also contains a residue (R39 in *T. elongatus*) which was shown to play an important role for the affinity of PSI to ferredoxin [74]. Besides its responsibility for the formation of a stable PSI-ferredoxin complex, PsaE is not important in electrostatic guiding of ferredoxin to its binding site, in contrast to PsaD, and most probably does not affect the geometry of ferredoxin binding [71]; for reviews see Sétif, 2001 [75] and Sétif and coworkers 2002 [64]. The idea that in cyanobacteria PsaE is involved in $NADP^+$ reduction via interaction with FNR in a ternary PSI-ferredoxin-FNR complex, based on kinetic data [76], was not confirmed [77]. In contrast, the interaction of PsaE with FNR has been shown in plants [78]. Studies on cyanobacterial mutants lacking PsaE point to another role of this subunit, which is obviously required for cyclic electron transfer around PSI [25, 79].

2.3.3
The Small Membrane-Integral Subunits

The low-molecular mass membrane-integral subunits, located at the periphery of the PsaA/PsaB heterodimer (Figure 2.1c), bind and stabilize the chlorophyll and carotenoid molecules of the PSI core antenna. Interactions with the chlorophyll molecules are always established by axial coordination of the central Mg^{2+} cations, by hydrophobic interactions, especially with the long phytyl chains, and, in many cases, by donating hydrogen bonds to oxygen atoms of the keto-carbonyl groups or the ester groups. In cyanobacteria, the chlorophylls which are exclusively coordinated by the small membrane-integral subunits may play a role in mediating excitation energy transfer between the peripheral antenna formed by the IsiA protein and the PSI core (see Chapter 6) [48], or between adjacent monomeric units in the PSI trimer. Protein–protein interactions responsible for the formation of the PSI trimer and for the stabilization of PSI–IsiA supercomplexes, should involve the small membrane-integral subunits. Detailed interactions between the PSI trimers and the rings of IsiA proteins cannot be resolved in the electron microscopy maps [46–48, 80].

2.3.3.1 Subunits PsaL, PsaI, and PsaM

Of these subunits, shown in Figure 2.2c, PsaL has a prominent function in cyanobacteria. It establishes the vast majority of inter-monomer contacts in the PSI trimer, followed by PsaI and the core subunit PsaB. Only a small number of interactions in the monomer-monomer interfaces involve PsaA and PsaM. Not only protein subunits, but also chlorophyll and carotenoids molecules, stabilize the PSI trimer by numerous pigment–protein and pigment–pigment contacts. The role of PsaL in the stabilization of PSI trimers was already proposed by mutagenesis studies, because no trimers can be detected in cyanobacterial PSI which lacks PsaL [81]. Formation of PSI trimers was shown to be important for the growth of cyanobacteria at low light intensities [10]. In the crystals of the PSI trimer, an electron density was found to be located between two adjacent monomers and suggested a Ca^{2+} ion because side chains of subunit PsaA from one monomer and of subunit PsaL from the two different monomers coordinate to it in a geometry typical for Ca^{2+} [17, 27]. Consistent with this suggestion, Ca^{2+} stimulates the formation of PSI trimers (P. Chitnis and J. Kruip, personal communication). Hydrogen bonds formed directly between the extra-membrane regions of the protein subunits, or mediated by water molecules, contribute to the stability of the PSI trimer; as well, each of the three transmembrane α-helices of PsaL participates in hydrophobic monomer-monomer interactions. The third of these transmembrane α-helices undergoes hydrophobic interactions in the trimerization domain with the first helix in a second monomer and the second helix in a third monomer. The presence of subunit PsaH in plant PSI, which was found to be located adjacent to PsaL and probably forms part of the docking site for the peripheral light-harvesting complex LHCII, explains why this complex cannot form trimers [22, 82]. Furthermore, plant PsaL has lost the C-terminal residues which stabilize the PSI trimer in cyanobacteria [22].

Within one monomer, the extra-membrane regions of PsaL, PsaI, and PsaM mainly interact with PsaB of the PSI core. Only PsaL shows a few contacts to PsaA; it is also connected by hydrogen bonds to PsaD. Within this group of three membrane-integral subunits, of which PsaI and PsaM contain a single transmembrane α-helix each, only PsaI interacts with both of the other two subunits. PsaI is also the only one of these subunits which does not provide an axial ligand to a chlorophyll, but it is remarkable that it forms van der Waals contacts with six chlorophyll molecules and interacts with three carotenoids of the same monomeric unit. Furthermore, its transmembrane α-helix stabilizes the PSI trimer by hydrophobic interactions with two chlorophylls of the adjacent monomer. Subunit PsaM was not reported for any preparation of higher plant PSI and is not present in the crystal structure of plant PSI [22]. The role of PsaM in axially coordinating an antenna chlorophyll was not completely clear in the electron density map of cyanobacterial PSI at 2.5 Å resolution [17], which suggested that a water molecule which is hydrogen bonded to residue R24 of PsaM is the axial ligand of this chlorophyll (M1). It can be concluded from the crystal structure that excitation energy transfer between this and the other chlorophylls of the same PSI monomer is slow [83]; it may instead be considered as part of the antenna chlorophyll network of the adjacent monomer [84]. PsaM is in van der Waals contact to two additional chlorophylls and a carotenoid molecule. The three chlorophyll molecules coordinated by PsaL are located close to the C_3 axis of the PSI trimer and may serve as linkers in excitation energy transfer between the chlorophylls of different monomeric units in the PSI trimer. Within the same monomer, PsaL interacts with seven additional chlorophylls, donating hydrogen bonds to three of them, and also with three carotenoids, of which one is located very close to the C_3 axis and is a prominent part of the monomer-monomer interface.

2.3.3.2 **Subunits PsaF, PsaJ and PsaX**

Of these subunits (Figure 2.2d) exposed to the lipid bilayer of the membrane, PsaF and PsaJ are located at positions which are roughly related by the pseudo-C_2 axis of PsaA/PsaB to PsaL and PsaI (Figure 2.1c). PsaF, PsaJ, and PsaX contain a single transmembrane α-helix each, but PsaF, the largest of these subunits, has the most unusual structure of the small membrane-integral subunits of PSI. It consists of two domains: an N-terminal domain at the lumenal side and a C-terminal domain which is mainly membrane-intrinsic, with the C-terminus located at the stromal side. The N-terminal domain contains two α-helices, *F-c* and *F-d,* as the most prominent features. These helices are oriented parallel to the membrane and located at a distance of 27 Å from the pseudo-C_2 axis of PsaA/PsaB, too far away from the docking site of cytochrome c_6 or plastocyanin to allow direct interaction with these proteins. This is in agreement with the observed functional difference between subunits PsaF in plants and cyanobacteria. It is only in plants that PsaF is involved in the docking of the electron donor plastocyanin to PSI [85–88], not in cyanobacteria [89]. The functional difference is due to an insertion of 20 amino acid residues in the N-terminal domain of PsaF in higher plants and green algae, leading to an extended helix-loop-helix motif on the lumenal side which is probably involved in plastocyanin docking [22]. A role of the α-helices *F-c* and *F-d* in cyanobacterial PSI may be to stabilize the structure of the lumenal surface (Figure 2.2a) [27]. In the C-terminal domain, located C-terminal of the transmembrane α-helix *F-f* and separated from it by a hydrophilic segment at the stromal side containing a short α-helix is a kinked hydrophobic α-helix, consisting of the segments *F-i* and *F-h* that form a V-shaped structure. The kink of this α-helix is located in the intramembrane space, at a depth of about one third of the membrane. The C-terminus, located in the stroma, interacts with PsaE and might be important for the stable assembly of this subunit into the stromal ridge. PsaF does not axially coordinate a chlorophyll, but is in van der Waals contact with 10 antenna chlorophylls, involving also a hydrogen bond to one of them and to 3 carotenoids, which demonstrates its function in shielding this pool of pigments from the lipid phase [27]. PsaF serves this function together with the smaller subunit PsaJ, which axially coordinates three chlorophylls and interacts with additional six chlorophylls and three carotenoids. PsaJ is tightly bound to PsaF by hydrogen bonds, a salt bridge, and hydrophobic interactions, all located in the extra-membrane regions. Because of the special location of PsaF and PsaJ at the periphery of the PSI trimers, these subunits and the antenna pigments stabilized by them are located relatively close to the ring of IsiA proteins in PSI-IsiA supercomplexes. The three chlorophylls which are axially coordinated by PsaJ were especially suggested to facilitate energy transfer between the peripheral antenna and the PSI trimer [48]; a specific role of PsaF and PsaJ for stabilizing the PSI-IsiA supercomplexes seems to be very likely. Interestingly, in mutant cyanobacteria lacking PsaF and PsaJ PSI-IsiA supercomplexes are induced, which consist of a central PSI trimer surrounded by an ordered and closed ring of 17 instead of 18 copies of IsiA [80]. These complexes seem to be less stable than the wild-type complex showing that PsaF and PsaJ facilitate formation of stable supercomplexes, but are not obligatory components of this process [90]. It was also speculated [27, 28] that PsaF might be important for the interaction of PSI with the large peripheral antenna system of phycobilisomes in cyanobacteria [91]. In mutants lacking PsaF and PsaJ, the IsiA protein is induced even at normal iron concentrations [92], which might be due to a loss of the docking site of phycobilisomes at PSI.

The presence of subunit in PsaX in PSI of *T. elongatus* was shown first by the high resolution crystal structure analysis [17], because the electron density map of this polypeptide (identified to be the twelfth subunit of *T. elongates*) PSI was compatible with an amino acid sequence homologous to PsaX from two other thermophilic cyanobacteria [93, 94]. The gene coding for PsaX was later identified in the genome of *T. elongatus* [95]. No corresponding gene was found in the genome of the mesophilic cyanobacterium *Synechocystis* sp. PCC6803 [96], supporting the idea that PsaX is a protein subunit specific to thermophilic cyanobacteria. In the electron density map of the PSI single crystals at 2.5 Å resolution [17], the six N-terminal amino acid residues could not be modeled suggesting that this part of the polypeptide, which is located in the stroma, adopts a flexible conformation. PsaX axially coordinates a chlorophyll molecule, is in van der Waals contact with additional four chlorophylls, donating a hydrogen bond to one of them, and is the only one of the membrane integral subunits of PSI which does not interact with any carotenoid. On the lumenal side, PsaX interacts with PsaF via a single hydrogen bond. Its attachment site at PSI is further characterized by hydrophobic interactions, a single hydrogen bond to PsaB, and an ionic interaction with the phosphodiester group of a lipid which is, in turn, hydrogen bonded to PsaB. PsaX may form further hydrophobic interactions with the fatty acid moieties of the same phospholipid molecule which, presumably due to their high flexibility, are only incompletely visible in the electron density map at 2.5 Å resolution [17]. It seems to be possible that this attachment site, including the binding of the specific lipid molecule, is unique to thermophilic cyanobacteria and strictly correlated with the presence of PsaX. The putative uniqueness of this site also includes the binding of antenna chlorophyll B33, which is part of a trimer of chlorophyll molecules (B31/B32/B33) stacked to each other in a parallel orientation and suggested to belong to the pool of "red chlorophylls" (see 2.5.1 and Jordan and coworkers, 2001 [17]). More genomes of thermophilic and mesophilic cyanobacteria have to be sequenced to discover whether PsaX is really required for stabilization of cyanobacterial PSI at higher temperatures [27], or if its presence is associated with specific spectral features of the PSI core antenna.

2.3.3.3 Subunit PsaK

Even in the electron density map of PSI at 2.5 Å resolution the side chains of the peripheral subunit PsaK could not be assigned to individual amino acid residues; the polypeptide backbone could only be traced imcompletely [17]. The lower quality of the map in this region is also attributed to high temperature factors, possibly indicating that PsaK is the most flexible subunit in PSI. PsaK contains two transmembrane α-helices, which were not visible in the electron density map at 4 Å resolution [21], connected by a loop on the stromal side. The N- and C-termini are located in the lumen, a topology which is consistent with insertion studies using the in vitro expressed protein [97]. This subunit is involved in the axial coordination of two chlorophylls, of which one is discussed to facilitate energy transfer between the PSI core and the peripheral antenna in PSI-IsiA supercomplexes [48]. In spite of the incomplete structural model of PsaK, van der Waals contacts to two additional chlorophylls and a single carotenoid molecule can be identified in the crystal structure. PsaA is the only protein subunit which interacts with PsaK. In PSI from plants, a subunit PsaG which is homologous to PsaK, is located at a position related by the pseudo-C_2 axis of PsaA/PsaB, in contact with PsaB [22]. Both PsaG and PsaK interact with proteins of the peripheral light-harvesting complex I (LHC-I), but in a completely different mode (see also Chapter 3).

2.4 The Electron Transfer Chain

With the exception of the [4Fe4S] clusters F_A and F_B, the cofactors which constitute the electron transfer chain of PSI (Figure 2.3a) are located in the center of PsaA/PsaB and bound to homologous amino acid residues of the core subunits. As a consequence, the arrangement of the PsaA/PsaB bound part of the electron transfer chain obeys the pseudo-C_2 axis of PsaA/PsaB and can be described to consist of two branches, each containing three chlorophyll molecules and a phylloquinone. The pseudo-symmetry also includes the [4Fe4S] cluster F_X, because it is bound to PsaA and PsaB by two pairs of homologous cysteine residues in these subunits. The two branches were called A- and B-branch (Figure 2.3a) to indicate whether PsaA or PsaB is the subunit which predominantly binds the cofactors [17, 27]. Each chlorophyll of the electron transfer chain is labeled eC, followed by A or B to indicate whether PsaA or PsaB is involved in axial liganding of the central Mg^{2+} ion, and numbered 1, 2, or 3 if it belongs to the first, second, or third pair

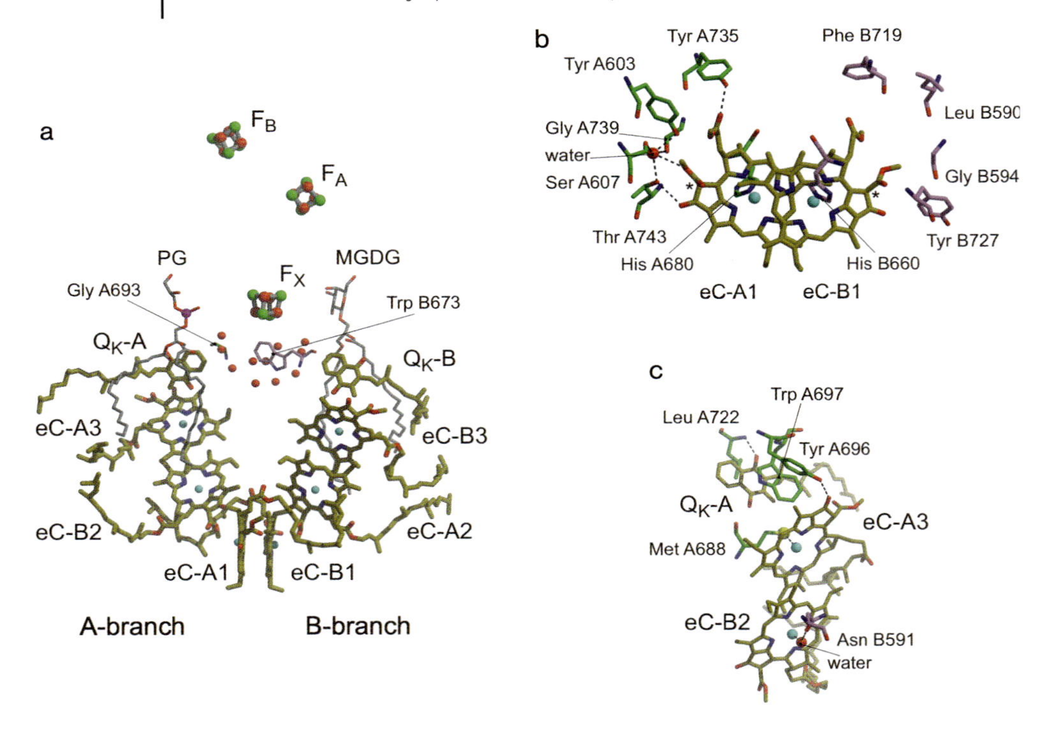

of chlorophylls, respectively, starting from the lumenal side. Correspondingly, the phylloquinone (vitamin K_1) molecule bound to PsaA is named Q_K-A; Q_K-B is the one bound to PsaB. The pseudo-symmetric arrangement of organic cofactors of the electron transfer chain in two branches is a structural feature which PSI shares with type-II photosynthetic reaction centers; this was already recognized when the resolution of the PSI structure was only 4 Å [41]. In type-II reaction centers, the terminal electron acceptors are two structurally corresponding quinone molecules in the two branches which have different roles, as one quinone (Q_B) follows the other (Q_A) in the sequence of electron transfer reactions. In other words, electron transfer from Q_A to Q_B proceeds across the pseudo-C_2 axis. This is different in PSI, where the the spectroscopically identified quinone A_1 is only an intermediate electron acceptor from which electron transfer has to proceed to the [4Fe4S] clusters. The structure of PSI and kinetic data suggest that both branches of the electron transfer chain may be involved in electron transfer, but it cannot be ruled out than only one branch is active. Whether electron transfer in PSI is uni- or bidirectional is still a matter of debate. It is of general interest for photosynthesis research because PSI can be regarded as intermediate between the type-I reaction centers from anoxygenic photosynthetic bacteria which have a homodimeric core, suggesting symmetric electron transfer along two branches and type-II reaction centers with a unidirectional electron pathway [98]. For reviews of this issue see

Figure 2.3 The electron transfer chain. **(a)** Arrangement of the cofactors of the electron transfer chain, view direction parallel to the membrane plane. Of the four organic cofactors in the A-branch, only eC-B2 is bound to PsaB and vice versa. The pair of chlorophyll *a′* eC-A1 and chlorophyll *a* eC-B1 is assigned to the primary donor P700. One or both of chlorophyll *a* molecules eC-A3 and eC-B3 are thought to be identical with the primary acceptor A_0, and similarly one or both of the phylloquinones Q_K-A and Q_K-B are the secondary acceptor A_1. Elements of asymmetry in the region surrounding the phylloquinones and [4Fe4S] cluster F_X have been indicated. These are Trp A693, which has glycine as its counterpart in PsaB (Gly B673), a cluster of six water molecules bound via a hydrogen bonding network to PsaB which corresponds to a cluster of five water molecules bound to PsaA, and a neutral monogalactosyldiacylglycerol (MGDG) located closer to Q_K-B which corresponds to a negatively charged phosphatidylglycerol (PG) situated closer to Q_K-A. **(b)** The chlorophyll *a′* / chlorophyll *a* heterodimer eC-A1/eC-B1 assigned to P700, view direction perpendicular to the planes of the chlorin rings. The long phytyl chains have been omitted for clarity. Two pairs of partially overlapping pyrrole rings are visible in the heterodimer. The carbon atoms $C13^2$, which have opposite configurations in chlorophyll *a′* and chlorophyll *a*, are labeled by asterisks. Of the local environment of P700, amino acid residues which axially coordinate the chlorophylls (His A680 and His B660) and which are involved in potential hydrogen bonds to these cofactors, are shown. Potential hydrogen bonding interactions are indicated by dotted lines and are in part mediated by a water molecule. Amino acid residues in PsaB which correspond to residues in PsaA that are involved in the hydrogen bonding network with eC-A1 are also shown. These residues in PsaB are incapable of forming hydrogen bonds to eC-B1 either because of their chemical structures or of their conformations. The hydrogen bond which seems to have the strongest effect on the electronic properties of P700 is formed between the side chain of Thr A743 and eC-A1, characterized by a distance of 3.0 Å between O of the threonine and $O13^1$ of the chlorophyll's keto group in ring V. **(c)** The remaining organic cofactors of the A-branch: chlorophyll *a* molecules eC-B2 and eC-A3, and phylloquinone Q_K-A. The arrangement of cofactors and interacting amino acid residues shown in this figure is very similar in the B-branch, the most striking difference being the different conformations of the phytyl chains of the phylloquinones (see (a)). The sulfur atom of the Met A688 side chain which coordinates eC-A3 is located at 2.6 Å from the central Mg^{2+} of this chlorophyll. The central Mg^{2+} of eC-B2 is coordinated by a water molecule at a distance of 2.0 Å, which in turn is hydrogen bonded to the side chain of Asn B591. The distances between the non-hydrogen atoms in the potential hydrogen bonds indicated by dotted lines are: Asn B591 $O_{\delta1} \cdots$ water O = 2.5 Å, Tyr A696 $O_\eta \cdots$ eC-A3 O (13^1-keto group) = 2.7 Å, Leu A722 N $\cdots$ Q_K-A O4 = 2.7 Å. The indole ring of Trp A697 and the quinone plane of Q_K-A are oriented almost paralell to each other, resulting in π-stacking with plane-to-plane distances between 3.0 and 3.5 Å.

Brettel and Leibl, 2001 [1], Rappaport and coworkers 2006 [99], and Redding and van der Est 2006 [100].

2.4.1
The First Pair of Chlorophylls

Located next to the lumenal surface, the pair eC-A1/eC-B1 was assigned to the primary donor P700 (Figure 2.3b). The axial ligands to these chlorophylls are histidine residues H680 in PsaA and H660 in PsaB of *T. elongatus*. Both chlorophyll molecules are stereochemically different: whereas eC-B1 is a chlorophyll *a*, eC-A1 is a chlorophyll *a* epimer, denoted chlorophyll *a*′, with inverted configuration at $C13^2$, the only asymmetric carbon atom in ring V. That P700 contains at least one chlorophyll *a*′ was already suggested by Watanabe and coworkers [101, 102], but was called into question until the structure of PSI at 2.5 Å resolution showed that P700 is, in fact, a chlorophyll *a*/chlorophyll *a*′ heterodimer [17]. The chlorin rings of the chlorophyll epimers are oriented parallel to each other and perpendicular to the membrane plane, resulting in a π-stacked arrangement with an interplanar distance of 3.6 Å. A molecular orbital study of P700 on the basis of the crystal structure [103] has shown that the strength of the electronic interaction of the chlorophylls is comparable to the special pair of bacteriochlorophylls in type-II bacterial photosynthetic reaction centers [35]. These calculations show that asymmetry in the electronic properties of P700 is already present in the isolated heterodimer, if interactions with the protein environment are not taken into account. The asymmetry can be traced not only to a specific structural feature like the stereochemical difference between eC-A1 and eC-B1, but it is also further enhanced by hydrogen-bonding interactions between PsaA and the chlorophyll *a*′, eC-A1, which have no counterpart in the B-branch. This theoretical study is compatible with results from ENDOR spectroscopy on solutions [104] and on single crystals of PSI [105] showing that in $P700^+$ about 85% of the spin density is localized on the chlorophyll which is bound to PsaB. Due to pronounced spin polarization effects, the distribution of the positive charge in $P700^+$ is estimated to be much less asymmetric, with less than 70% localized on eC-B1 [103]. However, this still contrasts with results from FTIR spectroscopy which are interpreted in terms of an essentially even distribution of the charge over the two chlorophylls in $P700^+$ [106, 107]. In context with the discussion on directionality of electron transfer in PSI, the asymmetry of P700 is of interest as it might be responsible for directing electron transfer either along the A- or the B-branch, according to the unidirectional model [17, 27]. On the other hand, the asymmetry could just be a byproduct of fine-tuning the redox and spectroscopic properties of P700. The redox potential of $P700^+$/P700 is much lower than that of monomeric chlorophyll *a*, the oxidation midpoint potential in PSI from *Chlamydomonas rheinhardtii* being +0.47 V [108]; and, consequently, the excited state P700* is strongly reducing. Dimerization of chlorophylls in P700 might be largely responsible for the negative shift of the redox potential [109], as concluded in analogy to work on the special pair of bacteriochlorophylls in type-II bacterial reaction centers [110]. The specific geometry of the chlorophyll *a*/*a*′ heterodimer also results in strong excitonic coupling which contributes to a red-shifted absorption, although the results of structure-based calculations strongly depend on the method which is used [83, 111–114]. Mutation of T739 in PsaA of *C. rheinhardtii* (corresponding to T743 in *T. elongatus*, which provides a hydrogen bond to the keto group in ring V of eC-A1), led to a blue shift by 6 nm in the P700 absorption. This might indicate the important role of this hydrogen bond in lowering the energy of the Q_y transition of eC-A1, which might be essential for effective trapping of excitation energy by P700 [108]. As the keto group in ring V is part of the conjugated π-electron system of chlorophyll *a*′, it is not surprising that a hydrogen bond to it affects the electronic properties of the pigment. In contrast, the role of chlorophyll *a′* in P700 and how is it synthesized remain open questions. Interestingly, $C13^2$ epimers of chlorophylls and bacteriochlorophylls are only present in type-I reaction centers of oxygenic and anoxygenic photosynthetic organisms (for review see Akiyama and coworkers, 2002 [115]). Of central importance to the understanding of oxygenic photosynthesis is the question: what makes the redox potentials of corresponding cofactors of the electron transfer chains in PSI and PSII so different? The redox potential for the oxidation of P680, the primary donor of PSII, is about 700 mV more positive than that of P700 [116, 117], in spite of the fact that both primary donors are constituted by chlorophyll molecules. In a theoretical approach, redox potentials were calculated for the (bacterio)chlorophyll pairs which were assigned to P680 in PSII, to P700 in PSI, and to P870 in the reaction center from the purple bacterium *Rhodobacter sphaeroides* by solving the linearized Poisson–Boltzmann equation for all atoms in the respective crystal structures [118]. In comparing PSI and PSII, it was found that the much more positive redox potential in P680 was mostly due to the presence of the Mn_4Ca cluster in PSII and to the different relative positions of the chlorophyll pairs with respect to the transmembrane α-helices which contain the axial ligands to the central Mg^{2+} ions of the pigments. Because of the only medium resolution of the presently available PSII structure and the uncertainties about the structure of the Mn_4Ca cluster in PSII, the details of this analysis must be taken with caution, but it demonstrates how a number of different relatively small contributions might add up to the observed dramatic differences in redox properties of the primary donors.

2.4.2
The Second and Third Pairs of Chlorophylls

The two pairs of chlorophyll *a* molecules, eC-A2/eC-B2 and eC-A3/eC-B3, are located close to each other and to P700 (Figure 2.3a and c), suggesting that they are involved in primary charge separation. The amino acid residues involved in axial coordination and hydrogen-bonding of these chlorophylls are strictly conserved in PSI from cyanobacteria, green algae, and higher plants. Although the detailed geometries are different, the overall arrangement of the six chlorophylls of the electron transfer chain in PSI slightly resembles that of the four (bacterio)chlorophylls and two (bacterio)pheophytins in type-II bacterial reaction centers [35] and PSII [40]. In analogy to the type-II reaction centers, the chlorophylls of the second pair eC-A2/eC-B2 are sometimes called accessory chlorophylls. It was suggested [19] that one or both chlorophylls of the third pair are identical with the spectroscopically identified primary acceptor A_0, but the rate of primary charge separation, estimated from experimental data to be about $(1\,ps)^{-1}$ (for review see Brettel and Leibl, 2001 [1]), is much faster than the optimal electron transfer rate from P700 to eC-A3 or eC-B3, calculated using the empirical relation proposed by Moser and Dutton [119] on the basis of the PSI structure at 4 Å resolution [117, 120]. Accordingly, one or both molecules of the pair eC-A2/eC-B2 should be intermediate acceptors. In each branch of the electron transfer chain, the second and third chlorophylls are in contact with each other; the shortest distance between atoms that are part of the conjugated π-electron systems is 3.8 Å, and are oriented roughly parallel to each other [17, 27]. Because of this, these chlorophylls might mutually influence their redox and spectroscopic properties, and their interaction, as suggested by the crystal structure, is in agreement with the contribution of more than one chlorophyll to the $(A_0^- - A_0)$ difference spectrum [121]. Strong excitonic coupling between the six chlorophylls of the electron transfer chain, and especially between the second and third chlorophyll in each branch [83, 114], could also be important for an increased spectral overlap between the chlorophylls of the core antenna and the electron transfer chain responsible for efficient flow of excitation energy to the reaction center. A delocalization of excitonic states over the six chlorophylls of the reaction center is also consistent with the proposal that one of the accessory chlorophylls, eC-A2 or eC-B2, may be the chlorophyll where electron transfer starts [122], in a mechanism analogous to that discussed for PSII [123, 124].

The chlorophylls of the second pair have water molecules as the axial ligands to Mg^{2+}. These are hydrogen-bonded to the amide side chains of asparagine residues, N604 in PsaA and N591 in PsaB of *T. elongatus,* used to determine the labeling of eC-A2 and eC-B2, respectively. Somewhat confusingly, eC-B2 belongs to the A-branch and eC-A2 is part of the B-branch, but this is because the second and the third chlorophylls are located in the interface between PsaA and PsaB, each being in van der Waals contact with both subunits. Coordination of Mg^{2+} by water, as found for the accessory chlorophylls, is also observed for 15 chlorophylls of the PSI core antenna [17]. A more surprising result in the crystal structure was that strictly conserved methionines, M688 in PsaA and M668 in PsaB of *T. elongatus,* axially coordinate the Mg^{2+} of the third chlorophylls eC-A3 and eC-B3 each by a sulfur atom. This coordination is unusual and the corresponding interaction expected to be very weak, as Mg^{2+} is a hard acid and the methionine sulfur is a soft base [17, 27]. As a weak ligand, the methionine side chain might be responsible for the relatively low redox potential of −1.05 V [117] of the A_0/A_0^- redox pair. The effects of site-directed mutations of the methionines serving as ligands to eC-A3 and eC-B3 have been studied in the green alga *C. rheinhardtii* [125, 126] and the cyanobacterium *Synechocystis* sp. PCC 6803 [127]. In the green alga, single mutations of the methionines corresponding to M688 in PsaA and to M668 in psaB of *T. elongatus* to histidine blocks electron transfer to the A-branch phylloquinone (Q_K-A) and strongly inhibits electron transfer to the B-branch phylloquinone (Q_K-B), respectively [125]. As only the B-branch mutant does not grow photoautotrophically, the authors conclude that the B-branch of the electron transfer chain is essential for photoautotrophic growth. Similar results are reported in another work, where the methionines were also mutated to leucine and serine residues [126], although limited photoautotrophic growth was also found in the B-branch mutants. Both of these studies indicate that, at least in green algae, the A- and the B-branches are active in electron transfer, but the B-branch seems to be more important. The blockage or inhibition of forward electron transfer from A_0^- in the histidine mutants could indicate the suggested responsibility of the methionine ligands for the low redox potential of A_0/A_0^-. Assuming that in the mutants, a nitrogen of the imidazole ring of histidine coordinates the central Mg^{2+} ion of the respective chlorophyll, this would be a strong interaction between a hard base and a hard acid which might result in a positive shift in the redox potential. In the extreme case, this could lead to endergonic forward electron transfer from A_0^- to A_1, and should strongly affect the kinetics of this reaction. This interpretation relies on the assumption that the histidine residues in the mutants have, in fact, substituted the methionines as axial ligands. In the cyanobacterium, the corresponding methionines were mutated to leucine residues. The B-branch mutant showed only minor changes in its physiological and spectroscopic properties, whereas the A-branch mutant showed slower growth rates, increased light sensitivity, reduced amounts of PSI, and less efficient forward electron transfer, in comparison with the wild type [127]. According to this study, electron transfer is asymmetric and proceeds primarily along the A-branch in cyanobacterial PSI. The results obtained with the mutants in the cyanobacterium and the green alga might be interpreted as contradictory, but another possibility is that they indicate a substantial difference in the directionality of electron transfer between prokaryotic and eukaryotic PSI.

Another difference between the second and third pairs of chlorophylls is that eC-A2 and eC-B2 are not involved in hydrogen-bonding, whereas eC-A3 and eC-B3 both accept a hydrogen bond via its keto group in ring V from the side chains of tyrosines Y696 in PsaA and Y676 in PsaB of *T. elongatus* [17, 27]. As discussed above (2.4.1), for eC-A1, this type of hydrogen bond should affect the conjugated π-electron system of the chlorophylls, and it might increase the redox potential of A_0/A_0^- with respect to a chlorophyll which does not have this hydrogen bond [128], similarly to an observation made for the pheophytin which is the primary acceptor in PSII [129]. Because of this, removing these hydrogen bonds, as was done in *C. rheinhardtii* single site mutants where the tyrosines interacting with eC-A3 and eC-B3 were substituted with phenylalanines [130], should reduce the change in Gibbs free energy for reduction of A_0 by P700* and thereby affect the rate, and/or the yield, of electron transfer. The results obtained by Li and coworkers can easily be interpreted in terms of the model of bidirectional electron transfer. According to this model, the rates of forward electron transfer measured for the reoxidation of the semiquinone radicals representing A_1^- were almost unaffected in both mutants, in comparison with the wild type, but the yields were significantly reduced for the branch affected by the mutation. For a detailed discussion, see Rappaport and coworkers, 2006 [99]. Irrespective of the importance of this study for the debate on the directionality of electron transfer in PSI, it does demonstrate the role of the hydrogen bonds to eC-A3 and eC-B3 in adjusting the redox properties of these chlorophylls for efficient forward electron transfer.

2.4.3 The Pair of Phylloquinones

One, or both, of the phylloquinones Q_K-A and Q_K-B are identical, with the secondary electron acceptor A_1. Reduction of A_1 by A_0^- is very fast, occurring within 30 ps, thereby preventing charge recombination as a result of re-reduction of $P700^+$ by A_0^-. In order to keep as much as possible of the reducing power of P700*, the A_1/A_1^- couple has the most negative redox potential of known quinone cofactors in nature, estimated to be <-700 mV [1, 131]. This redox potential is roughly 600–700 mV more negative than that of the quinones Q_A in type-II reaction centers from purple bacteria [132] and PSII [133]. Possible causes of the unusual redox potential can be found by comparing the protein-cofactor interactions into which the different quinones are involved (Figure 2.3c); (see also Chapters 4 and 12). Whereas, in each of the Q_A molecules, both carbonyl oxygens accept hydrogen bonds from the protein [35, 40], each of the phylloquinones in PSI forms only one hydrogen bond to the backbone NH groups of L722 in PsaA and L706 in PsaB of *T. elongatus* [17]. According to theoretical work, the lack of a single hydrogen bond in Q_K-A and Q_K-B may contribute to lowering their redox potentials [134]. Another contribution to a negative shift in redox potential by 50–150 mV [135] may be caused by the pronounced π-stacking which is found in the crystal structure of PSI between the quinones and the indole rings of W697 in PsaA and of W677 in PsaB [17]. Using quantum chemical methods, Kaupp proposed that a semiquinone radical anion which is formed upon reduction of a quinone would prefer a T-stacked arrangement with perpendicular ring systems and an N-H···π bond donated by the Trp side chain to the quinone headgroup; π-stacking would destabilize the semiquinone. Transient EPR spectroscopy [136], and modeling of the phylloquinone binding site based on the crystallographic structure at 4 Å resolution and EPR data [137], already suggested that the π-stacked arrangement is also present in the reduced state of the quinone. Reorientation of the quinone is probably prevented by sterical restraints, of which the most important are the hydrophobic interactions of the long phytyl chain of phylloquione with the protein. The idea that π-stacking plays a role in lowering the redox potential of the phylloquinone in PSI is supported by the fact that Q_A in bacterial reaction centers such as that from *Rhodopseudomonas viridis* [35] shows a weaker π-stacking with a tryptophan residue and that such an interaction is missing in PSII [40]. It was estimated that in PSI complexes from mutant cyanobacteria that cannot synthesize phylloquinone and bind plastoquinone in the Q_K-A and Q_K-B sites [138], the redox potential of the quinone is positively shifted by 135 mV [139]. This indicates that, indeed, the major contribution to the difference in quinone redox potential between PSI and type-II reaction centers must be caused by different protein environments. Structure based calculations suggest redox potentials of −531 mV for Q_K-A and of −686 mV for Q_K-B, and the authors ascribe a crucial role to the negatively charged F_X in causing the negative redox potentials [140]. The geometric counterpart to F_X in type-II reaction centers is the positively charged non-heme iron. A surprising result of the study by Ishikita and Knapp is the significant difference of circa 150 mV between the quinone redox potentials in both branches, which was attributed to different conformations of the polypeptide backbones and to a change in protonation state of D575 in PsaB upon quinone reduction, which induces asymmetry in the negatively charged state because this residue has a neutral glutamine as counterpart in PsaA (Q588).

That a fast (~10 ns) and a slow (~200 ns) phase for reoxidation of A_1^- exist in PSI has been known for quite a long time, but different interpretations of these biphasic kinetics are possible [1]. The observation of similar kinetics of A_1^- reoxidation in vivo in a green alga by optical spectroscopy led the authors to suggest that two phylloquinones in the symmetrically equivalent branches of the electron transfer chain within PSI are involved in electron transfer [141]. This hypothesis was supported by a mutagenesis study on PSI from the green alga *C. rheinhardtii* [142]. Mutation of the tryptophan in PsaA which is π-stacked to Q_K-A slowed the slower phase of A_1^- reoxidation, whereas the faster phase was slowed in the corresponding PsaB-side mutant. Within the bidirectional electron transfer scheme, this would mean that reoxidation of the quinones by the next acceptor in the electron transfer chain, F_X, is faster in the B- than in the A-branch. The time-resolution of EPR methods is insufficient to detect the fast kinetic phase which was observed by optical spectroscopy, but these techniques, when applied to the same mutants, which were used by Guergova-Kuras and coworkers [142], also showed that the slow phase is associated with the A-branch [143]. In more extensive mutagenesis studies on cyanobacterial PSI, amino acid residues were changed; they are part of the hydrogen-bonding network with which the quinones are involved. Based on optical spectroscopy and EPR spectroscopy on these mutants [144, 145], it was also concluded that the slow kinetic phase is associated with electron transfer along the A-branch and, because the relative amplitude of this phase is greater than 70%, the authors propose that electron transfer in PSI should be regarded as being strongly asymmetric in favor of the A-branch. They furthermore suggest that the asymmetry is determined by the asymmetric properties of P700.

As a possible origin of the different rates of forward electron transfer from A_1^-, different redox potentials for the two quinones were suggested [140, 146], requiring different local environments of these cofactors. Another possibility is different electronic couplings between each of the quinones and F_X. A striking element of asymmetry, which was suggested as being responsible for different rates of electron transfer from the two branches to F_X, is a tryptophan, W673 in PsaB of *T. elongatus,* which is not conserved in PsaA [17, 27]. The aromatic side chain of this residue is located in a position that suggests its suitability for a role as an intermediate electron acceptor between A_1 and F_X [147]; this should facilitate faster electron transfer along the B-branch. In a more recent study, mutation of the tryptophan to glycine was shown to slow the reoxidation of A_1^- [148]. The authors concluded that this mutation did not reduce the electronic coupling between Q_K-B and F_X, but affected the redox potential of the phylloquinone bound to PsaB. There are more asymmetric features in the region near the phylloquinones and F_X that might contribute to the observed difference in electron transfer rates. At a distance of circa 14 Å from Q_K-A, the negatively charged phosphodiester group of a phosphatidylgycerol is located, having the neutral galactosyl residue of a monogalactosyldiacylglycerol as its pseudosymmetric counterpart located at the same distance from Q_K-B [17, 27]. Another interesting and unusual feature is the presence of ordered water molecules in the intra-membrane space close to the quinones and F_X. These water molecules are arranged in two clusters, one involved in hydrogen-bonds with PsaA, the other with PsaB. The clusters differ in their geometries; the PsaA-bound cluster consists of five molecules, the PsaB-bound has six molecules. The water clusters could clearly play a role in affecting the protonation state of D575 in PsaB, the residue which was discussed by Ishikita and Knapp [140] as contributing to the difference in the quinone redox potentials and is hydrogen bonded to the water cluster in PsaB. The asymmetry induced by the water clusters might also contribute to differences in activation and reorganization energies of electron transfer along the two branches [28]. In fact, from the temperature dependence of the biphasic electron transfer from phylloquinone to F_X, an activation energy of 110 meV was deduced for the slower phase, whereas the faster phase was shown to be activationless [146].

2.4.4 The First [4Fe4S] Cluster, F_X

Because of its high electron density and its location on the pseudo-C_2 axis relating PsaA and PsaB (Figures 2.3a and 2.1c), the [4Fe4S] cluster F_X was already identified in the electron density map of cyanobacterial PSI at 6 Å resolution [16]. That the four strictly conserved cysteines, C578 and C587 in PsaA and C565 and C574 in PsaB of *T. elongatus,* are ligands to this clusters was suggested on the basis of the first amino acid sequences of PsaA/PsaB [34, 50]. The coordination of Fx by these cysteines was experimentally supported by mutagenesis studies [149, 150] and finally confirmed by the crystal structure at 2.5 Å resolution [17]. F_X is of central importance for the stability of the PSI core and for the stable attachment of the stromal subunits to the PSI core, as discussed above (2.3.1). The structural role of F_X was also shown in an elegant way by experiments using a mutant of the cyanobacterium *Synechococcus* sp. PCC 7002 which lacks an unusual rubredoxin, RubA [151, 152]. PSI complexes in this mutant lack F_X and the stromal subunits PsaC, PsaD, and PsaE, strongly suggesting that RubA is required for assembly of F_X into PSI, and also that stabilization of the stromal loop regions in PsaA and PsaB by F_X is necessary for the attachment of the stromal subunits to the PSI core.

Even when spectroscopic methods could be used to prove that F_X really is a [4Fe4S] cluster [153, 154], it remained unclear whether it is involved in forward electron transfer to the terminal [4Fe4S] clusters F_A and F_B [155]. The redox potential of F_X being <-700 mV [156, 157] probably makes it the most reducing iron-sulfur cluster in biology. The redox potential is similar to that of A_1, and much more negative than that of F_A and F_B, suggesting that forward electron transfer from F_X to F_A / F_B occurs in PSI. That F_X really is an intermediate acceptor in the electron transfer chain between A_1 and F_A / F_B was demonstrated by measuring identical reoxidation kinetics of A_1^- in the presence and the absence of F_A and F_B [158].

2.4.5 The Terminal [4Fe4S] Clusters, F_A and F_B

The [4Fe4S] clusters, F_A and F_B, are the terminal electron acceptors in PSI and the only cofactors of the electron transfer chain which are not bound to PsaA/PsaB

(Figures 2.3a and 2.1d). Removal of these clusters together with the three stromal subunits by chaotropic agents [59] has proved to be a helpful method in studying the role in electron transfer and the redox properties of the remaining cofactors of the electron transfer chain. Like F_X, F_A and F_B had already been identified in the electron density map of cyanobacterial PSI at 6 Å resolution [16]. But only results from mutagenesis studies involving the proposed cysteine ligands to the [4Fe4S] clusters, which identified the residues responsible for binding F_A and F_B [159, 160], helped to decide which of the two clusters in the stromal ridge is F_A and which is F_B. This was accomplished when the crystal structure at 2.5 Å resolution became available [17]. However, since the asymmetric arrangement of the two clusters, one located much closer (14.9 Å) to F_X than the other (22.0 Å) already visible in the lower resolution PSI crystal structures, suggested sequential electron transfer along the three [4Fe4S] clusters, and since a sequence of electron transfer reactions, $F_X \rightarrow F_A \rightarrow F_B \rightarrow$ ferredoxin, had already been concluded from the available experimental data in the late 1990s (for review see Golbeck, 1999 [161]), this assignment was not surprising, as it stated that F_B is more distal from F_X than F_A. Each electron transfer step from F_X to F_A and from F_A to F_B occurs in less than 500 ns [1]. For PSI from spinach, redox potentials of −540 mV and −590 mV were reported for F_A and F_B, respectively [117]. This would mean that electron transfer from F_A to F_B is energetically uphill and, in the absence of a soluble electron acceptor, light-induced charge separation in PSI should primarily lead to reduction of F_A (for a detailed discussion see Brettel, 1997 [117]). When ferredoxin or flavodoxin bind to PSI, the redox potentials of F_A and F_B may be adjusted to facilitate electron transfer to the soluble electron carriers [1]. EPR studies on PSI-ferredoxin cocrystals [60] suggest that the physico-chemical properties of F_B might indeed be altered upon binding of ferredoxin to PSI. Grotjohann and Fromme [28] propose that the physiological relevance of such a mechanism could be to minimize reduction of dioxygen by reduced F_B in the so-called Mehler reaction [162], leading to highly reactive superoxide O_2^-. The reasoning behind this is that F_B is located relatively close to the solvent-exposed surface of PSI, whereas F_A is deeply buried inside the protein and the electron transfer to dioxygen should be slow.

2.5 The Core Antenna System

Independently of interactions with different peripheral antenna proteins, PSI in cyanobacteria, green algae, and higher plants always harbor a core antenna system that, in the cyanobacterium *T. elongatus,* consists of 90 chlorophyll *a* and 22 carotenoid molecules, according to the crystal structure (Figure 2.4a and b) [17]. The function of these pigments is to absorb light and to transfer the excitation energy to the chlorophylls in the center of the complex, often called the reaction center, where the energy is used to initiate transmembrane charge separation. At room temperature, the core antenna of PSI is remarkably efficient in transferring excitation energy to the reaction center, which is expressed by a very low fluorescence yield of less than 1% [163] and a quantum yield of charge separation greater than 95% [164].

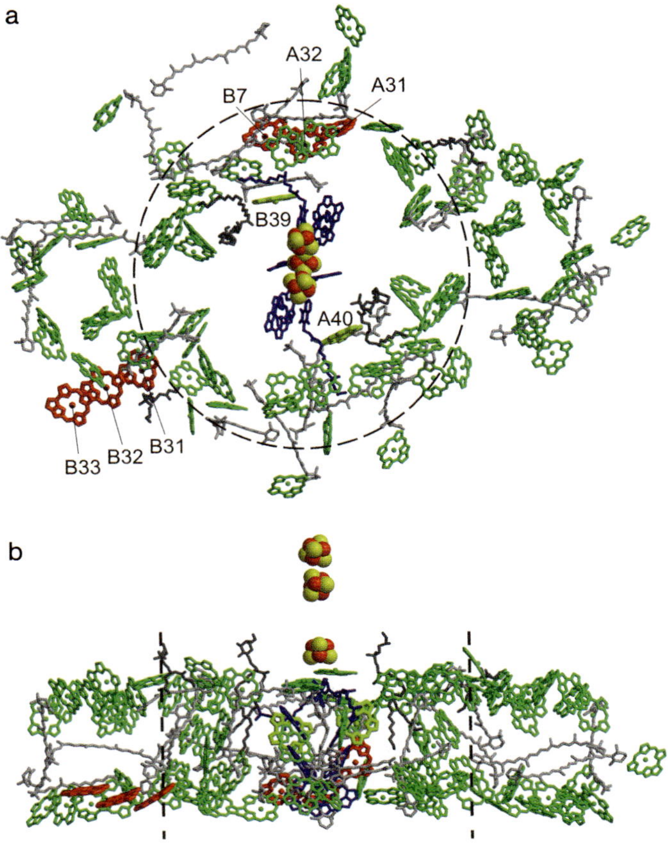

Figure 2.4 The PSI core antenna system. **(a)** Arrangement of the cofactors in a PSI monomer, shown in the same orientation as in Figure 2.1c. The antenna chlorophylls which have been mentioned in the text are highlighted by larger stick radii and labeled. The chlorophylls highlighted in red are assumed to belong to the "red chlorophylls"; the chlorophylls A40 and B39 shown in yellow are the "connecting chlorophylls" which are located closest to the electron transfer chain. The large ellipse centered at the pseudo-C_2 axis roughly indicates the boundary of the central domain of the PSI core antenna. **(b)** Same as in (a), but rotated 90° about the horizontal axis. The vertical broken lines indicate the boundary of the central antenna domain. Most of the chlorophylls in the peripheral domains are arranged in two layers, one at the lumenal and one at the stromal side.

2.5.1
The Antenna Chlorophylls

The majority of chlorophylls are axially coordinated by the PSI core subunits PsaA and PsaB, only ten are coordinated by the small membrane intrinsic subunits, and one is coordinated by the negatively charged phosphodiester group of a phosphatidylglycerol which is, in turn, bound to PsaA. By choosing this description, the role of the small subunits in stabilizing the complex structure of the PSI core antenna is clearly underestimated, as these proteins are in contact with many more chlorophylls, forming hydrogen bonds and hydrophobic interactions as mentioned above (2.3.3). The 79 chlorophylls which are coordinated by PsaA/PsaB follow the pseudo-C_2 symmetry of the PSI core in a loose sense [17, 83]. The antenna chlorophylls form a network with an average center-to-center distance between nearest neighbors of 9.9 Å [17, 83], facilitating efficient excitation energy transfer of the Förster type [165] and strong excitonic couplings within this network. Because the orientations of the chlorin rings of all chlorophylls are known from the X-ray structure, it is possible to assign the Q_y transition dipole moments of the electronic transition associated with the absorption band around 670 nm to the individual chlorophylls in the PSI crystal structure [17], and to model spectroscopic properties and kinetics of excitation energy transfer on the basis of excitonic coupling theory [83]. The structure based analysis shows that the arrangement of chlorophylls in PSI has been optimized for fast excitation energy transfer by the Förster mechanism between a single chlorophyll and two neighbors within a center-to-center distance of 12.5 Å on average. A side view of the arrangement of cofactors in PSI along the membrane plane and roughly perpendicular to the long axis of PsaA/PsaB (Figure 2.4b) shows that the core antenna system of PSI consists of a central domain which surrounds the cofactors of the electron transfer chain and two peripheral domains attached to it on both sides. In the central domain, which is well separated from the electron transfer chain, with the exception of two connecting chlorophylls A40 and B39 (Figure 2.4a), the chlorophylls are distributed over the full depth of the membrane. In contrast, most of the chlorophylls in the peripheral domains are organized in two layers, one located close the stromal surface, the other close to the lumenal surface of the membrane. Within the layers of the peripheral domains, excitation energy transfer can be expected to be very fast, and energy transfer between these layers should occur on average at slower rates [83]. The characteristic arrangement of the core antenna in these domains may favor fast energy transfer from the peripheral domain to the central domain, where numerous chlorophylls are in positions and orientations suitable for fast energy transfer to one of the reaction center chlorophylls, eC-A1 to eC-B3 [83]. Model calculations have not shown a crucial role for the connecting chlorophylls A40 and B39 in transferring energy to the reaction center [83, 166]. Another role suggested for these molecules is participation in excitonic coupling with the six reaction center chlorophylls with the effect of enhancing energy flow to the reaction center by increasing the spectral overlap between the core antenna and the chlorophylls of the electron transfer chain [167], as supported by analyses of mutants where the axial ligands to the connecting chlorophylls were altered [168].

A striking difference between the core antennae of PSI and PSII is that PSII almost completely lacks the central domain [36–40, 169] (see also Chapter 4), which may explain the lower efficiency of excitation energy transfer from the core antenna to the reaction center in PSII, as expressed by the only 85% quantum yield of charge separation [164]. Structure-based model calculations on the excitation energy transfer in the PSI trimer [84] indicate a 40% probability that excitation of chlorophyll in one PSI monomer migrates to a second monomer and is trapped there due to primary charge separation. This might increase the robust nature of PSI in trapping excitation energy by making it more tolerant of dysfunctional reaction centers in one or two monomers in the trimer [84]. This could be one reason why PSI trimers dominate when cyanobacteria grow under low light conditions [9].

The vast majority of antenna chlorophylls in PSI, the bulk chlorophylls, gives rise to a broad absorption band which shows a maximum at 680 nm [166]. A unique feature of PSI in eukaryotic and prokaryotic organisms is that it contains a number of chlorophylls that absorb at longer wavelengths than the primary donor P700. At room temperature, there is sufficient spectral overlap of these red chlorophylls with the antenna and the reaction center chlorophylls to enable efficient energy transfer, in line with a low fluorescence yield. At cryogenic temperatures, the fluorescence yield is increased more than 10-fold [163]; a significant decrease in P700 oxidation at wavelengths longer than 700 nm indicates that, under these conditions, the thermal energy is insufficient to allow uphill excitation energy transfer from the red chlorophylls to P700. The numbers and the spectral properties of red chlorophylls vary significantly between PSI from different organisms [166, 170]. This observation may be indicative of the suggested role of red chlorophylls in increasing the cross-section for absorption of red light by the PSI antenna [166, 171], which has been demonstrated to be responsible for a significant contribution to the total photon capture in plant PSI under different shade-light conditions [172]. Correspondingly, the variation in the content and absorption of red chlorophylls in different PSI complexes may be correlated to the different natural environments of oxygenic photosynthetic organisms. Other functions proposed for the red chlorophylls are to concentrate excitation energy close to P700, or to be involved in photoprotection (for reviews see Karapetyan and colleagues, 1999 [170]; Gobets and van Grondelle, 2001 [166]).

An interesting question is: why are red chlorophylls only found in PSI? The answer, given by Gobets and coworkers, 2001 [173], is that, in spite of an up to threefold increase in the time required for trapping of excitation energy due to charge separation as a result of the presence of red chlorophylls in PSI, these pigments lead to a decrease in the quantum yield of charge separation by less than 1%. In contrast, slowing down the trapping time by a factor of three in PSII would decrease the quantum efficiency by 20%. This means that the much higher efficiency of PSI is the reason why it can increase its absorption bandwidth during evolution by making use of red chlorophylls.

There may be several causes of red-shifted absorption for chlorophylls in PSI [174], of which excitonic coupling can be most safely deduced from the crystal structure [17]. That excitonic coupling within chlorophyll dimers or higher multimers contributes to the long-wavelength absorption in PSI was shown by spectroscopic methods [175–177], but other factors, such as interactions between chlorophylls and the protein, have to be taken into account in theoretical calculations of the PSI absorption spectrum based on the crystal structure, to explain the observed magnitudes in spectral shifts [111, 178]. Several suggestions have been made about which of the chlorophylls in the crystal structure of PSI from *T. elongatus* are the red ones (for a detailed discussion and review see Karapetyan and coworkers, 2006 [179]). It has been suggested by several authors [83, 84, 111] that the chlorophylls belonging to the chlorophyll trimer A31/A32/B7 (Figure 2.4a), located near the lumenal side, and in van der Waals contact with PsaL, belong to the red chlorophylls. Analysis of the excitation wavelength dependence on the fluorescence kinetics in PSI from two different cyanobacteria led to the assignment of a unique chlorophyll trimer, B31/B32/B33 (Figure 2.4a, b), located close to the lumenal surface in one of the peripheral domains of the PSI core antenna in *T. elongatus*, to the pool of the red-most absorbing chlorophylls (C719) in this complex [180]. This assignment is supported by a comparison of biochemical and spectroscopic data on PSI from *T. elongatus* and *Synechocystis* sp. PCC 6803. The latter cyanobacterium does not contain a pool of C719 chlorophylls; a lumenal loop region in PsaB of *T. elongatus* stabilized by PsaX, which is involved in hydrophobic interactions with chlorophyll B33, is too short to stabilize a chlorophyll in *Synechocystis* sp. PCC 6803. In addition to the incompleteness of the chlorophyll B33 binding site, there are more indications, such as the absence of PsaX in *Synechocystis* sp. PCC 6803 and the replacement of the axial ligand to B31, histidine H470 in PsaB of *T. elongatus*, by a serine in *Synechocystis* sp. PCC 6803, that the trimer B31/B32/B33 does not exist in *Synechocystis* sp. PCC 6803. Interestingly, the crystal structure of plant PSI shows that, of the chlorophyll trimer, B31 and B32 are conserved, whereas B33 is missing in the plant complex because the lumenal loop covering the chlorophylls from the lumenal side is, in fact, too short to stabilize a third chlorophyll [22]. This fits the arguments supporting the assignment of B31/B32/B33 in *T. elongatus* PSI to the red chlorophylls because the PSI core in plants does not have a pool of C719 chlorophylls. It is remarkable that the positions of 92 chlorophylls are identical in the plant PSI core and PSI from the cyanobacterium *T. elongatus*. The notable degree of conservation may be related to the highly optimized efficiency of the PSI core antenna. The variations in the core antenna during evolution may have occurred to adapt to environmental conditions by establishing specific pools of red chlorophylls not dependent on whether the organism was a prokaryote or eukaryote.

2.5.2
The Carotenoids

In the crystal structure of cyanobacterial PSI, 22 carotenoid molecules have been identified and modeled as β-carotene, of which only one is incomplete [17]. Carotenoids have several functions in photosynthesis. They can serve as antenna pigments which extend the spectral range available for light capturing, they are important for the structural integrity of protein–pigment complexes [181], they can be involved in electron transfer processes as in PSII (see also Chapter 4), and they have a central role in photoprotection [182]. Under conditions when the organism or the pigment-protein complex is over-exposed to light, there is increased probability that, from an excited singlet state of chlorophyll ($^1Chl^*$), a triplet state ($^3Chl^*$) is generated. $^3Chl^*$ is a dangerous species to the system because it can react with oxygen which is normally in its triplet state ($^3O_2$) to produce highly reactive singlet oxygen ($^1O_2^*$). As singlet oxygen can cause severe damage to the organism, quenching of $^3Chl^*$ is crucial for its survival. The crystal structure shows tight interactions of each carotenoid with at least one of the antenna chlorophylls (Figure 2.4a), which is necessary for effective quenching of excited chlorophyll triplet states by a carotenoid which produces a triplet carotenoid ($^3Car^*$) and a singlet chlorophyll in its ground state (^1Chl) in an electron exchange mechanism [183, 184]. The energy of $^3Car^*$ is too low to produce singlet oxygen, and $^3Car^*$ returns to its ground state by dissipating the excess energy as heat. Because of the very short lifetimes of the excited states of carotenoids, their distances to chlorophylls have to be short for efficient excitation energy transfer [185, 186]. That carotenoids are involved in light capturing in PSI was demonstrated by ultrafast spectroscopy which indicated an overall yield of about 90% for carotenoid-to-chlorophyll excitation energy transfer [187] in PSI, as opposed to an only ~35% yield in PSII [188]. The functional relevance of another feature of the carotenoid substructure, revealed by the crystal structure analysis of PSI, remains to be clarified. In good agreement with high-performance liquid chromatography analysis on PSI from the cyanobacterium *Thermosynechococcus vulcanus* [189], of the 21 complete carotenoids in the crystal structure, 16 are in the *all-trans* configuration, whereas 5 were found in different *cis* configurations. As none of either the *all-trans* or the *cis*-carotenoids are located close enough to a chlorophyll of the electron transfer chain for effective quenching of a chlorophyll triplet state, a specific photoprotective role of *cis*-carotenoids for the reaction center of PSI, as it was found for the photosynthetic reaction centers of purple bacteria [184], can be ruled out. It might be supposed that the *cis*-carotenoids should be located next to red chlorophylls because their energetically low-lying excited singlet states should have the highest probability to form excited triplet states. However, the distribution of carotenoid stereoisomers in the PSI structure does not provide much support for this idea. The role of carotenoid stereoisomers in PSI is, therefore, a subject for future studies.

Acknowledgements

I acknowledge and thank my colleagues who worked with me on the structure analysis of cyanobacterial PSI over a long period of time: Patrick Jordan, Petra Fromme, Olaf Klukas, Wolf-Dieter Schubert, Petra Jekow, Winfried Hinrichs, Ingrid Witt, Horst Tobias Witt and Wolfram Saenger. I thank my godson Lukas Nebel for supporting the literature search. The work on the PSI structure was funded by Deutsche Forschungsgemeinschaft (SFB 312 and SFB 498).

References

1 Brettel, K. and Leibl, W. (2001) Electron transfer in photosystem I. *Biochimica et Biophysica Acta*, **1507**, 100–14.

2 Rogers, L.J. (1987) Ferredoxins, flavodoxins and related proteins: structure, function and evolution, in *The Cyanobacteria* (eds P. Fay and C. Van Baalen), Elsevier, Amsterdam, pp. 35–67.

3 Navarro, J.A., Hervás, M. and De la Rosa, M.A. (1997) Co-evolution of cytochrome c_6 and plastocyanin, mobile proteins transferring electrons from cytochrome *b6f* to photosystem I. *Journal of Biological Inorganic Chemistry*, **2**, 11–22.

4 Hladik, J. and Sofrova, D. (1991) Does the trimeric form of the Photosystem 1 reaction center of cyanobacteria *in vivo* exist? *Photosynthesis Research*, **29**, 171–5.

5 Westermann, M., Neuschaefer-Rube, O., Morschel, E. and Wehrmeyer, W. (1999) Trimeric photosystem I complexes exist in vivo in thylakoid membranes of the *Synechocystis* strain BO9201 and differ in absorption characteristics from monomeric photosystem I complexes. *Journal of Plant Physiology*, **155**, 24–33.

6 Kruip, J., Bald, D., Boekema, E. and Rögner, M. (1994) Evidence for the existence of trimeric and monomeric photosystem I complexes in thylakoid membranes from cyanobacteria. *Photosynthesis Research*, **40**, 279–86.

7 Kruip, J., Chitnis, P.R., Lagoutte, B., Rögner, M. and Boekema, E. (1997) Structural organization of the major subunits in cyanobacterial photosystem 1. Localization of subunits PsaC, -D, -E, and -J. *Journal of Biological Chemistry*, **272**, 17061–9.

8 Kruip, J., Karapetyan, N.V., Terekhova, I.V. and Rögner, M. (1999) *In vitro* oligomerization of a membrane protein complex. Liposome-based reconstitution of trimeric photosystem I from isolated monomers. *Journal of Biological Chemistry*, **274**, 18181–8.

9 Fromme, P. (1998) Crystallization of photosystem i for structural analysis. *Habilitation*, Technical University Berlin, Berlin, Germany.

10 Fromme, P., Schlodder, E. and Jansson, S. (2003) Structure and function of the antenna system in photosystem I, in *Light-Harvesting Antennas in Photosynthesis* (eds B.R. Green and W.W. Parson), Kluwer Academic Publishers, Dordrecht, The Netherlands, pp. 253–79.

11 Boekema, E.J., Dekker, J.P., van Heel, M.G., Rögner, M., Saenger, W., Witt, I. and Witt, H.T. (1987) Evidence for a trimeric organization of the photosystem I complex from the thermophilic cyanobacterium *Synechococcus* sp. *FEBS Letters*, **217**, 283–6.

12 Kruip, J., Boekema, E.J., Bald, D., Boonstra, A.F. and Rögner, M. (1993) Isolation and structural characterization of monomeric and trimeric photosystem I complexes (P700•F_A/F_B and P700•F_X) from the cyanobacterium *Synechocystis* PCC 6803. *Journal of Biological Chemistry*, **268**, 23353–60.

13 Lelong, C., Boekema, E.J., Kruip, J., Bottin, H., Rögner, M. and Sétif, P. (1996) Characterization of a redox active crosslinked complex between cyanobacterial photosystem I and soluble ferredoxin. *EMBO Journal*, **15**, 2160–8.

14 Mühlenhoff, U., Kruip, J., Bryant, D.A., Rögner, M., Setif, P. and Boekema, E. (1996) Characterization of a redox-active cross-linked complex between cyanobacterial photosystem I and its physiological acceptor flavodoxin. *EMBO Journal*, **15**, 488–97.

15 Witt, I., Witt, H. T., Di Fiore, D., Rögner, M., Hinrichs, W., Saenger, W., Granzin, J., Betzel, C. and Dauter, Z. (1988) X-ray characterization of single crystals of the reaction center I of water splitting photosynthesis. *Berichte der Bunsen-Gesellschaft Physical Chemistry Chemical Physics*, **92**, 1503–6.

16 Krauss, N., Hinrichs, W., Witt, I., Fromme, P., Pritzkow, W., Dauter, Z., Betzel, C., Wilson, K.S., Witt, H.T. and Saenger, W. (1993) Three-dimesional structure of system I of photosynthesis at 6 Å resolution. *Nature*, **361**, 326–31.

17 Jordan, P., Fromme, P., Klukas, O., Witt, H.T., Saenger, W. and Krauß, N. (2001) Three-dimensional structure of cyanobacterial photosystem I at 2.5 Å resolution. *Nature*, **411**, 909–17.

18 Fromme, P. and Mathis, P. (2004) Unraveling the Photosystem I reaction center: a history, or the sum of many efforts. *Photosynthesis Research*, **80**, 109–24.

19 Krauß, N., Schubert, W.-D., Klukas, O., Fromme, P., Witt, H.T. and Saenger, W. (1996) Photosystem I at 4 Å resolution represents the first structural model of a joint photosynthetic reaction centre and core antenna system. *Nature Structural Biology*, **3**, 965–73.

20 Schubert, W.-D., Klukas, O., Krauß, N., Saenger, W., Fromme, P. and Witt, H.T. (1997) Photosystem I of *Synechococcus elongatus* at 4 Å resolution: comprehensive structure analysis. *Journal of Molecular Biology*, **272**, 741–69.

21 Klukas, O., Schubert, W.-D., Jordan, P., Krauß, N., Fromme, P., Witt, H.T. and Saenger, W. (1999) Photosystem I, an improved model of the stromal subunits PsaC, PsaD, and PsaE. *Journal of Biological Chemistry*, **274**, 7351–60.

22 Ben-Shem, A., Frolow, F. and Nelson, N. (2003) Crystal structure of plant photosystem I. *Nature*, **426**, 630–5.

23 Jolley, C., Ben-Shem, A., Nelson, N. and Fromme, P. (2005) Structure of plant photosystem I revealed by theoretical modeling. *Journal of Biological Chemistry*, **280**, 33627–36.

24 Amunts, A., Drory, O. and Nelson, N. (2007) The structure of a plant photosystem I supercomplex at 3.4 Å resolution. *Nature*, **447**, 58–63.

25 Golbeck, J.H. (1994) Photosystem I in cyanobacteria, in *The Molecular Biology of Cyanobacteria* (ed. D.A. Bryant), Kluwer Academic Publishers, Dordrecht, The Netherlands, pp. 319–60.

26 Chitnis, P.R. (1996) Photosystem I. *Plant Physiology*, **111**, 661–9.

27 Fromme, P., Jordan, P. and Krauß, N. (2001) Structure of photosystem I. *Biochimica et Biophysica Acta*, **1507**, 5–31.

28 Grotjohann, I. and Fromme, P. (2005) Structure of cyanobacterial Photosystem I. *Photosynthesis Research*, **85**, 51–72.

29 Fromme, P. and Grotjohann, I. (2006) Structural analysis of cyanobacterial photosystem I, in *Photosystem I: The Light-Driven Plastocyanin:Ferredoxin Oxidoreductase. Advances in Photosynthesis and Respiration*, Vol. **24** (ed. J.H. Golbeck), Springer, Dordrecht, The Netherlands, pp. 47–69.

30 Scheller, H.V., Jensen, P.E., Haldrup, A., Lunde, C. and Knoetzel, J. (2001) Role of subunits in eukaryotic Photosystem I. *Biochimica et Biophysica Acta*, **1507**, 41–60.

31 Knoetzel, J., Mant, A., Haldrup, A., Jensen, P.E. and Scheller, H.V. (2002) PSI-O, a new 10-kDa subunit of eukaryotic photosystem I. *FEBS Letters*, **510**, 145–8.

32 Loll, B., Raszewski, G., Saenger, W. and Biesiadka, J. (2003) Functional role of C^{a}–H···O hydrogen bonds between transmembrane a-helices in photosystem I. *Journal of Molecular Biology*, **328**, 737–47.

33 Saenger, W., Jordan, P. and Krauß, N. (2002) The assembly of protein subunits and cofactors in photosystem I. *Current Opinion in Structural Biology*, **12**, 244–54.

34 Fish, L., Kück, U. and Bogorad, L. (1985) Two partially homologous adjacent light-inducible maize chloroplast genes encoding polypeptides of the P700 chlorophyll a protein complex of hotosystem I. *Journal of Biological Chemistry*, **260**, 1413–21.

35 Deisenhofer, J., Epp, O., Sinning, I. and Michel, H. (1995) Crystallographic refinement at 2.3 Å resolution and refined model of the photosynthetic reaction centre from *Rhodopseudomonas viridis*. *Journal of Molecular Biology*, **246**, 429–57.

36 Zouni, A., Witt, H.T., Kern, J., Krauß, N., Saenger, W. and Orth, P. (2001) Crystal structure of photosystem II from *Synechococcus elongatus* at 3.8 Å resolution. *Nature*, **409**, 739–43.

37 Kamiya, N. and Shen, J.R. (2003) Crystal structure of oxygen-evolving photosystem II from *Thermosynechococcus vulcanus* at 3.7-Å resolution. *Proceedings of the*

National Academy of Sciences of the United States of America, **100**, 98–103.

38 Ferreira, K.N., Iverson, T.M., Maghlaoui, K., Barber, J. and Iwata, S. (2004) Architecture of the photosynthetic oxygen-evolving center. *Science*, **303**, 1831–8.

39 Biesiadka, J., Loll, B., Kern, J., Irrgang, K.-D. and Zouni, A. (2004) Crystal structure of cyanobacterial photosystem II at 3.2 Å resolution: a closer look at the Mn-cluster. *Physical Chemistry Chemical Physics*, **6**, 4733–6.

40 Loll, B., Kern, J., Saenger, W., Zouni, A. and Biesiadka, J. (2005) Towards complete cofactor arrangement in the 3.0 Å resolution structure of photosystem II. *Nature*, **438**, 1040–4.

41 Schubert, W.-D., Klukas, O., Saenger, W., Witt, H.T., Fromme, P. and Krauß, N. (1998) A common ancestor for oxygenic and anoxygenic photosynthetic systems: a comparison based on the structural model of photosystem I. *Journal of Molecular Biology*, **280**, 297–314.

42 Baymann, F., Brugna, M., Mühlenhoff, U. and Nitschke, W. (2001) Daddy, where did (PS)I come from? *Biochimica et Biophysica Acta*, **1507**, 291–310.

43 Grotjohann, I., Jolley, C. and Fromme, P. (2004) Evolution of photosynthesis and oxygen evolution: implications from the structural comparison of Photosystems I and II. *Physical Chemistry Chemical Physics*, **6**, 4743–53.

44 Hauska, G., Schoedl, T., Remigy, H. and Tsiotis, G. (2001) The reaction center of green sulfur bacteria. *Biochimica et Biophysica Acta*, **1507**, 260–77.

45 Neerken, S. and Amesz, J. (2001) The antenna reaction center complex of heliobacteria: composition, energy conversion and electron transfer. *Biochimica et Biophysica Acta*, **1507**, 278–90.

46 Bibby, T.S., Nield, J. and Barber, J. (2001) Iron deficiency induces the formation of an antenna ring around trimeric photosystem I in cyanobacteria. *Nature*, **412**, 743–5.

47 Boekema, E.J., Hifney, A., Yakushevska, A.E., Piotrowski, M., Keegstra, W., Berry, S., Michel, K.P., Pistorius, E.K. and Kruip, J. (2001) A giant chlorophyll–protein complex induced by iron deficiency in cyanobacteria. *Nature*, **412**, 745–8.

48 Nield, J., Morris, E.P., Bibby, T.S. and Barber, J. (2003) Structural analysis of the photosystem I supercomplex of cyanobacteria induced by iron deficiency. *Biochemistry*, **42**, 3180–8.

49 Barber, J., Nield, J., Duncan, J. and Bibby, T.S. (2006) Accessory chlorophyll proteins in cyanobacterial photosystem I, in *Photosystem I: The Light-Driven Plastocyanin:Ferredoxin Oxidoreductase. Advances in Photosynthesis and Respiration*, Vol. **24** (ed. J.H. Golbeck), Springer, Dordrecht, The Netherlands, pp. 47–69.

50 Cantrell, A. and Bryant, D.A. (1987) Molecular cloning and nucleotide sequence of the psaA and psaB genes of the cyanobacterium *Synechococcus* sp. PCC 7002. *Plant Molecular Biology*, **9**, 453–68.

51 Antonkine, M.L., Jordan, P., Fromme, P., Krauß, N., Golbeck, J.H. and Stehlik, D. (2003) Assembly of protein subunits within the stromal ridge of photosystem I. structural changes between unbound and sequentially PS I-bound polypeptides and correlated changes of the magnetic properties of the terminal iron sulfur clusters. *Journal of Molecular Biology*, **327**, 671–97.

52 Fromme, P., Schubert, W.D. and Krauß, N. (1994) Structure of Photosystem I: Suggestions on the docking sites for plastocyanin, ferredoxin and the coordination of P700. *Biochimica et Biophysica Acta*, **1187**, 99–105.

53 Frazão, C., Soares, C.M., Carrondo, M.A., Pohl, E., Dauter, Z., Wilson, K.S., Hervás, M., Navarro, J.A., De la Rosa, M. A. and Sheldrick, G. M. (1995) Ab initio determination of the crystal structure of cytochrome c_6 and comparison with plastocyanin. *Structure*, **3**, 1159–69.

54 Molina-Heredia, F.P., Díaz-Quintana, A., Hervás, M., Navarro, J.A. and De La Rosa, M.A. (1999) Site-directed mutagenesis of cytochrome c_6 from *Anabaena* species PCC 7119. Identification of surface residues of the hemeprotein involved in photosystem I reduction. *Journal of Biological Chemistry*, **274**, 33565–70.

55 Sun, J., Xu, W., Hervás, M., Navarro, J.A., De La Rosa, M.A. and Chitnis, P.R. (1999) Oxidizing side of the cyanobacterial photosystem I. Evidence for interaction between the electron donor proteins and a luminal surface helix of the PsaB subunit. *Journal of Biological Chemistry*, **274**, 19048–54.

56 Sommer, F., Drepper, F. and Hippler, M. (2002) The luminal helix l of PsaB is essential for recognition of plastocyanin or cytochrome c_6 and fast electron transfer to photosystem I in *Chlamydomonas reinhardtii*. *Journal of Biological Chemistry*, **277**, 6573–81.

57 Fromme, P., Melkozernov, A., Jordan, P. and Krauss, N. (2003) Structure and function of photosystem I: interaction with ist soluble electron carriers and external antenna systems. *FEBS Letters*, **555**, 40–4.

58 Sommer, F., Drepper, F., Haehnel, W. and Hippler, M. (2004) The hydrophobic recognition site formed by residues PsaATrp651 and PsaB-Trp627 of photosystem I in *Chlamydomonas reinhardtii* confers distinct selectivity for binding of plastocyanin and cytochrome c_6. *Journal of Biological Chemistry*, **279**, 20009–17.

59 Golbeck, J.H., Mehari, T., Parrett, K.G. and Ikegami, I. (1988) Reconstitution of the photosystem I complex from the P_{700} and F_X-containing reaction center core protein and the F_A/F_B polypeptide. *FEBS Letters*, **240**, 9–14.

60 Fromme, P., Bottin, H., Krauss, N. and Sétif, P. (2002) Crystallization and electron paramagnetic resonance characterization of the complex of photosystem I with its natural electron acceptor ferredoxin. *Biophysical Journal*, **83**, 1760–73.

61 Oh-Oka, H., Takahashi, Y., Kuriyama, K., Saeki, K. and Matsubara, H. (1988) The protein responsible for center A/B in spinach photosystem I: isolation with iron-sufur cluster(s) and complete sequence analysis, *Journal of Biochemistry*, **103**, 962–8.

62 Adman, E.T., Sieker, L.C. and Jensen, L.H. (1973) The structure of a bacterial ferredoxin. *Journal of Biological Chemistry*, **248**, 3987–96.

63 Fischer, N., Hippler, M., Sétif, P., Jacquot, J.-P. and Rochaix, J.-D. (1998) The PsaC subunit of photosystem I provides an essential lysine residue for fast electron transfer to ferredoxin. *EMBO Journal*, **17**, 849–58.

64 Sétif, P., Fischer, N., Lagoutte, B., Bottin, H. and Rochaix, J.-D. (2002) The ferredoxin docking site of photosystem I. *Biochimica et Biophysica Acta*, **1555**, 204–9.

65 Antonkine, M.L., Liu, G., Bentrop, D., Bryant, D.A., Bertini, I., Luchinat, C., Golbeck, J.H. and Stehlik, D. (2002) Solution structure of the unbound, oxidized photosystem I subunit PsaC, containing [4Fe–4S] clusters F_A and F_B: a conformational change occurs upon binding to Photosystem I. *Journal of Biological Inorganic Chemistry*, **7**, 461–72.

66 Jolley, C.C., Wells, S.A., Hespenheide, B.M., Thorpe, M.F. and Fromme, P. (2006) Docking of photosystem I subunit C using a constrained geometric simulation. *Journal of the American Chemical Society*, **128**, 8803–12.

67 Zanetti, G. and Merati, G. (1987) Interaction between photosystem I and ferredoxin. Identification by chemical cross-linking of the polypeptide which binds ferredoxin. *European Journal of Biochemistry*, **169**, 143–6.

68 Zilber, A.L. and Malkin, R. (1988) Ferredoxin cross-links to a 22 kDa subunit of photosystem I. *Plant Physiology*, **88**, 810–14.

69 Lelong, C., Sétif, P., Lagoutte, B. and Bottin, H. (1994) Identification of the amino acids involved in the functional interaction between photosystem I and ferredoxin from Synechocystis sp. PCC 6803 by chemical cross-linking. *Journal of Biological Chemistry*, **269**, 10034–9.

70 Chitnis, P.R., Xu, Q., Chitnis, V.P. and Neuchushtai, R. (1995) Function and organization of photosystem I polypeptides. *Photosynthesis Research*, **44**, 23–40.

71 Barth, P., Lagoutte, B. and Sétif, P. (1998) Ferredoxin reduction by photosystem I from *Synechocystis* sp. PCC 6803: toward an understanding of the respective roles of subunits PsaD and PsaE in ferredoxin binding. *Biochemistry*, **37**, 16233–41.

72 Falzone, C.J., Kao, Y.H., Zhao, J., Bryant, D.A. and Lecomte, J.T. (1994) Three-dimensional solution structure of PsaE from the cyanobacterium *Synechococcus* sp. strain PCC 7002, a photosystem I protein that shows structural homology with SH3 domains. *Biochemistry*, **33**, 6052–62.

73 Xia, Z., Broadhurst, R.W., Laue, E.D., Bryant, D.A., Golbeck, J.H. and Bendall, D.S. (1998) Structure and properties in solution of PsaD, an extrinsic polypeptide of photosystem I. *European Journal of Biochemistry*, **255**, 309–16.

74 Barth, P., Guillouard, I., Sétif, P. and Lagoutte, B. (2000) Essential role of a single arginine of photosystem I in stabilizing the electron transfer complex with ferredoxin. *Journal of Biological Chemistry*, **275**, 7030–6.

75 Sétif, P. (2001) Ferredoxin and flavodoxin reduction by photosystem I. *Biochimica et Biophysica Acta*, **1507**, 161–79.

76 van Thor, J.J., Geerlings, T.H., Matthijs, H.C. and Hellingwerf, K.J. (1999) Kinetic evidence for the PsaE-dependent transient ternary complex photosystem I/ferredoxin/ferredoxin:NADP$^+$ reductase in a cyanobacterium. *Biochemistry*, **38**, 12735–46.

77 Cassan, N., Lagoutte, B. and Sétif, P. (2005) Ferredoxin-NADP$^+$ reductase. Kinetics of electron transfer, transient intermediates, and catalytic activities studied by flash-absorption spectroscopy with isolated photosystem I and ferredoxin. *Journal of Biological Chemistry*, **280**, 25960–72.

78 Andersen, B., Scheller, H.V. and Møller, B.L. (1992) The PSI-E subunit of photosystem I binds ferredoxin: NADP$^+$ oxidoreductase. *FEBS Letters*, **311**, 169–73.

79 Yu, L., Zhao, J., Mühlenhoff, U., Bryant, D.A. and Golbeck, J.H. (1993) PsaE is required for in vivo cyclic electron flow around photosystem I in the cyanobacterium *Synechococcus* sp. PCC 7002. *Plant Physiology*, **103**, 171–80.

80 Kouril, R., Yeremenko, N., D'Haene, S., Yakushevska, A.E., Keegstra, W., Matthijs, H.C.P., Dekker, J.P. and Boekema, E.J. (2003) Photosystem I trimers from *Synechocystis* PCC 6803 lacking the PsaF and PsaJ subunits bind an IsiA ring of 17 units. *Biochimica et Biophysica Acta*, **1607**, 1–4.

81 Chitnis, V.P. and Chitnis, P.R. (1993) PsaL subunit is required for the formation of photosystem I trimers in the cyanobacterium *Synechocystis* sp. PCC 6803. *FEBS Letters*, **336**, 330–4.

82 Zhang, S.P. and Scheller, H.V. (2004) Light-harvesting complex II binds to several small subunits of photosystem I. *Journal of Biological Chemistry*, **279**, 3180–7.

83 Byrdin, M., Jordan, P., Krauss, N., Fromme, P., Stehlik, D. and Schlodder, E. (2002) Light harvesting in photosystem I: Modeling based on the 2.5-Å structure of photosystem I from *Synechococcus elongatus*. *Biophysical Journal*, **83**, 433–57.

84 Sener, M. K., Park, S., Lu, D., Damjanovic, A., Ritz, T., Fromme, P. and Schulten, K. (2004) Excitation migration in trimeric cyanobacterial photosystem I. *Journal of Chemical Physics*, **120**, 11183–95.

85 Haehnel, W., Jansen, T., Gause, K., Klosgen, R.B., Stahl, B., Michl, D., Huvermann, B., Karas, M. and Herrmann, R.G. (1994) Electron transfer from plastocyanin to photosystem I. *EMBO Journal*, **13**, 1028–38.

86 Farah, J., Rappaport, F., Choquet, Y., Joliot, P. and Rochaix, J.-D. (1995) Isolation of a psaF-deficient mutant of Chlamydomonas reinhardtii: efficient interaction of plastocyanin with the photosystem I reaction center is mediated by the PsaF subunit. *EMBO Journal*, **14**, 4976–84.

87 Hippler, M., Drepper, F., Farah, J. and Rochaix, J.-D. (1997) Fast electron transfer from cytochrome c6 and plastocyanin to photosystem I of *Chlamydomonas reinhardtii* requires PsaF. *Biochemistry*, **36**, 6343–9.

88 Hippler, M., Drepper, F., Haehnel, W. and Rochaix, J.D. (1998) The N-terminal domain of PsaF: precise recognition site for binding and fast electron transfer from cytochrome c_6 and plastocyanin to photosystem I of *Chlamydomonas reinhardtii*. *Proceedings of the National Academy of Sciences of the United States of America*, **95**, 7339–44.

89 Xu, Q., Yu, L., Chitnis, V.P. and Chitnis, P.R. (1994) Function and organization of photosystem I in a cyanobacterial mutant strain that lacks PsaF and PsaJ subunits. *Journal of Biological Chemistry*, **269**, 3205–11.

90 Kouril, R., Arteni, A.A., Lax, J., Yeremenko, N., D'Haene, S., Rögner, M., Matthijs, H.C.P., Dekker, J.P. and Boekema, E.J. (2005) Structure and functional role of supercomplexes of IsiA and Photosystem I in cyanobacterial photosynthesis. *FEBS Letters*, **579**, 3253–7.

91 Mullineaux, C.W. (1994) Excitation energy transfer from phycobilisomes to Photosystem I in a cyanobacterial mutant lacking Photosystem II. *Biochimica et Biophysica Acta*, **1184**, 71–7.

92 Jeanjean, R., Zuther, E., Yeremenko, N., Havaux, M., Matthijs, H.C.P. and Hagemann, M. (2003) A photosystem 1 psaFJ-null mutant of the cyanobacterium *Synechocystis* PCC 6803 expresses the isiAB operon under iron replete conditions. *FEBS Letters*, **549**, 52–6.

93 Koike, K., Ikeuchi, M., Hiyama, T. and Inoue, Y. (1989) Identifcation of photosystem I components from the cyanobacterium *Synechococcus vulcanus* by N-terminal sequencing. *FEBS Letters*, **253**, 257–63.

94 Ikeuchi, M., Nyhus, K.J., Inoue, Y. and Pakrasi, H.B. (1991) Identities of four low-molecular-mass subunits of the photosystem I complex from *Anabaena variabilis* ATCC 29413. Evidence for the presence of the psaI gene product in a cyanobacterial complex. *FEBS Letters*, **287**, 5–9.

95 Nakamura, Y., Kaneko, T., Sato, S., Ikeuchi, M., Katoh, H., Sasamoto, S., Watanabe, A., Iriguchi, M., Kawashima, K., Kimura, T., Kishida, Y., Kiyokawa, C., Kohara, M., Matsumoto, M., Matsuno, A., Nakazaki, N., Shimpo, S., Sugimoto, M., Takeuchi, C., Yamada, M. and Tabata, S. (2002) Complete genome structure of the thermophilic cyanobacterium *Thermosynechococcus elongatus* BP-1. *DNA Research*, **9**, 123–30.

96 Kaneko, T., Sato, S., Kotani, H., Tanaka, A., Asamizu, E., Nakamura, Y., Miyajima, N., Hirosawa, M., Sugiura, M., Sasamoto, S., Kimura, T., Hosouchi, T., Matsuno, A., Muraki, A., Nakazaki, N., Naruo, K., Okumura, S., Shimpo, S., Takeuchi, C., Wada, T., Watanabe, A., Yamada, M., Yasuda, M. and Tabata, S. (1996) Sequence analysis of the genome of the unicellular cyanobacterium *Synechocystis* sp. strain PCC6803. II. Sequence determination of the entire genome and assignment of potential protein-coding regions. *DNA Research*, **3**, 109–36.

97 Mant, A., Woolhead, C.A., Moore, M., Henry, R. and Robinson, C. (2001) Insertion of PsaK into the thylakoid membrane in a "horseshoe" conformation occurs in the absence of signal recognition particle, nucleoside triphosphates, or functional albino3. *Journal of Biological Chemistry*, **276**, 36200–6.

98 van der Est, A. (2006) Electron transfer involving phylloquinone in photosystem I, in *Photosystem I: The Light-Driven Plastocyanin:Ferredoxin Oxidoreductase. Advances in Photosynthesis and Respiration*, Vol. **24** (ed. J.H. Golbeck), Springer, Dordrecht, The Netherlands, pp. 387–411.

99 Rappaport, F., Diner, B.A. and Redding, K. (2006) Optical measurements of secondary electron transfer in photosystem I, in *Photosystem I: The Light-Driven Plastocyanin:Ferredoxin Oxidoreductase. Advances in Photosynthesis and Respiration*, Vol. **24** (ed. J.H. Golbeck), Springer, Dordrecht, The Netherlands, pp. 223–44.

100 Redding, K. and van der Est, A. (2006) The directionality of electron transport in photosystem I, in *Photosystem I: The Light-Driven Plastocyanin:Ferredoxin Oxidoreductase. Advances in Photosynthesis and Respiration*, Vol. **24** (ed. J.H. Golbeck), Springer, Dordrecht, The Netherlands, pp. 413–37.

101 Watanabe, T., Kobayashi, M., Hongu, A., Nakazato, M. and Hiyama, T. (1985) Evidence that a chlorophyll a′ dimer constitutes the photochemical reaction centre 1 (P700) in photosynthetic apparatus. *FEBS Letters*, **191**, 252–6.

102 Maeda, H., Watababe, T., Kobayashi, M. and Ikegami, I. (1992) Presence of two chlorophyll a molecules at the core of photosystem I. *Biochimica et Biophysica Acta*, **1099**, 74–80.

103 Plato, M., Krauß, N., Fromme, P. and Lubitz, W. (2003) Molecular orbital study of the primary elelctron donor P700 of photosystem I based on a recent X-ray single crystal structure analysis. *Chemical Physics*, **294**, 483–99.

104 Käß, H., Bittersmann-Weidlich, E., Andreasson, L.E., Bonigk, B. and Lubitz, W. (1995) ENDOR and ESEEM of the ^{15}N labeled radical cations of chlorophyll-a and the primary donor P_{700} in photosystem I. *Chemical Physics*, **194**, 419–32.

105 Käss, H., Fromme, P., Witt, H.T. and Lubitz, W. (2001) Orientation and electronic structure of the primary donor radical cation P_{700}^{+} in photosystem I: a single crystals EPR and ENDOR study. *Journal of Physical Chemistry B*, **105**, 1225–39.

106 Breton, J. (2001) Fourier transform infrared spectroscopy of primary electron donors in type I photosynthetic reaction centers. *Biochimica et Biophysica Acta*, **1507**, 180–93.

107 Breton, J., Xu, W., Diner, B.A. and Chitnis, P.R. (2002) The two histidine axial ligands of the primary electron donor chlorophylls (P700) in photosystem I are similarly perturbed upon $P700^{+}$ formation. *Biochemistry*, **41**, 11200–10.

108 Witt, H., Schlodder, E., Teutloff, C., Niklas, J., Bordignon, E., Carbonera, D., Kohler, S., Labahn, A. and Lubitz, W. (2002) Hydrogen bonding to P700: site-directed mutagenesis of threonine A739 of photosystem I in *Chlamydomonas reinhardtii*. *Biochemistry*, **41**, 8557–69.

109 Lubitz, W. (2006) EPR studies of the primary donor P700 in photosystem I, in *Photosystem I: The Light-Driven Plastocyanin:Ferredoxin Oxidoreductase. Advances in Photosynthesis and Respiration*, Vol. **24** (ed. J.H. Golbeck), Springer, Dordrecht, The Netherlands, pp. 245–69.

110 Müh, F., Lendzian, F., Roy, M., Williams, J.C., Allen, J.P. and Lubitz, W. (2002) Pigment-protein interactions in bacterial reaction centers and their influence on oxidation potential and spin density distribution of the primary donor. *Journal of Physical Chemistry B*, **106**, 3226–36.

111 Damjanovic, A., Vaswani, H.M., Fromme, P. and Fleming, G.R. (2002) Chlorophyll excitations in photosystem I of Synechococcus elongatus. *Journal of Physical Chemistry B*, **106**, 10251–62.

112 Sener, M.K., Lu, D., Ritz, T., Park, S., Fromme, P. and Schulten, K. (2002) Robustness and optimality of light harvesting in cyanobacterial photosystem I. *Journal of Physical Chemistry B*, **106**, 7948–60.

113 Madjet, M.E., Abdurahman, A. and Renger, T. (2006) Intermolecular coulomb couplings from ab initio electrostatic potentials: application to optical transitions of strongly coupled pigments in photosynthetic antennae and reaction centers. *Journal of Physical Chemistry B*, **110**, 17268–81.

114 Renger, T. and Schlodder, E. (2006) Modeling of optical spectra and light harvesting in photosystem I, in *Photosystem I: The Light-Driven Plastocyanin:Ferredoxin Oxidoreductase. Advances in Photosynthesis and Respiration*, Vol. **24** (ed. J.H. Golbeck), Springer, Dordrecht, The Netherlands, pp. 595–610.

115 Akiyama, M., Miyashita, H., Kise, H., Watanabe, T., Mimuro, M., Miyachi, S. and Kobayashi, M. (2002) Quest for minor but key chlorophyll molecules in photosynthetic reaction centers – unusual pigment composition in the reaction centers of the chlorophyll d-dominated cyanobacterium *Acaryochloris marina*. *Photosynthesis Research*, **74**, 97–107.

116 Rappaport, F., Guergova-Kuras, M., Nixon, P.J., Diner, B.A. and Lavergne, J. (2002) Kinetics and pathways of charge recombination in photosystem II. *Biochemistry*, **41**, 8518–27.

117 Brettel, K. (1997) Electron transfer and arrangement of the redox cofactors in photosystem I. *Biochimica et Biophysica Acta*, **1318**, 322–73.

118 Ishikita, H., Saenger, W., Biesiadka, J., Loll, B. and Knapp, E.-W. (2006) How photosynthetic reaction centers control oxidation power in chlorophyll pairs P680, P700, and P870. *Proceedings of the National Academy of Sciences of the United States of America*, **103**, 9855–60.

119 Moser, C.C. and Dutton, P.L. (1992) Engineering protein structure for electron transfer function in photosynthetic reaction centers. *Biochimica et Biophysica Acta*, **1101**, 171–6.

120 Klukas, O., Schubert, W.-D., Jordan, P., Krauß, N., Fromme, P., Witt, H.T. and Saenger, W. (1999) Localization of two phylloquinones, QK and QK*, in an improved electron density map of photosystem I at 4-Å resolution. *Journal of Biological Chemistry*, **274**, 7351–60.

121 Hastings, G., Hoshina, S., Webber, A.N. and Blankenship, R.E. (1995) Universality of energy and electron transfer processes in photosystem I. *Biochemistry*, **34**, 15512–22.

122 Müller, M.G., Niklas, J., Lubitz, W. and Holzwarth, A.R. (2003) Ultrafast transient absorption studies on photosystem I reaction centers from *Chlamydomonas reinhardtii*. 1. A new interpretation of the energy trapping and early electron transfer steps in photosystem I. *Biophysical Journal*, **85**, 3899–922.

123 Prokhorenko, V.I. and Holzwarth, A.R. (2000) Primary processes and structure of the photosystem II reaction center: a photon echo study. *Journal of Physical Chemistry B*, **104**, 11563–78.

124 Diner, B.A., Schlodder, E., Nixon, P.J., Coleman, W.J., Rappaport, F., Lavergne, J., Vermaas, W.F.J. and Chisholm, D.A. (2001) Site-directed mutations at D1-His198 and D2-His197 of photosystem II in *Synechocystis* PCC 6803: sites of primary charge separation and cation and triplet stabilization. *Biochemistry*, **40**, 9265–81.

125 Fairclough, W.V., Forsyth, A., Evans, M.C., Rigby, S.E., Purton, S. and Heathcote, P. (2003) Bidirectional electron transfer in Photosystem I: electron transfer on the PsaA side is not essential for phototrophic growth in *Chlamydomonas*. *Biochimica et Biophysica Acta*, **1606**, 43–55.

126 Ramesh, V.M., Gibasiewicz, K., Lin, S., Bingham, S.E. and Webber, A.N. (2004) Bidirectional electron transfer in photosystem I: accumulation of A_0^- in A-side or B-side mutants of the axial ligand to chlorophyll A_0. *Biochemistry*, **43**, 1369–75.

127 Cohen, R.O., Shen, G., Golbeck, J.H., Xu, W., Chitnis, P.R., Valieva, A.I., van der Est, A., Pushkar, Y. and Stehlik, D. (2004) Evidence for asymmetric electron transfer in cyanobacterial photosystem I: analysis of a methionine-to-leucine mutation of the ligand to the primary electron acceptor A_0. *Biochemistry*, **43**, 4741–54.

128 Krauß, N. (2003) Mechanisms for photosystems I and II. *Current Opinion in Chemical Biology*, **7**, 540–50.

129 Merry, S.A.P., Nixon, P.J., Barter, L.M.C., Schilstra, M.J., Porter, G., Barber, J., Durrant, J.R. and Klug, D. (1998) Modulation of quantum yield of primary radical pair formation in photosystem II by site-directed mutagenesis affecting radical cations and anions. *Biochemistry*, **37**, 17439–47.

130 Li, Y., van der Est, A., Lucas, M.G., Ramesh, V.M., Gu, F., Petrenko, A., Lin, S., Webber, A.N., Rappaport, F. and Redding, K. (2006) Directing electron transfer within Photosystem I by breaking H-bonds in the cofactor branches. *Proceedings of the National Academy of Sciences of the United States of America*, **103**, 2144–9.

131 Sétif, P. and Bottin, H. (1989) Identification of electron-transfer reactions involving the acceptor A_1 of photosystem I at room temperature. *Biochemistry*, **28**, 2689–97.

132 Prince, R.C., Leigh, J.S. and Dutton, P.L. (1976) Thermodynamic properties of the reaction center of *Rhodopseudomonas viridis*. In vivo measurement of the reaction center bacteriochlorophyll-primary acceptor intermediate electron carrier. *Biochimica et Biophysica Acta*, **440**, 622–36.

133 Krieger, A., Rutherford, A.W. and Johnson, G.N. (1995) On the determination of redox midpoint potential of the primary quinone electron acceptor, Q_A, in Photosystem II. *Biochimica et Biophysica Acta*, **1229**, 193–201.

134 O'Malley, P.J. (1999) Density functional calculated spin densities and hyperfine couplings for hydrogen bonded 1,4-naphthosemiquinone and phyllosemiquinone anion radicals: a model for the A_1 free radical formed in Photosystem I. *Biochimica et Biophysica Acta*, **1411**, 101–13.

135 Kaupp, M. (2002) The function of photosystem I. Quantum chemical insight into the role of tryptophan-quinone interactions. *Biochemistry*, **41**, 2895–900.

136 van der Est, A., Sieckmann, I., Lubitz, W. and Stehlik, D. (1995) Differences in the binding of the primary quinone acceptor in Photosystem I and reaction centres of *Rhodobacter sphaeroides*-R26 studied with transient EPR spectroscopy. *Chemical Physics*, **194**, 349–59.

137 Kamlowski, A., Altenberg-Greulich, B., van der Est, A., Zech, S.G., Bittl, R., Fromme, P., Lubitz, W. and Stehlik, D. (1998) The quinone acceptor A1 in photosystem I: binding site, and comparison to Q_A in purple bacteria reaction centers. *Journal of Physical Chemistry B*, **102**, 8278–87.

138 Semenov, A.Y., Vassiliev, I.R., van der Est, A., Mamedov, M.D., Zybailov, B., Shen, G., Stehlik, D., Diner, B.A., Chitnis, P.R. and Golbeck, J.H. (2000) Recruitment of a foreign quinone into the A_1 site of photosystem I. Altered kinetics of electron transfer in phylloquinone biosynthetic pathway mutants studied by time-resolved optical, EPR, and electrometric techniques. *Journal of Biological Chemistry*, **275**, 23429–38.

139 Shinkarev, V.P., Zybailov, B., Vassiliev, I.R. and Golbeck, J.H. (2002) Modeling of the $P700^+$ charge recombination kinetics with phylloquinone and plastoquinone-9 in the A_1 site of photosystem I. *Biophysical Journal*, **83**, 2885–97.

140 Ishikita, H. and Knapp, E.-W. (2003) Redox potential of quinones in both electron transfer branches of photosystem I. *Journal of Biological Chemistry*, **278**, 52002–11.

141 Joliot, P. and Joliot, A. (1999) In vivo analysis of the electron transfer within photosystem I: are the two phylloquinones involved? *Biochemistry*, **38**, 11130–6.

142 Guergova-Kuras, M., Boudreaux, B., Joliot, A., Joliot, P. and Redding, K. (2001) Evidence for two active branches for electron transfer in Photosystem I. *Proceedings of the National Academy of Sciences of the United States of America*, **98**, 4437–42.

143 Boudreaux, B., MacMillan, F., Teutloff, C., Agalarov, R., Gu, F., Grimaldi, S., Bittl, R., Brettel, K. and Redding, K. (2001) Mutations in both sides of the photosystem I reaction center identify the phylloquinone observed by electron paramagnetic resonance spectroscopy. *Journal of Biological Chemistry*, Vol. **276**, pp. 37299–306.

144 Xu, W., Chitnis, P., Valieva, A., van der Est, A., Pushkar, Y.N., Krzystyniak, M., Teutloff, C., Zech, S.G., Bittl, R., Stehlik, D., Zybailov, B., Shen, G. and Golbeck, J. H. (2003) Electron transfer in cyanobacterial photosystem I. I. Physiological and spectroscopic characterization of site-directed mutants in a putative electron transfer pathway from A_0 through A_1 to F_X. *Journal of Biological Chemistry*, **278**, 27864–75.

145 Xu, W., Chitnis, P.R., Valieva, A., van der Est, A., Brettel, K., Guergova-Kuras, M., Pushkar, Y.N., Zech, S.G., Stehlik, D., Shen, G., Zybailov, B. and Golbeck, J.H. (2003) Electron transfer in cyanobacterial photosystem I. II. Determination of forward electron transfer rates of site-directed mutants in a putative electron transfer pathway from A_0 through A_1 to F_X. *Journal of Biological Chemistry*, **278**, 27876–87.

146 Agalarov, R. and Brettel, K. (2003) Temperature dependence of biphasic forward electron transfer from the phylloquinones. A_1 in photosystem I: only the slower phase is activated. *Biochimica et Biophysica Acta*, **1604**, 7–12.

147 Ivashin, N. and Larsson, S. (2003) Electron transfer pathways in photosystem I reaction centers. *Chemical Physics Letters*, **375**, 383–7.

148 Ali, K., Santabarbara, S., Heathcote, P., Evans, M.C.W. and Purton, S. (2006) Bidirectional electron transfer in photosystem I: Replacement of the symmetry-breaking tryptophan close to the PsaB-bound phylloquinone (A_{1B}) with a glycine residue alters the redox properties of A_{1B} and blocks forward electron transfer at cryogenic temperatures. *Biochimica et Biophysica Acta*, **1757**, 1623–33.

149 Hallahan, B., Purton, S., Ivison, A., Wright, D. and Evans, M.C.W. (1995) Analysis of the proposed Fe-Sx binding region of Photosystem 1 by side directed mutation of PsaA in *Chlamydomonas rheinhardtii*. *Photosynthesis Research*, **46**, 257–64.

150 Vassiliev, I.R., Jung, Y.S., Smart, L.B., Schulz, R., McIntosh, L. and Golbeck, J. H. (1995) A mixed-ligand iron-sulfur cluster $C556S_{PsaB}$ or $C565S_{PsaB}$) in the F_X-binding site leads to a decreased quantum efficiency of electron transfer in photosystem I. *Biophysical Journal*, **69**, 1544–53.

151 Shen, G., Zhao, J., Reimer, S.K., Antonkine, M.L., Cai, Q., Weiland, S. M., Golbeck, J.H. and Bryant, D.A. (2002) Assembly of Photosystem I. I. inactivation of the *rubA* gene encoding a membrane-associated rubredoxin in the cyanobacterium *Synechococcus* sp. PCC 7002 causes a loss of photosystem I activity. *Journal of Biological Chemistry*, **277**, 20343–54.

152 Shen, G., Antonkine, M.L., van der Est, A., Vassiliev, I.R., Brettel, K., Bittl, R., Zech, S.G., Zhao, J., Stehlik, D., Bryant, D.A. and Golbeck, J.H. (2002) Assembly of Photosystem I. II. Rubredoxin is required for the *in vivo* assembly of F_X in *Synechococcus* sp. PCC 7002 as shown by optical and EPR spectroscopy. *Journal of Biological Chemistry*, **277**, 20355–66.

153 Petrouleas, V., Brand, J.J., Parrett, K.P. and Golbeck, J.H. (1989) A Mössbauer analysis of the low potential iron-sulfur center in photosystem I. Spectroscopic evidence that F_X is a 4Fe-4S cluster. *Biochemistry*, **28**, 8980–3.

154 McDermott, A.E., Yachandra, V.K., Guiles, R.D., Sauer, K., Parrett, K.G. and Golbeck, J.H. (1989) An EXAFS structuralmstudy of F_X, the low potential Fe-S center in photosystem I. *Biochemistry*, **28**, 8056–9.

155 Golbeck, J.H. (1992) Structure and function of photosystem I. *Annual Review of Plant Physiology and Plant Molecular Biology*, **43**, 293–324.

156 Ke, B., Dolan, E., Sugahara, K., Hawkridge, F.M., Demeter, S. and Shaw, E.R. (1977) Electrochemical and kinetic evidence for a transient electron acceptor in the photochemical charge separation in photosystem I. In: *Photosynthetic Organelles* [*Plant and Cell Physiology*, special issue no. *3* "Photosynthetic Organelles: Structure and Function"], 187–99.

157 Chamorovski, S. and Cammack, R. (1982) Direct determination of the midpoint potential of the acceptor X in chloroplast photosystem I by electrochemical reduction and ESR spectroscopy. *Photobiochem Photobiophys*, **4**, 195–200.

158 Lüneberg, J., Fromme, P., Jekow, P. and Schlodder, E. (1994) Spectroscopic characterization of PS I core complexes from thermophilic *Synechococcus* sp. Identical reoxidation kinetics of A_1^-, before and after removal of the iron-sulfur-clusters F_A and F_B. *FEBS Letters*, **338**, 197–202.

159 Zhao, J., Li, N., Warren, P.V., Golbeck, J.H. and Bryant, D.A. (1992) Site-directed conversion of a cysteine to aspartate leads to the assembly of a N iron-sulfur[3Fe-4S] cluster to PsaC of Photosystem I. The photoreduction of F_A is independent of F_B. *Biochemistry*, **31**, 5093–9.

160 Mehari, T., Qiao, F., Scott, M.P., Nellisi, D.F., Zhao, J., Bryant, D.A. and Golbeck, J.H. (1995) Modified ligands to F_A and F_B in photosystem I. I. Structural constraints for the formation of iron-sulfur clusters in free and rebound PsaC. *Journal of Biological Chemistry*, **270**, 28108–17.

161 Golbeck, J.H. (1999) A comparative analysis of the spin state distribution of *in vitro* and *in vivo* mutants of PsaC. A biochemical argument for the sequence of electron transfer in Photosystem I as $F_X \rightarrow F_A \rightarrow F_B \rightarrow$ ferredoxin/flavodoxin. *Photosynthesis Research*, **61**, 107–44.

162 Badger, M.R., von Caemmerer, S., Ruuska, S. and Nakano, H. (2000)

Electron flow to oxygen in higher plants and algae: rates and control of direct photoreduction (Mehler reaction) and rubisco oxygenase. *Philosophical Transactions of the Royal Society of London. Series B*, **355**, 1433–46.

163 Pålsson, L.O., Flemming, C., Gobets, B., van Grondelle, R., Dekker, J.P. and Schlodder, E. (1998) Energy transfer and charge separation in photosystem I: P700 oxidation upon selective excitation of the long-wavelength antenna chlorophylls of Synechococcus elongatus. *Biophysical Journal*, **74**, 2611–22.

164 Trissl, H.-W. and Wilhelm, C. (1993) Why do thylakoid membranes from higher plants form grana stacks? *Trends in Biochemical Sciences*, **18**, 415–19.

165 Förster, T. (1948) Zwischenmolekulare Energiewanderung und Fluoreszenz. *Annals of Physics (Leipzig)*, **2**, 55–75.

166 Gobets, B. and van Grondelle, R. (2001) Energy transfer and trapping in photosystem I. *Biochimica et Biophysica Acta*, **1507**, 80–99.

167 Gibasiewicz, K., Ramesh, V.M., Lin, S., Woodbury, N.W. and Webber, A.N. (2002) Excitation dynamics in eukaryotic PSI from *Chlamydomonas reinhardtii* CC2696 at 10K. Direct detection of the reaction center exciton states. *Journal of Physical Chemistry B*, **106**, 6322–30.

168 Gibasiewicz, K., Ramesh, V.M., Lin, S., Redding, K., Woodbury, N.W. and Webber, A.N. (2003) Excitonic interactions in wild-type and mutant PSI reaction centers. *Biophysical Journal*, **85**, 2547–59.

169 Fromme, P., Kern, J., Loll, B., Biesiadka, J., Saenger, W., Witt, H.T., Krauss, N. and Zouni, A. (2002) Functional implications on the mechanism of the function of photosystem II including water oxidation based on the structure of photosystem II. *Philosophical Transactions of the Royal Society of London. Series B*, **357**, 1337–44.

170 Karapetyan, N.V., Holzwarth, A.R. and Rögner, M. (1999) The photosystem I trimer of cyanobacteria: molecular organization, excitation dynamics and physiological significance. *FEBS Letters*, **460**, 395–400.

171 Trissl, H.-W. (1993) Long-wavelength absorbing antenna pigments and heterogeneous absorption bands concentrate excitons and increase absorption cross section. *Photosynthesis Research*, **35**, 247–63.

172 Rivadossi, A., Zucchelli, G., Garlaschi, F.M. and Jennings, R.C. (1999) The importance of PS I chlorophyll red forms in light-harvesting by leaves. *Photosynthesis Research*, **60**, 209–15.

173 Gobets, B., Kennis, J.T.M., Ihalainen, J.A., Brazzoli, M., Croce, R., van Stokkum, I.H.M., Bassi, R., Dekker, J.P., van Amerongen, H., Fleming, G.R. and van Grondelle, R. (2001) Excitation energy transfer in dimeric Light-Harvesing Complex I: a combined streak camera/fluorescence upconversion study. *Journal of Physical Chemistry B*, **105**, 10132–9.

174 Fleming, G.R. and van Grondelle, R. (1997) Femtosecond spectroscopy of photosynthetic light-harvesting systems. *Current Opinion in Structural Biology*, **7**, 738–48.

175 Gobets, B., van Amerongen, H., Monshouwer, R., Kruip, J., Rögner, M., van Grondelle, R. and Dekker, J.P. (1994) Polarized site-selected fluorescence spectroscopy of isolated Photosystem I particles. *Biochimica et Biophysica Acta*, **1188**, 75–85.

176 Frese, R.N., Palacios, M.A., Azzizi, A., van Stokkum, I.H.M., Kruip, J., Rögner, M., Karapetyan, N.V., Schlodder, E., van Grondelle, R. and Dekker, J.P. (2002) Electric field effects on red chlorophylls, β-carotenes and P700 in cyanobacterial Photosystem I complexes. *Biochimica et Biophysica Acta*, **1554**, 180–91.

177 Zazubovich, V., Matsuzaki, S., Johnson, T.W., Hayes, J.M., Chitnis, P.R. and Small, G.J. (2002) Red antenna states of photosystem I from cyanobacterium *Synechococcus elongatus*: a spectral hole burning study. *Chemical Physics*, **275**, 47–59.

178 Yang, M., Damjanovic, A., Vaswani, H.M. and Fleming, G.R. (2003) Energy transfer in photosystem I of cyanobacteria *Synechococcus elongatus*: model study with structure-based semiempirical Hamiltonian and experimental spectral density. *Biophysical Journal*, **85**, 140–58.

179 Karapetyan, N.V., Schlodder, E., van Grondelle, R. and Dekker, J.P. (2006) The long wavelength chlorophylls in photosystem I, in *Photosystem I: The Light-Driven Plastocyanin:Ferredoxin Oxidoreductase. Advances in Photosynthesis and Respiration*, Vol. **24** (ed. J.H. Golbeck), Springer, Dordrecht, The Netherlands, pp. 177–92.

180 Gobets, B., van Stokkum, I.H.M., van Mourik, F., Dekker, J.P. and van Grondelle, R. (2003) Excitation wavelength dependence of the fluorescence kinetics in photosystem I particles from *Synechocystis* PCC 6803 and *Synechococcus elongatus*. *Biophysical Journal*, **85**, 3883–98.

181 Horn, R. and Paulsen, H. (2002) Folding *in vitro* of light-harvesting chlorophyll-*a/b* protein is coupled with pigment binding. *Journal of Molecular Biology*, **318**, 547–56.

182 Cogdell, R.J. (1985) Carotenoids in photosynthesis. *Pure and Applied Chemistry*, **57**, 723–8.

183 Dexter, D. (1953) A theory of sensitized luminescence in solids. *Journal of Chemical Physics*, **21**, 836–50.

184 Koyama, Y. (1991) Structures and functions of carotenoids in photosynthetic systems. *Journal of Photochemistry and Photobiology B*, **9**, 265–80.

185 Koyama, Y., Kuki, M., Andersson, P.O. and Gillbro, T. (1996) Singlet excited states and the light-harvesting function of carotenoids in bacterial photosynthesis. *Photochemistry and Photobiology*, **63**, 243–56.

186 Chynwat, V. and Frank, H.A. (1995) The application of the energy gap law to the S_1 energies and dynamics of carotenoids. *Chemical Physics*, **194**, 237–44.

187 Kennis, J.T.M., Gobets, B., van Stokkum, I.H.M., Dekker, J.P., van Grondelle, R. and Fleming, G.R. (2001) Light harvesting by chlorophylls and carotenoids in the photosystem I core complex of *Synechococcus elongatus*: a fluorescence upconversion study. *Journal of Physical Chemistry B*, **105**, 4485–94.

188 van Dorssen, R.J., Breton, J., Plijter, J.J., Satoh, K., van Gorkom, H.J. and Amesz, J. (1987) Spectroscopic properties of teh reaction center and of the 47 kDa chlorophyll protein of Photosystem II. *Biochimica et Biophysica Acta*, **893**, 267–74.

189 Ashikawa, I., Miyata, A., Koike, H., Inoue, Y. and Koyama, Y. (1986) Light-induced structural change of ß-carotene in thylakoid membranes. *Biochemistry*, **25**, 6154–60.

3
A Glimpse into the Atomic Structure of Plant Photosystem I

Alexey Amunts, Omri Drory, and Nathan Nelson

3.1
Introduction

Photosynthesis underpins the survival of virtually all higher-life forms. It is the principle producer of both oxygen and organic matter on earth. By producing oxygen and assimilating carbon dioxide into an organic matter, oxygenic photosynthesis determines, to a large extent, the composition of the atmosphere while providing essential food and fuel. The primary step in the photosynthesis process is driven by four multi-subunit membrane protein complexes: Photosystem I (PSI), Photosystem II (PSII), cytochrome b_6f complex and ATP synthase. Photosystem I generates the most negative redox potential in nature and thus determines much of the global amount of enthalpy in living systems. The plant PSI supercomplex contains approximately 200 cofactors, and, because of its ability to combine the core-complex of PSI with the four peripheral antenna proteins, is much larger than its cyanobacterial counterpart [1–4]. Nevertheless, it performs its photochemical activity with an unprecedented quantum yield of close to one [5, 6]. The vast complexity of PSI that belies its efficiency is one of the most intriguing wonders in nature. Only a high-resolution structure of the supercomplex will provide a yardstick for a theoretical approach to solve this enigma.

Biochemically, PSI can be defined as a plastocyanin-ferredoxin oxidoreductase [7–9]. Photochemically, PSI operates by a reversible photooxidation of a primary electron donor P700, which launches a multistep electron transfer via a series of redox cofactors of the reaction center (RC). The excitation energy for the function of the primary electron donor in the RC is delivered via a core antenna made up of chlorophyll (Chl) and carotenes. In contrast to most other photosynthetic systems (where the primary antenna and reaction centers are located on distinct complexes), PSI has a combined system in which the light-harvesting chlorophylls are associated with the same proteins that bind the redox cofactors of the electron transfer in the RC [10]. Plant PSI consists of two separate complexes of the RC and the light-harvesting complex (LHC-I) [1, 2]. The medium-resolution structure of plant PSI revealed the mode of interaction between these two

Photosynthetic Protein Complexes: A Structural Approach. Edited by P. Fromme

ISBN: 978-3-527-31730-1

loosely bound subcomplexes [3, 4]. The structure has tuned the imagination of theoreticians into a more plausible architecture of the system and has provided evidence for the evolutionary steps that enabled the development of terrestrial life about 1.5 billion years ago. However, it only marginally helped to solve the enigma of the high quantum yield efficiency of the system [11]. Toward this goal, research is taking place to try to improve the resolution of the PSI crystals and to reach a state that will be more useful in achieving the desired explanation.

3.2 The General Architecture of Plant PSI

The model of plant PSI is based on the data obtained at 4.4 Å [4], the very recent 3.4 Å resolution [12], and the theoretical model that was based on the 4.4 Å resolution structure [13]. A view from the stroma on a plant PSI (Figure 3.1a) gives the impression that it has a twofold symmetry between the left side (PsaA) and the right (PsaB) of the core of the PSI–LHC-I supercomplex. Moreover, the two LHC-I dimers, Lhca1/Lhca4 and Lhca2/Lhca3, were superficially ascribed to this notion. However, a look at the membrane plane (Figure 3.1b) shows that only chlorophylls that belong to the electron transport chain (ETC) and some of their adjacent chlorophylls maintain pseudo-twofold symmetry. This arrangement has far reaching biochemical and evolutionary consequences [14]. The view from the stroma reveals that the PSI-core and LHC-I form two distinct and loosely associated halves, with a deep cleft between them. The four antenna proteins assemble into two dimers that are arranged in a series, creating a half-moon shaped belt that docks to the PSI-core's PsaF side. The LHC-I belt contributes a mass of 170 kDa out of approximately 600 kDa. LHC-I is composed of four proteins, the nuclear gene products (Lhca1-Lhca4) that are 20–24 kDa polypeptides and belong to the LHC family of chlorophyll a/b binding proteins. The archetype of this family, and the most abundant membrane protein in nature, are the major LHC-II proteins (Lhcb1 and Lhcb2); its structures were recently revealed by X-ray crystallography at 2.7 Å and 2.5 Å resolutions [15, 16] (see also Chapter 2). The main difference between the cyanobacterial and plant PSI-core is that the latter is monomeric, both in vitro and in vivo [17]. This opens up the possibility of lateral interactions with other membrane proteins in an asymmetric fashion [14]. Assignment of the four LHC-I monomers and subunits K and G was based on several biochemical studies. SDS-PAGE and mass-spectrometry analysis confirm that plant PSI crystals contain four different LHC-I proteins, namely Lhca1-4 [4, 18, 19]. No direct evidence based on electron densities was provided at 4.4 Å resolution, but a theoretical model calculated by Jolley et al. [13] gave credence not only to the general architecture of plant PSI, but also suggested the position and orientation of most of the chlorophyll molecules in the complex. This model remains to be verified by a higher resolution structure of plant PSI.

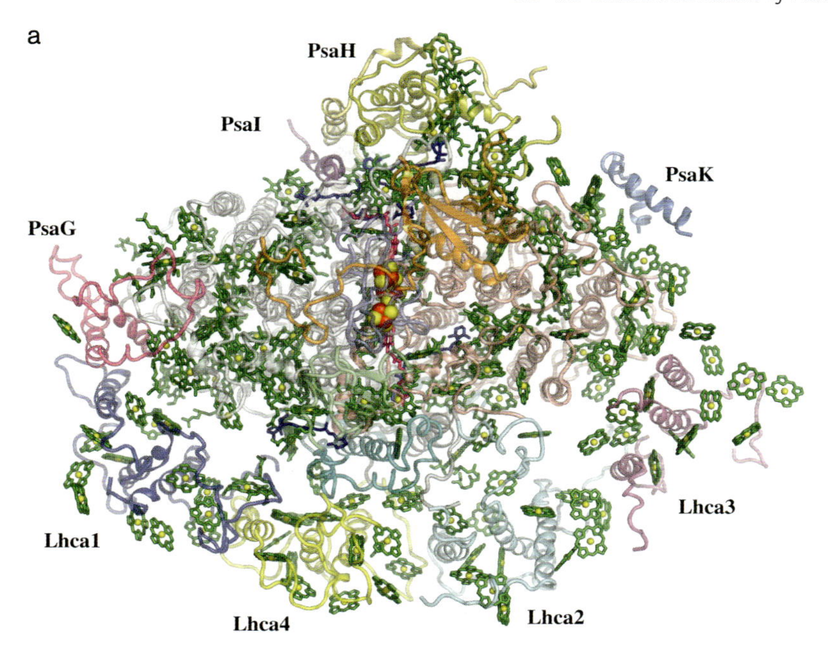

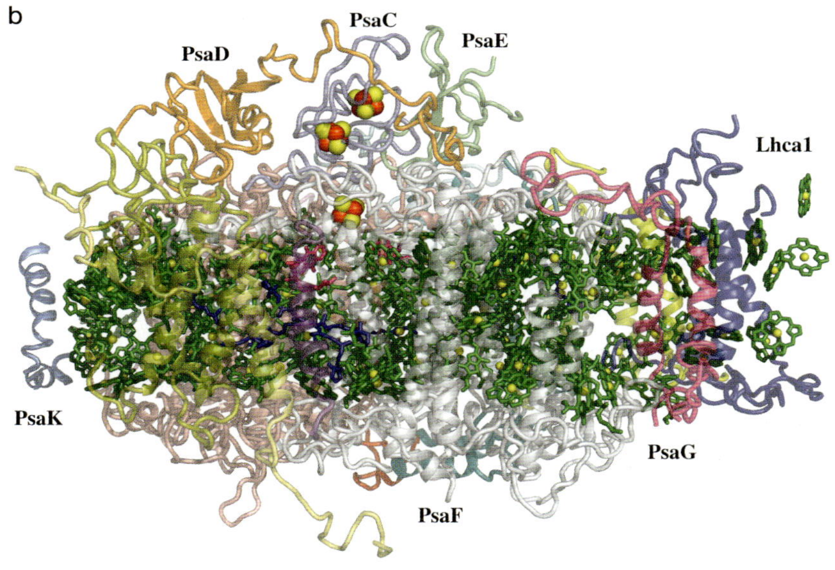

Figure 3.1 The overall structure of plant PSI. The model of plant photosystem I at 3.4 Å resolution [12]. **(a)** View from the stromal side of the thylakoid membrane. **(b)** View perpendicular to the membrane normal. Each subunit is colored differently (PsaA – wheat; PsaB – gray; PsaC – light blue; PsaD – orange; PsaE – light green; PsaF – teal; PsaG – pink; PsaH – pale yellow; PsaI – purple; PsaJ – dark gray; PsaK – blue; PsaL – bright yellow; Lhca1 – green; Lhca2 – cyan; Lhca3 – hot pink; Lhca4 – lemon) and represented in transparent cartoon. The Fe-S clusters are shown as spheres (Fe atoms in red, S atoms in yellow). Chlorophylls are shown in green.

3.3 The Core Complex

The RC of plant PSI contains 12 polypeptides (PsaA, PsaB, PsaC, PsaD, PsaE, PsaF, PsaG, PsaH, PsaI, PsaJ, PsaK, and PsaL) and approximately 100 chlorophyll molecules [7, 8]. The plant RC retains the location and orientation of the electron transfer components and all cyanobacterial transmembrane helices, except those of subunits X and M that are not present in plants (Figure 3.2). In addition, two RC proteins are exclusively present in plants and green algae – subunits PsaG and PsaH. These subunits are 10 kDa membrane proteins, where PsaH contains a single transmembrane helix adjacent to PsaL (Figures 3.1 and 3.2). The position and shape of PsaH conform well to its proposed role as a docking site for LHC-II [20]. On the opposite side of the RC, PsaG with its two tilting transmembrane helices contributes most of the contact surface area associated with LHC-I [4]. PsaG evolved by a gene duplication of PsaK and the new gene product was assembled onto PsaB. Its biogenesis involves a unique sequence of events that

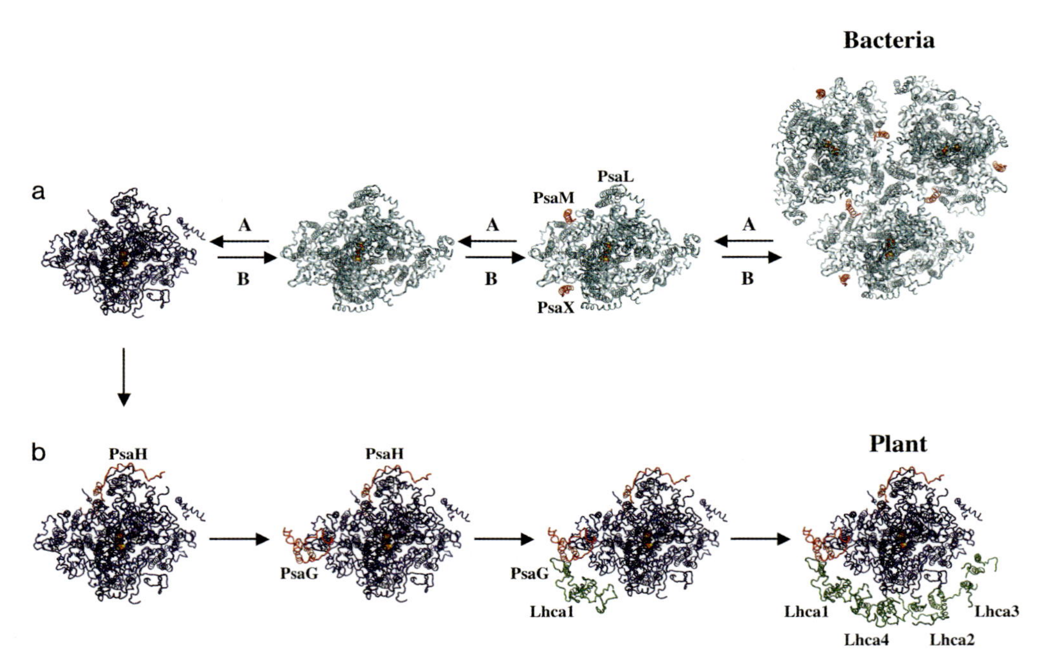

Figure 3.2 The two possible scenarios for plant PSI evolution. Scenario **(a)** suggests that the ancestral, bacterial PSI functioned as a trimer. During the course of evolution, the monomeric complex with modified PsaL was naturally selected. In the continuous process, this complex became free of PsaM and PsaX (red). The additional plant PsaH (red) has prevented the trimer formation, while the added PsaG (red) served as a template for the light-harvesting proteins (green). Scenario **(b)** predicts primordial PSI as a monomeric complex. Addition of PsaM and PsaX, as well as extension of PsaL, resulted in formation of the trimeric structure.

places the loop situated between the two transmembrane helices at the stromal side of PSI [4, 21]. On the luminal side, the most noticeable distinction between plant and cyanobacterial reaction centers is the helix-loop-helix motif contributed by the longer N-terminus domain of plant PsaF (Figure 3.1b). This domain enables the more efficient plastocyanin binding in plants and, as a result, two orders of magnitude faster electron transfer from this copper protein to P_{700} [22–24].

The two new eukaryotic subunits, PsaG and PsaH, were probably crucial for the evolution of algae that inhabited the surface of the oceans about 1.2 billion years ago and eventually the evolution of plants which settled on land. In addition, the modification of PsaL played a role in the advancement towards monomeric PSI [14]. The protruding C-terminus of the cyanobacterial PsaL is essential for the PSI trimer formation [10, 25]. A forced model of trimeric plant PSI showed that not only is this part of PsaL missing in plants, but also that the added subunit H prevents any possibility of forming a trimeric structure [14]. The evolutionary origin of PsaH is not clear. Concomitantly with the shortening of PsaL and the emergence of PsaH, PsaK underwent gene duplication to form the eukaryotic PsaK and PsaG (Figure 3.2). The new subunit (PsaG) found a binding site on the opposite side of the RC and served as a template for the evolving LHC-I complex (Figure 3.2). A smaller change in the eukaryotic RC made it ready for vertical interactions with LHC-II, and perhaps with the cytochrome b_6f complex, as improved efficient horizontal interactions with plastocyanin and ferredoxin [5]. The above-mentioned evolutionary scenario may be subjected to major modifications. One of them is the possibility that the ancestral PSI acted as a monomer with the onset of oxygenic photosynthesis [26, 27]. This alternate scenario, depicted in Figure 3.2, predicts an ancestral monomer in ancient cyanobacteria. It contained the basic PSI-RC subunits excluding X and M and a short version of PsaL. The emergence of algae caused a limitation for the light available for the cyanobacteri and prompted them to extend the C-terminus of PsaL and to add PsaM and PsaX to their RC. This allowed them to assemble the trimeric form of PSI that may be more efficient for dim light harvesting [26, 27].

The two scenarios for the evolution of PSI describe the main events over the last 3.5 billon years in the evolution of the system. The events that led to the development of cyanobacterial PSI from the homodimeric RC that is currently present in green and other photosynthetic bacteria are not clear [28]. Photosystem I complex isolated from the cyanobacterium *Gloeobacter violaceus* PCC 7421 contains only eight subunits: PsaA, PsaB, PsaC, PsaD, PsaE, PsaF, PsaL, and PsaM [29]. A novel subunit, PsaZ, was also discovered, but PsaI, PsaJ, PsaK, and PsaX were absent. It is not clear whether this PSI represents a primordial PSI-like RC, or if it evolved from a reaction center containing the whole complement of cyanobacterial subunits. As shown in Figure 3.2, PsaM is situated in the interface between individual subunits in the trimer formation. Accordingly, it lost its function in the monomeric plant PSI and eventually disappeared during the chloroplast evolution. It is likely that the surface exposed by the absence of PsaM serves now for binding of other membrane complexes [14]. PsaX had to be removed to make space for the assembly of PsaG and Lhca1 (Figure 3.2). Additionally, PsaX may not be present in all cyanobacteria – it has so far been identified in thermophilic cyanobacteria. While PsaI and PsaJ function primarily as space-filling modules, PsaK seems to function in a variety of interactions; it is the most mobile of all the other RC subunits. Consequently, it was poorly resolved in the 2.5 Å structure of the cyanobacterial PSI [10], and even less resolved in the plant PSI–LHC-I supercomplex [4]. Presumably, PsaK might also undergo extensive post-translational modifications.

3.4 Light-Harvesting Complex of Higher Plants

Photosynthesis provides plants with almost all the required energy for their life processes. Plants proliferate on most of the surface of earth, and they should operate efficiently under stress conditions of temperature, humidity, and light intensity. In general, light is a rather dilute energy source and, even on a bright day, the radiation flux of solar energy that a chlorophyll absorbs is only 10 photons per second [30]. The same variety of the plant may have to complete its life cycle under radiation flux that is a thousand fold lower. The evolution of light-harvesting antennas permitted photosynthetic organisms to synchronize the light-harvesting and photosynthetic electron transport under a wide variety of light intensities. The modular and adjustable LHC-I was designed to operate efficiently under ever-changing light intensities [31]. The binding of LHC-I to the RC is asymmetric and much stronger on the G-pole than on the K-pole of the core (Figure 3.1). Lhca1 is strongly attached to the core through the helix bundle formed between its transmembrane helix C and the two tilted helices of PsaG, as well as due to the close interaction of its stromal loop with the novel Loop1 of PsaB. The structure suggests that Lhca1 acts as an anchor point for facilitating the binding of other LHC-I monomers and dimers (and their isoforms) at varying stoichiometries, depending on environmental conditions. The other LHC-I proteins interact with the core mainly through small binding surfaces at their stromal exposed regions (Figure 3.1). Lhca4 binds to PsaF and Lhca2 associates weakly with PsaJ. The interaction between Lhca subunits and PSI core is strongly cooperative [32]. The loss of Lhca4 destabilizes the whole LHC-I and leads to the greatest diminishment of the antennas [33]. In addition, Lhca3 has probably some structural peculiarity as compared to the other Lhca complexes; this is supported by recent spectroscopic measurement [34].

Even though the light-harvesting chlorophyll *a*/*b* binding proteins that constitute the peripheral antennas of PSI and PSII (LHC-II) share sequence and structural homology, their oligomeric states vary considerably [35, 36]. Whereas LHC-I proteins assemble into dimers [18, 37], the light-harvesting proteins that associate with PSII form either trimers or monomers as minor antenna members CP24, CP26 and CP29 [38]. Lhca1-4 and LHC-II bind around 13 chlorophyll $a + b$ molecules each and possess the LHC-II general fold [4, 15, 39]. The absorption peak of the bulk chlorophylls of LHC-I proteins is also shifted to lower energies, in comparison with LHC-II [4, 32, 40]. Dimerization in LHC-I is mediated by relatively small contact surfaces at the luminal side by the C-terminus and at the stromal side of the N-terminal domain of the Lhca proteins [4, 41]. The dimer form and the relatively long distance between the RC and LHC-I required additional chlorophylls to facilitate excitation transfer between the LHC-I monomers and between LHC-I and the RC. Such "gap chlorophylls" are situated in positions that allowed a synchronized excitation flow between the two complexes [11, 36]. The current structure of plant PSI does not yet suggest a plausible hypothesis for the pathways and rates of excitation transfer from LHC-I to the reaction pathway [12].

The high degree of flexibility may be important for the assembly of LHC-I proteins in its half-moon shape. This may contribute to the efficient energy-transfer along the LHC-I belt and between it and the RC. A view of chlorophyll arrangement within the plant holo-complex, presented in Figure 3.3, reveals that most LHC-I chlorophylls are contained within a narrow belt with Mg-Mg distances of 20–30 Å from the nearest RC chlorophyll. There are, however, three contact regions where much smaller inter-pigment distances (10–15 Å) are observed. The denser contact regions are located at the two poles of the LHC-I belt near PsaK and PsaG, with an additional minor one located at the center of the LHC-I belt in the vicinity of PsaF. Recent biochemical, proteomic, and genomic studies of the PSI antenna indicate the presence of five to six distinct Lhca proteins [33, 42, 43]. However, they are only present in substoichiometrical amounts and may play a role for higher plant light harvesting under changing environmental conditions. Lhca5 was shown to interact with PSI via the Lhca2/3 site; it is also proposed the formation of a homodimer in the Lhca1/4 binding site in the absence of Lhca1 and Lhca4 [44]. Whether and how it contributes to the remarkable efficiency of plant PSI remains to be shown.

3.5
Interaction with Electron Donors and Acceptors

The binding mode of ferredoxin and ferredoxin-NADP-reductase (FNR) to plant PSI is not clear. A relatively strong interaction between the two proteins was demonstrated [45, 46] and a crystal structure of the ferredoxin-FNR complex has been determined [47]. It has been reported that FNR in higher plant chloroplasts is localized peripherally on the stromal side of thylakoid membranes through association with a membrane-intrinsic protein [48, 49]–either cytochrome (cyt) b_6f complex [50–52], NAD(P)H dehydrogenase complex [53], or PSI [54]. The localization of FNR at the cyt b_6f and/or NAD(P)H dehydrogenase complexes implies a role in cyclic electron transport. It has been shown that endogenously bound FNR interacts with the PsaE subunit of plant PSI, as demonstrated by cross-linking experiments using two different types of cross-linkers, the identification of the products by Western blotting, and the use of monospecific antibodies [54]. Currently, detergent-isolated and purified PSI [55] contains no bound FNR; yet, for efficient linear electron transport, it is likely that FNR is bound to PSI at a close

Figure 3.3 The arrangement of chlorophyll, Fe-S cluster, phylloquinone, and beta-carotene molecules. **(a)** View from the stromal side of the thylakoid membrane. **(b)** View perpendicular to the membrane normal. The Fe-S clusters are shown in transparent spheres (Fe atoms – red; S atoms – yellow), all other prosthetic groups are shown as sticks (phylloquinones – blue; beta-carotenes – magenta; chlorophylls – CPK; Mg atoms – yellow). Sixty-five out of 168 chlorophylls are shown with orientation of the Qx and Qy transition dipolar moments. Two phylloquinones, three 4Fe4S clusters, and five beta-carotenes could be shown in this model at 3.4 Å resolution.

proximity to the electron accepting site of ferredoxin. Plausible binding sites for FNR and the ferredoxin-FNR complex were explored, and suggest that PSI likely binds FNR on the stromal surface of PsaA (Figure 3.4). This binding mode is primarily hydrophobic, and the complex might be released upon detergent treatment. Attempts to move FNR towards PsaE caused conflict with the N-terminus of PsaA and left no space for the oxidized ferredoxin to associate close enough, so that a fast electron transfer from F_B would be possible.

Figure 3.4 shows the suggested model of the position of FNR on PSI and the movement of ferredoxin from a proposed binding site at close proximity to PsaC toward a complex with FNR for donation of its electron to FAD. The proposed positions of oxidized and reduced ferredoxin, as well as FNR, cause no steric hindrance and may provide the right environment for rapid electron transport at a very negative redox potential in the presence of molecular oxygen. The distance between F_B and the ferredoxin Fe-S is only 13 Å – similar to the distance between F_X and F_A. This architecture should enable electron transport with stoichiometric amounts of PSI, ferredoxin, and FNR. In contrast, when the same reaction takes place in solution, it requires a large excess of ferredoxin and FNR [7], and enormous excess of ferredoxin in the cyclic phosphorylation that was measured in vitro.

Another advantage of the proposed architecture of PSI-ferredoxin-FNR complex is that under conditions when NADP is not required, the reduced ferredoxin is free to leave its binding site, even when FNR is still bound, and donate the electron to cytochrome b_6f complex for the cyclic photophosphorylation.

Plastocyanin (Pc) is the universal PSI electron donor, although cytochromes can replace it in cyanobacteria and algae (see also Chapter 8). Under certain physiological conditions, cyt c_6 can alternate with Pc as an electron carrier between cyt f and P700 [56] in cyanobacteria. Recent studies indicate that Pc is the only mobile electron donor to higher plant PSI [57–59]. The Pc binding site of PSI is comprised mainly of a hydrophobic interaction with PsaA and B; it is facilitated and controlled by charge-charge interactions with PsaF. As a result, electron transfer from Pc to PSI is two orders of magnitude faster in plants than in cyanobacteria, and the release of oxidized Pc limits the electron transfer rates [22–24, 60, 61]. This was attributed to more efficient Pc binding in plants, mediated by the extra 18 amino acid residues in the plant PsaF N-terminus. Site-directed mutagenesis studies have identified several lysine residues in this N-terminal region [23], as well as an acidic patch on plastocyanin [62, 63] as vital for the formation of the PSI–Pc complex. In the plant PSI structure, this extra N-terminal domain forms an amphipathic helix-loop-helix motif on the luminal side of the thylakoid membrane [4]; which, in comparison to the cyanobacterial structure [10], is the only alteration in the Pc binding pocket. The model of interaction of between PSI and Pc (Figure 3.4c) brings a conserved cluster of negatively charged residues (Asp42, Glu43, Asp44, and Glu45) of Pc into contact with the positively charged N-terminal domain of PsaF. Thus, Pc interacts with hydrophobic surface of PsaA and PsaB and a positively charged site of PsaF [4, 64].

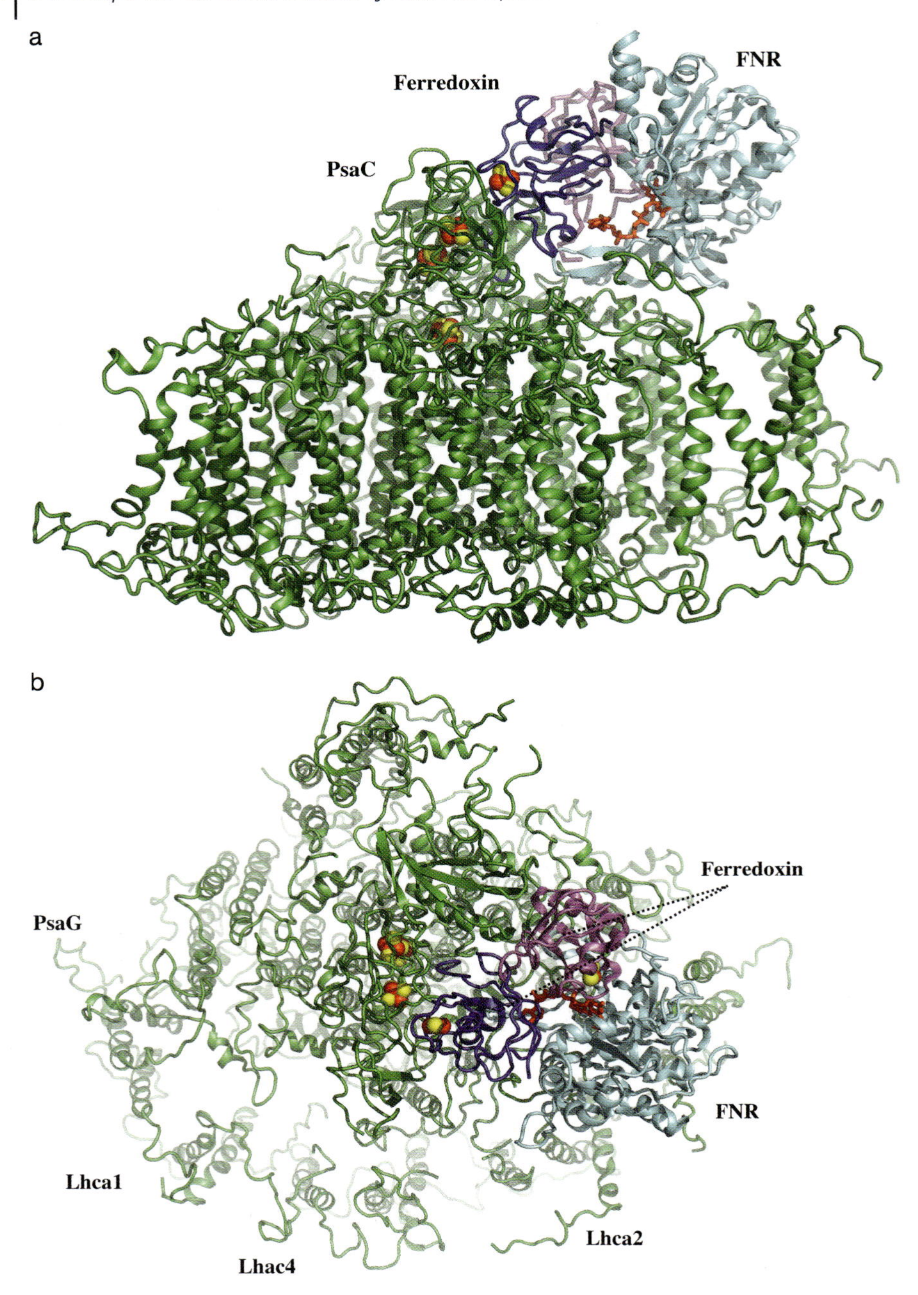
a
FNR
Ferredoxin
PsaC
b
Ferredoxin
PsaG
FNR
Lhca1
Lhca2
Lhac4

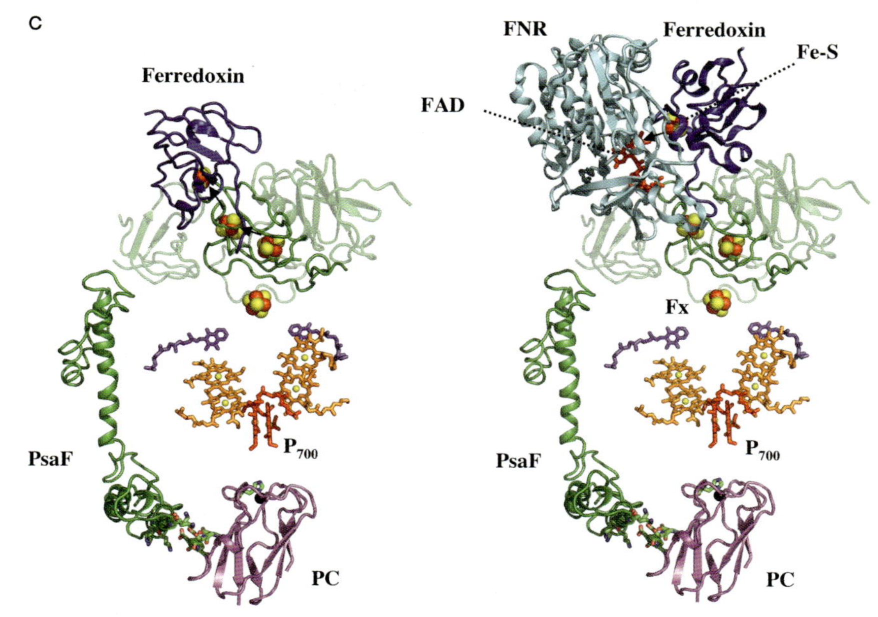

Figure 3.4 Interaction with electron donor and acceptors. These are proposed models of interaction between plant PSI and ferredoxin, ferredoxin-FNR complex, and plastocyanin. **(a)** View along the membrane plane and **(b)** view from the stroma of the PSI-ferredoxin-FNR complex. Oxidized ferredoxin (blue) accepts an electron from F_B, and then moves to a different location (magena) to serve as an electron donor to FAD of FNR. **(c)** Proposed architecture of supra-complex between PSI, plastocyanin, ferredoxin, and FNR. Negatively charged residues of Pc (cyan, copper in black) form electrostatic interactions with the positively charged N-terminal domain of PsaF (green). Electrons are transported to P_{700} (red) and then to Fx (yellow-red spheres). On the left, oxidized ferredoxin (blue) is bound onto its proposed binding site on PSI (green). Fe-S of ferredoxin accepts an electron from F_B (yellow-red spheres) of PSI. On the right, the reduced ferredoxin moves towards the bound FNR (cyan) to donate its electron to FAD. The ferredoxin-FNR complex [47] was fitted to the structure of plant PSI. A site on PsaA was found to be compatible with the requirements of rapid, ferredoxin mediated, electron transfer between F_B and FAD (red). At this position, no steric interferences for the movements of ferredoxin were detected.

3.6
The Modality of PSI

The onset of oxygenic photosynthesis took place about 3.5 billion years ago, under harsh conditions, in cyanobacteria-like organisms that lived in the deep-water habitats of ancient oceans [28, 65]. They enjoyed anaerobic conditions that protected them from singlet oxygen damage and relied upon wide water shielding from excess light intensities. However, avoidance of the water surface limited the light intensities under which they could proliferate. Oxygen production by those organisms caused "the big bang of evolution" [66] about 2.5 billion years ago, a time when most organisms had vanished and the rest had to adopt drastic changes. The next significant event, initiated by further elevated oxygen concentration in the atmosphere, was the evolution of eukaryotic cells about 1.5 billion years ago [67]. Anaerobic organisms with a primitive network of endomembranes established symbiosis with various eubacteria that resulted in the emergence of eukaryotic cells [68]. Today, several variants of eukaryotes exist; two forms are most common [69]. Both contain mitochondria that originated from aerobic endosymbionts that were able to utilize oxygen for ATP formation. Some underwent further symbiosis to form organisms related to the current algae. Eukaryotic cells are inherently inefficient; they use the plentiful energy provided by photosynthesis to run an energy economy full of slips [70]. Therefore, without the modality of plant oxygenic photosynthesis, advanced living organisms could not proliferate on land. The main obstacle for photosynthetic organisms to inhabit the surface of the oceans and land was the unbearable light intensity that they might encounter [5, 28]. Here, the evolution of PSI modality was critical.

Plant oxygenic photosynthesis operates under an enormous variety of environmental conditions. Plants can grow at light intensities around $1\,\mu E\,m^{-2}\,s^{-1}$ and can tolerate light intensities of over $1000\,\mu E\,m^{-2}\,s^{-1}$. How do they cope with ever-changing environmental conditions? On the one hand, plants developed a highly sophisticated antenna system to maximize light harvesting under low-light intensities; on the other hand, they developed a protective means against high-light intensities. Lost quanta can inflict damage on the photosynthetic complexes; therefore, the two photosystems have evolved to operate with the highest possible quantum yield. However, even though PSI operates with an almost perfect quantum yield of 1.0, PSII operates with a lower quantum yield of about 0.85 [6, 71]. Since sunlight is abundant, the difference in quantum yield may seem unimportant; however, the less efficient PSII pays a penalty for the lost quanta by exhibiting more significant damages and a high turnover.

What are the mechanisms of high quantum yield and the stability of PSI? Reaction centers evolved to be highly compact structures through which water and protons are not allowed to pass. Because they are photochemical machines, large conformational changes at the reaction centers are not permitted. This arrangement is inherently inflexible and is strictly conserved during long evolutionary periods. Attesting to the antiquity of the chlorophyll arrangement in the PSI-core, the crystal model of plant PSI reveals that the majority of the RC chlorophylls retained the same position and tilting angle as in the PSI from the cyanobacterium *TS. elongatus* [4, 10]. Furthermore, a sequence alignment of PsaA from plants and *Gloeobacter Violaceus* [29] supports the notion that the designing principles of the chlorophyll arrangement in PSI have remained unchanged for more than 3 billion years. Since the majority of chlorophylls in RC are coordinated by histidines and, according to 16S rRNA analysis, *Gloeobacter* ,is the species closest to the ancestral cyanobacteria, this attests to the antiquity of the chlorophyll arrangement in PSI.

The modality of PSI is a property of its peripheral parts. The utilization of LHC monomers, their assembly to dimers, and their organization as a half crescent play a major role in the adaptability of PSI and its ability to be effective under different light intensities [40]. In addition, PSI serves as a template for the formation of supra-complexes (see also Chapter 6). The most important one exists during state-transitions where LHC-II trimer presumably bind to PSI in state II [72, 73]. State transitions in higher plants are limited. In state II, additional light harvesting by PSI does not exceed 20% [74], which is just about what one LHC-II trimer per PSI complex could contribute. Cross-linking experiments analyzing the PSI–LHC-II interactions showed that LHC-II is situated in close proximity to the PsaI, L and H subunits [20, 75]. However, attempts to fit LHC-II trimer or monomer structures to PSI in a way that satisfied the cross-linking data were only partially successful [28]. None of the configurations tested could fit LHC-II into the super-complex such that it would be in simultaneous contact with the three subunits and provide efficient energy transfer to the photosystem. A model for the binding of a LHC-II trimer on the PsaA side of plant PSI was proposed [28, 76] – supported by a single particle analysis of state II-enriched PSI [77]. In green algae, state transitions may also increase the numbers of LHC-I complexes/PSI-core [78].

The modality of plant PSI consists of many more possibilities, including interactions with more LHC-I complexes, cytochrome b6f complexes, F-ATPase, and several membrane chaperons. Structural models for the horizontal interactions between PSI and LHC-II or LHC-I complexes, and vertical interactions with plastocyanin and ferredoxin and/or FNR, have been presented. These models have to be examined by co-crystallization of PSI with soluble and membrane factors. A large number of studies were implemented to understand how the biological significance of plant PSI is matched by its structural elements. However, this is just the beginning of the journey toward understanding the molecular basis of this keystone photochemical super-complex.

Acknowledgements

We gratefully acknowledge the ESRF for synchrotron beam time and staff scientists of the ID14, ID 29 and ID 23 stations clusters for their assistance. The Israel Science Foundation supported this work through Grant 356/06.

References

1 Mullet, J.E., Burke, J.J. and Arntzen, C.J. (1980) Chlorophyll proteins of Photosystem I. *Plant Physiology*, **65**, 814–22.

2 Mullet, J.E., Burke, J.J. and Arntzen, C.J. (1980) A developmental study of Photosystem I peripheral chlorophyll proteins. *Plant Physiology*, **65**, 823–7.

3 Ben-Shem, A., Nelson, N. and Frolow, F. (2003) Crystallization and initial X-ray diffraction studies of higher plant photosystem I. *Acta Crystallographica Section D – Biological Crystallography*, **59**, 1824–7.

4 Ben-Shem, A., Frolow, F. and Nelson, N. (2003) Crystal structure of plant photosystem I. *Nature*, **426**, 630–5.

5 Nelson, N. and Ben-Shem, A. (2004) The complex architecture of oxygenic photosynthesis. *Nature Reviews Molecular Cell Biology*, **5**, 971–82.

6 Nelson, N. and Yocum, C. (2006) Structure and function of photosystems I and II. *Annual Review of Plant Biology*, **57**, 521–65.

7 Bengis, C. and Nelson, N. (1975) Purification and properties of the photosystem I reaction center from chloroplasts. *Journal of Biological Chemistry*, **250**, 2783–8.

8 Bengis, C. and Nelson, N. (1977) Subunit structure of chloroplast photosystem I reaction center. *Journal of Biological Chemistry*, **252**, 4564–9.

9 Nelson, N. and Ben-Shem, A. (2002) Photosystem I reaction center: past and future. *Photosynthesis Research*, **73**, 193–206.

10 Jordan, P., Fromme, P., Witt, H.T., Klukas, O., Saenger, W. and Krauss, N. (2001) Three-dimensional structure of cyanobacterial photosystem I at 2.5 angstrom resolution. *Nature*, **411**, 909–17.

11 Sener, M.K., Jolley, C., Ben-Shem, A., Fromme, P., Nelson, N., Croce, R. and Schulten, K. (2005) Comparison of the light-harvesting networks of plant and cyanobacterial photosystem I. *Biophysical Journal*, **89**, 1630–42.

12 Amunts, A., Drory, O. and Nelson, N. (2007) The structure of a plant photosystem I supercomplex at 3.4 Å resolution. *Nature*, **447**, 58–63.

13 Jolley, C., Ben-Shem, A., Nelson, N. and Fromme, P. (2005) Structure of plant photosystem I revealed by theoretical modeling. *Journal of Biological Chemistry*, **280**, 33627–36.

14 Ben-Shem, A., Frolow, F. and Nelson, N. (2004) Evolution of Photosystem I – from symmetry through pseudosymmetry to asymmetry. *FEBS Letters*, **564**, 274–80.

15 Liu, Z., Yan, H., Wang, K., Kuang, T., Zhang, J., Gui, L., An, X. and Chang, W. (2004) Crystal structure of spinach major light-harvesting complex at 2.72 A resolution. *Nature*, **428**, 287–92.

16 Standfuss, J., Terwisscha van Scheltinga, A.C., Lamborghini, M. and Kuhlbrandt, W. (2005) Mechanisms of photoprotection and nonphotochemical quenching in pea light-harvesting complex at 2.5 A resolution. *EMBO Journal*, **24**, 919–28.

17 Scheller, H.V., Jensen, P.E., Haldrup, A., Lunde, C. and Knoetzel, J. (2001) Role of subunits in eukaryotic photosystem I. *Biochimica et Biophysica Acta*, **1507**, 41–60.

18 Jansson, S., Andersen, B. and Scheller, H.V. (1996) Nearest-neighbor analysis of higher-plant photosystem I holocomplex. *Plant Physiology*, **112**, 409–20.

19 Croce, R., Morosinotto, T., Castelletti, S., Breton, J. and Bassi, R. (2002) The Lhca antenna complexes of higher plants photosystem I. *Biochimica et Biophysica Acta*, **1556**, 29–40.

20 Lunde, P., Jensen, P.E., Haldrup, A., Knoetzel, J. and Scheller, H.V. (2000) The PSI-H subunit of photosystem I is essential for state transitions in plant photosynthesis. *Nature*, **408**, 613–15.

21 Zygadlo, A., Robinson, C., Scheller, H.V., Mant, A. and Jensen, P.E. (2006) The properties of the positively charged loop region in PSI-G are essential for its "spontaneous" insertion into thylakoids and rapid assembly into the photosystem I complex. *Journal of Biological Chemistry*, **281**, 10548–54.

22 Hippler, M., Reichert, J., Sutter, M., Zak, E., Altschmied, L., Schroer, U., Herrmann, R.G. and Haehnel, W. (1996) The plastocyanin binding domain of photosystem I. *EMBO Journal*, **15**, 6374–84.

23 Hippler, M., Drepper, F., Haehnel, W. and Rochaix, J.D. (1998) The N-terminal domain of PsaF: precise recognition site for binding and fast electron transfer from cytochrome c6 and plastocyanin to photosystem I of Chlamydomonas reinhardtii. *Proceedings of the National Academy of Sciences of the United States of America*, **95**, 7339–44.

24 Hippler, M., Drepper, F., Rochaix, J.D. and Muhlenhoff, U. (1999) Insertion of the N-terminal part of PsaF from Chlamydomonas reinhardtii into photosystem I from Synechococcus elongatus enables efficient binding of algal plastocyanin and cytochrome c6. *Journal of Biological Chemistry*, **274**, 4180–8.

25 Chitnis, P.R., Purvis, D. and Nelson, N. (1991) Molecular cloning and targeted mutagenesis of the gene psaF encoding subunit III of photosystem I from the cyanobacterium Synechocystis sp. PCC

6803. *Journal of Biological Chemistry*, **266**, 20146–51.

26 Schubert, W.D., Klukas, O., Saenger, W., Witt, H.T., Fromme, P. and Krauss, N.A. (1998) A common ancestor for oxygenic and anoxygenic photosynthetic systems: a comparison based on the structural model of photosystem I. *Journal of Molecular Biology*, **280**, 297–314.

27 Fromme, P., Jordan, P. and Krauss, N. (2001) Structure of photosystem I. *Biochimica et Biophysica Acta*, **1507**, 5–31.

28 Nelson, N. and Ben-Shem, A. (2005) The structure of photosystem I and evolution of photosynthesis. *Bioessays*, **27**, 914–22.

29 Inoue, H., Tsuchiya, T., Satoh, S., Miyashita, H., Kaneko, T., Tabata, S., Tanaka, A. and Mimuro, M. (2004) Unique constitution of photosystem I with a novel subunit in the cyanobacterium Gloeobacter violaceus PCC 7421. *FEBS Letters*, **578**, 275–9.

30 Blankenship, R.E. (2002) *Molecular Mechanisms of Photosynthesis*, Blackwell Science, Oxford, UK.

31 Bailey, S., Walters, R.G., Jansson, S. and Horton, P. (2001) Acclimation of Arabidopsis thaliana to the light environment: the existence of separate low light and high light responses. *Planta*, **213**, 794–801.

32 Morosinotto, T., Castelletti, S., Breton, J., Bassi, R. and Croce, R. (2002) Mutation analysis of Lhca1 antenna complex. Low energy absorption forms originate from pigment-pigment interactions. *Journal of Biological Chemistry*, **277**, 36253–61.

33 Klimmek, F., Ganeteg, U., Ihalainen, J.A., van Roon, H., Jensen, P.E., Scheller, H.V., Dekker, J.P. and Jansson, S. (2005) Structure of the higher plant light harvesting complex I: in vivo characterization and structural interdependence of the Lhca proteins. *Biochemistry*, **44**, 3065–73.

34 Mozzo, M., Morosinotto, T., Bassi, R. and Croce, R. (2006) Probing the structure of Lhca3 by mutation analysis. *Biochimica et Biophysica Acta*, **1757**, 1607–13.

35 Pichersky, E. and Jansson, S. (1996) *The Light Reactions in Oxygenic Photosynthesis*, Kluwer Academic Publishers, Dordrecht, The Netherlands, pp. 507–21.

36 Melkozernov, A.N., Barber, J. and Blankenship, R.E. (2006) Light harvesting in photosystem I supercomplexes. *Biochemistry*, **45**, 331–45.

37 Schmid, V.H., Cammarata, K.V., Bruns, B.U. and Schmidt, G.W. (1997) In vitro reconstitution of the photosystem I light-harvesting complex LHCI-730: heterodimerization is required for antenna pigment organization. *Proceedings of the National Academy of Sciences of the United States of America*, **94**, 7667–72.

38 Barber, J. (2002) Photosystem II: a multisubunit membrane protein that oxidises water. *Current Opinion in Structural Biology*, **12**, 523–30.

39 Durnford, G., Deane, J.A., Tan, S., McFadden, G.I., Gantt, E. and Green, B.R. (1999) A phylogenetic assessment of the eukaryotic light-harvesting antenna proteins, with implications for plastid evolution. *Journal of Molecular Evolution*, **48**, 59–68.

40 Ben-Shem, A., Frolow, F. and Nelson, N. (2004) Light-harvesting features revealed by the structure of plant photosystem I. *Photosynthesis Research*, **81**, 239–50.

41 Schmid, V.H., Paulsen, H. and Rupprecht, J. (2002) Identification of N- and C-terminal amino acids of Lhca1 and Lhca4 required for formation of the heterodimeric peripheral photosystem I antenna LHCI-730. *Biochemistry*, **41**, 9126–31.

42 Ganeteg, U., Klimmek, F. and Jansson, S. (2004) Lhca5 – an LHC-type protein associated with photosystem I. *Plant Molecular Biology*, **54**, 641–51.

43 Storf, S., Jansson, S. and Schmid, V.H. (2005) Pigment binding, fluorescence properties, and oligomerization behavior of Lhca5, a novel light-harvesting protein. *Journal of Biological Chemistry*, **280**, 5163–8.

44 Lucinski, R., Schmid, V.H., Jansson, S. and Klimmek, F. (2006) Lhca5 interaction with plant photosystem I. *FEBS Letters*, **580**, 6485–8.

45 Nelson, N. and Neumann, J. (1969) Interaction between ferredoxin and ferredoxin nicotinamide adenine dinucleotide phosphate reductase in pyridine nucleotide photoreduction and some partial reactions. I. Inhibition of ferredoxin nicotinamide adenine dinucleotide phosphate reductase by ferredoxin. *Journal of Biological Chemistry*, **244**, 1926–31.

46 Nelson, N. and Neumann, J. (1969) Interaction between ferredoxin and ferredoxin nicotinamide adenine dinucleotide phosphate reductase in pyridine nucleotide photoreduction and some partial reactions. II. Complex formation between ferredoxin and ferredoxin nicotinamide adenine dinucleotide phosphate reductase and its relevance to pyridine nucleotide photoreduction. *Journal of Biological Chemistry*, **244**, 1932–6.

47 Kurisu, M., Kusunoki, E., Katoh, T., Yamazaki, K., Teshima, Y., Onda, Y., Kimata-Ariga, Y. and Hase, T. (2001) Structure of the electron transfer complex between ferredoxin and ferredoxin-NADP(+) reductase. *Nature Structural Biology*, **8**, 117–21.

48 Vallejos, R.H., Ceccarelli, E. and Chan, R. (1984) Evidence for the existence of a thylakoid intrinsic protein that binds ferredoxin-NADP+ oxidoreductase. *Journal of Biological Chemistry*, **259**, 8048–51.

49 Matthijs, H.C., Coughlan, S.J. and Hind, G. (1986) Removal of ferredoxin:NADP+ oxidoreductase from thylakoid membranes, rebinding to depleted membranes, and identification of the binding site. *Journal of Biological Chemistry*, **261**, 12154–8.

50 Clark, R.D., Hawkesford, M.J., Coughlan, S.J., Bennett, J. and Hind, G. (1984) Association of ferredoxin-NADP$^+$ oxidoreductase with the chloroplast cytochrome b-f complex. *FEBS Letters*, **174**, 137–42.

51 Zhang, H. and Whitelegge, J.P. and Cramer, W.A. (2001) Ferredoxin:NADP+ oxidoreductase is a subunit of the chloroplast cytochrome b6f complex. *Journal of Biological Chemistry*, **276**, 38159–65.

52 Okutani, S., Hanke, G.T., Satomi, Y., Takao, T., Kurisu, G., Suzuki, A. and Hase, T. (2005) Three maize leaf ferredoxin:NADPH oxidoreductases vary in subchloroplast location, expression, and interaction with ferredoxin. *Plant Physiology*, **139**, 1451–9.

53 Quiles, M.J., Garcia, A. and Cuello, J. (2000) Separation by blue-native PAGE and identification of the whole NAD(P)H dehydrogenase complex from barley stroma thylakoids. *Plant Physiology and Biochemistry*, **38**, 225–32.

54 Andersen, B., Scheller H.V. and Moller, B.L. (1992) The PSI-E subunit of photosystem I binds ferredoxin:NADP+ oxidoreductase. *FEBS Letters*, **311**, 169–73.

55 Amunts, A., Ben-Shem, A. and Nelson, N. (2005) Solving the structure of plant photosystem I–biochemistry is vital. *Photochemical & Photobiological Sciences*, **4**, 1011–15.

56 Kerfeld, A.A. and Krogmann, D.W. (1998) Photosynthetic cytochromes c in cyanobacteria, algae and plants. *Annual Review of Plant Physiology and Plant Molecular Biology*, **49**, 397–425.

57 Molina-Heredia, F.P., Wastl, J., Navarro, J.A., Bendall, D.S., Hervas, M., Howe, C.J. and De la Rose, M.A. (2003) Photosynthesis: a new function for an old cytochrome? *Nature*, **424**, 33–4.

58 Weigel, M., Pesaresi, P. and Leister, D. (2003) Tracking the function of the cytochrome c6-like protein in higher plants. *Trends in Plant Science*, **8**, 513–17.

59 Weigel, M., Varotto, C., Pesaresi, P., Finazzi, G., Rappaport, F., Salamini, F. and Leister, D. (2003) Plastocyanin is indispensable for photosynthetic electron flow in Arabidopsis thaliana. *Journal of Biological Chemistry*, **278**, 31286–9.

60 Finazzi, G., Sommer, F. and Hippler, M. (2005) Release of oxidized plastocyanin from photosystem I limits electron transfer between photosystem I and cytochrome b6f complex in vivo. *Proceedings of the National Academy of Sciences*, **102**, 7031–6.

61 Golding, A.J., Joliot, P. and Johnsin, G.N. (2005) Equilibration between cytochrome f and P700 in intact leaves. *Biochimica et Biophysica Acta*, **1706**, 105–9.

62 Young, S., Sigfridsson, K., Olesen, K. and Hansson, O. (1997) The involvement of the two acidic patches of spinach plastocyanin in the reaction with photosystem I. *Biochimica et Biophysica Acta*, **1322**, 106–14.

63 Olesen, K., Ejdeback, M., Crnogorac, M.M., Kostic, N.M. and Hansson, O.

(1999) Electron transfer to photosystem 1 from spinach plastocyanin mutated in the small acidic patch: ionic strength dependence of kinetics and comparison of mechanistic models. *Biochemistry*, **38**, 16695–705.

64 Sommer, F., Drepper, F. and Hippler, M. (2002) The luminal helix l of PsaB is essential for recognition of plastocyanin or cytochrome c6 and fast electron transfer to photosystem I in Chlamydomonas reinhardtii. *Journal of Biological Chemistry*, **277**, 6573–81.

65 Blank, C.E. (2004) Evolutionary timing of the origins of mesophilic sulphate reduction and oxygenic photosynthesis: a phylogenomic dating approach. *Geobiology*, **2**, 1–20.

66 Barber, J. (2004) Engine of life and big bang of evolution: a personal perspective. *Photosynthesis Research*, **80**, 137–55.

67 Martin, W., Rotte, C., Hoffmeister, M., Theissen, U., Gelius-Dietrich, G., Ahr, S. and Henze, K. (2003) Euglena gracilis rhodoquinone:ubiquinone ratio and mitochondrial proteome differ under aerobic and anaerobic conditions. *IUBMB Life*, **55**, 193–204.

68 Lopez-Garcia, P. and Moreira, D. (2006) Selective forces for the origin of the eukaryotic nucleus. *Bioessays*, **28**, 525–33.

69 Simpson, A.G. and Roger, A.J. (2004) The real “kingdoms” of eukaryotes. *Current Biology*, **14**, 693–6.

70 Nelson, N., Sacher, A. and Nelson, H. (2002) The significance of molecular slips in transport systems. *Nature Reviews Molecular Cell Biology*, **3**, 876–81.

71 Trissl, H.W. and Wilhelm, C. (1993) Why do thylakoid membranes from higher plants form grana stacks? *Trends in Biochemical Sciences*, **18**, 415–19.

72 Allen, J.F., Bennett, J., Steinback, K.E. and Arntzen, C.G. (1981) Chloroplast protein phosphorylation couples plastoquinone redox state to distribution of excitation energy between photosystems. *Nature*, **291**, 25–9.

73 Kyle, J., Staehelin, L.A. and Arntzen, C.J. (1983) Lateral mobility of the light-harvesting complex in chloroplast membranes controls excitation energy distribution in higher plants. *Archives of Biochemistry and Biophysics*, **222**, 527–41.

74 Allen, J.F. (2003) Cyclic, pseudocyclic and noncyclic photophosphorylation: new links in the chain. *Trends in Plant Science*, **8**, 15–19.

75 Zhang, S. and Scheller, H.V. (2004) Light-harvesting complex II binds to several small subunits of photosystem I. *Journal of Biological Chemistry*, **279**, 3180–7.

76 Amunts, A. and Nelson, N. (2008) Functional organization of a plant photosystem I – evolution of a highly efficient photochemical machine. *Plant Physiology and Biochemistry*, **46**, 228–37 (in press).

77 Kouril, R., Zygadlo, A., Arteni, A.A., de Wit, C.D., Dekker, J.P., Jensen, P.E., Scheller, H.V. and Boekema, E.J. (2005) Structural characterization of a complex of photosystem I and light-harvesting complex II of Arabidopsis thaliana. *Biochemistry*, **44**, 10935–40.

78 Subramanyam, R., Jolley, C., Brune, D.C., Fromme, P. and Webber, A.N. (2006) Characterization of a novel Photosystem I-LHCI supercomplex isolated from Chlamydomonas reinhardtii under anaerobic (State II) conditions. *FEBS Letters*, **580**, 233–8.

4
Structure and Function of Photosystem II

Jian-Ren Shen, Takahiro Henmi, and Nobuo Kamiya

4.1
Introduction

Photosystem II (PSII) is the site of oxygen production, leading to the oxygenic atmosphere required for the survival and evolution of all oxygenic life on earth. In view of its importance, the structure and function of PSII has been studied extensively in the past several decades and has been well summarized in a recently published book solely devoted to it [1]. PSII is a large, multi-membrane protein complex embedded in the thylakoid membranes from cyanobacteria to higher plants. The heart of PSII is largely conserved from cyanobacteria to higher plants. In its native state, PSII is surrounded by a similar, or even larger, protein complex that functions to harvest light-energy and to transfer it to the PSII reaction center. In cyanobacteria and red algae, this light-harvesting complex is the phycobilisome that consists of phycocyanin and phycoerythrin, as well as linker proteins (see also Chapter 11); whereas in green algae and higher plants, a chlorophyll *a*/*b* binding protein complex called light-harvesting complex II (Chapter 11) functions to collect the light energy. These light-harvesting proteins can be stripped from the PSII core complex without affecting the oxygen-evolving activity significantly from both cyanobacteria and higher plants. Structural studies are thus possible using the purified PSII core complex only. This chapter focuses on the structure and function of the PSII core complex (herein referred to as PSII); for the structure of light-harvesting complex II of higher plants and phycobilisome antenna of cyanobacteria; please refer to Chapters 10 and 11.

The PSII core complex from cyanobacteria contains 17 transmembrane protein subunits and 3 extrinsic proteins (for review, see Chapters 3–6 in [1]). The transmembrane subunits include the D1 and D2 reaction center subunits encoded by the *psbA* and *psbD* genes with deduced molecular masses of 38.2 kDa and 39.4 kDa, respectively; and CP47, CP43 chlorophyll-binding subunits encoded by the *psbB* and *psbC* genes with deduced molecular masses of 56.4 kDa and 51.6 kDa (from cyanobacterial genes). In addition, 13 subunits with molecular

Photosynthetic Protein Complexes: A Structural Approach. Edited by P. Fromme

ISBN: 978-3-527-31730-1

masses below 10 kDa are present; they are referred to as low-molecular mass (LMM) subunits and include the PsbE (α-subunit of cytochrome (Cyt) *b*559), PsbF (β-subunit of Cyt *b*559), PsbH, PsbI, PsbJ, PsbK, PsbL, PsbM, PsbTc, PsbX, PsbY, PsbZ, and Ycf12 (where Ycf represents hypothetical chloroplast reading frame) subunits encoded by the corresponding genes, respectively [2–5]. Among these subunits, 12 contain a single transmembrane helix, and only PsbZ has two transmembrane helices. In addition to these transmembrane subunits, three extrinsic proteins are bound to the lumenal side. In cyanobacteria, these three extrinsic proteins are the 33 kDa and 12 kDa proteins encoded by the *psbO* and *psbU* genes, and Cyt *c*550 encoded by the *psbV* gene (deduced molecular mass: 15.1 kDa), whereas in green algae and higher plants the 23 kDa and 17 kDa proteins encoded by the *psbP* and *psbQ* genes, respectively, were found to replace the 12 kDa protein and Cyt *c*550 [6]. These extrinsic proteins, combining with the extra-membrane loop regions of D1, D2, CP47, CP43, and the Mn_4Ca cluster, constitute the oxygen-evolving complex (OEC) of PSII. In the cyanobacterial genome, two genes homologous to the plant *psbP* and *psbQ* genes have been discovered, and their products were indeed found to associate with purified PSII in some cases, but in sub-stoichiometric amounts. [7, 8]. PSII purified from thermophilic cyanobacteria that have been used for structural studies appears to not contain these subunits. Therefore, their structure and functions will not be discussed further here.

The crystal structure of PSII core was analyzed with PSII purified from two species of thermophilic cyanobacteria, *Thermosynechococcus elongatus (T. elongatus)* and *T. vulcanus*, because of their high stability in the purified form that is needed for crystallization. No reports have been published on the crystal structure analysis of PSII from mesophilic cyanobacteria and higher plants. The first crystal structure was reported by Zouni and coworkers [9] with PSII purified from *T. elongatus*, a thermophilic cyanobacterium grown at 50–55 °C, at a 3.8 Å resolution. This report provided the overall structure of PSII and formed the basis for further detailed structural and functional analysis of PSII; however, the structure was built with Cα atoms of some major PSII subunits without providing the detailed structure of amino acid side chains or the location of most of the LMM subunits. Subsequently, Kamiya and Shen [10] reported the structure of PSII from *T. vulcanus*, a close relative of *T. elongatus*, at a 3.7 Å resolution, which included the orientation of some of the amino acid side chains as well as the structure of 12 kDa protein, an extrinsic subunit whose structure had not been resolved previously. More detailed structures were reported by Ferreira *and coworkers* [11] at a 3.5 Å resolution and more recently, by Loll *and coworkers* [12] at a 3.0 Å resolution; both used PSII from *T. elongatus*. In this chapter we describe the PSII structure and function based on these recent structural analyses.

4.2 The Structure of Protein Subunits

4.2.1 Overall Structure

PSII purified from thermophilic cyanobacteria exists as a dimer both in solution and in crystallized form. The overall structure of PSII dimer at 3.0 Å resolution is shown in Figure 4.1 [12]. The two monomers in PSII are related by a local twofold symmetry axis in its center; each monomer is composed of 36 transmembrane helices (TMHs). Among them, 10 are assigned to the D1, D2 subunits located in the center of the monomer (Figure 4.1b). The D1 and D2 subunits are surrounded by the CP47 and CP43 subunits in each side, each with 6 TMHs. In addition, 10 TMHs are attributed to the LMM subunits of PsbE, F, H, I, J, K, L, M, Tc, and PsbY subunits [14], and 2 TMHs to PsbZ. The remaining two TMHs, designated X1 and X3 (Figure 4.1b), are not assigned in the current structure due to the limited resolution.

A large extra-membrane region, found at the lumenal side of PSII, is formed by the three extrinsic proteins PsbO, PsbU, and PsbV, together with the large, hydrophilic loops of D1, D2, CP47, and CP43 protruding from the membrane surface (Figure 4.1a). The Mn_4Ca cluster is located at the immediate surface of the membrane facing the lumenal side and is thus deeply covered by the large extra-membrane region. In contrast, a relatively small extra-membrane region is found at the cytoplasmic side (stromal side in the case of chloroplasts) of PSII, which may be related to the fact that no distinct function has been assigned in the peripheral region of the stromal side except the location of Q_A and Q_B, two plastoquinone electron acceptors immediately underneath the cytoplasmic membrane surface (see 4.3.3). The lack of a large extra-membrane region in the cytoplasmic side may facilitate the association of phycobilisome at the membrane surface required for collecting the light energy in cyanobacteria. In higher plants, phycobilisome is replaced by light-harvesting complex II, a transmembrane protein complex. The lack of an extra-membrane region in the stromal side may, therefore, facilitate the stacking of multiple thylakoid membranes. This phenomenon is known to occur in green algal and higher plants containing light-harvesting complex II, but not in cyanobacteria.

4.2.2 Structure of the D1 and D2 Reaction Center Subunits

The D1 and D2 proteins virtually bind all of the PSII electron transfer components and are thus designated as PSII reaction center (RC) subunits. This was shown by isolation of a complex containing these two subunits, together with the α-, β-subunits of cytochrome *b*559 and PsbI that retains the capability of light-induced charge separation [15]. The overall structure of D1 and D2 is similar to the L and M subunits of bacterial RC [16, 17]. Each of the D1 and D2 subunits contains five

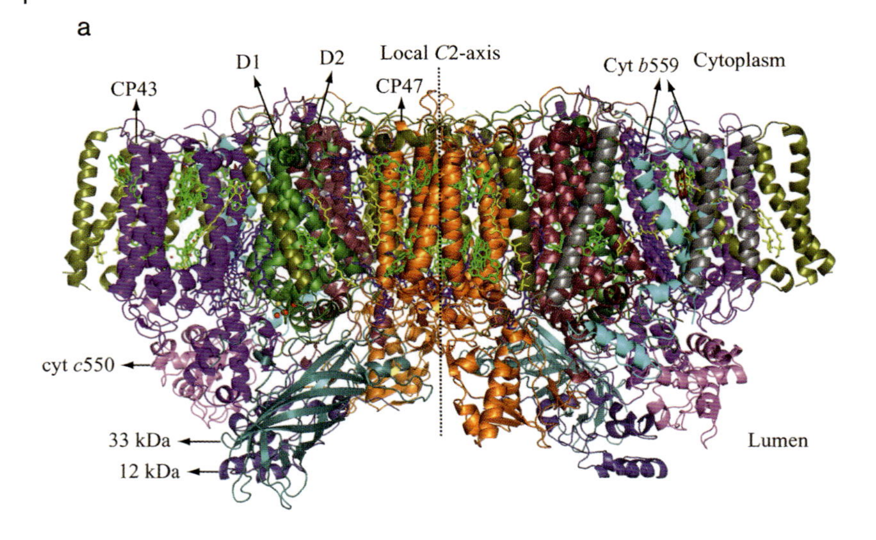

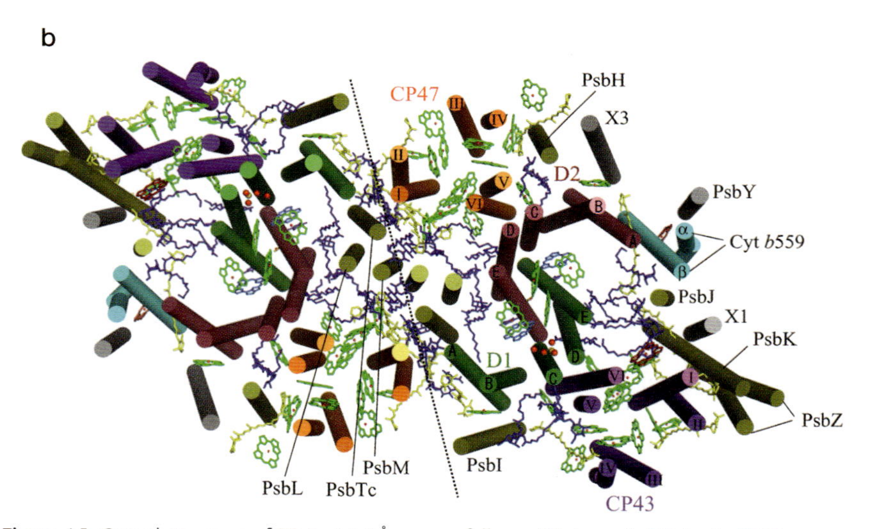

Figure 4.1 Crystal structure of PS II at 3.0 Å resolution from *Thermosynechococcus elongatus* [12]. **(a)** Side view of PS II dimer along with the membrane plane. **(b)** The arrangement of TMHs and cofactors seen as a top view of PS II dimer perpendicular to the membrane plane from the lumenal side. For clarity, the extrinsic proteins and hydrophilic regions of membrane-spanning subunits were omitted. For both panels **(a)** and **(b)**, major protein subunits are colored as follows: D1 (green), D2 (red), CP47 (orange), CP43 (purple), α- and β-subunits of Cyt *b*559 (cyan), and the three extrinsic proteins 33 kDa (cyan), Cyt *c*550 (pink), 12 kDa (deep blue). Cofactors are shown as: Chls (green), carotenoids (yellow), lipids (deep blue), Q_A and Q_B, Pheo (blue), Mn, Fe and heam ring of Cyt (red), and Ca (yellow). This figure and all of the following figures were drawn from PDB data 2AXT with PyMol [13].

TMHs, helices *A-E* (Figure 4.2a). The organization of these five TMHs is similar between D1 and D2; they are related by a pseudo-*C*2-symmetry with the symmetric axis passing through the non-heme iron that is located at the center of PSII monomer. The five helices of D1 and D2 are tilted against each other in the cytoplasmic side, but separated in the lumenal side. The PSII-RC chlorophylls (Chls) are associated with these two subunits in a region close to the lumenal surface, whereas two pheophytin (Pheo) molecules are associated with D1/D2 subunits toward the cytoplasmic side. At the lumenal side, there are two main helices for each of the D1 and D2 subunits. One is in the C-D loop and the other is in the C-terminal region of both subunits. The Mn_4Ca cluster is located in a region close to the *cd* helix as well as the C-terminal loop of D1 (see 4.3.4). In addition, a short helix is present in the A-B loop of both subunits, and an additional short helix is present in the C-terminal loop region of D2. At the cytoplasmic side, there are two helices for each of the D1 and D2 subunits; one is in the N-terminal region and the other is in the D-E loop. The *de* helix in the D-E loop is involved in the binding of Q_B in D1 and that of Q_A in D2 (see 4.3.3).

4.2.3
Structure of the CP47 and CP43 Subunits

CP47 is located at the D2 side while CP43 is located at the D1 side, each with six TMHs (Figure 4.1). Together, with several LMM subunits, they form a belt of TMHs that surround the D1/D2 RC heterodimer. The 6 TMHs in CP47 and CP43 are symmetrically related, and are organized into three clusters, each consisting of two TMHs that form a trimer of dimer structure in each subunit. This trimer-of-dimers organization is similar to the N-terminal region of Chl-binding RC proteins PsaA and PsaB of PSI [18] (see also Chapters 2 and 13). However, large hydrophilic regions protrude into the lumenal side from both CP47 and CP43 (Figure 4.2b) whose structures are not homologous to the PsaA and PsaB subunits in PSI. The spaces between each cluster of TMHs in both CP47 and CP43 are filled with hydrophobic Chls, β-carotenes (car), and lipids (Figure 4.1b) (see 4.4.1). Two main functions of CP47 and CP43 are, thus, the harvest of light energy by these Chls and the transfer of the excitation energy to the PSII reaction center. Because of this, and also because CP47 and CP43 surround the PSII-RC directly, they are called intrinsic light-harvesting proteins–this distinguishes them from the large amount of light-harvesting complex II and additional chlorophyll-binding proteins (so called cp's), associated in the periphery of higher plant PSII. Another important function associated with CP47 and CP43 is the maintenance of assembly of the PSII core, since loss of either CP47 or CP43 leads to the disassembly of PSII as well as loss of oxygen-evolution [19–22]. Note, however, that the loss of CP43 has been reported to have less effect on the remaining PSII complex as a small amount of PSII without oxygen-evolving activity was still assembled in the CP43-deletion mutant, whereas virtually no PSII was assembled in the CP47-deletion mutant [19–23]. In particular, various residues in the large lumenal hydrophilic loop regions of both CP47 and CP43 have been shown to be required for

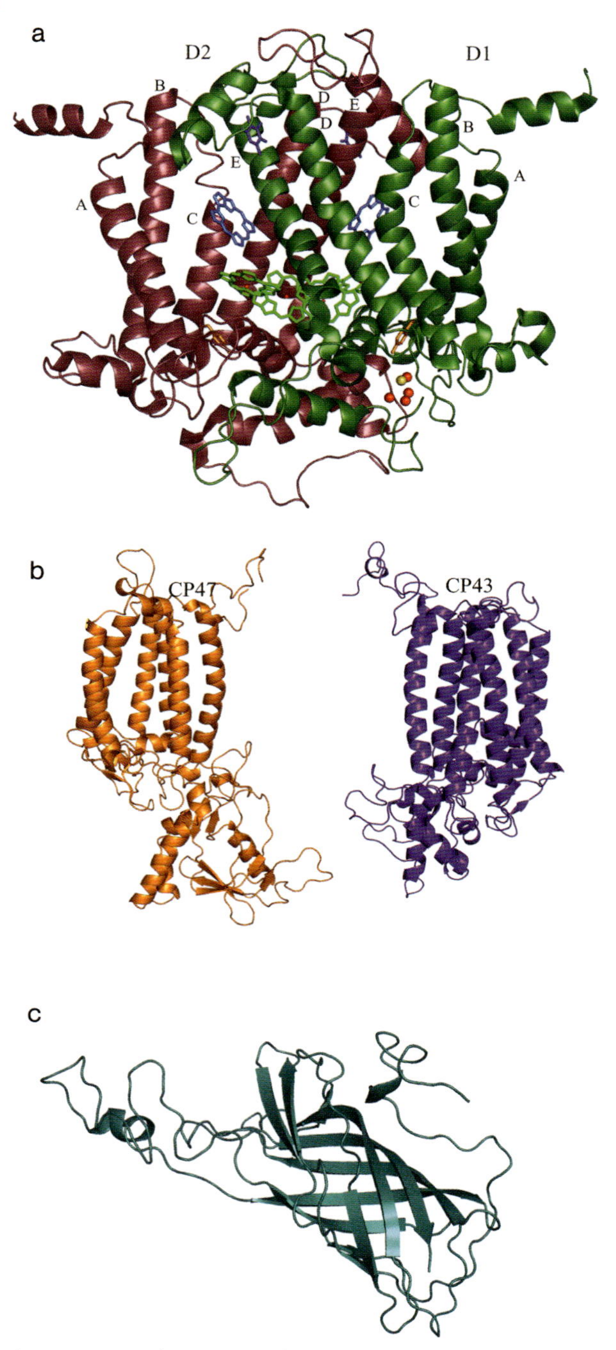

Figure 4.2 Crystal structures of individual components of *T. elongatus* PS II at 3.0 Å resolution [12]. **(a)** Structure of D1/D2 subunits from a side view along with the membrane plane. The TMHs are labeled as A-E. Cofactors are shown as Chls (green), Pheo (blue), Mn (red), Ca (yellow), Tyr_Z and Tyr_D (orange). **(b)** Structure of CP47, CP43 subunits from a side view. **(c)** Structure of the extrinsic 33 kDa protein.

binding of the extrinsic proteins and for the assembly of a stable Mn_4Ca cluster ([22], and references cited therein). In support of this, the crystal structures now show that CP43 even directly provides a ligand for the Mn_4Ca cluster (see 4.3.4).

4.2.4
Structure of the LMM Subunits

In addition to the 4 major subunits, 13 LMM subunits were shown to be present in PSII: PsbE, PsbF, PsbH, PsbI, PsbJ, PsbK, PsbL, PsbM, PsbTc, PsbX, PsbY, PsbZ, and Ycf12 [2–5]. Among them, the location of 11 subunits has been determined in the crystal structure (Figure 4.1b) [11–14]. PsbE and PsbF are the α- and β-subunits of Cyt *b*559, with molecular masses of 9.0 kDa and 4.5 kDa respectively. They are located adjacent to TMH-*A* of D2. Three TMHs are found surrounding Cyt *b*559: PsbJ, PsbY, and X3, where X3 has unknown identity due to limited resolution. Adjacent to PsbJ, two TMHs are present, PsbK and unknown X1. PsbK is closely associated to CP43. PsbZ is also closely related to CP43, but PsbZ is the only LMM subunit that contains two TMHs. In the middle of PSII dimer, there are three TMHs at each side of the monomer which are assigned to PsbM, PsbL, and PsbTc respectively; these subunits may thus function to stabilize the dimeric structure.

Among the 11 LMM subunits identified, PsbY was assigned as an unknown helix X2 in the 3.0 Å structure but was identified by crystal structure analysis of a PsbY-deletion mutant [14]. This helix was found in the 3.7 Å [10] and 3.0 Å structure [12], but not in the 3.5 Å structure [11]. Such findings suggest that it is associated with PSII weakly and may be lost during purification in some cases. In addition, there are two TMHs (X1 and X3) whose identities have yet to be determined. They should correspond to PsbX and Ycf12, since their presence in PSII has been confirmed [5, 22]. In fact, X1 has been suggested to be Ycf12, and X3 was suggested to be PsbX, based on their sequences [24] and the incomplete side chain structure in the present crystal structure [5].

The functions of LMM subunits have been studied by deletion mutagenesis. It has been established that Cyt *b*559 is absolutely required for the assembly of the PSII protein complex–deletion of either the *psbE* or *psbF* gene results in an obligate photohetrotrophic growth as well as disassembly of PSII in cyanobacteria (for review, see [4]). For most of the other LMM subunits, however, deletion of the corresponding gene in cyanobacteria often did not lead to a loss of photoautotrophic growth; rather, such deletions gave rise to an impaired growth rate with their PSII activities lowered to varying extents. Interestingly, deletion of the same LMM subunit seems to have greater effect on PSII in eukaryotic alga and higher plants than what is observed in cyanobacteria (reviewed in [4]). The causes for these apparent differences are not yet clear. Possible sources of differences may include a different assembly process of PSII and a slightly different composition of LMM subunits, between higher plants and cyanobacteria. The assembly of PSII in eukaryotic alga and higher plants may be more complicated or need a higher degree of cooperation among different subunits (as a result of localization of some genes into the nuclear genome) than that in cyanobacteria, resulting in harsher effects caused by subunit deletion. In addition, a PsbW subunit is present in higher plant PSII but not in cyanobacteria, suggesting a possible slightly different structural organization of PSII between higher plant and cyanobacteria [3, 4]. From a structural point of view, the functions of LMM subunits in maintaining the binding environment of Chls, carotenoids, lipids, and their roles in maintaining the structural integrity and dimeric conformation of PSII can now be deduced. For example, PsbI and unassigned X3 form a cap for $Chlz_{D1}$ and $Chlz_{D2}$, respectively, thus shielding them from the solution environment outside the membrane. In addition, PsbH, PsbK, and PsbZ are in contact with other Chls. The side chains of PsbI, PsbJ, PsbK, PsbT, PsbZ, and X1 are in contact with carotenoids; PsbH, PsbI, PsbJ, PsbL, PsbM, and PsbT are involved in lipid binding. These possible functions of LMM subunits should be examined in further structural and functional studies, taking the structural information into consideration.

4.2.5 Structure of the Extrinsic Proteins

Three extrinsic proteins of 33 kDa, 12 kDa, and Cyt *c*550 encoded by *psbO, psbU* and *psbV* genes are located at the lumenal side. Among these extrinsic proteins, the 33 kDa protein is attached to the lumenal surface at the D2/CP47 side. It has a characteristic cylindrical shape (Figure 4.2c) similar to those of outer membrane proteins family [25–27]. Cyt *c*550 is attached to the lumenal surface at the CP43/D1 side, with its overall structure being similar to other *c*-type cytochromes [28]. The 12 kDa protein is composed of mainly α-helices, and is located between the 33 kDa protein and Cyt *c*550. Additionally, it is approximately 30 Å away from the lumenal surface of the membrane and may be in contact with some of the loop regions of D2, CP47, and CP43 that protrude from the lumenal surface (Figure 4.1a). The contact of the 12 kDa protein with these membrane proteins, however, is not strong enough to hold it onto PSII since binding of the 12 kDa protein to PSII requires the presence of both the 33 kDa protein and Cyt *c*550 [29, 30]. These three extrinsic proteins together with the large extra-membrane loop regions of D1, D2, CP47, and CP43, form a large hydrophilic domain at the lumenal side that covers the Mn_4Ca cluster located close to the membrane at the lumenal surface. This structural organization suggests that both the 33 kDa protein and Cyt *c*550 function to stabilize the Mn_4Ca cluster. Indeed, release of either the 33 kDa protein or Cyt *c*550 has been found to result in a destabilized Mn_4Ca cluster which was deactivated rapidly in the dark [31, 32]. Absence of either of these two subunits even leads to a higher efficiency of photoactivation, a process of light-induced reassembly of the Mn_4Ca cluster [31, 32]. This suggests that Mn and/or Ca could access to the Mn_4Ca cluster more freely in the absence of these subunits.

Two genes, homologous to *psbP* and *psbQ*, of higher plants that encode the extrinsic 23 kDa and 17 kDa proteins of green algal and higher plant PSII have been found in the cyanobacterial genome; the products of these genes have been confirmed in PSII purified from *Synechocystis* sp. PCC 6803 with a His-tag attached [8]. The presence of PsbP and PsbQ proteins were also confirmed in the thylakoids of thermophilic cyanobacteria [7]. In the present structure, however, the PsbP and PsbQ proteins were not found, and they were also absent in the purified PSII of thermophilic cyanobacteria. These proteins are thus released during the purification process, suggesting that they are weakly bound to PSII during assembly.

4.3 The Electron Transfer Chain (ETC)

4.3.1 Overview

There are two sets of similar electron transfer cofactors arranged approximately symmetrically along the non-crystallographic pseudo-C2 axis, one at the D1 side and one at the D2 side (Figure 4.3a). The active electron transfer chain is located at the D1 side, since the Mn_4Ca cluster is bound to the D1 side only. The charge separation occurs at four Chls bound to the D1/D2 subunits (Figure 4.3b), which are designated RC-Chls. Among these four Chls, two Chls, P_{D1} and P_{D2}, are located toward the lumenal side with their head groups arranged parallel to each other and perpendicular to the membrane plane. P_{D1} and P_{D2} are considered as the R.C. Chls of PSII, namely P680 (however, see 4.3.2). The other two Chls surround P_{D1} and P_{D2} respectively; they are designated Chl_{D1} and Chl_{D2} and are called accessory Chls because of their similar location as the bacterial accessory Chls. These two accessory Chls are tilted against the membrane plane by approximate 30°. These Chls form a cluster in a region from the middle to the lumenal side of the membrane. The electrons generated in the initial charge separation events at the RC-Chls are transferred to a pheophytin molecule ($Pheo_{D1}$) at the D1 side, with a center-to-center distance of 10.5 Å from Chl_{D1} to the cytoplasmic membrane surface. The electrons from $Pheo_{D1}$ are then transferred to Q_A, the first bound plastoquinone accepter, 13.1 Å away from the center of $Pheo_{D1}$, and then to the second bound quinone acceptor Q_B.

At the donor side, tyrosine of D1-161 donates electrons directly to the RC-Chls; this is designated Tyr_Z and is located 13.6 Å from the center of P_{D1}. A symmetrically located Tyr in the D2 side, namely, D2-Tyr-160 or Tyr_D, is found which although not directly involved in electron transfer, may play a role in photoassembly of the Mn_4Ca cluster. The Mn_4Ca cluster is located 5.4 Å away from D1-Tyr_Z, which sequentially extracts electrons from water and donates them to Tyr_Z. After four electrons are removed from two molecules of water, one molecule of oxygen is formed. This process is accomplished by the release of four protons, whose exact pattern is still under debate. During this process, the configuration of the Mn_4Ca cluster is changed through the S-state cycle (where $i=0-4$) as defined by the Kok-model [33, 34].

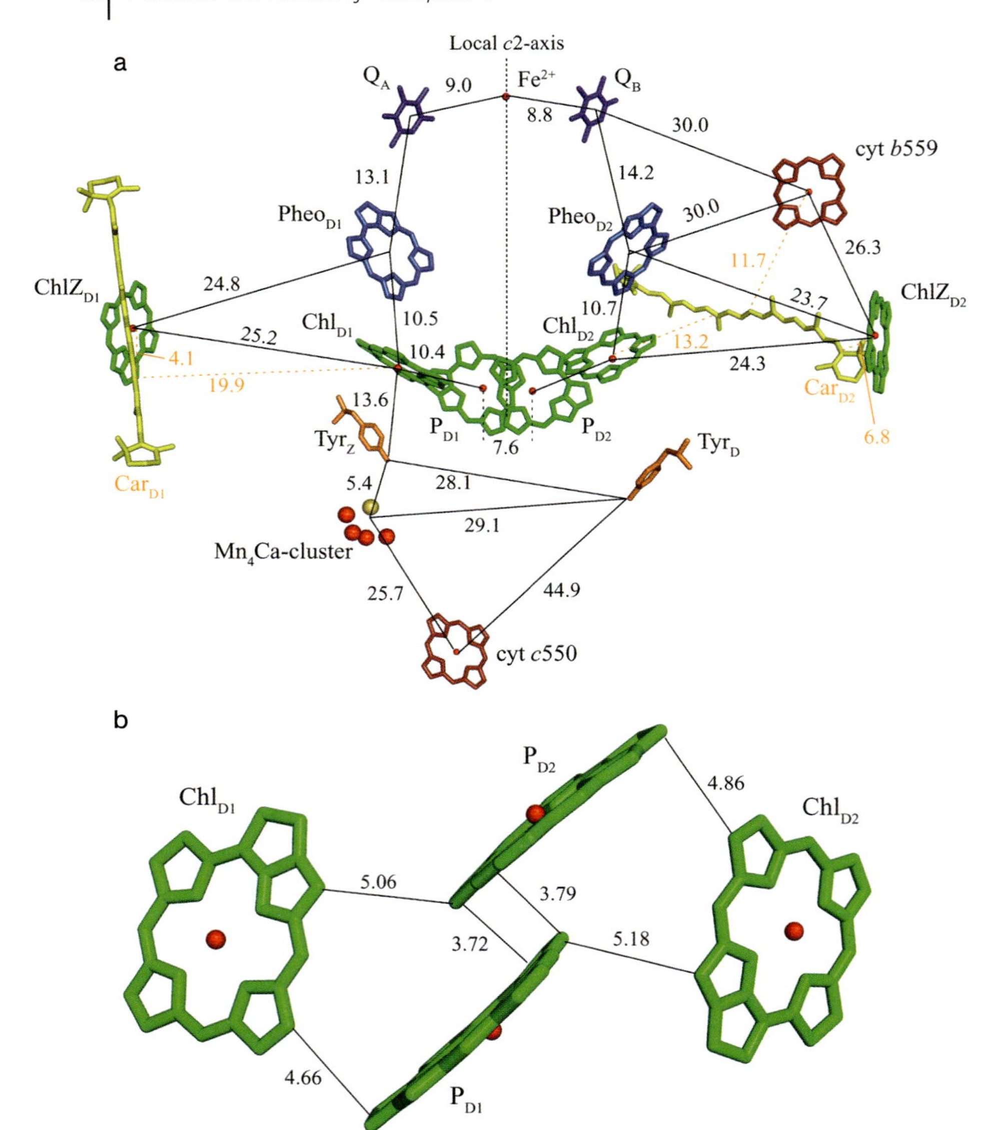

Figure 4.3 (a) Arrangement of the electron transfer chain and other cofactors in PSII-RC from a side view. Distances between the centers of each component were shown in Å. **(b)** An enlarged picture of the four Chls located in the center of PSII-RC. Distances between the edges of each component were shown in Å.

4.3.2
The Photosystem II Reaction Center Chlorophylls (P680)

One of the most distinct features of PSII is its unusually high midpoint redox potential ($E_m \approx 1.1 \sim 1.3$ V) generated by its RC-Chls (P680) [35–37] that enables oxidation of water. Six Chls are bound to the D1/D2 subunits, four of which form a cluster at the center of ETC and constitute the PSII-RC-Chls, whereas the other two Chls are located at the periphery of D1/D2 ($ChlZ_{D1}$, $ChlZ_{D2}$). The two peripheral Chls $ChlZ_{D1}$ and $ChlZ_{D2}$ are coordinated by D1-His118 and D2-His117, respectively, and are located far (~24 Å) from PSII-RC-Chls and thus do not participate in the electron transfer reactions directly under normal conditions (Figure 4.3a) (for discussions on the roles of $ChlZ_{D1}$ and $ChlZ_{D2}$, see 4.3.5). Among the four Chls of RC, P_{D1} and P_{D2} are coordinated by D1-His198 and D2-His197, respectively, whereas no corresponding His residues have been found as coordinating ligands for Chl_{D1} and Chl_{D2}. Instead, their head groups are found in a hydrophobic environment surrounded by D1-Thr179 and D2-Leu205 for Chl_{D1}, and D2-Ile178 and D1-Phe206 for Chl_{D2} [12]. Previous kinetic studies suggested that a multimer or tetramer model involving these four Chls as the site of the initial light-induced charge separation [38–40]. Indeed, when the first medium-resolution PSII structures were reported [9, 10], the center-to-center distances between these four Chls were assigned to be 10–11 Å, which suggested that the interactions among these four Chls are rather similar, in agreement with requirements imposed by the tetramer model. The higher resolution structure has yielded a center-to-center distance of 7.6 Å between P_{D1} and P_{D2}, and 10.3 Å, 10.4 Å, respectively, between P_{D1}-Chl_{D1}, and between P_{D2}-Chl_{D2} (Figure 4.3a). Furthermore, the edge-to-edge distance between P_{D1}-P_{D2} is apparently shorter than that between P_{D1}-Chl_{D1} and between P_{D2}-Chl_{D2} (Figure 4.3b), suggesting that the excitonic coupling between P_{D1}-P_{D2} is rather stronger than those between P_{D1}-Chl_{D1} and between P_{D2}-Chl_{D2}. This suggests a rather similar organization between P_{D1}/P_{D2} and the bacteriochlorophyll *a* homodimer of the bacterial RC. The question is, then, which pigment is the real initial charge separator? By comparing PSII to the bacterial RC, it has long been believed that the initial charge separation occurs in the P_{D1}-P_{D2} homodimer, with the charge of $P680^{+}\cdot$ largely localized at P_{D1}. If this is the case, the electron generated at P_{D1} in the initial charge separation event is transferred to Chl_{D1} and subsequently to $Pheo_{D1}$. This consideration, however, has been questioned by studies on the localization of the triplet state [41] which have shown a remarkable difference between the orientations of the chlorin ring where the Chl triplet resides and the normal of the membrane plane (or the chlorin ring of P_{D1}). Later studies indicate that at low temperatures, below that of liquid nitrogen, the Chl triplet is accumulated at Chl_{D1} [40, 42, 43]. Thus, at least at low temperature, the initial charge separation may take place at Chl_{D1}, which transfers the electron to $Pheo_{D1}$ and extracts an electron from P_{D1}, thereby forming $P680^{+*}$.

Recent femtosecond transition absorption studies identified a 5.5-ps component at room temperature that may reflect the lifetime of the primary charge separation in intact PSII cores [44]. In these studies, the authors found that Pheo is reduced in the first charge separation step, leading to the conclusion that Chl_{D1} is the place for primary charge separation also at room temperature. Indeed, this has been suggested to occur in isolated D1/D2/Cyt *b*559 complex earlier [40, 43, 45, 46] and also from a study using pigment-modified RCs [47]. Thus, the location of the triplet state at Chl_{D1} at low temperature can be explained by the assumption that reduction of Chl_{D1} by P_{D1} is limited at low temperature after the first charge separation step. At ambient temperature, the reduction of Chl_{D1} by P_{D1} takes place rapidly, which may have prevented identification of positive charges at Chl_{D1} in the first step. If this is the case, nature has chosen to adopt a totally different approach, namel using the accessory Chl *a* monomer as the RC in PSII in order to fulfill the requirement of the high redox potential for water oxidation, rather than modifying the properties of Chl *a* dimer found in the similar position as that of bacterial RC.

The cause of the unusually high redox potential of PSII-RC as compared with bacterial RC (Em ≈ 500 mV, [48]) and PSI (Em ≈ 500 mV, [49]) can now be analyzed based on the X-ray structures of all these complexes. Based on the current PSII structure, the mid-point redox potentials have been calculated to be 1206 mV, 1222 mV for P_{D1} and P_{D2}, and 1262 mV, 1320 mV for Chl_{D1} and Chl_{D2}, respectively [50, 51]. These data suggest that the redox potential of P_{D1}/P_{D2} is slightly lower than that of Chl_{D1}/Chl_{D2}, indicating that the charge-separated state would be stabilized at P_{D1}/P_{D2} rather than Chl_{D1}/Chl_{D2} [39], in agreement with the preferential detection of the triplet state at P_{D1}/P_{D2} at room temperature. Among the major factors that contribute to the up-shift of the redox potential of PSII-RC, the charges of the Mn_4Ca cluster were found to up-shift the E_m of P_{D1}/P_{D2} by 210/100 mV [51]. To compensate for the positive charges on the Mn_4Ca cluster, there are more acidic than basic residues on the D1 side. Such a composition, in turn, down shifts the E_m of P_{D1} by 135 mV without affecting the E_m of P_{D2} significantly. The TMH-*D* of D1 and residues in the lumenal *cd* loop of D1 contributed to an up-shift of 223/195 mV for P_{D1}/P_{D2} relative to the bacterial RC. In addition, the protein subunits of PSII, other than D1/D2, contributed to an up-shift of 170–200 mV in the redox potential of P_{D1}/P_{D2}. Similar effects were found on the E_m of Chl_{D1}/Chl_{D2} [51]. The combination of these factors yielded the characteristic high potential of PSII-RC relative to those of bacterial RC, PSI, and also of free Chl *a* monomer in organic solvents ($E_m \approx 800 \sim 850$ mV, [52, 53]).

4.3.3
The Quinone Acceptors and Non-Haem Iron

The first quinone acceptor Q_A is located at the D2 subunit 13.1 Å away from $Pheo_{D1}$ and acts as a one-electron acceptor (Figure 4.3a). Its binding pocket is at the cytoplasmic membrane surface, under the coverage of helix *de* of D2 in the cytoplasmic side. Its quinone plane is sandwiched between D2-Trp253 and D2-Leu267, with its keto oxygens hydrogen bonded to D2-His214 and D2-Phe261. The isoprenoid chain is partially modeled, but not fully resolved, in the current structure [12].

The second quinone acceptor, Q_B, is located on the D1 subunit in a symmetric position relative to Q_A, defined by the pseudo-*C*2 axis. Q_B acts as a two-electron acceptor which, after accepting two electrons from Q_A^-, binds two protons and dissociates from its binding site, and then replaced by free plastoquinone (PQ). The residues forming the Q_B-binding pocket, possibly interacting with the quinone plane, involve D1-Phe255, D1-Lue271, D1-His215, and D1-Ser264 or D1-Phe265. These residues are distributed at the beginning of THM-*E*, the stromal *de* helix and the end region of TMH-*D* of D1. The binding of Q_B is generally weaker than Q_A, thus the content of Q_B in the purified PSII core used for crystallization was lower (circa 0.5 per PSII) than that of Q_A due to a partial loss of Q_B during the purification procedure. This has prevented a precise determination of the binding pocket for Q_B, especially for its isoprenoid tail. Nevertheless, the Q_B-binding site was found to open into a large cavity, with the wall of the cavity formed by lipophilic phyto chains of Chls, acyl chains of four lipids and two carotenoids [12]. Furthermore, the cavity has two openings perpendicular to each other, one towards the cytoplasmic side and the other one facing the membrane interior. Since the quinone head group is located on the opposite side of the membrane-facing opening, it was suggested that PQ/PQH_2 diffuses through the membrane-facing opening between the Q_B-binding site and the plastoquinone pool of the thylakoid membrane [12]. The membrane-facing opening is surrounded by TMHs of PsbJ and Cyt *b*559; these two subunits may, therefore, regulate the PQ diffusion. An effect of PsbJ on electron transfer between Q_A, Q_B, and the plastoquinone pool has been reported [54]. However, the function of the cytoplasmic side-facing opening, which is larger than the membrane-facing opening is not clear.

The non-heme iron is located in between Q_A and Q_B (Figure 4.3a), and is coordinated by D1-His25, D1-His272, D2-His214, and D2-His268, which are analogous to the purple bacterial RC. In addition, a bicarbonate was found to provide the fifth and/or sixth ligand to the non-heme iron by FTIR measurements of PSII [55]. This bicarbonate was placed at a site interacting with D1-Tyr246 and D2-Tyr244 in the 3.0 Å structure [12], but was proposed to interact with D2-Lys264 and D2-Tyr244 in the 3.5 Å structure [11]. The non-heme iron does not participate directly in the electron transfer; on the other hand, various functions, such as an essential structural role in maintaining the integrity of the acceptor side, regulation of the electron transfer rate on the acceptor side, and possible protective functions against harmful active oxygen species generated at the PSII-RC, have been proposed for it (for review, see [56]). The redox potential of the non-haem is affected variously by pH and by molecules such as 3-(3,4-dichlorophenyl)-1,1-dimethylurea (DCMU) that bind and interfere with the Q_B-site.

4.3.4
The Mn_4Ca Cluster

The structure and function of the Mn_4Ca cluster has been the topic of extensive studies using biochemical, molecular biological, spectroscopic and X-ray crystallographic analysis (see also Chapter 5). The crystal structure of PSII at 3.0 Å resolution [12] yielded an overall shape like a hook for the four Mn atoms, with three Mn atoms forming the triangle of the hook and a fourth Mn located in the knob of the hook (Figure 4.4). This overall shape is basically similar to the Y-shaped structures reported at lower resolutions [9, 10]) and the 3 + 1 model suggested from spectroscopic studies ([57, 58], Chapter 5). However, the Ca atom was only resolved at higher resolution in the 3.5 Å [11] and 3.0 Å [12] structures of PSII. This Ca is located at the top of the triangle formed by the three Mn atoms, leading to a pseudo-pyramid structure with the Ca and three Mn atoms occupying each corner of the pyramid (Figure 4.4). The distances between Mn atoms and those between Mn-Ca could not be determined accurately from the current crystal structure, and thus have to adopt the restraints imposed by EXAFS studies. Such studies have yielded Mn–Mn distances of 2.7 and 3.3 Å, and possibly Ca–Mn distances of 3.3–3.4 Å (for review, see [58]). The short distance of 2.7 Å suggests a di-μ-oxo bridge between Mn–Mn, whereas the longer distance of 3.3 Å suggests a mono-μ-oxo bridge between Mn–Mn; however, no oxygen atoms could be identified by X-ray structure analysis due to the limited resolution.

The ligands for the Mn_4Ca cluster were identified as D1-Glu189, D1-Asp170, D1-Glu333, D1-His337, D1-Asp342, and the C-terminal residue of D1-Ala344 in the 3.0 Å structure (Figure 4.4). In addition, CP43-Glu354 was also found to be close enough to coordinate the Mn atom(s). Among these residues, D1-Asp170 and D1-Asp342 were arranged as mono-dentate ligands, whereas other carboxylic

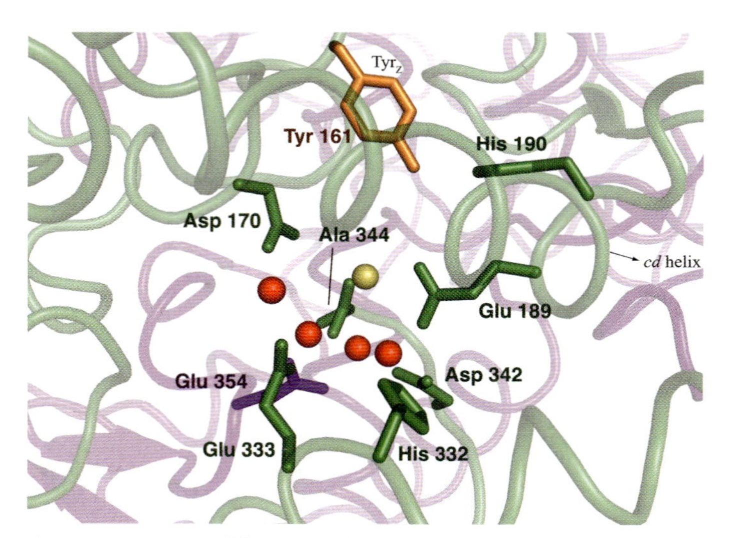

Figure 4.4 Structure of the Mn_4Ca cluster and its ligand environment. Mn atoms are shown in red, Ca in yellow, Tyr_Z in orange, D1 in green, and CP43 in purple. The side chains of amino acid residues that may provide ligands to the Mn_4Ca cluster are drawn and labeled, whereas the backbone structures of D1 and CP43 are shown in light colors.

residues could act as bidentate ligands to the metal cluster. One of the carboxylate ligands from both D1-Glu189 and D1-Ala344 was close to Ca^{2+}, rather than Mn, although the ligation of D1-Ala344, the C-terminal residue of D1, to Ca^{2+} has been questioned by FTIR studies [59, 60]. Thus, a total of 9–10 ligands were found for the four Mn atoms from the amino acids and one or two ligands for Ca^{2+}. The remaining ligands may be provided by water molecules and bicarbonate, a molecule that was found to affect the oxygen-evolving activity [1]. It should be mentioned here that, although the overall shape and ligands of the Mn_4Ca cluster are essentially similar between the structures of 3.0 Å resolution [12] and 3.5 Å resolution [11], there are some differences regarding the precise positions of each metal atom and also the ligand structure of each residue. Especially in the 3.5 Å structure, fewer residues were reported to act as bidentate ligands. This results in a lower number of protein ligands (7–8 ligands compared with 9–10 ligands in the 3.0 Å structure) for the four Mn atoms, leading to a lower saturation level for the ligands. This raises different requirements for additional ligands other than amino acids, namely water or mono- and di-μ-oxo bridges in the two structures, and will probably lead to a different atomic structure of the metal cluster. In fact, a recent study using polarized extended X-ray absorption fine structure (EXAFS) measurement on PSII single crystals has yielded structural models different from either of those derived from the crystal structure analysis [61]. Thus, the precise structure of the Mn_4Ca cluster needs to be determined at a higher resolution in order to solve these discrepancies.

4.3.5
Carotenoids and the Secondary Electron Transfer Pathway

Seven carotenoid molecules were found in the 3.5 Å structure, whereas 11 were identified in the 3.0 Å structure; all of these carotenoids have an all-trans configuration. These numbers are lower compared to the 22 carotenoids found in cyanobacterial PSI [18], which probably coincides with the lower number of Chls in PSII than in PSI (96 Chls were found in PSI, [18]). Among the 11 carotenoids in PSII, 3 are located in CP43, 5 in CP47, 1 in a space surrounded by CP43 and PsbJ, PsbK, PsbZ, and X1, and one each in D1 and D2. Four of the five carotenoids in CP47 are clustered in the monomer-monomer interface, whereas other carotenoids are located at the periphery of PSII. Almost all of these carotenoids are in close contact with Chl *a* molecules, making them capable of rapidly transferring excitation energy to the nearby Chl *a* molecules, as well as quenching the triplet state of Chl *a* in order to protect PSII from excessive light-induced damage.

The two carotenoids associated with PSII-RC have special functions, one associated with D1 and the other one with D2. Their presence in the isolated D1/D2-RC complex has been reported earlier from biochemical [62] and various spectroscopic studies [63–65]. Car_{D2} is located close to $ChlZ_{D2}$ and Cyt *b*559 with an orientation approximately parallel to the membrane, resulting in a distance of 6.8 Å between Car_{D2}-$ChlZ_{D2}$, 11.7 Å between Car_{D2}-Cyt *b*559, and 13.2 Å between Car_{D2}-Chl_{D2} at the closest positions of each molecule pair (Figure 4.3a). These distances suggest that Car_{D2} is capable of mediating electron transfer from $ChlZ_{D2}$ and Cyt *b*559 to the PSII-RC, a path that has been designated as secondary, or side electron transfer reactions which operates when the Mn_4Ca cluster is not functional, such as at low temperature [64, 66, 67]. This side reaction may also become important in protecting the PSII-RC from photo-damage under strong light illumination where electron donation from the Mn_4Ca may not be rapid enough to compensate for the charge generated at RC. On the other hand, Car_{D1} is located close to $ChlZ_{D1}$ with an orientation nearly perpendicular to the membrane plane. The distance between Car_{D1}-$ChlZ_{D1}$ is 4.1 Å, and that between Car_{D1}-Chl_{D1} is 19.9 Å, at their closest positions (Figure 4.3a). The longer distance between Car_{D1} and the RC-Chl suggests that this carotenoid is not efficient in mediating the possible secondary electron transfer from $ChlZ_{D1}$ to RC; the close location of Car_{D1} with $ChlZ_{D1}$ suggests that it is optimized for quenching the $ChlZ_{D1}$ triplet state as well as singlet oxygen that could be produced by the triplet state of RC (^{3}P680) [64]. Indeed, Car_{D1} is in close position with a cluster of four carotenoids associated with CP47 in the neighboring monomer, which is then connected to the bulk Chl *a* of CP47, supporting the role of Car_{D1} as an energy quencher [12, 68]. Also in agreement with this is the slightly higher redox potential of Car_{D1} than Car_{D2}, calculated from the current structure [68], making Car_{D2} a preferential carrier mediating electron transfer from $ChlZ_{D2}$ and/or Cyt *b*559 to PSII-RC. The preferential location of the secondary electron transfer pathway on the D2 side involving Cyt *b*559 and $ChlZ_{D2}$ has been reported based on spectroscopic as well as mutagenesis studies. Some earlier reports, however, also suggested the involvement of $ChlZ_{D1}$ at the D1 side in the secondary electron transfer reactions (for review, see [67]).

4.3.6
Location and Possible Functions of Lipids

Fourteen lipids were found in the 3.0 Å structure of PSII, which is significantly more than the four lipids found in PSI [18]. The 14 lipids of PSII included 4 digalactosyldiacylglycerol (DGDG), 6 monogalactosyldiacylglycerol (MGDG), 3 sulfoquinovosyldiacylglycerol (SQDG), and 1 phosphatidyldiacylglycerol (PG). Among these lipids, 11 constitute a belt surrounding the PSII-RC, separating it from the antenna and some LMM subunits. The remaining three lipids are found at the monomer-monomer interface.

A number of studies have suggested some specific functions for lipids in PSII, either using mutants defective in synthesizing the specific lipid molecules, or by examining the effects of treatments with lipases and phospholipases on isolated thylakoid membranes or PSII particles (for reviews, see [69, 70]). For example, loss of SQDG was found to not only increase the sensitivity of PSII to the herbicide DCMU, but also to inhibit the electron transfer from water to Tyr_Z due to a decreased stability of the Mn_4Ca cluster and the oxygen-evolving machinery [71–73]. In particular, several lines of evidence have indicated that PG is required for maintaining the integrity of the Q_B-binding site. The loss of PG, either by mutagenesis in vivo, or by phospholipase treatment in vitro, has been shown to inhibit the electron transfer from Q_A to Q_B and to destabilize the dimeric structure of PSII [69, 74, 75]. Whether or not these functions of SQDG and PG can be correlated with the positions of specific molecules identified in the crystal structure of PSII is not clear. Since multiple functions have been suggested for PG, and also since the only PG molecule identified in the crystal structure is neither close enough to Q_A and Q_B nor located in the monomer-monomer interface, the previously suggested functions for PG may not be accomplished by this single PG molecule. This suggests that there may be additional PG molecules associated with PSII. Indeed, it has been found that there are 2–3 PG molecules that can not be decomposed by phospholipase A_2 in the purified PSII dimer used for crystallization (Shen *and coworkers*, unpublished results), suggesting the presence of more than one PG molecule in PSII.

4.4 The Energy Transfer Pathways

4.4.1 The Location of Chlorophylls and the Energy Transfer Pathways within CP47 and CP43

CP47 binds 16 Chls and CP43 binds 13 Chls [12]. Most of these Chls are arranged in two layers, one toward the stromal surface and the other toward the lumenal surface (Figure 4.5a), which is similar to the N-terminal domains of PsaA and PsaB in PSI [18]. There are eight Chls each in CP47 and CP43 in the layer close to the cytoplasmic surface, seven Chls in CP47, and four Chls in CP43 in the layer close to the lumenal surface. The arrangement of Chls in a layer close to the stromal surface may facilitate the absorption of light-energy, as well as energy transfer from phycobilisome antenna attached on the cytoplasmic surface of PSII. The Chl layer close to the lumenal side, however, should be less involved in excitation energy transfers from phycobilisomes and energy absorption (as the light intensity reaching to the lumenal layer would become weaker); this may account for why the numbers of Chls present in the lumenal layer are fewer than those in the cytoplasmic layer, especially in the case of CP43. Nevertheless, the Chls in the lumenal side layer may be necessary to fulfill the spaces between TMHs in both CP47 and CP43, in order to maintain their stable structures. In fact, mutagenesis of a single His residue that is coordinated to a Chl has been found to produce a significant effect on PSII oxygen-evolving activity, in most cases in both CP47 and CP43 (for review, see [22]).

The Chls within the two layers of both CP47 and CP43 are arranged closely with each other, with their Mg-Mg distances ranging between 8.1–13.6 Å, enabling a fast equilibrium of excitation energy within each layer [76, 77]. The distances between the two layers, however, are above 16 Å, rendering a direct energy transfer between the Chls in two layers less efficient. This is overcome by a single Chl each in CP47 and CP43 located in the middle of the two pigment layers (Figure 4.5a);

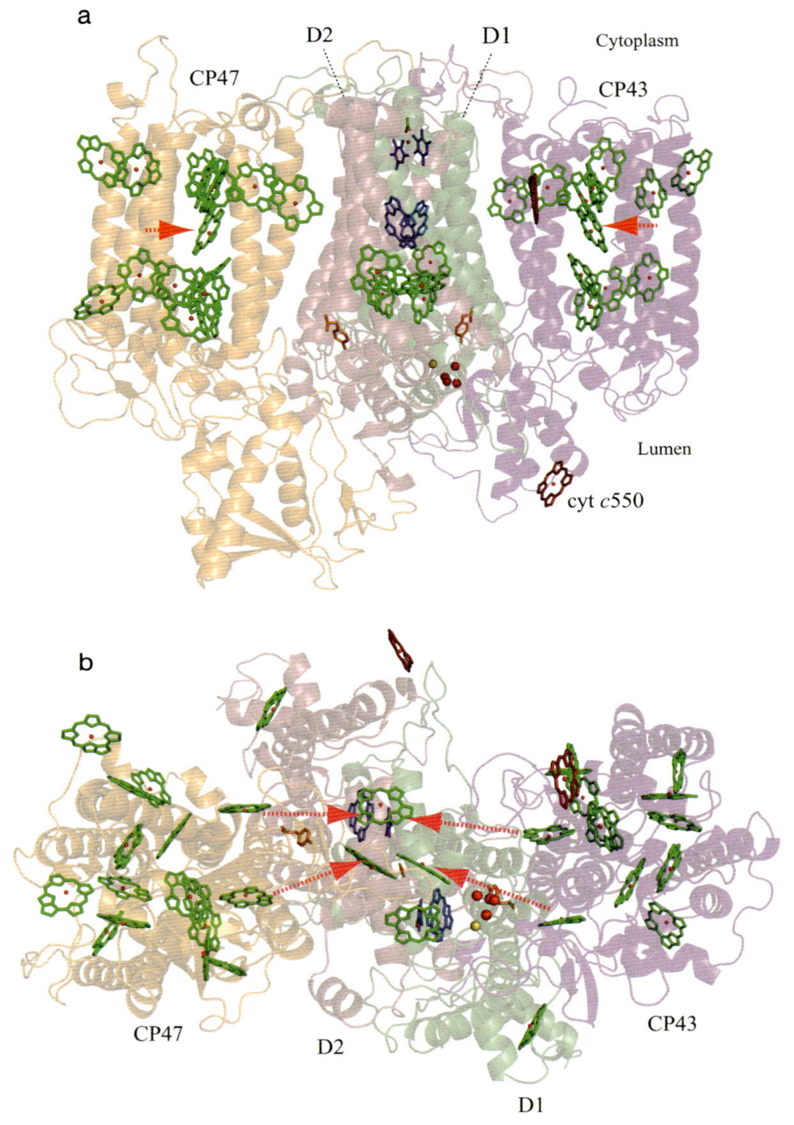

Figure 4.5 **(a)** Arrangement of Chls and other cofactors in a PSII monomer of *T. elongatus* from a side view. The protein backbone structures were shown in light colors. The arrows in CP47 and CP43 indicate two Chls located in the middle between the two Chl layers in the cytoplasmic and lumenal side. Two Pheo molecules and QA, QB are shown in blue, and heam rings of Cyt c550, Cyt b559 in red, TyrZ, TyrD in orange, Mn in red, Ca in yellow. **(b)** Same as **(a)** with a top view on to the membrane plane from the lumenal side. Arrows indicate four Chls located between CP47 and RC, and between CP43 and RC that are suggested as mediating energy transfer from bulk Chls of CP47 and CP43 to PSII-RC.

the distances between this Chl and each of the two layers are 9–10 Å. This Chl is thus important for mediating energy transfer between the two layers, enabling the fast equilibrium of excitation energy also possible between the two layers.

4.4.2
The Energy Transfer from CP47 and CP43 to PSII-RC

The excitation energy absorbed by CP47 and CP43 is transferred to PSII-RC by excitation energy transfers. From the crystal structure, there are two Chls each in CP47 and CP43 located between the bulk antenna Chls and the RC-Chls (P_{D1}, P_{D2}, Chl_{D1}, and Chl_{D2}) (Figure 4.5b). It is suggested that the four Chls in CP47 and CP43 mediate the energy transfer from the bulk antenna of CP47 and CP43 to PSII-RC. Indeed, the orientations of the head groups of these four Chls have been shown to be nearly optimized for mediating such energy transfer, based on Monte Carlo simulations of orientations of Chls with kinetic models for excitation energy transfer [77, 78]. These simulations also indicate that orientations of the two peripheral Chls, $ChlZ_{D1}$, and $ChlZ_{D2}$ are not optimized for transferring energy to RC, in agreement with their roles in secondary energy transfer and triplet quenching as described previously. Thus, natural selection also worked efficiently at the structural level of the functional molecules around PSII-RC.

4.5
Concluding Remarks and Perspectives

Our understanding of the structure and function of PSII has been advanced greatly, owing to the elucidation of its three-dimensional structure. Based on this, the reaction mechanisms of electron transfer, energy transfer, and water oxidations are analyzed in much more detail with the powerful biophysical techniques such as X-ray absorption spectroscopy, FTIR, EPR, and so on. The resolution of current crystal structure, however, is not high enough to reveal the full structure of PSII at a real atomic level. For example, there are two TMHs whose identities are not clear; in particular, the structure of Mn_4Ca cluster is not fully resolved. There are some discrepancies in the ligand structure, as well as in the positions of the metal ions in the current structures. Furthermore, the position of Cl^-, which is an essential co-factor required for oxygen-evolution and is thought to be close to the Mn_4Ca cluster, has not been determined. Thus, the goal in the near future is to improve the resolution of the crystal in order to reveal the full picture of PSII at the atomic level. Crystals with higher resolution are also a prerequisite for understanding the structural changes at the Mn_4Ca cluster accompanying electron transfer reactions. With the higher resolution crystals, it will be possible to analyze the structure of the Mn_4Ca cluster at various intermediates; in fact, one such intermediate, the S_2-state, has been successfully trapped with the current single crystals at low temperature [79, 80], which will enable X-ray structural analysis of PSII at the S_2-state possible. At the same time, the functions of most of the LMM subunits are not clear at present; crystal structure analysis of cyanobacterial mutants lacking a specific subunit is required. This is, indeed, an effective approach to clarify the function of their subunits. Finally, the current structure has been obtained with cyanobacterial PSII. Since there are some differences in the subunit compositions of PSII between prokaryotes and eukaryotes, especially with regard to the subunits involved in the oxygen-evolving complex, structural analysis for PSII from eukaryotes, including higher plants, are required to fully reveal the roles of these subunits as well as the evolutionary process of PSII from prokaryotes to eukaryotes.

References

1 Wydrzynski, T.J. and Satoh, K. (eds) (2005) *Photosystem II, The Light-Driven Water: Plastoquinone Oxidoreductase*, Springer, Dordrecht, The Netherlands.

2 Zouni, A., Kern, J., Loll, B., Fromme, P., Witt, H.T., Orth, P., Krauß, N., Saenger, W. and Biesiadka, J. (2001) Biochemical characterization and crystal structure of water oxidizing photosystem II from *Synechococcus elongatus*, in *PS2001 Proceedings, 12th International Congress on Photosynthesis*, S5-003, CSIRO, Australia, pp. 1–8.

3 Shi, L.X. and Schröder, W.P. (2004) The low molecular mass subunits of the photosynthetic supracomplex, photosystem II. *Biochimica et Biophysica Acta*, **1608**, 75–96.

4 Thornton, L.E., Roose, J.L., Pakrasi, H.B. and Ikeuchi, M. (2005) The low molecular weight proteins of photosystem II, in *Photosystem II, The Light-Driven Water:Plastoquinone Oxidoreductase* (eds T.J. Wydrzynski and K. Satoh), Springer, Dordrecht, The Netherlands, pp. 121–38.

5 Kashino, Y., Takahashi, T., Inoue-Kashino, N., Ban, A., Ikeda, Y., Satoh, K. and Sugiura, M. (2007) Ycf12 is a core subunit in the photosystem II complex. *Biochimica et Biophysica Acta*, **1767**, 1269–75.

6 Bricker, T.M. and Burnap, R.L. (2005) The extrinsic proteins of photosystem II, in *Photosystem II, The Light-Driven Water: Plastoquinone Oxidoreductase* (eds T.J. Wydrzynski and K. Satoh), Springer, Dordrecht, The Netherlands, pp. 95–120.

7 Kashino, Y., Koike, H., Yoshio, M., Egashira, H., Ikeuchi, M., Pakrasi, H.B. and Satoh, K. (2002) Low-molecular-mass polypeptide components of a photosystem II preparation from the thermophilic cyanobacterium *Thermosynechococcus vulcanus*. *Plant and Cell Physiology*, **43**, 1366–73.

8 Kashino, Y., Lauber, W.M., Carroll, J.A., Wang, Q., Whitmarsh, J., Satoh, K. and Pakrasi, H.B. (2002) Proteomic analysis of a highly active photosystem II preparation from the cyanobacterium *Synechocystis* sp. PCC 6803 reveals the presence of novel polypeptides. *Biochemistry*, **41**, 8004–12.

9 Zouni, A., Witt, H.T., Kern, J., Fromme, P., Krauß, N., Saenger, W. and Orth, P. (2001) Crystal structure of photosystem II from *Synechococcus elongatus* at 3.8 Å resolution. *Nature*, **409**, 739–43.

10 Kamiya, N. and Shen, J.R. (2003) Crystal structure of oxygen-evolving photosystem II from *Thermosynechococcus vulcanus* at 3.7-Å resolution. *Proceedings of the National Academy of Sciences of the United States of America*, **100**, 98–103.

11 Ferreira, K.N., Iverson, T.M., Maghlaoui, K., Barber, J. and Iwata, S. (2004) Architecture of the photosynthetic oxygen-evolving center. *Science*, **303**, 1831–8.

12 Loll, B., Kern, J., Saenger, W., Zouni, A. and Biesiadka, J. (2005) Towards complete cofactor arrangement in the 3.0 Å resolution structure of photosystem II. *Nature*, **438**, 1040–4.

13 Delano, W.L. (2002) *The PyMOL Molecular Graphics System* (2002) on World Wide Web www.pymol.org.

14 Kawakami, K., Iwai, M., Ikeuchi, M., Kamiya, N. and Shen, J.-R. (2007) Location of PsbY in oxygen-evolving photosystem II revealed by mutagenesis and X-ray crystallography. *FEBS Letters*, **581**, 4983–7.

15 Nanba, O. and Satoh, K. (1987) Isolation of a Photosystem II reaction center consisting of D-1 and D-2 polypeptides and cytochrome b-559, *Proceedings of the National Academy of Sciences of the United States of America*, **84**, 109–12.

16 Michel, H. and Deisenhofer, J. (1988) Relevance of the photosynthetic reaction center from purple bacteria to the structure of photosystem II. *Biochemistry*, **27**, 1–7.

17 Deisenhofer, J., Epp, O., Sinning, I. and Michel, H. (1995) Crystallographic refinement at 2.3 Å resolution and refined model of the photosynthetic reaction center from *Rhodopseudomonas viridis*. *Journal of Molecular Biology*, **246**, 429–57.

18 Jordan, P., Fromme, P., Witt, H.T., Klukas, O., Saenger, W. and Krauß, N. (2001) Three-dimensional structure of cyanobacterial photosystem I at 2.5 Å resolution. *Nature*, **411**, 909–16.

19 Vermaas, W.F.J., Williams, J.G.K. and Arntzen, C.J. (1987) Sequencing and modification of *psbB*, the gene encoding the CP47 protein of photosystem II, in the cyanobacterium *Synechocystis* 6803. *Plant Molecular Biology*, **8**, 317–26.

20 Vermaas, W.F.J., Ikeuchi, M. and Inoue, Y. (1988) Protein composition of the photosystem II core complex in genetically engineered mutants of the cyanobacterium *Synechocystis* sp. PCC 6803. *Photosynthesis Research*, **17**, 97–113.

21 Dzelzkalns, V.A. and Bogorad, L. (1988) Molecular analysis of a mutant defective in photosynthetic oxygen evolution and isolation of a complementing clone by a novel screening procedure. *The EMBO Journal*, **7**, 333–8.

22 Eaton-Rye, J.J. and Putnam-Evans, C. (2005) The CP47 and CP43 core antenna components, in *Photosystem II, The Light-Driven Water:Plastoquinone Oxidoreductase* (eds T.J. Wydrzynski and K. Satoh), Springer, Dordrecht, The Netherlands, pp. 45–70.

23 Rögner, M., Chisholm, D.A. and Diner, B.A. (1991) Site-directed mutagenesis of the *psbC* gene of photosystem II: isolation and functional characterization of CP43-less photosystem II core complexes. *Biochemistry*, **30**, 5387–95.

24 Nakamura, Y., Kaneko, T., Sato, S., Ikeuchi, M., Katoh, H., Sasamoto, S., Watanabe, A., Iriguchi, M., Kawashima, K., Kimura, T., Kishida, Y., Kiyokawa, C., Kohara, M., Matsumoto, M., Matsuno, A., Nakazaki, N., Shimpo, S., Sugimoto, M., Takeuchi, C., Yamada, M. and Tabata, S. (2002) Complete genome structure of the thermophilic cyanobacterium *Thermosynechococcus elongatus* BP-1. *DNA Research*, **9**, 123–30.

25 Pautsch, A. and Schulz, G.E. (1998) Structure of the outer membrane protein A transmembrane domain. *Nature Structural and Molecular Biology*, **5**, 1013–17.

26 Pautsch, A. and Schulz, G.E. (2000) High-resolution structure of the OmpA membrane domain. *Journal of Molecular Biology*, **298**, 273–82.

27 Vogt, J. and Schulz, G.E. (1999) The structure of the outer membrane protein OmpX from *Escherichia coli* reveals possible mechanisms of virulence. *Structure with Folding and Design*, **7**, 1301–9.

28 Sawaya, M.R., Krogmann, D.W., Serag, A., Ho, K.K., Yeates, T.O. and Kerfeld, C.A. (2001) Structures of cytochrome c_{549} and cytochrome c_6 from the cyanobacterium *Arthrospira maxima*. *Biochemistry*, **40**, 9215–25.

29 Shen, J.-R. and Inoue, Y. (1993) Binding and functional properties of two new extrinsic components, cytochrome *c*550 and a 12 kDa protein, in cyanobacterial photosystem II. *Biochemistry*, **32**, 1825–32.

30 Enami, I., Kikuchi, S., Fukuda, T., Ohta, H. and Shen, J.-R. (1998) Binding and functional properties of four extrinsic proteins of photosystem II from a red alga, *Cyanidium caldarium* as studied by release-reconstitution experiments. *Biochemistry*, **37**, 2787–93.

31 Burnap, R.L., Qian, M. and Pierce, C. (1996) The manganese stabilizing protein of photosystem II modifies the *in vivo* deactivation and photoactivation kinetics of

the H_2O oxidation complex in *Synechocystis* sp. PCC6803. *Biochemistry*, **35**, 874–82.

32 Shen, J.-R., Qian, M., Inoue, Y. and Burnap, R.L. (1998) Functional characterization of *Synechocystis* sp. PCC 6803 Δ*psb*U and Δ*psb*V mutants reveals important roles of cytochrome *c*-550 in cyanobacterial PSII. *Biochemistry*, **37**, 1551–8.

33 Kok, B., Forbush, B. and McGloin, M. (1970) Cooperation of charges in photosynthetic oxygen evolution. I. a linear four step mechanism. *Photochemistry and Photobiology*, **11**, 457–75.

34 Joliot, P. (2003) Period-four oscillations of the flash-induced oxygen formation in photosynthesis. *Photosynthesis Research*, **76**, 65–72.

35 Jursinic, P. and Govindjee (1977) Temperature dependence of delayed light emission in the 6-340 microsecond range after a single flash in chloroplasts. *Photochemistry and Photobiology*, **26**, 617–28.

36 Klimov, V.V., Allakhverdiev, S.I., Demeter, S. and Krasnovsky, A.A. (1979) Photoreduction of pheophytin in photosystem II of chloroplasts as a function of redox potential of the medium. *Doklady Akademii Nauk SSSR*, **249**, 227–30.

37 Rappaport, F., Guergova-Kuras, M., Nixon, P.J., Diner, B.A. and Lavergne, J. (2002) Kinetics and pathways of charge recombination in photosystem II. *Biochemistry*, **41**, 8518–27.

38 Durrant, J.R., Klug, D., Kwa, S.L.S., Grondelle, R.V., Porter, G. and Dekker, J.P. (1995) A multimer model for P680, the primary electron donor of photosystem II, *Proceedings of the National Academy of Sciences of the United States of America*, **92**, 4798–802.

39 Dekker, J.P. and Grondelle, R. (2000) Primary charge separation in photosystem II. *Photosynthesis Research*, **63**, 195–208.

40 Prokhorenko, V.I. and Holzwarth, A.R. (2000) Primary processes and structure of the photosystem II reaction center: a photon echo study. *The Journal of Physical Chemistry B*, **104**, 11563–78.

41 Van Mieghem, F.J.E. and Rutherford, A.W. (1993) Comparative spectroscopy of photosystem-II and purple bacterial reaction centers. *Biochemical Society Transactions*, **21**, 986–91.

42 Kamlowski, A., Frankemöller, L., Van der Est, A., Stehlik, D. and Holzwarth, A.R. (1996) Evidence for delocalization of the triplet state ^{3}P680 in the D1/D2/cyt *b*559-complex of photosystem II. *Berichte der Bunsen-Gesellschaft Physical Chemistry Chemical Physics*, **100**, 2045–51.

43 Diner, B.A., Schlodder, E., Nixon, P.J., Coleman, W.J., Rappaport, F., Lavergne, J., Vermaas, W.F.J. and Chisholm, D.A. (2001) Site-directed mutations at D1-His198 and D2-His197 of photosystem II in *Synechocystis* PCC 6803: site of primary charge separation and cation and triplet stabilization. *Biochemistry*, **40**, 9265–81.

44 Holzwarth, A.R., Müller, M.G., Reus, M., Nowaczyk, M., Sander, J. and Rögner, M. (2006) Kinetics and mechanism of electron transfer in intact photosystem II and in the isolated reaction center: pheophytin is the primary electron acceptor. *Proceedings of the National Academy of Sciences of the United States of America*, **103**, 6895–900.

45 Barter, L.M.C., Durrant, J.R. and Klug, D.R. (2003) A quantitative structure-function relationship for the photosystem II reaction center: supermolecular behavior in natural photosynthesis. *Proceedings of the National Academy of Sciences of the United States of America*, **100**, 946–51.

46 Raszewski, G., Saenger, W. and Renger, T. (2005) Theory of optical spectra of photosystem II reaction centers: location of the triplet state and the identity of the primary electron donor. *Biophysical Journal*, **88**, 986–98.

47 Germano, M., Gradinaru, C.C., Shkuropatov, A.Y., van Stokkum, I.H.M., Shuvalov, V.A., Dekker, J.P., van Grondelle, R. and van Gorkom, H.J. (2004) Energy and electron transfer in photosystem II reaction centers with modified pheophytin composition. *Biophysical Journal*, **86**, 1664–72.

48 Williams, J.C., Alden, R.G., Murchison, H.A., Peloquin, J.M., Woodbury, N.W. and Allen, J.P. (1992) Effects of mutations near the bacteriochlorophylls in reaction centers from *Rhodobacter sphaeroides*. *Biochemistry*, **31**, 11029–37.

49 Brettel, K. (1997) Electron transfer and arrangement of the redox cofactors in photosystem I. *Biochimica et Biophysica Acta*, **1318**, 322–71.

50 Ishikita, H., Loll, B., Biesiadka, J., Saenger, W. and Knapp, E.W. (2005) Redox potentials of chlorophylls in the photosystem II reaction center. *Biochemistry*, **44**, 4118–24.

51 Ishikita, H., Saenger, W., Biesiadka, J., Loll, B. and Knapp, E.W. (2006) How photosynthetic reaction centers control oxidation power in chlorophyll pairs P680, P700, and P870. *Proceedings of the National Academy of Sciences of the United States of America*, **103**, 9855–60.

52 Fujita, I., Davis, M.S. and Fajer, J. (1978) Anion radicals of pheophytin and chlorophyll *a*: their role in the primary charge separations of plant photosynthesis. *Journal of the American Chemical Society*, **100**, 6280–2.

53 Watanabe, T. and Kobayashi, M. (1991) Electrochemistry of chlorophylls, in *Chlorophylls* (ed. H. Scheer), CRC Press, Boca Raton, pp. 287–315.

54 Ohad, I., Dal Bosco, C., Herrmann, R.G. and Meurer, J. (2004) Photosystem II proteins PsbL and PsbJ regulate electron flow to the plastoquinone pool. *Biochemistry*, **43**, 2297–308.

55 Berthomieu, C. and Hienerwadel, R. (2001) Iron coordination in photosystem II: interaction between bicarbonate and the Q_B pocket studied by Fourier transform infrared spectroscopy. *Biochemistry*, **40**, 4044–52.

56 Petrouleas, V. and Crofts, A.R. (2005) The iron-quinone acceptor complex, in *Photosystem II, The Light-Driven Water: Plastoquinone Oxidoreductase* (eds T.J. Wydrzynski and K. Satoh), Springer, Dordrecht, The Netherlands, pp. 177–206.

57 Peloquin, J.M., Campbell, K.A., Randall, D.W., Evanchik, M.A., Pecoraro, V.L., Armstrong, W.H. and Britt, R.D. (2000) ^{55}Mn ENDOR of the S_2-state multiline EPR signal of photosystem II: implications on the structure of the tetranuclear Mn cluster. *Journal of the American Chemical Society*, **122**, 10926–42.

58 Yachandra, V.K. (2005) The catalytic manganese cluster: organization of the metal ions, in *Photosystem II, The Light-Driven Water:Plastoquinone Oxidoreductase* (eds T.J. Wydrzynski K. and Satoh), Springer, Dordrecht, The Netherlands, pp. 235–60.

59 Kimura, Y., Mizusawa, N., Yamanari, T., Ishi, A. and Ono, T.A. (2005) Structural changes of D1 C-terminal α-carboxylate during S-state cycling in photosynthetic oxygen evolution. *Journal of Biological Chemistry*, **280**, 2078–83.

60 Strickler, M.A., Walker, L.M., Hiller, W. and Debus, R.J. (2005) Evidence from biosynthetically incorporated strontium and FTIR difference spectroscopy that the C-terminus of the D1 polypeptide of photosystem II does not ligate calcium. *Biochemistry*, **44**, 8571–7.

61 Yano, J., Kern, J., Sauer, K., Latimer, M.J., Pushkar, Y., Biesiadka, J., Loll, B., Saenger, W., Messinger, J., Zouni, A. and Yachandra, V.K. (2006) Where water is oxidized to dioxygen: structure of the photosynthetic Mn_4Ca cluster. *Science*, **314**, 821–5.

62 Satoh, K. (1996) Introduction to the Photosystem II reaction center–isolation and biochemical and biophysical characterization, in *Oxygenic Photosynthesis: The Light Reactions* (eds D.R. Ort and C.F. Yocum), Kluwer Academic Publishers, Dordrecht, pp. 193–211.

63 Telfer, A., De Las Rivas, J. and Barber, J. (1991) β-carotene within the isolated photosystem II reaction center: photooxidation and irreversible bleaching of this chromophore by oxidized P680. *Biochimica et Biophysica Acta*, **1060**, 106–14.

64 Telfer, A. (2002) What is β-carotene doing in the photosystem two reaction center? *Philosophical Transactions of the Royal Society of London Series B*, **357**, 1431–9.

65 Telfer, A., Frolov, D., Barber, J., Robert, B. and Pascal, A. (2003) Oxidation of the two β-carotene molecules in the photosystem II reaction center. *Biochemistry*, **42**, 1008–15.

66 Hanley, J., Deligiannakis, Ym, Pascal, A., Faller, P., and Rutherford A.W. (1999) Carotenoid oxidation in photosystem II. *Biochemistry*, **38**, 8189–95.

67 Faller, P., Fufezan, C. and Rutherford, A.W. (2005) Side-path electron donors: cytochrome *b*559, chlorophyll Z and β-carotene, in *Photosystem II, The Light-Driven Water:Plastoquinone Oxidoreductase* (eds T.J. Wydrzynski and K. Satoh), Springer, Dordrecht, The Netherlands, pp. 347–65.

68 Ishikita, H., Loll, B., Biesiadka, J., Kern, J., Irrgang, K.D., Zouni, A., Saenger, W. and Knapp, E.W. (2006) Function of two beta-carotenes near the D1 and D2 proteins in photosystem II dimers. *Biochimica et Biophysica Acta*, **1767**, 79–87.

69 Sato, N. (2004) Roles of the acidic lipids sulfoquinovosyl diacylglycerol and phosphatidylglycerol in photosynthesis: their specificity and evolution. *Journal of Plant Research*, **117**, 495–505.

70 Jones, M.R. (2007) Lipids in photosynthetic reaction centers: structural roles and functional holes. *Progress in Lipid Research*, **46**, 56–87.

71 Minoda, A., Sato, N., Nozaki, H., Okada, K., Takahashi, H., Sonoike, K. and Tsuzuki, M. (2002) Role of sulfoquinovosyl diacylglycerol for the maintenance of photosystem II in *Chlamydomonas reinhardtii*. *European Journal of Biochemistry*, **269**, 2353–8.

72 Minoda, A., Sonoike, K., Okada, K., Sato, N. and Tsuzuki, M. (2003) Decrease in the efficiency of the electron donation to tyrosine Z of photosystem II in an SQDG-deficient mutant of *Chlamydomonas*. *FEBS Letters*, **553**, 109–12.

73 Sato, N., Aoki, M., Maru, Y., Sonoike, K., Minoda, A. and Tsuzuki, M. (2003) Involvement of sulfoquinovosyl diacylglycerol in the structural integrity and heat-tolerance of photosystem II. *Planta*, **217**, 245–51.

74 Kruse, O., Hankamer, B., Konczak, C., Gerle, C., Morris, E., Radunz, A., Scmid, G.H. and Barber, J. (2000) Phosphatidylglycerol is involved in the dimerization of photosystem II. *Journal of Biological Chemistry*, **275**, 6509–14.

75 Gombos, Z., Värkonyi, M., Hagio, M., Iwaki, M., Kovacs, L., Masamoto, K., Itoh, S. and Wada, H. (2002) Phosphatidylglycerol requirement for the function of electron acceptor plastoquinone Q_B in the photosystem II reaction center. *Biochemistry*, **41**, 3796–802.

76 Vasil'ev, S., Orth, P., Zouni, A., Owens, T.G. and Bruce, D. (2001) Excited-state dynamics in photosystem II: Insights from the x-ray crystal structure. *Proceedings of the National Academy of Sciences of the United States of America*, **98**, 8602–7.

77 Vasil'ev, S. and Bruce, D. (2004) Optimization and evolution of light harvesting in photosynthesis: the role of antenna chlorophyll conserved between photosystem II and photosystem I. *Plant Cell*, **16**, 3059–68.

78 Vasil'ev, S., Shen, J.-R., Kamiya, N. and Bruce, D. (2004) The orientations of core antenna chlorophylls in photosystem II are optimized to maximize the quantum yield of photosynthesis. *FEBS Letters*, **561**, 111–16.

79 Matsuoka, H., Furukawa, K., Kato, T., Mino, H., Shen, J.-R. and Kawamori, A. (2006) g-Anisotropy of the S_2-state manganese cluster in single crystals of cyanobacterial photosystem II studied by W-band electron paramagnetic resonance spectroscopy. *The Journal of Physical Chemistry B*, **110**, 13242–7.

80 Teutloff, C., Kessen, S., Kern, J., Zouni, A. and Bittl, R. (2006) High-field (94-GHz) EPR spectroscopy on the S(2) multiline signal of photosystem II. *FEBS Letters*, **580**, 3605–9.

5
Current Models and Mechanism of Water Splitting

Robert M. McCarrick and R. David Britt

5.1
Introduction

5.1.1
Photosystem II

Oxygenic photosynthesis in plants and bacteria plays a central role in nature in that it forms the energetic base for most life on earth [1]. The overall process is separated into two sets of reactions: the light reactions in which light energy is used as the driving force to oxidize water, reduce ferrodoxin, and provide a proton gradient that helps provide the energy for ATP production (see Chapter 9); and the dark, or carbon fixation reactions in which NADPH is used to reduce atmospheric CO_2 for incorporation into simple sugars.

The focus of this chapter is the structure and mechanism of the oxygen-evolving complex (OEC) in Photosystem II (PSII), which directly carries out the remarkable water-oxidation reaction [2–5] (for more details on the structure of Photosystem II see Chapter 4). PSII is located in the chloroplast thylakoid membranes of higher plants and algae and in the plasma membranes of cyanobacteria. The groundbreaking discovery that showed a cyclic relationship between the absorption of photons and oxygen production was carried out by Pierre Joliot [6]. In these experiments, a series of short flashes was given to a preparation containing dark-adapted chloroplast thylakoid membranes, and the production of oxygen was monitored as a function of the flash number. These experiments showed that the initial release of oxygen occurred following the third flash and then subsequently after each fourth flash. Results from subsequent experiments were interpreted by Bessel Kok and colleagues, leading to the development of the S state, or Kok cycle, depicted in Figure 5.1, where S_1 is the dark stable state and S_4 spontaneously decomposes to S_0, releasing one molecule of oxygen [7].

This S state cycle is driven in four of the five steps by the photoexcitation of the primary pigment, P680. The resulting P680* excited state decays via electron transfer to an adjacent pheophytin that then reduces a bound quinone, Q_A, which,

Photosynthetic Protein Complexes: A Structural Approach. Edited by P. Fromme

ISBN: 978-3-527-31730-1

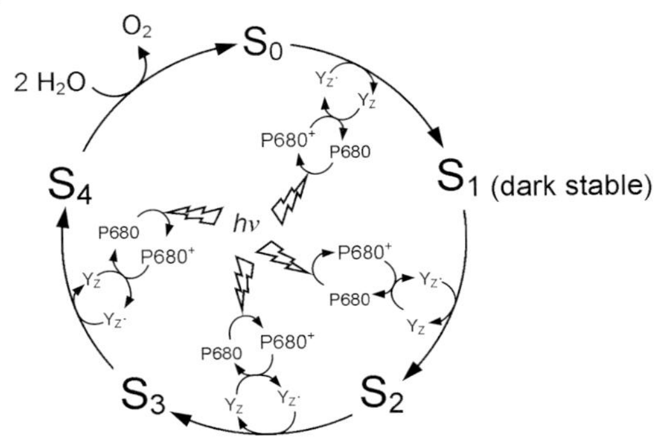

Figure 5.1 The Kok cycle. The Kok cycle was developed to interpret the pattern of O_2 release as a function of flash. Oxygen is released on the third flash and then subsequently on each fourth flash. The S_1 state is the dark-stable, resting state of PSII and S_4 spontaneously decays to S_0, releasing oxygen [7].

in turn, reduces a mobile quinone, Q_B. Once doubly reduced, Q_B is released into a membrane quinone pool. The oxidized $P680^+$ molecule is reduced on the time scale of nanoseconds (20–40 ns depending on the S state) by a redox active tyrosine residue, Y_Z, which is in turn reduced by the OEC [8]. Following the buildup of four oxidative equivalents on the OEC, O_2 is formed and released, yielding the S_0 state.

5.2.1 The Oxygen-Evolving Complex

The OEC consists of a tetranuclear Mn cluster and one atom each of Ca^{2+} and Cl^- [9]. Early experiments provided clues to the composition of the OEC. Algae grown under Mn-depleted conditions were shown not to be able to carry out photosynthesis [10]. More targeted experiments in 1955 showed that Mn was specifically required for the production of oxygen [11]. Similarly to the previous study, algae grown under Mn depleted conditions were shown to be deficient only in the production of oxygen. While the stoichiometry of Mn to PSII was initially controversial, electron paramagnetic resonance (EPR) and atomic absorption spectroscopies were able to determine that the ratio of Mn to P680 was 4 : 1 [12–14]. Some redox enzymes can function with different metals bound to the active site, but no other metal has been shown to be capable of replacing Mn within the OEC [9]. The development of a preparative method which removed two extrinsic polypeptides (the 17 and 23 kDa polypeptides) and destroyed high affinity Ca^{2+} binding (while leaving the Mn cluster intact) led to the initial discovery that Ca^{2+} was an essential cofactor for water oxidation. Only when Ca^{2+} was added back in high concentrations to these preparations was oxygen-evolving activity regained [15, 16]. Unlike many Ca^{2+} containing enzymes where Ca^{2+} can be functionally replaced by several cations, only Sr^{2+} has been shown to restore activity, albeit at a much lower rate [16]. There has also been extensive research into the role of Cl^- in water oxidation. In contrast to Ca^{2+} and Mn, several anions can functionally replace Cl^-. The replacement of Cl^- with Br^- has little effect on the rate of oxygen-evolution, while replacement with NO_2^-, NO_3^-, and I^- leads to the retention of activity at diminished rates [17–20].

5.2 The Structure of the OEC

5.2.1 Early Spectroscopic Results from Chloroplast Membranes

Determining the structure of the OEC has been the target of a large number of researchers, exploiting every applicable spectroscopic probe. Two spectroscopic techniques have historically provided the most significant results: electron paramagnetic resonance (EPR) spectroscopy and X-ray absorption spectroscopy (XAS). In EPR spectroscopy, a magnetic field splits the energy levels of paramagnetic electron states and spin transitions are driven using microwave radiation. The results obtained can give insight into the structural and chemical environment of the electron spin. However, one limitation of this technique is the requirement of unpaired electrons with typically a non-integer total spin for the use of conventional continuous wave and pulsed techniques.

The EPR spectrum of isolated chloroplast membranes from spinach poised in the S_2 state by a high intensity flash of light, exhibits a multiline EPR signal with 18–20 resolved hyperfine lines, centered at $g = 2$, shown in Figure 5.2a [21, 22]. To understand the origin of this EPR spectrum, the origin of mononuclear Mn(II) system with an electronic spin (S) value of 5/2 can be considered first. In a typical mononuclear Mn(II) EPR spectrum, the inner allowed $m_S = 1/2$ to $-1/2$ spin transitions give rise to an EPR signal split into six lines by the hyperfine interaction with the $I = 5/2$ ^{55}Mn nucleus (the number of transitions is equal to $2I + 1$) [23]. The splitting between each of these transitions is dictated by A, the magnitude of the hyperfine coupling. The system becomes more complex when more than one Mn nuclei are present. Biological Mn ions are always high spin with Mn(II), Mn(III), and Mn(IV) ions presenting S values of 5/2, 2, and 3/2 respectively [24]. In Mn dimers, such as the extensively examined $Mn(III)(IV)O_2bpy_4$ compound (spectrum shown in Figure 5.2b), electronic exchange coupling between the two Mn atoms is facilitated through the μ-oxo bridges [25]. This exchange coupling (J) is a measure of the degree of spin exchange between the two Mn ions. For these di-μ-oxo bridged Mn compounds, J is typically negative and between −100 and $-300\,cm^{-1}$, indicating strong antiferromagnetic behavior [26]. For both Mn(II)(III) and Mn(III)(IV) di-μ-oxo bridged dimers, the antiferromagnetic coupling gives a ground spin state with the lowest net coupled spin, 1/2. The above mentioned synthetic model compounds contain two ^{55}Mn nuclei; there is a total of 36 allowed

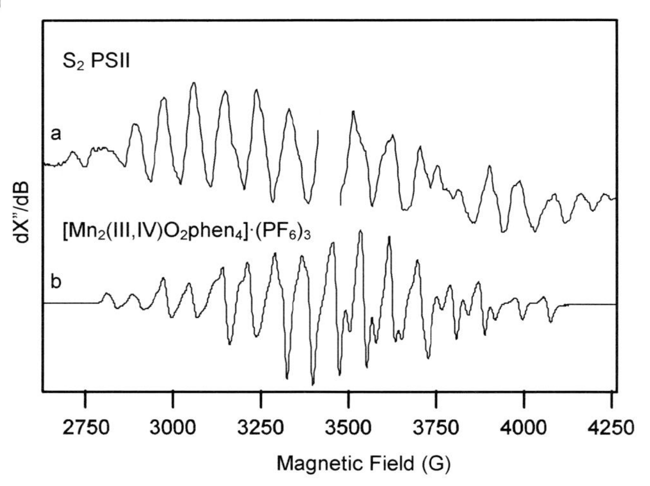

Figure 5.2 The typical multiline EPR signals associated with PSII in the S_2 state and $[Mn_2(III)(IV)O_2Phen_4]\cdot(PF_6)_3$. Shown in the figure are continuous wave-field swept EPR spectra. The data were collected at liquid helium cryogenic temperatures. Data collected in the Britt laboratory.

EPR transitions; the 6 transitions from the coupling of the net unpaired electron to the first ^{55}Mn nucleus are each further split into 6 transitions by the coupling to the second ^{55}Mn nucleus. Due to fairly broad linewidths, many of these transitions overlap, giving rise to 20 or fewer observed spectral features. In the case of the EPR spectrum of $Mn(III)(IV)O_2bpy_4$, 16 major observable features are present. The broadness of the EPR spectrum of an antiferromagnetic Mn dimer is primarily determined by the hyperfine coupling, of which Mn(II) tends to have larger hyperfine couplings than Mn(III) or Mn(IV) and more importantly, the projection factors for the ions, which are greater for a Mn(II)(III) dimer (7/3 and −4/3) than a Mn(III)(IV) dimer (2 and −1). These projection factors dictate the extent of the hyperfine coupling that is observed in the dimer, in relation to what would have been present in a mononuclear compound.

The multiline EPR signal associated with the S_2 state of the OEC, first discovered by Dismukes and colleagues [21, 22], is shown in Figure 5.2a. The increased linewidth, 1900 G, of this signal, compared to the EPR spectra of mixed valent Mn(III)(IV) model compounds, was apparent.

Mn(II)(III) EPR spectra are typically broader than those of Mn(III)(IV) dimers as discussed above. The possibility of a Mn(II)(III) dimer within the OEC was given as one possible explanation for the increased broadness of the multiline signal in PSII as compared to Mn(III)(IV) synthetic dimers [21]. Alternatively, a broader spectrum could result from coupling of the electron spin to more than two ^{55}Mn nuclei. Due to the additional number of EPR transitions (216 for a Mn trimer and 1296 for a Mn tetramer), it would be expected that a trimer or tetramer with similar hyperfine interactions would yield an EPR spectrum with a larger overall spectral width. Therefore, the possibility of a $Mn(III)_3(IV)$ or $Mn(III)(IV)_3$ tetranuclear origin for the multiline signal was given as an explanation for the S_2 state multiline signal [21, 22].

Concomitant with the EPR studies described above, XAS experiments on chloroplast membranes by Klein and coworkers gave a more direct glimpse at the structural features present in the OEC [27]. In contrast to electronic spin state requirements of EPR, XAS can be employed regardless of the electronic state of the metal center. In XAS, the absorption of X-ray photons as a function of energy is monitored,either directly or via fluorescence resulting from the filling of the core electron hole created by the incident x-ray photon. Due to the requirement for a high intensity light source spanning a broad X-ray energy range, these experiments are conducted at synchrotron radiation facilities. The element specific XAS techniques are able to provide information about oxidation states and the structural features present around the absorbing atom. In K-edge EXAFS (extended X-ray absorption fine structure) spectroscopy, an ejected electron from the 1s orbital of the target element, in this case Mn, is scattered by nearby atoms, leading to a modulation in the absorption coefficient as a function of X-ray energy. The resulting spectra can then be analyzed using Fourier transform techniques, giving a radial distribution of electron density out from the target atom [28]. The first series of EXAFS experiments on isolated spinach chloroplasts showed similar results to Mn dimer compounds with di-μ-oxo bridges. Particularly, both sets of data show a peak at approximately 2.7 Å, the Mn–Mn distance in the di-μ-oxo bridged compounds.

Another XAS experiment is X-ray absorption near edge structure (XANES) spectroscopy, where one studies the large rise in absorption observed when the X-ray ejects core electrons into unoccupied orbitals and unbound continuum states (a 1s electron in the case of K-edge XAS). The energy at which this absorption edge occurs is highly sensitive to the oxidation state of the target atom. The XANES spectra of dark adapted (S_1) spinach chloroplast membranes indicated that the average oxidation state of the cluster was higher than Mn(II) [29]. Due to the pronounced feature that is present in a typical Mn(II) XANES spectrum, as seen in Figure 5.3a, this oxidation state is easy to differentiate from Mn(III) and Mn (IV) in a given spectrum. The combination of XANES and EXAFS led to the proposal that the Mn within the OEC was more oxidized than Mn(II) in the S_1 state and that the structural environment was similar to that of Mn model compounds with oxo bridging. The combination of these results, along with the discovery of the multiline EPR signal for the first time, shed light on structural features present in the OEC.

5.2.2
Spectroscopic Studies of Isolated PSII

The development of methods for efficiently isolating highly active PSII preparations from spinach, by Yocum and colleagues (the BBY preparation) led to an increase in the output of important papers on the OEC and water splitting

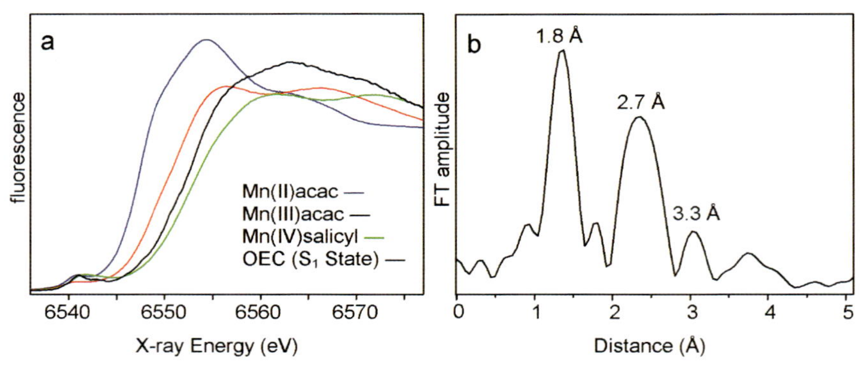

Figure 5.3 XAS spectra of PSII and Mn model compounds. **(a)**: XANES spectra of dark adapted (S_1) PSII and representative Mn(II), Mn(III) and Mn(IV) models. Spectra were obtained with permission form Yano, Yachandra and colleagues. **(b)**: uncorrected, Fourier transformed EXAFS data of dark adapted PSII with the corresponding distances indicated. Spectra collected as part of the first author's PhD research.

[30, 31]. In particular, the new preparations enabled Klein and coworkers to obtain much improved XAS results. EXAFS spectra of isolated PSII showed three distinct peaks of electron as shown in Figure 5.3b [32].

While XAS is a highly useful spectroscopic technique, in that it is capable of isolating the spectra of a particular element from the rest of the elements within the protein, the interpretation of spectra from multinuclear species, such as the OEC, is complicated because one records an average spectrum for all of the selected atoms. Specifically, the Mn EXAFS spectrum from the OEC is an average of all four individual Mn environments within the cluster. Nevertheless, three well resolved peaks were observed that are most consistent with Mn–O ligation for the 1.8 Å peak, two Mn–Mn distances for the 2.7 Å interaction, and one Mn–Mn distance with the possibility of a contribution from a Mn–Ca distance for the 3.3 Å interaction [32, 33]. The first two shells are very similar to that observed for Mn(III)(IV) dimers, while the third shell provided additional support for additional Mn or Ca interactions. While these results represented a great improvement over those obtained with chloroplast membranes and very precise distance measurements of the various interactions within the cluster were obtained, no specific structures were initially proposed.

5.2.3
The Dimer of Dimers Structure

The first specific proposal for the structure of the OEC came from Klein and coworkers, and used the higher resolution XAS data as well as the EPR spectra of the S_2 state to propose a dimer of dimers arrangement of the Mn atoms within the cluster [21, 22, 34].

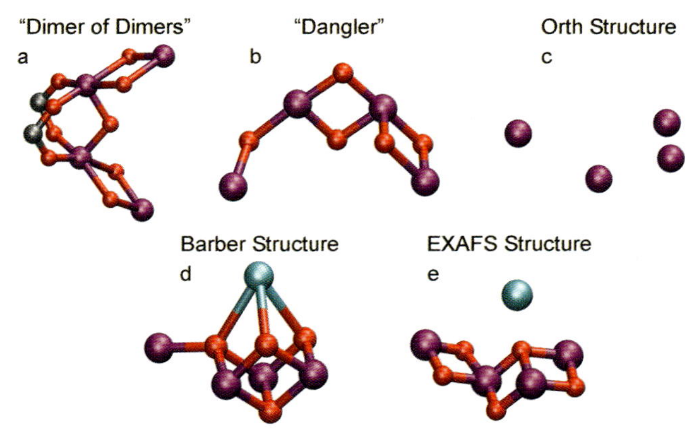

Figure 5.4 OEC structural models. Shown are five structural models of the OEC, presented in chronological order of the initial proposal [34, 57–60].

This structure was consistent with the XAS data in that there were Mn–O distances at 1.8 Å, one 2.7 Å Mn-Mn interaction for each of the dimers, and a longer 3.3 Å Mn–Mn distance linking the two dimers as seen in Figure 5.4a. This structure was also capable of roughly explaining the origin of the S_2 multiline signal. If the oxidation states for the cluster were either $(III)_3(IV)$ or $(III)(IV)_3$ for the S_2 state, as will be further discussed in Section 5.3.2, the structure would contain one spin 1/2 Mn(III)(IV) dimer and one spin 0 Mn(III)(III) or (IV)(IV) dimer, for a net spin 1/2 system.

While this dimer of dimers structure was in solid agreement with the XAS results and the early interpretations of the S_2 multiline signal, it was incapable of explaining a second EPR signal that was discovered and eventually attributed to a higher spin state of the OEC. This so called $g = 4.1$ signal was first suggested as arising from Fe, but was later shown to arise from the Mn cluster [35–37]. The most compelling evidence that suggested the multiline and $g = 4.1$ signals originated from the same cluster was a series of experiments in which ammonia treated oriented membrane PSII samples exhibited a multiline signal at both $g = 2$ and $g = 4.1$ [38]. This $g = 4.1$ multiline could only be successfully modeled if it was assumed that the unpaired electron spin was coupled to the entire Mn cluster. The interpretation that this signal originated from a spin 3/2 or 5/2 state of the cluster cast doubt on the dimer of dimers structure. Using the electron coupling scheme detailed above, there is no arrangement of exchange couplings or oxidation state distributions of atoms that could give either a ground or low lying excited 3/2 or 5/2 spin state; the dimer of dimers structure lost favor within the research community. The $g = 4.1$ signal has recently been shown, using higher frequency Q-band EPR, to indeed originate from a spin 5/2 state of the cluster, confirming the proposal from an earlier study [39, 40].

An interesting aspect of the presence of the two EPR signals associated with the S_2 state is the series of conditions that have been shown to affect the relative intensities of the two signals. High concentrations (5% by volume) of small alcohols lead to complete elimination of the g = 4.1 signal and an increase in the intensity of the multiline signal [41]. In contrast, high concentrations of sucrose, the presence of F^-, Sr^{2+}, or near infrared illumination, lead to an increase in the g = 4.1 signal [35, 36, 42]. While no specific explanation of these results has been provided, they do suggest that the exchange couplings between the Mn ions in the cluster are extremely sensitive to a variety of conditions.

In addition to the EPR signals associated with the S_2 state, subsequent discoveries allowed the S_0 and S_1 states to be studied using EPR spectroscopy. A multiline signal similar to that of the S_2 multiline signal was shown by Messinger and colleagues to be present in the S_0 state. The signal was produced in two ways: with the reduction of dark adapted PSII with hydroxylamine, to reduce the centers to the S_0 state and with three laser flashes to advance the OEC from S_1 to S_0 [43, 44]. Just as the S_2 multiline signal was shown to be enhanced by the addition of methanol, the S_0 signal has only been detected clearly in samples which were supplemented with methanol. The appearance of this EPR signal is similar to the S_2 multiline, but spread over a larger spectral range, which led to the initial proposal that it originates from a Mn(II)(III)$(IV)_2$ oxidation state of the cluster. That interpretation has since been reevaluated, and Lubitz and Messinger now favor a $Mn(III)_3$(IV) distribution of oxidation states for the S_0 state [45, 46].

While the OEC in the S_1 state does not exhibit an EPR signal in conventional perpendicular mode EPR spectroscopy, it does have a low field (g = 4.8) parallel mode signal associated with a paramagnetic, non-Kramer spin system, that was discovered by Klein and colleagues [47, 48]. Another parallel mode EPR signal attributed to the S_1 state was later shown to originate from multiple ^{55}Mn nuclei by Britt and coworkers.[49, 50] This signal is present in bacterial PSII, and in spinach PSII that has been depleted of the 17 and 23 kDa extrinsic polypeptides, where it contains approximately 18 lines and is centered at g = 12 [49, 50]. Recently, a model compound containing all Mn(IV) has been shown to exhibit a similar parallel mode signal, representing the first spectroscopic model for the S_1 state [51].

5.2.4
The Monomer Trimer or "Dangler" Structure

The examination of an EPR signal can yield a great deal of information, particularly with regard to the hyperfine couplings between the electrons and the surrounding nuclei. However, simulations of the S_2 multiline signal attempting to model these parameters were controversial because several research groups arrived at varying values for the hyperfine coupling parameters, as well as the underlying J couplings between the four Mn atoms [52–56]. This ambiguity stems from the nature of the system: 1296 hyperfine lines overlap extensively to give a relatively information-poor spectrum that can be satisfactorily modeled with many different parameter sets. More direct spectroscopic evidence for a structure different from the dimer of dimers arrangement was provided by a series of ^{55}Mn ESE-ENDOR experiments conducted by Britt and colleagues [57]. In this pulsed EPR technique, the magnetic field is held constant at a position where the targeted EPR transition is in resonance with the frequency of the microwave pulses. A separate radio frequency (rf) pulse is applied and its effect on the EPR signal is monitored as a function of the radio frequency. The resulting ENDOR spectrum is essentially a nuclear magnetic resonance (NMR) spectrum of the nuclei in the vicinity of the unpaired electron spin, including the contribution of the electron-nuclear hyperfine interaction, which can be determined with high accuracy. For the S_2 multiline signal, the fact that both the EPR and the ENDOR spectra can be simultaneously examined and simulated gives much greater confidence in the interpretation than with the EPR spectrum alone. Specifically, the precise determination of the hyperfine parameters allowed the researchers to test the various structures that were deemed consistent with EXAFS results. Using a projection factor approach, which relates the hyperfine parameters for a mononuclear species to those observed in multinuclear coupled systems, it was determined that a monomer-trimer arrangement was most consistent with the experimental data. As mentioned above, the exchange coupling between two Mn atoms with di-μ-oxo bridges tends to be antiferromagnetic. This representation becomes much more complicated when examining compounds with a nuclearity greater than two. In the case of a trimer, three J couplings arise. Since the multiline signal was shown to arise from a tetranuclear cluster, six possible J couplings exist. From the EPR and ^{55}Mn ESE-ENDOR data, it was determined that there were two simple coupling schemes that satisfied the experimental data well, both of which show three of the couplings were zero, two at ~100 cm^{-1}, and one around 15 cm^{-1}. Both coupling schemes best fit a monomer-trimer arrangement for the Mn in the OEC, one of which is shown in Figure 5.4b [57]. This was colloquially termed the Dangler model.

5.2.5 X-Ray Diffraction and XAS of PSII Crystals

Shortly after the publication of the proposed monomer-trimer structure for the OEC from the ^{55}Mn ESE-ENDOR study, the first crystal structure of PSII from *Synechococcus elongatus* was published by Zouni and coworkers [58]. This structure represented a significant accomplishment, owing to the difficulty in crystallizing large multisubunit membrane proteins. In this first structure, the relative locations of many of the polypeptides and various cofactors were clear, but the low resolution (3.8 Å) was not sufficient to yield an atomic resolution model for the OEC. In the electron density map, a pear-shaped feature was present that could fit with four Mn atoms in an arrangement depicted in Figure 5.4c. These atoms appeared to be in a monomer-trimer arrangement, supporting the results from the ^{55}Mn ESE-ENDOR studies. Following this initial structure, Shen and colleagues published a structure at slightly better resolution (3.7 Å), again showing a pear-shaped region of electron density for the OEC [59] (for the structure of

Photosystem II see also Chapter 4). While these results were significant, much uncertainty still remained. While all four Mn atoms seemed to be present, there was no electron density detected for either Ca^{2+} or Cl^-. Several years after the first crystal structure for PSII, a structure at 3.5 Å resolution was presented which included the model for the OEC shown in Figure 5.4d [60]. Barber and colleagues (for the first time) were able to locate the Ca^{2+} atom by examining the difference electron density map generated by acquiring diffraction data from above and below the Ca^{2+} XANES edge. In this class of diffraction study, known as anomalous diffraction, the positions in the electron density map that show the greatest difference between the two data sets are likely to correspond to that element, specifically Ca in this case. Additionally, the enhanced resolution of the structure allowed the researchers to identify potential protein ligands to the OEC. The arrangement of these putative amino acid ligands in the vicinity of the OEC were used to constrain the positions of the Mn and Ca within the electron density map of the OEC. A number of specific ligand positions to the Mn atoms were assigned, with the suggestion that the remaining ligand sites were occupied by waters to bring each Mn to 5 or 6 coordinate. However, uncertainty in the position of these nearby amino acids has cast doubt on their use to constrain the metal ion geometry and assign a specific ligand arrangement.

While this structure provides a significant step toward understanding the structural properties of the OEC, a subsequent XAS study by Yachandra and coworkers revealed a significant problem with all of the published crystal structure data [61]. Since XAS probes both the oxidation state of the metal examined and its local structural environment, it is an ideal technique for determining the extent of possible radiation damage. Therefore, XANES and EXAFS spectra were collected on single PSII crystals under temperature and X-ray dosage conditions comparable to those used to collect the diffraction data. In the XANES spectra, full reduction of all of the Mn in the OEC to Mn^{2+} was observed. In addition, the EXAFS spectrum showed a complete loss of the spectral features characteristic to the OEC, giving rise to one strong feature similar to that of hydrated Mn(II) [61]. At roughly two thirds of the photon dosage used to collect the crystallographic data, 80% of the Mn had been reduced to Mn^{2+} [61]. These results show that, for all of the published crystal structures, the electron density map observed for the OEC represents a time dependent average that is biased toward the radiation damaged structure. Indeed, an examination of the agreement of the structural model with the experimental data shows a partial occupancy in Ca^{2+} and two of the four Mn atoms. However, the similarity between these structural models and those obtained using other techniques, does suggest a certain degree of reliability. In addition, there is still the potential to get a more reliable structure in the near future. The XAS results showed that X-ray induced reduction of the cluster occurred at a slower rate at liquid He temperatures. Groups are undoubtedly working on obtaining structures at lower temperatures and using multiple crystals with shorter x-ray exposure to help mitigate the radiation damage problem.

X-ray diffraction techniques are tremendously useful for determining the overall structure of proteins. However, XAS methods can also give more accurate local structural data. Specifically, the bond lengths determined from a typical EXAFS experiment are far more precise than those obtained by X-ray diffraction [28]. However, XAS experiments are most often conducted on solutions of the protein, so the spherical electron density profile contains no directional information. Several studies have used two dimensionally oriented PSII membranes and have yielded some directional information, with respect to the membrane normal; these techniques are still relatively limited [62, 63]. With the progress that has been achieved towards obtaining high quality PSII crystals, single crystal XAS experiments can now be conducted in parallel with X-ray diffraction experiments. Recently, a series of structures by Yachandra was published using low temperature orientation selective XAS [64]. In oriented crystals, a significant dichroism was present in the characteristic EXAFS features (the same features that are observed in EXAFS spectra of randomly oriented PSII particles). However, even this additional information did not allow the researchers to unambiguously assign a single structure for the OEC. Three arrangements of the Mn atoms were presented for which simulated spectra accurately match the experimental data. Each of these XAS derived structures was inserted into the crystallographically determined structure from Barber and coworkers with the second XAS model, shown in Figure 5.4e, giving the best agreement with the electron density map and the position of the surrounding ligands. However, Yachandra and coworkers note that, due to the significant radiation damage that occurs in the measurement of the diffraction data, the ligand assignments were only tentative. Even with the ambiguity, the structures derived from the orientation dependent EXAFS data may be the most reliable to date. This is due not only to the highly accurate distance measurements and directional parameters that are obtained from EXAFS spectroscopy, but also to the fact that, at the low temperatures and low doses of X-rays used, these experiments are carried out under non-destructive conditions.

5.2.6
Quantum Mechanical Structural Models

In addition to the variety of spectroscopic techniques that have been utilized to resolve the elusive structure of the OEC, the increased power of computers, along with advancements in quantum mechanical modeling, have made it possible to begin examining systems as complex as the OEC. For example, in recent studies from Batista and coworkers, quantum mechanical/molecular mechanical modeling (QM/MM) methods have been used to create a model for the structure of the OEC and the surrounding amino acids [65–67]. Computational models can be used to predict electronic properties such as oxidation state and localization of unpaired spin density in addition to the geometric structural properties. Through comparison to experimentally derived observables, these computationally derived parameters are used to judge the quality of the QM model. Thus tested, a successful computational model of the OEC could provide insight into the chemical properties that govern the water-oxidation mechanism. In the case of Batista and coworkers, a simulation of the expected EXAFS spectrum from the QM model was compared with an experimental EXAFS spectrum. While the fit was not exact, the general features were conserved. In fact, the simulated spectra based on the geometry of the quantum mechanical model better fit the experimental spectra than that which was simulated based on the crystallographically derived model of Barber and colleagues. However, it should be noted that when simulating the EXAFS spectra based on the crystallographic parameters, the researchers did not include water molecules that were absent in the electron density, but could, nonetheless, be bound to the OEC in the crystal structure, as was noted in Section 5.2.5. As the results from spectroscopic and diffractions studies on PSII converge with those predicted from quantum mechanical modeling, a complete picture will emerge that will be very useful in determining how the structure and chemistry of the OEC directs water oxidation.

5.3
Mechanisms of Water Oxidation

5.3.1
Energetic Considerations

The oxidation of water to molecular oxygen is a remarkable chemical reaction in nature. The fact that plants and bacteria use sunlight as the energetic driving force for the reaction gives it added significance. Despite the necessity of oxygen for humans and other aerobes, it is merely a byproduct of photosynthesis. The products that are crucial for the photosynthetic organisms are the protons which help fuel a pH gradient, and electrons which provide reducing equivalents for the carbon fixation reactions, while providing reducing equivalents for the carbon fixation reactions. The reaction requires a potential of approximately 0.9 V per electron at the pH on the luminal side of PSII [24].

S States:

$$S_0 \xrightarrow{h\nu} S_1 \xrightarrow{h\nu} S_2 \xrightarrow{h\nu} S_3 \xrightarrow{h\nu} S_4 \xrightarrow{h\nu} S_0$$

$$2H_2O \longrightarrow O_2 + 4H^+ \quad E_m = 0.9\,V/e^-$$

In the initial step, P680 accepts exciton energy from the antenna pigments, creating an excited state, P680*. This excited state decays via the reduction of an adjacent pheophytin molecule (pheo). The resulting $P680^+$ cation has a reduction potential of 1.2 V, which is sufficient for the oxidation of the redox active tyrosine cofactor, Y_Z, which has a reduction potential of approximately 1.0 V [8]. Substantial rearrangements of the cluster and the energetic penalties of burying an additional positive charge within a hydrophobic membrane protein represent kinetic and thermodynamic hurdles. In order to avoid the buildup of excess charge, it is widely accepted that protons are released to achieve charge balance, keeping the oxidation potential of the OEC in the various S states within a range accessible to oxidation by Y_Z [68].

In Section 5.2.6, a series of quantum mechanical studies targeting the structural properties of the OEC were discussed. In addition, Siegbahn and coworkers have used DFT to examine models for the various S States. The relative free energy for each S state led the researchers to divide the Kok cycle (Figure 5.1) into two energetically favorable transitions for S_0-S_1 and S_1-S_2, and two energetically unfavorable steps in S_2-S_3 and S_3-S_4 which are compensated for by the large favorability of the S_4-S_0 step [69]. For the S_4 state, the researchers propose an oxyl radical and not a Mn=O oxo species, which is in agreement with results that will be discussed in Section 5.3.2. However, a potential problem with the studies of Siegbahn and colleagues is their use of the highest electronic spin state, the pure ferromagnetically coupled state, for their calculations. As discussed in Section 5.2.1, this state is quite different from the antiferromagnetically coupled ground state. More recent studies have used the "broken symmetry" approach that more accurately reflects certain properties the ground state of the system [65].

5.3.2
S State Advancement

Due to the relative ease of advancing PSII in a step-wise fashion using either short, high intensity laser flashes or extended, low temperature illuminations, several of the intermediates in the Kok cycle have been extensively characterized. In Section 5.2.2, the ability of XAS spectroscopy to elucidate structural information through EXAFS experiments was discussed. As was also discussed, in XANES experiments on samples poised in the S_1 state, a comparison with Mn model compounds suggests that there is no Mn(II) present in the sample [29]. While the detection of Mn(II) present within a sample is relatively easy, it is more difficult to differentiate between Mn(III) and Mn(IV). However, with careful analysis of the XANES data, the S_1 state of the enzyme has been assigned to $Mn(III)_2(IV)_2$ [29, 70]. While most researchers agree with this assignment, others prefer an all Mn(III) oxidation state for the dark stable S_1 state [21, 71]. When the same XANES experiments are conducted on the once oxidized S_2 state, there is universal agreement that the absorption of a photon by P680 leads to the oxidation of a Mn(III) to Mn(IV), as determined by a shift of the edge to a higher energy [72]. This interpretation is also consistent with the observed EPR and ^{55}Mn ESE-ENDOR data [21, 22, 57].

A significant controversy arises when considering the nature of OEC oxidation during the S_2–S_3 transition. Based on XANES spectra of samples that received two saturating flashes, Klein and coworkers interpret the observed edge shift to be too small to indicate Mn oxidation, while Dau and colleagues have suggested that metal-centered oxidation does occur [73–75]. Using K_β emission, which is very sensitive to the oxidation state of a metal, Klein and coworkers obtained further results that were suggestive of the absence of Mn centered oxidation on the S_2–S_3 transition [76]. Thus far, EPR signals attributed to the S_3 state have not been unambiguously assigned to a specific origin and therefore have not contributed significantly to the understanding of the S_3 state [77–81]. While the large array of pigment molecules within PSII greatly complicates optical experiments, it is possible to carry out ultraviolet (UV) absorption studies. Experiments by Van Gorkom and coworkers showed an increase in absorption in the UV range upon both the S_1–S_2 and S_2–S_3 transitions. Similar difference spectra are observed between Mn(II)(III) and Mn(III)(IV) model compounds, and thus these results have provided evidence for Mn oxidation on both the S_1–S_2 and S_2–S_3 transitions [82–84]. Difference Fourier transform infrared (FTIR) spectroscopy (n + 1 minus n flash) allows minute differences in various S state vibrational spectra to be distinguished despite the inherent complexity of spectral features from such a large, multi-protein complex as PSII. In these experiments, no changes that can be directly attributed to the oxidation of Mn have been observed for the S_2–S_3 transition [85–87]. However, it is difficult to conclusively determine that absence of a significant change in regions expected to contain vibrational modes from Mn ligands signifies that no oxidation is taking place, particularly since oxidation could be occurring at a Mn which is ligated by a residue that is not contributing to the observed spectra [86]. In fact, recent experiments have suggested that amino acid residues that appear to ligate Mn in the crystal structures do not show significant changes in the FTIR spectra as a function of S state [85, 87].

A compelling series of experiments that has directly examined the Mn cluster as a function of S state used time-resolved room temperature XANES methods [88, 89]. In a typical XANES experiment, the inherently low signal to noise demands a great deal of signal averaging and low temperatures. However, Dau and colleagues were able to develop methods that allowed for the time resolved measurement of XANES data following laser flashes at room temperature [89]. Due to the shape of a typical XANES spectrum, shifts in the edge position to a higher energy result in a decrease in absorption at any given position in the rise of the edge, whereas a decrease in the energy of the edge position leads to an increase in absorption. The results showed that on the S_0–S_1, S_1–S_2 and S_2–S_3 transitions, a decrease in absorption monitored at several positions in the edge was observed, indicating that Mn oxidation occurred. These results represent the most direct evidence that Mn oxidation does in fact occur on the S_2–S_3 transition.

A popular goal of Photosystem II research has long been the isolation and characterization of the short lived ($t_{1/2} = \sim 1.0$ ms) S_4 state [8]. Recently, two different approaches have given a glimpse into the nature of this elusive intermediate. Clausen and Junge yielded a candidate for the S_4 state by utilizing very high O_2 pressure, 20 bar or ~130 times ambient O_2 pressure, to trap the enzyme in the final state prior to oxygen release via product feedback inhibition [90, 91]. By monitoring the changes in the UV-Vis spectra as a function of the oxygen pressure during the S_3–S_4–S_0 transition, a model for the S_4 state was developed that is best described as S_2-H_2O_2. This suggested that the oxidation of water to oxygen occurs in two steps – the oxidation of the OEC to the S_4 state, followed by the reduction of the OEC back to an S_2 like state by the substrate water molecules, forming a peroxo species. That state would then decay back to S_0, concurrent with O_2 release [91]. A more recent set of experiments used delayed chlorophyll fluorescence, a technique in which fluorescence occurs as a result of the equilibrium between the charge separated state in the reaction center and the antenna chlorophylls [90].

These experiments tentatively supported the previous assignment of an S_2-H_2O_2 state for S_4.

In the same series of time resolved X-ray absorption experiments discussed above, the nature of the transition between the S_3 and S_0 states were probed. As expected, the S_3–S_0 transition showed an increase in absorption due to the reduction of the OEC. However, the more interesting aspect was the kinetic behavior that was observed. Following the third laser flash, a 250 μs lag was observed during which no change in the X-ray absorption was detected [89]. This was remarkable for two reasons: first, it suggests that no Mn oxidation occurs on the S_3–S_4 transition and second, it suggested that some chemical process occurred on the time scale of 250 μs prior to the oxidation of water and reduction of the cluster to S_0. Interestingly, following this lag phase, the reduction of the cluster shows monophasic exponential behavior. This suggested that the oxidation of water and reduction of the cluster occur in one concerted step, in contrast to the prediction by Junge. In addition to the kinetic behavior of the traces obtained at energy positions on the edge, data were collected below the edge, at the energy where a 1s-3d transition may occur. This feature is extremely sensitive to the degree of mixing of the ligand p and metal d orbitals since the 1s–3d transition is not formally allowed and only becomes prominent when such d-p mixing occurs. A high valent Mn=O species is predicted to have a significant degree of d-p mixing of the orbitals and would, therefore, be detected by an increase in this feature. The lack of an increase in absorption of this region suggests that no Mn=O species is present as the reactive species in the S_4 state. These data point to the possibility of ligand, potentially a bound water molecule, centered oxidation on the S_3–S_4 transition. In a parallel set of experiments, the delayed chlorophyll florescence was observed as a function of laser flash. In these experiments, a kinetic component following the third flash was observed with a time of approximately 250 μs. The pH dependence and deuterium isotope effect of this component suggested that this lag phase corresponds to a deprotonation event. The overall model, based on these experiments, for the S_3–S_0 transition is as follows: the absorption of a photon by P680 leads to the formation of an S_3–Yz· state, which is stable for ~250 μs, until a deprotonation somewhere near the OEC occurs. Subsequently, a resultant shift in redox potential leads to the oxidation of water and O–O bond formation, releasing oxygen and producing the S_0 state.

5.3.3
Water Binding and Proton Release

The previous section examined the water-oxidation reaction from the standpoint of the Mn cluster. Two important aspects to consider, with respect to the mechanism of water oxidation, are the point at which the two substrate water molecules bind and when the protons are released. It has long been known that water was the oxygen donor and molecular oxygen was released on the S_3–S_0 transition, based on a series of mass spectrometry and bare platinum electrode experiments [6, 7, 92]. However, the question of when the waters bind is much more difficult to answer. Various EPR techniques have been able to detect potential substrate water molecules bound to the OEC in the S_2 state. In a proton and deuterium ESEEM and ENDOR examination by Britt and coworkers, two distinct classes of protons are observed that were suggestive of a directly Mn-ligated water and a second water that could potentially be bound to Ca^{2+} [93]. In a series of experiments from Pace and colleagues, ^{17}O ($I = 5/2$) labeled water was used and the observation of very weak ESEEM signal was attributed to a strongly hyperfine coupled ^{17}O [94]. While the researchers cited the signal as spectroscopic evidence for a bound water molecule, the quality of the experimental data is not sufficient to justify a rigorous interpretation. A mass spectrometry approach has been used by Wydrzynski and coworkers to more precisely probe the point at which waters bind. In these experiments, the detection of ^{18}O in the O_2 product was monitored as a function of the delay of ^{18}O labeled water introduction following a series of laser flashes [95, 96]. Two different classes of water molecules were observed: one which had resolvable exchange rates for each of the S states, and one which exchanged too rapidly to detect in the S_0 and S_1 states. This suggested that water molecules were bound as early as the S_0 state, but exchange much more slowly following the oxidation of the OEC to the S_2 state and the latter states. It should be mentioned that the QM/MM modeling discussed in Section 5.2.6 included two water molecules, one bound to Ca^{2+} and one bound to Mn, with differing exchange rates, in accordance with the mass spectrometry and ESEEM data [65]. While these experiments represent a direct probe of water binding, the experimental limits, with respect to time resolution, coupled with the difficulties associated with the complexity of water interactions with PSII, lead to difficulties in rigorously interpreting the data. The exchange rate of a bound water or hydroxide can change quite drastically with changes in the oxidation state of the metal and protonation/deprotonation events [97, 98]. All of these effects are very difficult to examine in simple model systems, and are even more so for multinuclear metal clusters of ill defined structure as found in PSII.

It can be assumed that the release of protons from the OEC not only helps fuel the electrochemical gradient that provides the energy for ATP synthesis, but also aids in reducing the energetic costs of having a highly charged metal center buried within a hydrophobic membrane protein. However, the pattern of proton release as a function of S-state has remained controversial. Some researchers have suggested that the proton release stoichiometry should be 1:1:1:1, one proton for each of the four photon driven steps [99]. However, a series of experiments in which highly sensitive electrodes were used to determine the pH as a function of flash showed that the likely proton release stoichiometry was 1:0:1:2 [100]. This proposal was further elaborated in a recent letter to *Science* clarifying the time resolved XANES and delayed chlorophyll fluorescence results discussed above [5.3.2]. In this model, it was suggested that one proton is released for each electron that is removed from the OEC. In this so-called I scheme, the OEC cycles from I_0 to I_8 while superimposed on the S state cycle (Figure 5.5).

At each I state advancement, alternating proton and electron loss occurs. This culminates in the formation of the S_3-$Y_Z\cdot$ state (I_6) where a proton is first removed forming the I_7 state, coinciding with the 250 μs lag phase (described earlier, in

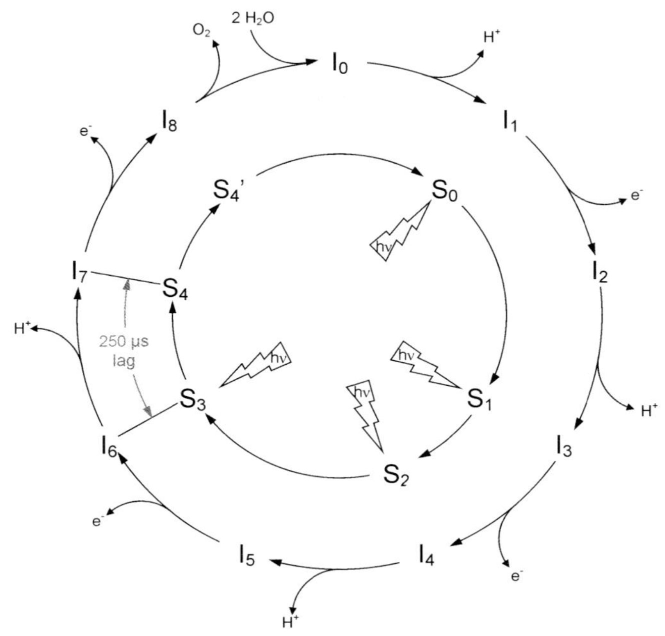

Figure 5.5 The "I" scheme of oxygen evolution. An adaptation of the I scheme superimposed on the Kok, S state cycle. In the I scheme, alternating protons and electrons are removed from the cluster, culminating with the release of oxygen following the I_8 state. (Figure reproduced from reference 101.)

Section 5.3.2) followed by extraction of an electron from somewhere other than the Mn cluster, yielding the I_8 state, which decomposes to the I_0/S_0 state.

5.3.4 Hydrogen Atom Abstraction Model (Babcock Model)

In their hydrogen atom abstraction proposal for the mechanism of water oxidation, Babcock and colleagues used the prevailing dimer of dimers model to develop a specific model for how the OEC might evolve oxygen (Figure 5.6) [99]. While the dimer of dimers model has now been largely eliminated as a possible structure of the OEC, the mechanism can still be adapted to the more recent monomer-trimer structures. The central feature of this mechanism is tyrosine radical based hydrogen atom abstraction. The researchers used the dissociation energies for the total water-oxidation reaction to justify the possibility that the mechanism proceeded via the abstraction of a hydrogen atom at each step by the photogenerated radical, $Y_Z\cdot$. By their calculations, the overall hydrogen atom abstraction reaction from two water molecules would be energetically favorable by 23 kcal/mole. Starting with two water molecules bound to two Mn atoms, each photon leads to the loss of a

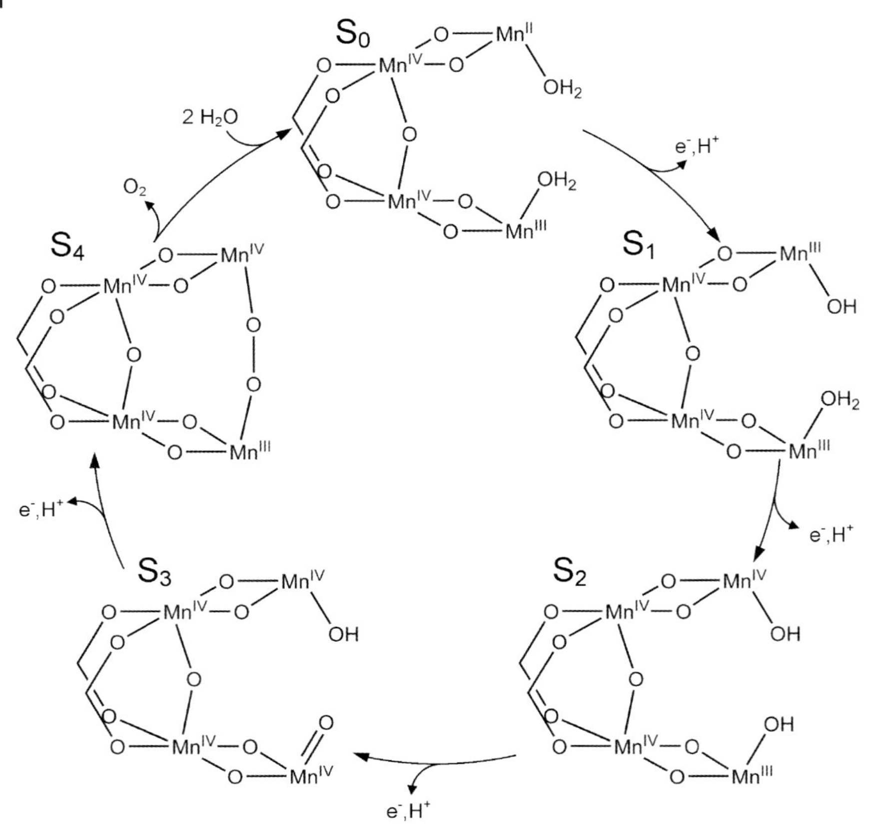

Figure 5.6 The Hydrogen Abstraction (HA)/Babcock model of oxygen evolution. The HA/Babcock model invokes the loss of a hydrogen atom to Y_Z on each S state advancement. At the final stage, the O–O bond forms on water molecules bound to adjacent Mn atoms. Figure reproduced from reference 99.

hydrogen atom and the resulting oxidation of the corresponding Mn atom. This results in the formation of a Mn(IV)=O species which acts as an electrophile and is the subject of a nucleophilic attack from a hydroxide bound to the adjacent Mn, forming the O–O bond in the S_4 state.

While it can certainly be adapted to the more current structures, the hydrogen atom abstraction model has fallen out of favor. Such a mechanism might be expected to exhibit a very strong isotope effect, yet water-oxidation has been shown to have a relatively weak H/D isotope effect [102]. In addition, electrochromic shift experiments examining the rates and nature of proton release as a function of pH, H/D ratio and temperature have suggested that Y_Z is not a hydrogen atom abstractor from the OEC; rather, it is likely hydrogen bonded to an adjacent histidine residue throughout the Kok cycle [103, 104]. There is also evidence refuting this model from the crystallographic data. Barber and colleagues contend that there is no obvious proton path between the OEC and Y_Z [60]. However, this could arise from an inability to resolve any water molecules between the OEC and Y_Z that could be present. Furthermore, the likelihood that there is no Mn oxidation after the S_3 state argues against a Mn(IV)=O as the reactive species in S_4.

5.3.5
Brudvig/Pecoraro Mechanisms for Water Oxidation

Two subsequent models also proposed a high valent Mn=O species in the S_4 state, like the Babcock model, but with the second water bound to Ca^{2+} [68, 105]. Both of these models specifically address the roles of Ca^{2+} and Cl^-, two essential cofactors in water oxidation. In the Pecoraro model, the energetics of hydrogen bond dissociation of free water and water bound to various metal compounds were surveyed to justify why Mn is specifically needed (Figure 5.7). In addition, a justification for the Ca^{2+} requirement, and the fact that it can only be functionally replaced with Sr^{2+}, contrary to most Ca^{2+} containing enzymes, was presented [68]. In this model, the Lewis acidity of highly valent Mn atoms provides a large decrease in the hydrogen–oxygen bond dissociation energy, as observed in a series of model compounds. In contrast, the differing Lewis acidity of Ca^{2+} poises the second bound water for a nucleophilic attack on the adjacent high valent Mn=O.

Recent ESEEM and ENDOR studies from Britt and colleagues have shown that there are two classes of protons (and deuterons in ^2H exchanged samples) that fit the predicted binding scheme from the Pecoraro mechanism, having one water bound Mn and another to Ca^{2+} [106]. In this mechanism, protons are lost on the S_1-S_2 and S_2-S_3 transitions, leading to two hydroxide ligands on Mn and Ca. On the S_3–S_4 transition, O–O bond formation occurs, producing an OEC bound peroxo species. The Cl^- ion, bound to the adjacent Mn atom, is then displaced by the recently formed O–OH species, which moves to the other Mn, where it is further oxidized, forming O_2. This suggests that Cl^- is a placeholder, keeping water from binding to the second Mn atom. A recent paper from Dau and colleagues suggests, however, based on Br EXAFS results, that Cl^- might actually bind as far as 5 Å from the Mn cluster [107]. In contrast, ESEEM experiments in which acetate or azide are present, two ions known to compete with Cl^- for binding at the OEC, have detected deuterium nuclei, in the case of labeled acetate, and nitrogen nuclei for labeled azide coupled directly to the OEC. These results suggest the possibility that Cl^- binds either directly to the Mn cluster, or in close proximity, supporting an intimate role for Cl^- in the water-oxidation reaction [108, 109].

The Brudvig model (Figure 5.8) was inspired by this group's work with a binuclear Mn model compound that is capable of forming an O-O bond in the presence of sodium hypochlorite as an external oxidant [105]. This reaction mechanism was proposed to involve a high valent Mn=O species. This was used as the justification for a similar reactive species in the OEC at the S_4 state. Like the Pecoraro mechanism, the Brudvig mechanism requires the binding of one of the substrate water molecules to Ca^{2+}. However, in this case, Cl^- is postulated to bridge the Mn and

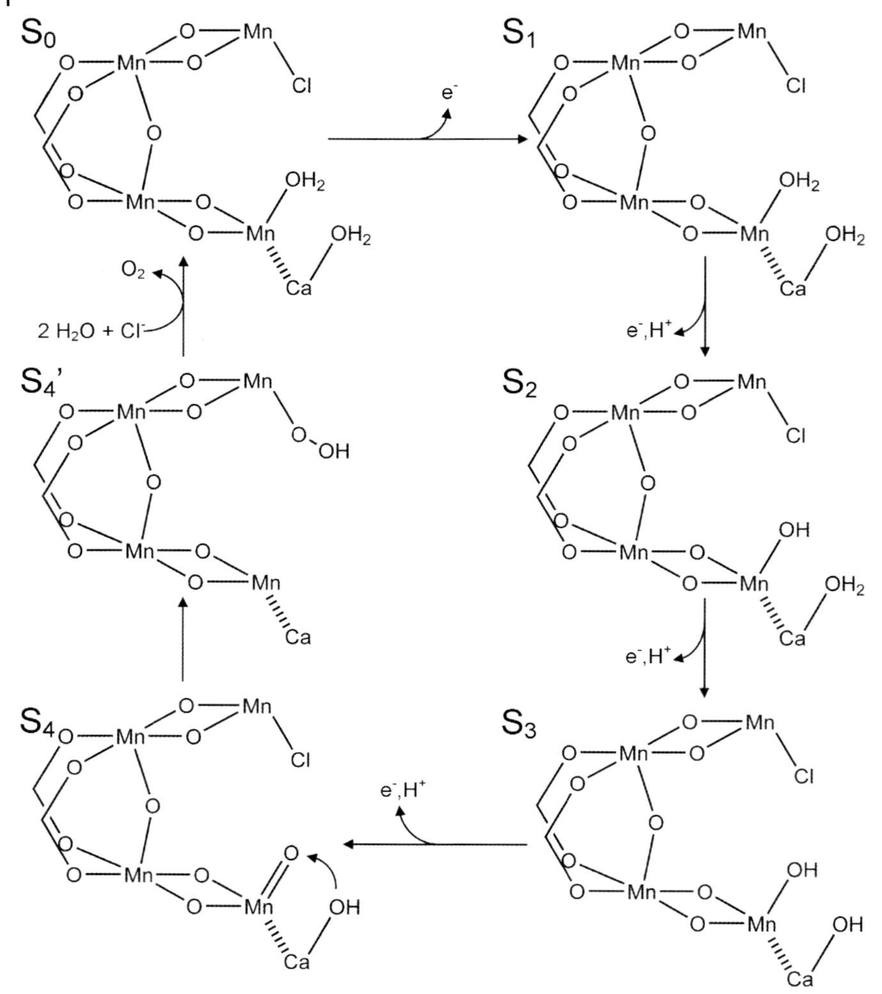

Figure 5.7 The Pecoraro model of oxygen evolution. The Pecoraro model contains both Ca^{2+} and Cl^- and was the first to do so. The role of Ca^{2+} is to bind one of the substrate water molecules and activate it for a nucleophilic attack on a water bound to an adjacent Mn. The Cl^- ion occupies a ligand position on a second Mn and is released in the final step of water oxidation. (Figure reproduced from reference 68.)

Ca atoms that bind the substrate water molecules. The Cl^- ion tunes the redox potential of the water-bound to the Ca ion as a function of S states, ensuring proper reactivity.

This mechanistic proposal has been explored by the recent QM/MM calculations described in Section 5.2.6 [65, 66]. For this QM model of the S_2 state, computed exchange rates for waters bound to Ca and Mn match those determined by mass spectrometry measurements discussed in Section 5.3.3. In the preliminary analy-

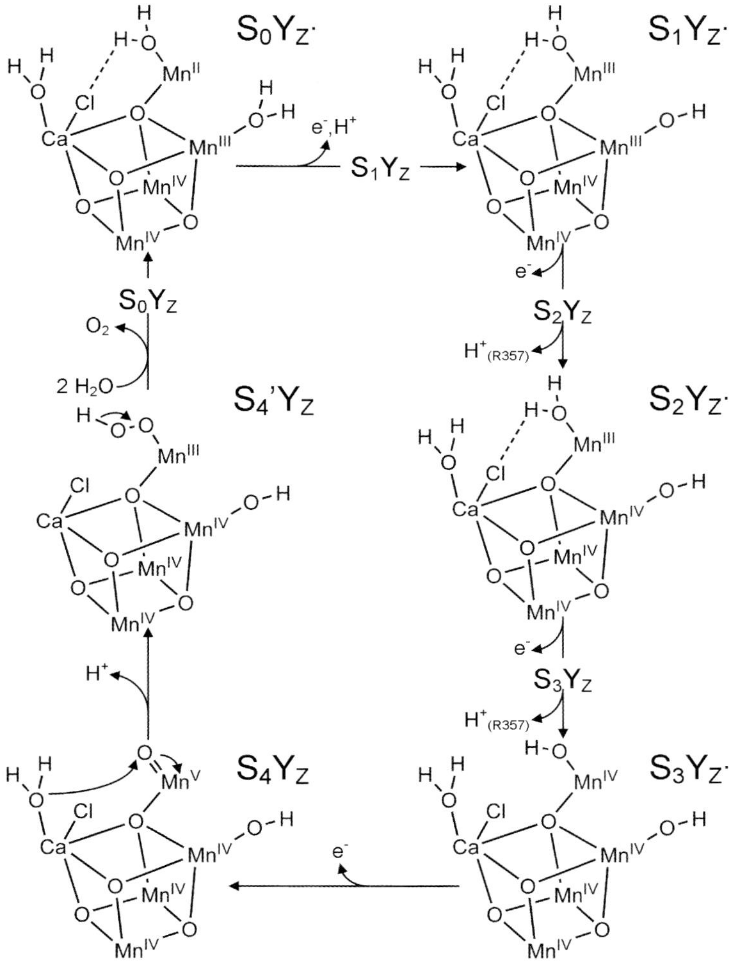

Figure 5.8 The Brudvig model of oxygen evolution. In the Brudvig model, one substrate water molecule is bound to Ca^{2+} and the other to Mn. Like the Pecoraro mechanism, the water bound to Mn is the subject of a nucleophilic attach from the hydroxide ion bound to Ca^{2+}. The role of the Cl^- ion is to control the reduction potential of two Mn by acting as a bridging ligand. (Figure reproduced from reference 67.)

sis of the QM/MM model for the various S states, the models are in general agreement with the Brudvig model for water oxidation.

5.3.6 Messinger Mechanism

Another recent mechanism (Figure 5.9), described by Messinger and colleagues, invokes a different arrangement of water. In this proposal, the O–O bond forms

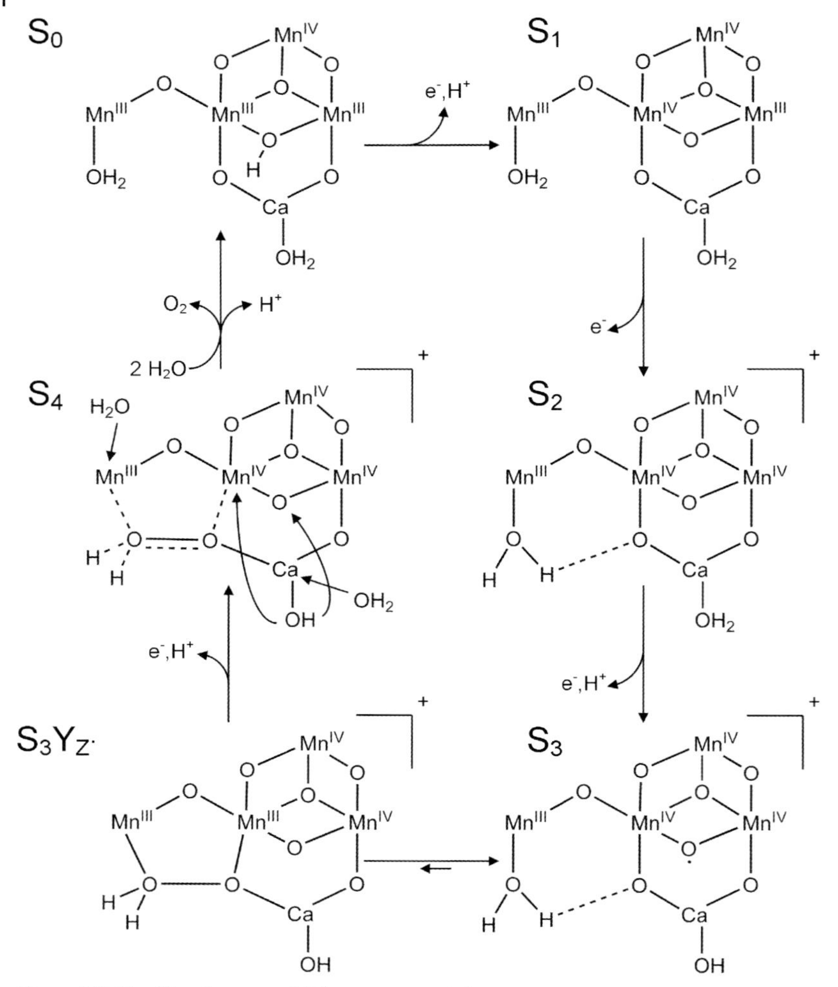

Figure 5.9 The Messinger model for oxygen evolution. Contrary to the other three main models of water oxidation, Messinger and colleagues implicate a bridging oxygen radical, instead of a terminal Mn=O as the reactive species in the S_4 state. (Figure reproduced from reference 110.)

between a bridging oxygen atom and a terminal hydroxide bound to Ca^{2+}. The primary rationale for this choice in the mechanism hinges on the kinetic data that has been obtained for water exchange, discussed in Section 5.3.3 [95, 110]. Messinger and colleagues suggest that the rapid exchanging water is bound to Ca^{2+}, with the other, more slowly exchanging water, occupying a bridging position between two Mn atoms. In addition, this mechanism follows the interpretation of the XANES data discussed in Section 5.3.2, and leads to the prediction that oxidation on the S_2–S_3 transition occurs at the bridging oxygen position invoked to form the O–O bond, and not Mn as has been proposed in the other mechanisms as well as the time resolved XANES data. This mechanism has support from quantum mechanical calculations conducted by Siegbahn and colleagues in which the stability of an oxygen centered radical was demonstrated and suggested to be more likely than a Mn=O reactive species in S_4 [69].

5.3.7
Agreement between Proposed Mechanisms and Spectroscopic Data

A significant number of spectroscopic results have been obtained since the the specific mechanisms for water oxidation were first proposed. One common feature in most of the mechanisms discussed is the presence of a high-valent Mn=O species in the S_4 state. Yet, the time resolved XANES, and 1s–3d transition measurements discussed earlier (in Section 5.3.2), suggest that no Mn=O species is present at S_4, and, in fact, that there is no Mn centered oxidation at all following the formation of the S_3 state [89]. When all of the spectroscopic data are combined, the oxidation of water appears to proceed as follows: on the S_0 to S_1 transition, the time resolved and static XANES measurements suggest that a Mn centered oxidation occurs. Based on the EPR interpretation of the S_0 oxidation state and the XANES interpretation of the S_1 state, the oxidation is likely of a Mn(III) to Mn(IV) accompanied by release of a proton. With respect to substrate water molecules, both waters are in rapid exchange in both in the S_0 and S_1 states following oxidation. In the S_1–S_2 transition, based on wide agreement with the XANES results, a different Mn(III) is oxidized to Mn(IV); perhaps without the release of a proton [72, 88, 100, 101]. This oxidation also substantially slows the rate of water exchange [95]. While the S_2–S_3 transition still remains controversial, the most compelling evidence suggests that Mn-centered oxidation does occur, to give $Mn(III)(IV)_3$. On this transition, it is suggested that one proton might be released [88, 89, 100, 101]. Following the formation of the S_3-$Y_Z\cdot$ state, in approximately 250 μs, a deprotonation occurs prior to the formation of the S_4 state, which is perhaps better described as an S_3-oxidized ligand state [89, 101]. Once this state is achieved, the OEC is reduced back to the S_0 state and O_2 is released together with a proton.

The past several years have seen a remarkable increase in the understanding of the OEC; however, many questions remain unanswered. Pressing issues such as the precise structure of the OEC, the surrounding protein and substrate ligands at each state within the Kok cycle, and the chemical nature of the 250 μs component of the S_4 state, will likely become clearer in the coming years.

References

1 Blankenship, R.E. (2006) Photosynthesis: the light reactions, in *Plant Physiology*, 4th edn (ed Lincoln Taiz, Eduardo Zeiger), Sinauer Associates, pp. 125–58.

2 Barber, J. (2003) Photosystem II: the engine of life. *Quarterly Reviews of Biophysics*, **36**, 71–89.

3 Bricker, T.M. and Ghanotakis, D.F. (1996) Introduction to oxygen evolution and the oxygen-evolving complex. *Advances in Photosynthesis*, **4**, 113–36.

4 Britt, R.D. (1996) Oxygen evolution. *Advances in Photosynthesis*, **4**, 137–64.

5 Nelson, N. and Yocum, C.F. (2006) Structure and function of photosystems I and II. *Annual Review of Plant Biology*, **57**, 521–65.

6 Joliot, P., Barbieri, G. and Chabaud, R. (1969) Model of the System II photochemical centers. *Photochemistry and Photobiology*, **10**, 309–29.

7 Kok, B., Forbush, B. and McGloin, M. (1970) Cooperation of charges in photosynthetic oxygen evolution. I. A linear four step mechanism. *Photochemistry and Photobiology*, **11**, 457–75.

8 Babcock, G.T., Barry, B.A., Debus, R.J., Hoganson, C.W., Atamian, M., McIntosh, L., Sithole, I. and Yocum, C.F. (1989) Water oxidation in photosystem II: from radical chemistry to multielectron chemistry. *Biochemistry*, **28**, 9557–65.

9 Debus, R.J. (1992) The manganese and calcium ions of photosynthetic oxygen evolution. *Biochimica et Biophysica Acta, Bioenergetics*, **1102**, 269–352.

10 Pirson, A. (1937) A study of the nutrition and metabolism of Fontinalis and Chlorella. *Zeitschrift fuer Botanik*, **31**, 193–267.

11 Kessler, E. (1955) On the role of manganese in the oxygen-evolving system of photosynthesis. *Archives of Biochemistry and Biophysics*, **59**, 527–9.

12 Astier, C., Styring, S., Maison-Peteri, B. and Etienne, A.L. (1986) Preparation and characterization of thylakoid membranes and photosystem II particles from the facultative phototrophic cyanobacterium Synechocystis 6714. *Photobiochemistry and Photobiophysics*, **11**, 37–47.

13 Ghanotakis, D.F., Demetriou, D.M. and Yocum, C.F. (1987) Isolation and characterization of an oxygen-evolving photosystem II reaction center core preparation and a 28 kDa Chl-a-binding protein. *Biochimica et Biophysica Acta, Bioenergetics*, **891**, 15–21.

14 Yamada, Y., Tang, X.S., Ito, S. and Sato, K. (1987) Purification and properties of an oxygen-evolving photosystem II reaction-center complex from spinach. *Biochimica et Biophysica Acta, Bioenergetics*, **891**, 129–37.

15 Ghanotakis, D.F., Topper, J.N. and Yocum, C.F. (1984) Structural organization of the oxidizing side of photosystem II. Exogenous reductants reduce and destroy the manganese complex in photosystems II membranes depleted of the 17- and 23-kDa polypeptides. *Biochimica et Biophysica Acta, Bioenergetics*, **767**, 524–31.

16 Ghanotakis, D.F., Babcock, G.T. and Yocum, C.F. (1984) Calcium reconstitutes high rates of oxygen evolution in polypeptide depleted photosystem II preparations. *FEBS Letters*, **167**, 127–30.

17 Homann, P.H. (1988) Structural effects of chloride and other anions on the water oxidizing complex of chloroplast photosystem II. *Plant Physiology*, **88**, 194–9.

18 Ono, T., Nakayama, H., Gleiter, H., Inoue, Y. and Kawamori, A. (1987) Modification of the properties of S2 state in photosynthetic O2-evolving center by replacement of chloride with other anions. *Archives of biochemistry and biophysics*, **256**, 618–24.

19 Kelley, P.M. and Izawa, S. (1978) The role of chloride ion in photosystem II. I. Effects of chloride ion on photosystem II electron transport and on hydroxylamine inhibition. *Biochimica et Biophysica Acta, Bioenergetics*, **502**, 198–210.

20 Sandusky, P.O. and Yocum, C.F. (1984) The chloride requirement for photosynthetic oxygen evolution. Analysis of the effects of chloride and other anions on amine inhibition of the oxygen-evolving complex. *Biochimica et Biophysica Acta, Bioenergetics*, **766**, 603–11.

21 Dismukes, G.C. and Siderer, Y. (1981) Intermediates of a polynuclear manganese center involved in photosynthetic oxidation of water. *Proceedings of the National Academy of Sciences of the United States of America*, **78**, 274–8.

22 Dismukes, G.C. and Siderer, Y. (1980) EPR spectroscopic observations of a manganese center associated with water oxidation in spinach chloroplasts. *FEBS Letters*, **121**, 78–80.

23 Reed, G.H. and Markham, G.D. (1984) EPR of manganese(II) complexes with enzymes and other proteins. *Biological Magnetic Resonance*, **6**, 73–142.

24 Law, N.A., Caudle, M.T. and Pecoraro, V.L. (1998) Manganese redox enzymes and model systems: properties, structures, and reactivity, *Advances in Inorganic Chemistry*, **46**, 305–440.

25 Cooper, S.R., Dismukes, G.C., Klein, M.P. and Calvin, M. (1978) Mixed valence interactions in di-m-oxo bridged manganese complexes. Electron paramagnetic resonance and magnetic susceptibility studies. *Journal of the American Chemical Society*, **100**, 7248–52.

26 Mukhopadhyay, S., Mandal, S.K., Bhaduri, S. and Armstrong, W.H. (2004) Manganese clusters with relevance to Photosystem II. *Chemical Reviews (Washington, DC, United States)*, **104**, 3981–4026.

27 Kirby, J.A., Robertson, A.S., Smith, J.P., Thompson, A.C., Cooper, S.R. and Klein, M.P. (1981) State of manganese in the photosynthetic apparatus. 1. Extended x-ray absorption fine structure studies on chloroplasts and di-m-oxo-bridged dimanganese model compounds. *Journal of the American Chemical Society*, **103**, 5529–37.

28 Penner-Hahn, J.E. (2004) X-ray absorption spectroscopy. *Comprehensive Coordination Chemistry II*, **2**, 159–86.

29 Kirby, J.A., Goodin, D.B., Wydrzynski, T., Robertson, A.S. and Klein, M.P. (1981) State of manganese in the photosynthetic apparatus. 2. X-ray absorption edge studies on manganese in photosynthetic membrane. *Journal of the American Chemical Society*, **103**, 5537–42.

30 Ghanotakis, D.F., Babcock, G.T. and Yocum, C.F. (1984) Structural and catalytic properties of the oxygen-evolving complex. Correlation of polypeptide and manganese release with the behavior of Z+ in chloroplasts and a highly resolved preparation of the PS II complex. *Biochimica et Biophysica Acta, Bioenergetics*, **765**, 388–98.

31 Berthold, D.A., Babcock, G.T. and Yocum, C.F. (1981) A highly resolved, oxygen-evolving photosystem II preparation from spinach thylakoid membranes. EPR and electron-transport properties. *FEBS Letters*, **134**, 231–4.

32 Yachandra, V.K., Guiles, R.D., McDermott, A., Britt, R.D., Dexheimer, S.L., Sauer, K. and Klein, M.P. (1986) The state of manganese in the photosynthetic apparatus. 4. Structure of the manganese complex in photosystem II studied using EXAFS spectroscopy. The S1 state of the oxygen-evolving photosystem II complex from spinach. *Biochimica et Biophysica Acta, Bioenergetics*, **850**, 324–32.

33 George, G.N., Prince, R.C. and Cramer, S.P. (1989) The manganese site of the photosynthetic water-splitting enzyme. *Science (Washington, DC, United States)*, **243**, 789–91.

34 Yachandra, V.K., DeRose, V.J., Latimer, M.J., Mukerji, I., Sauer, K. and Klein, M.P. (1993) Where plants make oxygen: a structural model for the photosynthetic oxygen-evolving manganese cluster. *Science (Washington, DC, United States)*, **260**, 675–9.

35 Zimmermann, J.L. and Rutherford, A.W. (1986) Electron paramagnetic resonance properties of the S2 state of the oxygen-evolving complex of photosystem II. *Biochemistry*, **25**, 4609–15.

36 Zimmermann, J.L. and Rutherford, A.W. (1984) EPR studies of the oxygen-evolving enzyme of photosystem II. *Biochimica et Biophysica Acta, Bioenergetics*, **767**, 160–7.

37 Casey, J.L. and Sauer, K. (1984) EPR detection of a cryogenically photogenerated intermediate in photosynthetic oxygen evolution. *Biochimica et Biophysica Acta, Bioenergetics*, **767**, 21–8.

38 Kim, D.H., Britt, R.D., Klein, M.P. and Sauer, K. (1992) The manganese site of the photosynthetic oxygen-evolving complex probed by EPR spectroscopy of oriented photosystem II membranes: the g = 4 and g = 2 multiline signals. *Biochemistry*, **31**, 541–7.

39 Haddy, A., Lakshmi, K.V., Brudvig Gary, W., Harry, F. and A. (2004) Q-band EPR of the S2 state of photosystem II confirms an S = 5/2 origin of the X-band g = 4.1 signal. *Biophysical Journal*, **87**, 2885–96.

40 Astashkin, A.V., Kodera, Y. and Kawamori, A. (1994) Pulsed EPR study of manganese g = 4.1 signal in plant photosystem II, *Journal of Magnetic Resonance, Series B*, **105**, 113–9.

41 Force, D.A., Randall, D.W., Lorigan, G.A., Clemens, K.L. and Britt, R.D. (1998) ESEEM studies of alcohol binding to the manganese cluster of the oxygen evolving complex of photosystem II. *Journal of the American Chemical Society*, **120**, 13321–33.

42 Boussac, A., Girerd, J.-J. and Rutherford, A.W. (1996) Conversion of the spin state of the manganese complex in Photosystem II induced by Near-Infrared Light. *Biochemistry*, **35**, 6984–9.

43 Messinger, J., Robblee, J., Yu, W.O., Sauer, K., Yachandra, V.K. and Klein, M.P. (1997) The S0 state of the oxygen-evolving complex in photosystem II is paramagnetic: detection of an EPR multiline signal. *Journal of the American Chemical Society*, **119**, 11349–50.

44 Messinger, J., Nugent, J.H. and Evans, M.C.W. (1997) Detection of an EPR multiline signal for the S0 state in photosystem II. *Biochemistry*, **36**, 11055–60.

45 Kulik, L.V., Lubitz, W. and Messinger, J. (2005) Electron spin-lattice relaxation of the S0 state of the oxygen-evolving complex in Photosystem II and of dinuclear manganese model complexes. *Biochemistry*, **44**, 9368–74.

46 Kulik, L.V., Epel, B., Lubitz, W. and Messinger, J. (2005) 55Mn pulse ENDOR at 34 GHz of the S0 and S2 states of the oxygen-evolving complex in Photosystem II. *Journal of the American Chemical Society*, **127**, 2392–3.

47 Dexheimer, S.L. and Klein, M.P. (1992) Detection of a paramagnetic intermediate in the S1 state of the photosynthetic oxygen-evolving complex. *Journal of the American Chemical Society*, **114**, 2821–6.

48 Yamauchi, T., Mino, H., Matsukawa, T., Kawamori, A. and Ono, T.-A. (1997) Parallel polarization electron paramagnetic resonance studies of the S1-state manganese cluster in the photosynthetic oxygen-evolving system. *Biochemistry*, **36**, 7520–6.

49 Campbell, K.A., Gregor, W., Pham, D.P., Peloquin, J.M., Debus, R.J. and Britt, R.D. (1998) The 23 and 17 kDa extrinsic proteins of Photosystem II modulate the magnetic properties of the S1-State manganese cluster. *Biochemistry*, **37**, 5039–45.

50 Campbell, K.A., Peloquin, J.M., Pham, D.P., Debus, R.J. and Britt, R.D. (1998) Parallel polarization EPR detection of an S1-State \"Multiline\" EPR signal in Photosystem II particles from synechocystis sp. PCC 6803. *Journal of the American Chemical Society*, **120**, 447–8.

51 Hsieh, W.-Y., Campbell Kristy, A., Gregor, W., David Britt, R., Yoder Derek, W., Penner-Hahn James, E. and Pecoraro Vincent, L. (2004) The first spectroscopic model for the S1 state multiline signal of the OEC. *Biochimica et Biophysica Acta*, **1655**, 149–57.

52 Zheng, M. and Dismukes, G.C. (1996) Orbital configuration of the valence electrons, ligand field symmetry, and manganese oxidation states of the photosynthetic water oxidizing complex: analysis of the S2 state multiline EPR signals. *Inorganic Chemistry*, **35**, 3307–19.

53 Hasegawa, K., Ono, T., Inoue, Y. and Kusunoki, M. (1999) Spin-exchange interactions in the S2-state manganese tetramer in photosynthetic oxygen-evolving complex deduced from g = 2 multiline EPR signal. *Chemical Physics Letters*, **300**, 9–19.

54 Hasegawa, K., Kusunoki, M., Inoue, Y. and Ono, T.-a. (1998) Simulation of S2-State multiline EPR signal in oriented Photosystem II membranes: structural implications for the manganese cluster in an oxygen-evolving complex. *Biochemistry*, **37**, 9457–65.

55 Aahrling, K.A. and Pace, R.J. (1995) Simulation of the S2 state multiline electron paramagnetic resonance signal of Photosystem II: a multifrequency approach. *Biophysical Journal*, **68**, 2081–2090.

56 Lakshmi, K.V., Eaton, S.S., Eaton, G.R. and Brudvig, G.W. (1999) Orientation of the tetranuclear manganese cluster and tyrosine Z in the O2-evolving complex of Photosystem II: an EPR study of the S2YZ.bul. state in oriented acetate-inhibited Photosystem II membranes. *Biochemistry*, **38**, 12758–67.

57 Peloquin, J.M., Campbell, K.A., Randall, D.W., Evanchik, M.A., Pecoraro, V.L., Armstrong, W.H. and Britt, R.D. (2000) 55Mn ENDOR of the S2-state multiline EPR signal of Photosystem II: implications on the structure of the tetranuclear Mn cluster. *Journal of the American Chemical Society*, **122**, 10926–42.

58 Zouni, A., Witt, H.-T., Kern, J., Fromme, P., Krauss, N., Saenger, W. and Orth, P. (2001) Crystal structure of photosystem II from Synechococcus elongatus at 3.8 .ANG. resolution. *Nature (London)*, **409**, 739–43.

59 Kamiya, N. and Shen, J.-R. (2003) Crystal structure of oxygen-evolving photosystem II from Thermosynechococcus vulcanus at 3.7-A resolution. *Proceedings of the National Academy of Sciences of the United States of America*, **100**, 98–103.

60 Ferreira, K.N., Iverson, T.M., Maghlaoui, K., Barber, J. and Iwata, S. (2004) Architecture of the photosynthetic oxygen-evolving center. *Science (Washington, DC, United States)*, **303**, 1831–8.

61 Yano, J., Kern, J., Irrgang, K.-D., Latimer, M.J., Bergmann, U., Glatzel, P., Pushkar, Y., Biesiadka, J., Loll, B., Sauer, K., Messinger, J., Zouni, A. and Yachandra, V.K. (2005) X-ray damage to the Mn4Ca complex in single crystals of photosystem II: a case study for metalloprotein crystallography. *Proceedings of the National Academy of Sciences of the United States of America*, **102**, 12047–52.

62 Cinco Roehl, M., Robblee John, H., Messinger, J., Fernandez, C., McFarlane Holman Karen, L., Sauer, K. and Yachandra Vittal, K. (2004) Orientation of calcium in the Mn4Ca cluster of the oxygen-evolving complex determined using polarized strontium EXAFS of photosystem II membranes. *Biochemistry*, **43**, 13271–82.

63 Cinco, R.M., Robblee, J.H., Rompel, A., Fernandez, C., Visser, H., Messinger, J., McFarlane, K., YachAndra, V.K., Sauer, K. and Klein, M.P. (2000) Structure of the calcium cofactor in oxygen evolution probed by polarized strontium EXAFS on oriented photosystem II membranes, *Book of Abstracts, 219th ACS National Meeting* (March 26–30, 2000), INOR-064, San Francisco, CA.

64 Yano, J., Kern, J., Sauer, K., Latimer, M.J., Pushkar, Y., Biesiadka, J., Loll, B., Saenger, W., Messinger, J., Zouni, A. and Yachandra, V.K. (2006) Where water is oxidized to dioxygen: structure of the photosynthetic Mn4Ca Cluster. *Science (Washington, DC, United States)*, **314**, 821–5.

65 Sproviero, E.M., Gascon, J.A., McEvoy, J.P., Brudvig, G.W. and Batista, V.S. (2006) QM/MM models of the O2-evolving complex of Photosystem II. *Journal of Chemical Theory and Computation*, **2**, 1119–34.

66 Sproviero, E.M., Gascon, J.A., McEvoy, J.P., Brudvig, G.W. and Batista, V.S. (2006) Characterization of synthetic oxomanganese complexes and the inorganic core of the O2-evolving complex in photosystem II: evaluation of the DFT/B3LYP level of theory. *Journal of Inorganic Biochemistry*, **100**, 786–800.

67 McEvoy, J.P., Gascon, J.A., Batista, V.S. and Brudvig, G.W. (2005) The mechanism of photosynthetic water splitting. *Photochemical and Photobiological Sciences*, **4**, 940–9.

68 Pecoraro, V.L., Baldwin, M.J., Caudle, M.T., Hsieh, W.-Y. and Law, N.A. (1998) A proposal for water oxidation in photosystem II. *Pure and Applied Chemistry*, **70**, 925–9.

69 Siegbahn, P.E. and Lundberg, M. (2005) The mechanism for dioxygen formation in PSII studied by quantum chemical methods. *Photochemical and Photobiological Sciences*, **4**, 1035–43.

70 Riggs, P.J., Yocum, C.F., Penner-Hahn, J.E. and Mei, R. (1992) Reduced derivatives of the manganese cluster in the photosynthetic oxygen-evolving complex. *Journal of the American Chemical Society*, **114**, 10650–1.

71 Zheng, M. and Dismukes, G.C. (1992) Photosynthetic water oxidation: what have we learned from the multiline EPR signals? *Research in Photosynthesis*, Vol **II** (ed. N. Murata), Kluwer Academic Publ., Dordrecht, pp. 305–8.

72 Yachandra, V.K., Guiles, R.D., McDermott, A.E., Cole, J.L., Britt, R.D., Dexheimer, S.L., Sauer, K. and Klein, M.P. (1987) Comparison of the structure of the manganese complex in the S1 and S2 states of the photosynthetic O2-evolving complex: an x-ray absorption spectroscopy study. *Biochemistry*, **26**, 5974–81.

73 Iuzzolino, L., Dittmer, J., Dorner, W., Meyer-Klaucke, W. and Dau, H. (1998) X-ray absorption spectroscopy on layered photosystem II membrane particles suggests manganese-centered oxidation of the oxygen-evolving complex for the S0-S1, S1-S2, and S2-S3 transitions of the water oxidation cycle. *Biochemistry*, **37**, 17112–19.

74 Roelofs, T.A., Liang, W., Latimer, M.J., Cinco, R.M., Rompel, A., andrews, J.C., Sauer, K., YachAndra, V.K. and Klein, M.P. (1996) Oxidation states of the manganese cluster during the flash-induced S-state cycle of the photosynthetic oxygen-evolving complex. *Proceedings of the National Academy of Sciences of the United States of America*, **93**, 3335–40.

75 Ono, T., Noguchi, T., Inoue, Y., Kusunoki, M., Matsushita, T. and Oyanagi, H. (1992) X-ray-detection of the period-4 cycling of the manganese cluster in photosynthetic water oxidizing enzyme. *Science*, **258**, 1335–7.

76 Messinger, J., Robblee, J.H., Bergmann, U., Fernandez, C., Glatzel, P., Visser, H., Cinco, R.M., McFarlane, K.L., Bellacchio, E., Pizarro, S.A., Cramer, S.P., Sauer, K., Klein, M.P. and Yachandra, V.K. (2001) Absence of Mn-centered oxidation in the S2 -> S3 transition: implications for the mechanism of photosynthetic water oxidation. *Journal of the American Chemical Society*, **123**, 7804–20.

77 Hallahan, B.J., Nugent, J.H.A., Warden, J.T. and Evans, M.C.W. (1992) Investigation of the origin of the \"S3\" EPR signal from the oxygen-evolving complex of photosystem 2: the role of tyrosine Z. *Biochemistry*, **31**, 4562–73.

78 Ioannidis, N., Nugent, J.H. and Petrouleas, V. (2002) Intermediates of the S3 state of the oxygen-evolving complex of Photosystem II. *Biochemistry*, **41**, 9589–600.

79 Geijer, P., Morvaridi, F. and Styring, S. (2001) The S3 state of the oxygen-evolving complex in Photosystem II Is converted to the S2YZ.bul. State at Alkaline pH. *Biochemistry*, **40**, 10881–91.

80 Ioannidis, N. and Petrouleas, V. (2000) Electron paramagnetic resonance signals from the S3 state of the oxygen-evolving complex. A broadened radical signal induced by low-temperature near-infrared light illumination. *Biochemistry*, **39**, 5246–54.

81 Boussac, A. and Rutherford, A.W. (1992) The origin of the split S3 EPR signal in calcium-depleted photosystem II: histidine versus tyrosine. *Biochemistry*, **31**, 7441–5.

82 Dekker, J.P., Van Gorkom, H.J., Wensink, J. and Ouwehand, L. (1984) Absorbance difference spectra of the successive redox states of the oxygen-evolving apparatus of photosynthesis. *Biochimica et Biophysica Acta, Bioenergetics*, **767**, 1–9.

83 Dekker, J.P., Van Gorkom, H.J., Brok, M. and Ouwehand, L. (1984) Optical characterization of photosystem II electron donors. *Biochimica et Biophysica Acta, Bioenergetics*, **764**, 301–9.

84 Dekker, J.P. (1992) Optical studies on the oxygen-evolving complex of photosystem II. *Manganese Redox Enzymes*, 85–103.

85 Strickler, M.A., Hillier, W. and Debus, R.J. (2006) No evidence from FTIR difference spectroscopy that glutamate-189 of the D1 polypeptide ligates a Mn ion that undergoes oxidation during the S0 to S1, S1 to S2, or S2 to S3 transitions in Photosystem II. *Biochemistry*, **45**, 8801–11.

86 Kimura, Y. and Ono, T.-a. (2006) Structural and functional studies of photosynthetic oxygen evolving Mn cluster by means of FTIR spectroscopy. *Seibutsu Butsuri*, **46**, 124–9.

87 Debus, R.J., Strickler, M.A., Walker, L.M. and Hillier, W. (2005) No evidence from FTIR difference spectroscopy that Aspartate-170 of the D1 polypeptide ligates a manganese ion that undergoes oxidation during the S0 to S1, S1 to S2, or S2 to S3 transitions in Photosystem II. *Biochemistry*, **44**, 1367–74.

88 Haumann, M., Muller, C., Liebisch, P., Iuzzolino, L., Dittmer, J., Grabolle, M., Neisius, T., Meyer-Klaucke, W. and Dau, H. (2005) Structural and oxidation state changes of the photosystem II manganese complex in four transitions of the water oxidation cycle (S0 → S1, S1 → S2, S2 → S3, and S3,4 → S0) characterized by X-ray absorption spectroscopy at 20 K and room temperature. *Biochemistry*, **44**, 1894–908.

89 Haumann, M., Liebisch, P., Muller, C., Barra, M., Grabolle, M. and Dau, H. (2005) Photosynthetic O2 formation tracked by time-resolved x-ray experiments. *Science*, **310**, 1019–21.

90 Clausen, J., Junge, W., Dau, H. and Haumann, M. (2005) Photosynthetic water oxidation at high O2 backpressure monitored by delayed chlorophyll fluorescence. *Biochemistry*, **44**, 12775–9.

91 Clausen, J. and Junge, W. (2004) Detection of an intermediate of photosynthetic water oxidation, *Nature (London, United Kingdom)*, **430**, 480–3.

92 Radmer, R. and Ollinger, O. (1985) Photosynthesis studies using mass spectrometric techniques. *Gas Enzymolology: Proceedings of a Symposium*, 91–104.

93 Aznar, C.P. and Britt, R.D. (2002) Simulations of the 1H electron spin echo-electron nuclear double resonance and 2H electron spin echo envelope modulation spectra of exchangeable hydrogen nuclei coupled to the S2-state photosystem II manganese cluster. *Philosophical Transactions of the Royal Society of London, Series B: Biological Sciences*, **357**, 1359–66.

94 Evans, M.C.W., Nugent, J.H.A., Ball, R.J., Muhiuddin, I. and Pace, R.J. (2004) Evidence for a direct manganese-oxygen ligand in water binding to the S2 state of the photosynthetic water oxidation complex. *Biochemistry*, **43**, 989–94.

95 Hillier, W. and Wydrzynski, T. (2004) Substrate water interactions within the Photosystem II oxygen evolving complex, *Physical Chemistry Chemical Physics*, **6**, 4882–9.

96 Hendry, G. and Wydrzynski, T. (2002) The two substrate-water molecules are already bound to the oxygen-evolving complex in the S2 state of Photosystem II. *Biochemistry*, **41**, 13328–34.

97 Richens, D.T. (2005) Ligand substitution reactions at inorganic centers. *Chemical Reviews (Washington, DC, United States)*, **105**, 1961–2002.

98 Helm, L. and Merbach, A.E. (2005) Inorganic and bioinorganic solvent exchange mechanisms. *Chemical Reviews (Washington, DC, United States)*, **105**, 1923–59.

99 Hoganson, C.W. and Babcock, G.T. (1997) A metalloradical mechanism for the generation of oxygen from water in photosynthesis. *Science*, **277**, 1953–6.

100 Schlodder, E. and Witt, H.T. (1999) Stoichiometry of proton release from the catalytic center in photosynthetic water oxidation Reexamination by a glass electrode study at pH 5.5–7.2. *Journal of Biological Chemistry*, **274**, 30387–92.

101 Dau, H. and Haumann, M. (2006) Response to photosynthetic oxygen production. *Science (Washington, DC, United States)*, **312**, 1471–2.

102 Haumann, M., Bogershausen, O., Cherepanov, D., Ahlbrink, R. and Junge, W. (1997) Photosynthetic oxygen evolution: H/D isotope effects and the coupling between electron and proton transfer during the redox reactions at the oxidizing side of Photosystem II. *Photosynthesis Research*, **51**, 193–208.

103 Ahlbrink, R., Haumann, M., Cherepanov, D., Boegershausen, O., Mulkidjanian, A. and Junge, W. (1998) Function of tyrosine z in water oxidation by Photosystem II: electrostatical promoter instead of hydrogen abstractor. *Biochemistry*, **37**, 1131–42.

104 Haumann, M., Mulkidjanian, A. and Junge, W. (1999) Tyrosine-Z in oxygen-evolving Photosystem II: a hydrogen-bonded tyrosinate. *Biochemistry*, **38**, 1258–67.

105 Limburg, J., Szalai, V.A. and Brudvig, G.W. (1999) A mechanistic and structural model for the formation and reactivity of a MnV:O species in photosynthetic water oxidation. *Journal of the Chemical Society, Dalton Transactions: Inorganic Chemistry*, 1353–62.

106 Britt, R.D., Campbell, K.A., Peloquin, J.M., Gilchrist, M.L., Aznar, C.P., Dicus, M.M., Robblee, J. and Messinger, J. (2004) Recent pulsed EPR studies of the Photosystem II oxygen-evolving complex: implications as to water oxidation mechanisms. *Biochimica Et Biophysica Acta-Bioenergetics*, **1655**, 158–71.

107 Haumann, M., Barra, M., Loja, P., Loescher, S., Krivanek, R., Grundmeier, A., andreasson, L.-E. and Dau, H. (2006) Bromide does not bind to the Mn4Ca complex in its S1 state in Cl–depleted and Br–reconstituted oxygen-evolving Photosystem II: evidence from X-ray absorption spectroscopy at the Br K-edge. *Biochemistry*, **45**, 13101–7.

108 Yu, H., Aznar, C.P., Xu, X.Z. and Britt, R.D. (2005) Evidence that azide occupies the chloride binding site near the manganese cluster in photosystem II. *Biochemistry*, **44**, 12022–9.

109 Clemens, K.L., Force, D.A. and Britt, R.D. (2002) Acetate binding at the photosystem II oxygen evolving complex: an S-2-state multiline signal ESEEM study. *Journal of the American Chemical Society*, **124**, 10921–33.

110 Messinger, J. and Renger, G. The Royal Society Chemistry (2008) Photosynthetic water splitting. *Primary Processes of Photosynthesis: Basic Principles and Apparatus, Comprehensive Series in Photochemical and Photobiological Sciences*, (in press).

6
Supercomplexes of Photosystems I and II with External Antenna Complexes in Cyanobacteria and Plants

Jan P. Dekker and Egbert J. Boekema

6.1
Introduction

Many chapters in this book describe impressive progress in determining the structure of almost all components of the electron transport chains in the photosynthetic apparatus of organisms performing oxygenic photosynthesis. Chlorophyll-containing photosystems which have structures known at intermediate and high resolutions include trimeric Photosystem I (PSI) of the cyanobacterium *Thermosynechococcus elongatus* [1] (see also Chapter 2), a supercomplex of PSI and light-harvesting complex I (LHC-I) from pea [2] (see also Chapter 3), dimeric Photosystem II (PSII) of *T. elongatus* [3–5] and *T. vulcanus* [6] (see also Chapter 4), and trimeric light-harvesting complex II (LHC-II) of spinach [7] and pea [8] (see also Chapter 10).

A good understanding of the process of photosynthesis does not, however, only require a deep knowledge of the structures of the individual photosystems and their light-harvesting complexes; it also requires knowledge of how the various complexes work together in the photosynthetic membrane. This is particularly relevant for the light-harvesting function and its regulation. During evolution, the various photosynthetic organisms have adopted different strategies to develop and survive in very diverse environmental conditions such as: high light; low light; and strongly fluctuating light; conditions in which the incoming light is spectrally altered due to shading by water or other photosynthetic organisms; and certain types of stress conditions. Photosynthetic organisms have adapted to these varying conditions by making use of different types of light-harvesting systems and by using fixed and flexible connections between reaction center and light-harvesting complexes in the form of supercomplexes. This, in turn, also determines the shape and organization of the photosynthetic membranes.

This chapter presents a review of the current knowledge on supercomplexes involving PSI and PSII from oxygenic photosynthetic organisms. Because the numbers and types of light-harvesting complexes vary considerably within the organisms performing oxygenic photosynthesis, this chapter will discuss those

Photosynthetic Protein Complexes: A Structural Approach. Edited by P. Fromme

ISBN: 978-3-527-31730-1

complexes in relation to type of light-harvesting complex. These can be classified in three main categories: complexes belonging to the so-called LHC superfamily; complexes belonging to the core antenna family; and, phycobiliproteins [9]. It is stressed that in many cases, the supercomplexes are quite fragile and are easily disrupted into at least two parts. A reason for this is that the supercomplex does not have to be a stable organizational unit under all physiological conditions; a rearrangement of parts of supercomplexes and a change of the overall macromolecular organization may occur in certain conditions to increase function and viability. This chapter is restricted to discussions of supercomplexes with membrane-bound antenna complexes, that is, complexes with light-harvesting proteins from the LHC and core antenna families.

6.2 Supercomplexes with Proteins from the LHC Superfamily

The LHC superfamily of proteins includes not only the most common proteins involved in light harvesting, but also proteins that clearly have other functions [9]. Most well-known is the major LHC-II complex of green plants, whose structure is now known at 2.5 Å [7, 8] (see also Chapter 10). This protein has three transmembrane α-helices, binds chlorophyll *a*, chlorophyll *b* and the carotenoids lutein, violaxanthin, and neoxanthin, occurs in trimers in vivo, is the major light-harvesting protein in green plants and several types of eukaryotic algae, and usually harvests light for PSII [10]. In special conditions, however, LHC-II may also harvest light for PSI or play a direct role in photoprotection, and thus can be a switch between light harvesting and photoprotection [11]. Most organisms contain a number of related, often monomeric, proteins, including proteins of this family that are bound to PSI, which are therefore denoted as light-harvesting complex I (LHC-I). In green plants, PSI binds four LHC-I proteins, and the structure of this supercomplex is known at 3.4 Å resolution [2]. Red algae, cryptophytes, dinoflagellates, and diatoms contain very similar proteins which bind chlorophyll *c* instead of chlorophyll *b*, or none of these at all, while they also bind different types of xanthophylls.

A special member of this family is PsbS. This protein has four transmembrane α-helices, occurs probably only in higher plants, and contains no pigments, or only a few. Therefore, this protein can not be regarded as a light-harvesting protein, but it is in a still unknown way involved in photoprotection, because mutants lacking this protein lack the rapidly reversible high-energy quenching (qE), a physiologically important mechanism by which excess excitation energy is harmlessly converted into heat [12, 13]. qE is a major contributor to the non-photochemical quenching (NPQ) of chlorophyll fluorescence. Other members of the LHC family are the early light-induced proteins (ELIPs) induced in many species in response to high light or other types of stress [14], and a number of partial homologues with only one transmembrane α-helix that have been found in cyanobacteria [15].

In this review, attention will be focused on the binding of light-harvesting proteins of the LHC family to PSI or PSII.

6.2.1
PSII-LHC-II Supercomplexes

In grana membranes of green plants and green algae, a variable number of proteins of the LHC superfamily have been shown to associate with dimeric PSII core complexes to form the so-called PSII-LHC-II supercomplexes. Such complexes were first observed in electron microscopic images of grana membranes from spinach solubilised by the mild detergent β-dodecylmaltoside [16]. These rectangular supercomplexes, in the following denoted as C_2S_2 supercomplexes (see below), contain all, or almost all, PSII core proteins. Additionally, at each side, they contain one trimeric LHC-II complex consisting of the Lhcb1 and Lhcb2 gene products, one CP29 monomer (the Lhcb4 gene product), and one CP26 monomer (the Lhcb5 gene product) [17]. EM micrographs of partially unfolding grana membranes clearly reveal the presence of rectangular supercomplexes in the membranes [18], indicating that the supercomplexes occur as such in the membranes and represent native organizations of PSII and LHC-II. A three-dimensional structure of this supercomplex was constructed by Nield and Barber based on a low-resolution structure from spinach obtained by cryoelectron microscopy and on the high-resolution crystal structures of the (cyanobacterial) PSII core, LHC-II and the extrinsic PsbP and PsbQ proteins [19]. Very similar structures of PSII–LHC-II complexes were observed in the green alga *Chlamydomonas reinhardtii* [20] and the liverwort *Marchantia polymorpha* [21]. The C_2S_2 PSII–LHC-II supercomplex was also observed to be the organizational unit of PSII and LHC-II in a barley mutant lacking PSI [22] and was, therefore, suggested to form the unit in which the smallest possible amount of LHC complexes is bound to PSII in grana.

It was noted that the CP29–LHC-II–CP26 peripheral antenna structure needs a PSII core dimer for its binding. CP29 binds to one core monomer, CP26 to the other and LHC-II to both (Figure 6.1). This may explain why small supercomplexes of a PSII core monomer and one or two peripheral antenna proteins were never found. It was shown by analysis of mutants that at least one small PSII core protein enhances the binding of the CP29–LHC-II–CP26 unit. This protein is PsbZ, a small PSII protein with two transmemebrane α-helices. Mutants without PsbZ appeared to have strongly reduced CP26 levels and very unstable PSII–LHC-II supercomplexes [24].

In grana membranes, there are usually considerably more than two trimeric LHC-II complexes per PSII dimer, especially in low-light growing conditions. In addition, there is a third monomeric LHC protein (CP24, the Lhbc6 gene product). Two additional binding sites for trimeric LHC-II and one additional binding site for a monomeric LHC-II protein were, indeed, found after a very short and mild detergent treatment with α-dodecylmaltoside of PSII grana membranes from spinach [25–27]. The binding sites of the trimers were designated S, M, and L (for strongly, moderately and loosely bound LHC-II,) (Figure 6.1). In *Arabidopsis*

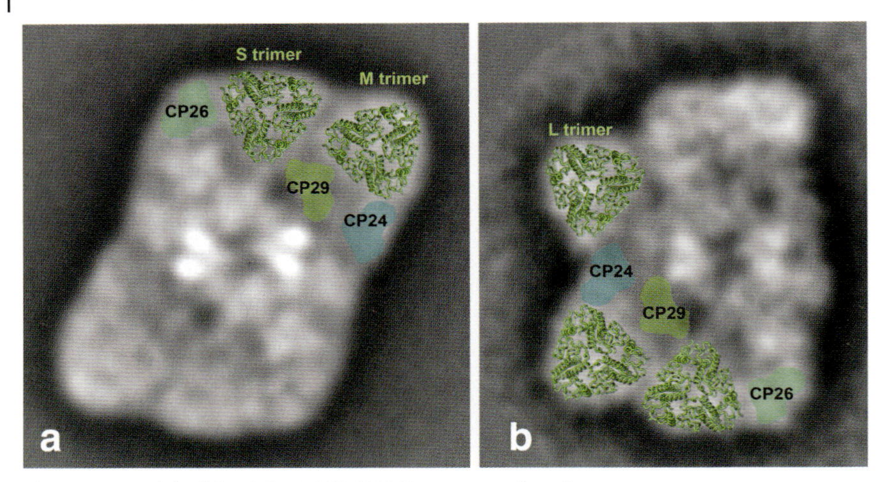

Figure 6.1 Model of the $C_2S_2M_2$ PSII-LHC-II supercomplex of *Arabidopsis thaliana* and of the C_2S_2ML supercomplex of spinach. The atomic model of trimeric LHC-II has been used for fitting of the S-, M- and L-trimers. The $C_2S_2M_2$ map iscurrently the best resolved supercomplex with a resolution of about 15 Å (adapted from [23]).

thaliana and *Marchantia polymorpha,* S and M trimers were found in the same position, but L trimers could not be detected [21, 28]. In *Chlamydomonas reinhardtii,* LHC-II binds only to the PSII core dimer at the S position, which is most likely due to the absence of a CP24 homologue in this organism. Based on an up-to-date supercomplex map (the $C_2S_2M_2$ supercomplex map from *Arabidopsis*), the M trimer is rotationally shifted by about 25° compared to the S trimer. In plants, the S trimer consists predominantly of the Lhcb1 and Lhcb2 gene products; the M trimer most likely consists of the Lhcb1 and Lhcb3 gene products [29]. The additional binding of, at maximum, four LHC-II trimers at the rectangular C_2S_2 supercomplexes does not mean that all available LHC-II trimers are directly bound to PSII. In normal growth conditions, about half of all LHC-II trimers are not directly connected to PSII and occur in the membranes in LHC-II-only regions. A small number of these unconnected trimers have been shown to form supercomplexes of seven LHC-II trimers [30].

In some conditions, the PSII-LHC-II supercomplexes organize themselves into semi-crystalline arrays in grana membranes. The formation of regular lattices of PSII in rows is made possible by the rectangular (C_2S_2) or diamond-like shape ($C_2S_2M_2$) of the supercomplexes. The way these supercomplexes organize themselves into larger complexes could be made visible by analysis of solubilised megacomplexes, dimers of PSII-LHC-II supercomplexes. To date, six types of megacomplexes have been observed [29, 31]. Details of the rows in grana membranes were visualized by image analysis of negatively stained specimens; it appeared that in most of the detected types of rows, the supercomplexes were connected to each other as in the megacomplexes [29]. In spinach, the majority of these rows were shown to consist of an asymmetric C_2S_2M repeating unit [32]. A few crystalline arrays consisting of a C_2S_2 repeating unit were also observed. In membranes obtained from a PSI-less barley mutant, the crystalline arrays consisted exclusively of a C_2S_2 repeating unit, consistent with the absence of CP24 and the M-trimer in this mutant [22]. In *Arabidopsis* membranes, only rows consisting of a $C_2S_2M_2$ repeating unit were observed [28]. Isolated supercomplexes of *Arabidopsis* contained a larger number of occupied M binding sites than those of spinach [28]; so, it seems that the binding of LHC-II at the M site is stronger in *Arabidopsis* than in spinach.

Analysis of supercomplexes isolated from *Arabidopsis* plants expressing antisense constructs to Lhcb gene products or from plants with a T-DNA knockout in an *lhcb* gene provided information of the specific roles in the supramolecular organization of PSII and LHC-II. Thus far, supercomplexes and crystalline arrays in grana membranes were analyzed from mutants lacking Lhcb4 (CP29), Lhcb5 (CP26), Lhcb6 (CP24), and Lhcb2 and Lhcb1 (trimeric LHC-II).

No intact supercomplexes could be isolated from plants lacking CP29 and crystalline arrays were not detected [33]. This suggests that CP29 occupies a unique position in the PSII macrostructure and that its presence is essential for the formation of PSII-LHC-II supercomplexes. In all other investigated mutants, both supercomplexes and crystalline arrays were observed. From plants lacking CP26, supercomplexes and crystalline arrays were observed with empty CP26 binding sites but with an otherwise normal appearance [33]. Because earlier investigations of PSII-LHC-II supercomplexes from spinach revealed many complexes with empty CP26 binding sites [26, 27], these results indicate that the presence of CP26 is not required for the formation of PSII-LHC-II macroorganization. Plants lacking CP24 only give rise to C_2S_2 supercomplexes and to crystalline arrays in membranes with the C_2S_2 supercomplex as a repeating unit [31]. This indicates that the presence of CP24 is required for the binding of LHC-II at the M site. In all mutants, the binding site appeared to be unique and could not be occupied by other monomeric proteins of the LHC family. The plants lacking CP29 or CP26 showed a rather normal photosynthetic performance [34], though in the field the fitness was lower [35]. Plants lacking CP24, however, showed a stronger phenotype [31], with decreased abilities of state transitions (see Section 6.2.3) and rapidly reversible non-photochemical quenching (see Section 6.2). In particular, these plants lacked the so-called psi-type circular dichroism signal [36], reflecting long-range interactions between chlorophylls. The decrease of the signal indicates a change in the overall macrostructure and stresses the importance of the presence of the M trimer for proper photosynthetic functioning.

In plants with antisense constructs against Lhcb2, not only was the synthesis of Lhcb2 almost completely abolished, but also that of the strongly related Lhcb1 protein [34]. In these plants, strongly increased levels of Lhcb5 and, to a lesser extent, Lhcb3 were observed, resulting in supercomplexes and crystalline arrays with trimers of Lhcb5 and Lhcb3 at the S and M sites [37, 38]. This replacement is unique because expression of antisense constructs to other LHC-types of proteins, including those of PSI (see Section 6.2.2), did not lead to increased synthesis of other proteins, stressing the importance of the organization of PSII and LHC-II in supercomplexes.

LHC-II does not only bind to PSII to form PSII-LHC-II supercomplexes, it is also primarily responsible for the formation of grana stacks in green plants. The stromal surface of LHC-II is not only very flat and overall negatively charged, but it also contains one domain of four positively charged residues. The interactions between the positively charged residues of one LHC-II with the negatively charged surface of an LHC-II in the opposing membrane, combined with the electrostatic screening of the negative surface with divalent cations, are probably the most important factors in determining the stacking of the thylakoid membranes [8]. Because of the stacking, the proteins with large extrinsic groups at the stromal surface, PSI and the ATP synthase complex, are excluded from the stacks, resulting in a spatial separation of PSI and PSII (Figure 6.2). This separation prevents spill-over of excitation energy from PSII to PSI and allows the formation of a very large light-harvesting antenna for PSII [39]. It is possible that interactions between the lumenal extensions of the extrinsic proteins of PSII involved in water oxidation in opposing membranes also influence the shape of the grana [29].

Supramolecular associations of PSII and other peripheral light-harvesting proteins of the LHC superfamily have not been studied until very recently when the first map of supercomplexes of a PSII core dimer and three or four monomeric Chl *a*/*c* proteins of the cryptophyte *Rhodomonas* CS24 became available [40]. The projection of the largest complex is shown in Figure 6.3 suggests that the four Chl *a*/*c* proteins are bound at similar positions as the CP24, CP29, one monomer of the S-LHC-II trimer and CP26 proteins in green plants.

6.2.2
PSI-LHC-I Supercomplexes

A landmark in the growing knowledge of the structure of the native PSI complex from green plants was the completion in 2003 of a 4.4 Å structure of this complex by Nelson and coworkers [41, 42], which was recently improved to 3.4 Å resolution [2] (see also Chapter 3). In fact, this is the only available crystal structure of a reaction center complex and its peripheral light-harvesting antenna. While the structures of the reaction center core part clearly resembled that of cyanobacterial PSI [1] and the four LHC-I proteins (the Lhca1-4 gene products) of LHC-II [7, 8], it was a surprise to find many chlorophyll molecules between the core and peripheral antenna proteins and between the four LHC-I proteins. These chlorophylls should facilitate the excitation energy transfer between the core and antenna parts of PSI; however, studies on the dynamics of energy transfer suggest a kinetic constraint within the peripheral antenna [43].

It is now clear that in vivo, green plant PSI only occurs as a monomer. Dimers and larger aggregates were sometimes found after detergent solubilization [44], but these aggregates always contained up-and-down oriented complexes, thought to have been caused by artificial aggregation after solubilization. In cyanobacteria,

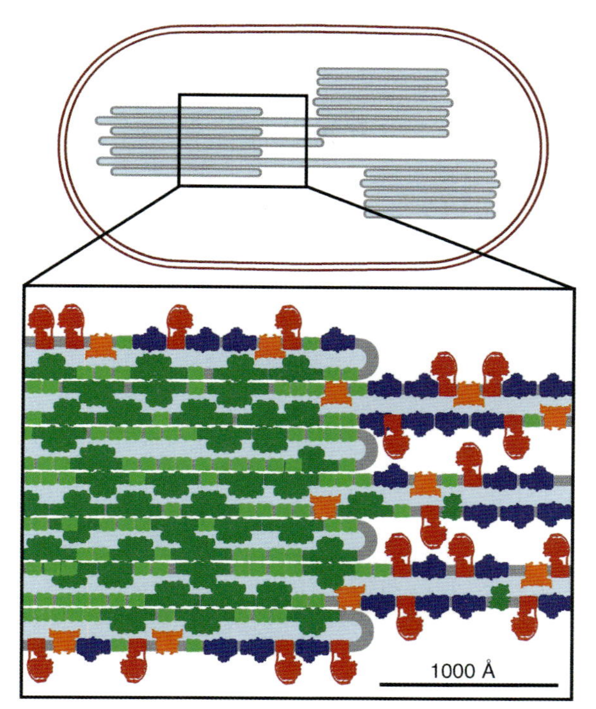

Figure 6.2 Model of the thylakoid membrane organization within the chloroplast, based on atomic models of the main protein components and thin sectioning electron microscopy. The model shows a chloroplast with three stacks of grana membranes, connected by stroma membranes. In 3D all membranes are interconnected and enclose the lumen (depicted in light blue). The closer view below shows a stack of grana domains on the left, with dimeric PSII complexes in dark green and LHC-II trimers in bright green. On the right, there are some interconnecting stroma domains with PSI in dark blue, the cytochrome b_6f complex in orange, ATP synthase in red and PSII monomers in dark green.

PSI frequently occurs as a trimer [45], which is mediated by the presence of the PSI-L protein [46]. It was shown by Ben-Shem *et al.* [47] that the presence of the plant-specific PsaH protein on top of PsaL prevents the trimerization in plants. It is possible that two other plant specific proteins (PsaO and PsaP) also contribute to the lack of trimerization. The impossibility of forming trimers in plants is related to the binding of LHC-II in conditions where additional light harvesting for PSI is needed (see Section 6.2.3).

PSI–LHC-I complexes have also been isolated from *Arabidopsis* plants expressing antisense constructs to Lhca gene products and from plants with a T-DNA knockout in an *lhca* gene [48]. It appeared that in most cases, the binding position of a lacking Lhca polypeptide is not occupied by another Lhca protein and that the levels of Lhca2 and Lhca3 were mutually considerably lower in the absence of the other protein, just as those of Lhca1 and Lhca4. Such findings confirm the idea

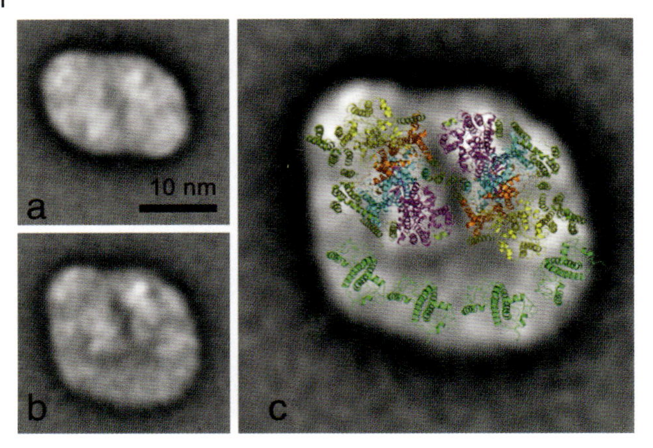

Figure 6.3 Electron micrisocopy projection of a complex of a PSII core dimer and four monomeric LHC complexes from the cryptophyte *Rhodomonas* CS24.

that these proteins form heterodimers [49] and that these heterodimers form the building blocks of the binding of the Lhca proteins to PSI [50].

Arabidopsis plants contain two other Lhca gene products, known as Lhca5 and Lhca6 [35]. The Lhca5 protein has now been analyzed in more detail. Normally, it is expressed in sub-stoichiometric amounts with respect to Lhca1-4, but under certain stress conditions its content increases compared to that of Lhca1-4 [51]. It was suggested that Lhca5 binds to the native PSI–LHC-I complex at the Lhca2/3 site and that in complexes without Lhca1-4 it can bind directly to the PSI core at the Lhca1/4 site [52]. Time-resolved fluorescence measurements of PSI particles from plants with antisense constructs to Lhca4, which lacked Lhca4, Lhca1, Lhca2, and Lhca3, were consistent with a particle consisting of the PSI core complex and one Lhca protein with the spectral characteristics of Lhca5 [53].

The PSI core complex of the green alga *Chlamydomonas reinhardtii* binds more Lhca proteins than that of green plants [54, 55]. However, only low-resolution structures are available and considerable differences are reported for the structure of the PSI–LHC-I complex of *Chlamydomonas*. Barber and coworkers [56] suggested that six LHC-I proteins bind to the PSI core, of which four are bound at the same positions as the four LHC-I proteins in green plants and two at the other side of the complex. It has also been suggested that 9 or 10 LHC-I proteins bind to the PSI core complex [29, 57], in line with biochemical data [58], of which 8 or 9 bind in 2 rows at the PsaF/J side of the complex; one is bound in the vicinity of the subunits PsaL, PsaH, PsaA, and PsaK (Figure 6.4).

In several other types of organisms, it is suspected that proteins of the LHC superfamily bind to PSI [57]; structural studies that confirm this have rarely been performed. A complex of PSI and chlorophyll *a*/*c* binding LHC proteins from the cryptophyte *Rhodomonas* CS24 has been studied and shows a PSI–LHC supercomplex with a very different shape from those of green plants and algae [40], see

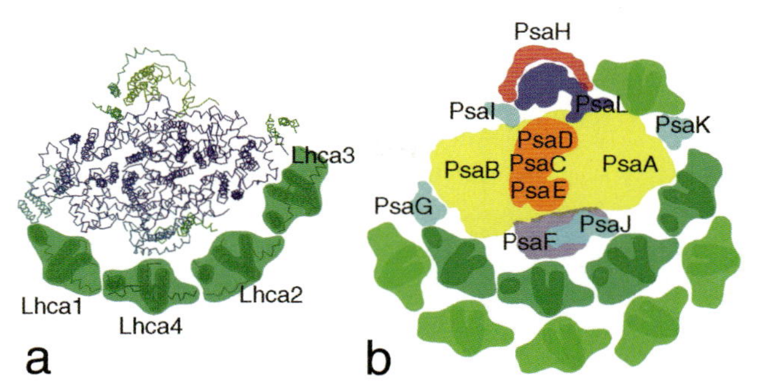

Figure 6.4 A model of the core and peripheral antenna subunit arrangement in the PSI–LHC-I complexes in green plants **(a)** and *Chlamydomonas reinhardtii* **(b)**.

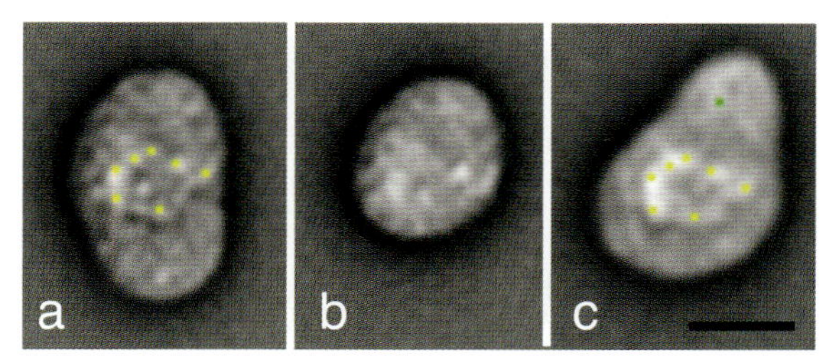

Figure 6.5 A comparison of the electron microscopy projections of the complex of PSI and LHC from the cryptophyte *Rhodomonas* CS24 **(a, b)** and the PSI–LHC-II supercomplex from *Arabidopsi thaliana* **(c)**. Similar positions in the core complex parts are marked yellow; the center of the LHC-II trimer in the PSI–LHC-II supercomplex is marked green. In the full cryptophyte complex of A, there are peripheral LHC copies bound on two sides of the core complex; in the smaller complex of B it appears that the lower peripheral antenna is lacking. The space bar is 10 nm.

Figure 6.5. The PSI–LHC supercomplex seems to bind LHC proteins at both sides of the complex and is only slightly smaller than the PSI–LHC-I–LHC-II complex from green plants (see Section 6.2.3). The total number of bound LHC complexes is, therefore, expected to be close to six. The precise number will depend on the type of association of the core and peripheral antenna proteins.

6.2.3 PSI-LHC-I-LHC-II Supercomplexes

Green plants and algae have been shown to be able to balance the energy distribution to PSI and PSII by a move of a mobile part of the peripheral antenna from PSII to PSI and back again. This process is known as a state transition [59–61] and is triggered by the redox state of the plastoquinone pool and the cytochrome b_6f complex. The signalling pathway includes the activation of a kinase that phosphorylates LHC-II. Phosphorylated LHC-II functionally uncouples from PSII, and reversibly connects to PSI. A phosphatase can dephosphorylate LHC-II, after which it will move back to PSII. The state in which a part of LHC-II is bound to PSI is called state 2; the state in which all LHC-II is bound to PSII is called state 1. In plants, the presence of the PsaH subunit is required for the binding of phosphorylated LHC-II to PSI [23]. It was later shown that LHC-II does not need to be phosphorylated to bind to PSI and that the LHC-II binding pocket on PSI consists of the PsaH, PsaI and PsaL subunits [62].

It has been very difficult to isolate and purify the PSI–LHC-I–LHC-II complex of state 2. Due to the reversible binding of LHC-II to the PSI–LHC-I complex, it is likely that such complexes are too labile to be purified by common chromatography procedures. Researchers have solubilized *Arabidopsis* membranes prepared in states 1 and 2 by the very mild detergent digitonin, analyzed the solubilized fractions immediately by single particle EM without further purification [63], and found a pear-shaped complex in state 2 only that consists of one PSI–LHC-I complex [2] and one LHC-II trimer [7] (Figures 6.5c and 6.6). The LHC-II trimer appears to be attached at a well-defined position close to the subunits PsaA, PsaH, PsaL, and PsaK. It is possible that the subunits PsaO and PsaP, which were not present in the 3.4 Å crystal structure [2], also contribute to the binding of LHC-II to PSI. It has been suggested [63] that there may be a second binding site for LHC-II to PSI at the symmetry-related site near PsaH, PsaB, and PsaG; there is no experimental evidence to support this suggestion.

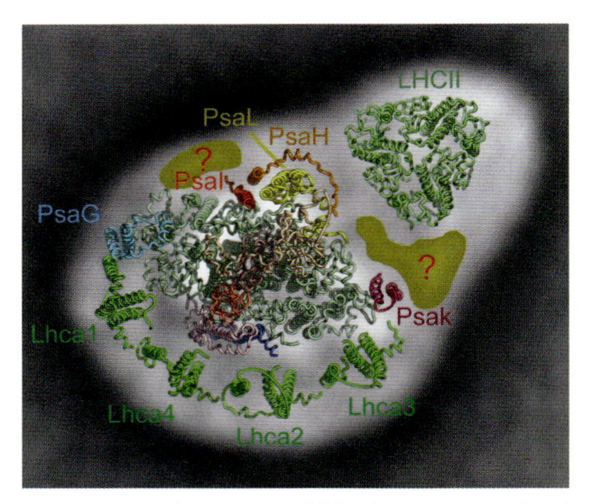

Figure 6.6 Pseudo-atomic model for the plant PSI–LHC-II supercomplex viewed from the stromal side of the membrane based on the electron microscopy data of the PSI–LHC-II supercomplex from *Arabidopsis thaliana* obtained by single particle averaging (improved from [23]). The high-resolution structures of PSI [2] and trimeric LHC-II [7] have been fitted. Positions of the four peripheral antenna subunits Lhca1-4 and the small subunits PsaG, -H, -I, -K and -L closest to the LHC-II trimer have been indicated. Densities in the interface of PSI and LHC-II and close to PsaH and –I (dark yellow) are not covered by the atomic structures of PSI and LHC-II and likely consist of additional subunits, which may include the PsaN and/ or PsaO subunits.

Unstacking the thylakoid membranes results in a response that resembles the state transitions. Such unstacking results in a redistribution of excitation energy from PSII to PSI, just as in the transition from state 1 to state 2. The relative increase of the flow of excitation energy to PSI can be caused by the disintegration of the PSII-LHC-II supercomplexes into separated LHC-II and monomeric PSII complexes [64], leading to a smaller probability that excitations reach PSII, and by the binding of one or more LHC-II trimers to PSI, which leads to an increased probability that excitations reach PSI. Time-resolved fluorescence measurements on stacked and unstacked thylakoid membranes from spinach have suggested a fast (~30 ps) energy transfer phase from LHC-II to PSI in unstacked membranes [65]. These kinetics are identical to those interpreted before as the main kinetics of energy transfer between LHC-II trimers in LHC-II aggregates [66], and suggest that PSI–LHC-I-LHC-II supercomplexes also exist in unstacked thylakoid membranes.

In the green alga *Chlamydomonas reinhardtii*, an analysis of detergent-solubilized PSI–LHC-II particles resulted in two types of particles [52]. These particles were identical at the side of the complex where most of the LHC-I proteins bind, but different at the opposite side near PsaH (Figure 6.4). A later analysis of the membranes used by Germano *et al.* [54], and of membranes poised in state 1 and state 2, revealed that the membranes used by Germano *et al.* were in a state just between state 1 and state 2, and that the smaller particle was dominant in state 1 and the larger in state 2 (A.E. Yakushevska *et al.*, unpublished observations). It was suggested that in state 2 the PSII protein CP29 (Lhcb4) is bound to the tip of the PSI–LHC-I complex [56, 67], and that of all PSII proteins CP29 has the largest number of phosphorylation sites [68]. Because the size of the additional mass of the larger particle is consistent with that of a monomeric LHC type of protein, the conclusion can be drawn that the additional mass in the larger particle is CP29. Note that a particle with one or more additional trimeric LHC-II complexes was not yet observed, and that the LHC-II binding site of plant PSI is already occupied by an LHC-I type of protein in PSI from *Chlamydomonas*. Perhaps, because of the crucial role of CP29 in the formation of PSII-LHC-II supercomplexes (see Section 6.2.1), it is not a coincidence that CP29, instead of, for instance, CP26, binds to the PSI–LHC-I complex from *Chlamydomonas*. A preferred binding of CP29 to PSI would prevent the formation of PSII-LHC-II supercomplexes, and thus the association of LHC-II to PSII. So it is possible that the large extent of the state transitions in *Chlamydomonas* is not primarily caused by a much larger antenna size of PSI, but rather by a much smaller antenna size of PSII because of the extensive phosphorylation and binding to PSI of a protein that plays a dominant role in the association of PSII and LHC-II in supercomplexes.

6.3 Supercomplexes with Proteins from the Core Complex Family

This family of proteins consists of the PSII core antenna proteins CP47 and CP43, the cyanobacterial iron-stress-induced protein A (IsiA), and prochlorophyte chlorophyll *a/b* proteins (Pcbs). The Pcbs originate from a rather diverse group of cyanobacteria that completely, or almost completely, lack phycobiliproteins, but do have the chlorophyllide *a* oxygenase enzyme required for the biosynthesis of chlorophyll *b* [9]. This group also includes the unusual cyanobacterium *Acaryochloris marina* that contains almost exclusively chlorophyll *d*. The proteins of the core antenna family all have six transmembrane α-helices. High-resolution structures are available for CP47 and CP43 as constituents of the PSII core complex [4, 5]. In this chapter, the focus is on the supercomplexes that the PSI and PSII core complexes have been shown to form with IsiA and Pcb proteins.

6.3.1 PSI-IsiA and PSI-Pcb Supercomplexes

A break-through in the understanding of the association of proteins from the core antenna family with PSI came in 2001 when two groups found a supercomplex of trimeric PSI encircled by a ring of 18 IsiA proteins in two cyanobacterial species grown in the absence of iron: *Synechocystis* PCC 6803 [69, 70] and *Synechococcus* PCC 7942 [71]. It was later found that a very similar structure of trimeric PSI and 18 Pcb proteins occurs in low-light adapted *Prochlorococcus* SS120 cells [72], in *Prochlorothrix hollandica* [73] and in *Acaryochloris marina* grown in the absence of iron [74]. A 20 Å structure was reported of the PSI-IsiA supercomplex from *S.* 6803 by cryoelectron microscopy [75, 76] and was used to produce a structural model based on the crystal structures of the related PSI and CP43 complexes from *T. elongatus*. Spectroscopic measurements have indicated that the chlorophylls and carotenoids of IsiA harvest light for PSI in a very efficient way [77–80]; while the structural model indicates several routes of energy transfer between IsiA and the central PSI complex.

The possibility of obtaining a 20 Å structure of the PSI-IsiA supercomplex implies that there is a preferred binding of the six IsiA complexes at each monomer. It was suggested that organization of the IsiA and PSI chlorophylls is optimized to get efficient energy transfer, such as by the three chlorophylls bound to the PsaJ protein and, possibly, by hitherto unrecognized linker chlorophylls [43, 76]. The role of PsaF and PsaJ in the binding of IsiA to the PSI core trimer was investigated in a PsaF/PsaJ double mutant of *Synechocystis* PCC 6803 [81]. The data show that PSI-IsiA supercomplexes can still be formed, but the IsiA ring consists in this mutant of 17 units, instead of 18 as in PSI-IsiA complexes from wild-type cells. A ring of 17 units around a trimer implies that every PSI monomer in the trimeric unit faces IsiA in a different way. These data indicate that PsaF and PsaJ are not needed for the binding of IsiA to PSI, and that the size of the ring around the photosystem is determined primarily by the circumference of the PSI trimer; the specific PSI-IsiA interactions are secondary. The circumference of the wild-type PSI trimer is larger than that of the mutant (PsaF and PsaJ subunits have four transmembrane α-helices).

In normal growth conditions of cyanobacteria with a sufficient amount of iron in the medium, the *isiA* gene is not expressed and PSI occurs predominantly as core trimers and PSII as core dimers. The PSI to PSII ratio is generally very high in these conditions (at least five), and phycobilisomes attached to the stromal surface of the membrane harvest most of the light for PSII. However, in many aquatic ecosystems the amount of iron is extremely low due to the poor solubility of Fe^{3+} in water. This poses a significant problem for the biosynthesis of the photosynthetic apparatus and respiration complexes, because the PSI, PSII, the cytochrome b_6f, and the NDH-1 complexes all need iron for essential electron transfer processes. In addition, a number of enzymes required for the biosynthesis of phycobilisomes cannot function without iron. This problem is most prominent for PSI because PSI needs the largest number of iron ions. Cyanobacteria contain a very efficient response mechanism for a lack of iron during growth, of which the induction of the *isiA* gene forms an essential part [82]. In early stages of iron deficiency, the cyanobacterium responds to this condition by lowering the PSI to PSII ratio, and the induction of IsiA as a light-harvesting complex bound to PSI is therefore a good means to restore the balance of the flow of excitation energy to the two photosystems.

Iron depletion is not the only inducer of the *isiA* gene; high light and other types of oxidative stress do as well [83, 84]. It is possible that this role is related to a second function that was proposed for IsiA – a role as a dissipater of excitation energy, a photoprotective role [85, 86]. A dual role as light-harvester and photoprotector is also considered for green plant LHC-II [11]. Because there is a whole range of gradations in the extent of iron depletion or of oxidative stress in general, the amount of available IsiA complexes in the cells can vary substantially. It is to be expected, therefore, that there are more associations of PSI and IsiA than the well-resolved $(PSI)_3(IsiA)_{18}$ supercomplex. Indeed, an analysis of *Synechocystis* PCC 6803 cells grown for short and long times in the absence of iron has indicated that many types of PSI-IsiA supercomplexes can exist [87]. A limited iron deficiency resulted, for instance, in PSI core trimers with only a few attached IsiA complexes, whereas pronounced iron deficiency was shown to result in PSI monomers surrounded by rings of 12, 13, or 14 units, or by double rings of 12 and 19, 13 and 20, and 14 and 21 units. Also PSI monomers with partial single or double rings were frequently found [87]. In the absence of the PsaL subunit, PSI trimers can no longer be formed. If mutant cells without this subunit are grown without iron, most of the IsiA accumulates in incomplete rings at the PsaFJ side of the complex [88, 89]. However, also in these cells, complete rings and double rings can be formed, suggesting that the PsaL protein is not necessary for the binding of IsiA. The formation of PSI monomers upon growth in iron-deficient conditions is probably related to the strongly decreased number of PSI complexes in the cells and the strongly increased amount of IsiA, thus preventing the formation of PSI core trimers [86]. In carefully grown cyanobacteria under conditions where iron is very limited but not totally absent, it is possible that PSI trimers with a double ring of 18 and 25 IsiA units can be formed (I.M. Folea, D. Chauhan, R. Kouřil, E.J. Boekema and P. Fromme, unpublished observations, see Figure 6.7).

It was found that growth of *Synechocystis* PCC 6803 in a medium without iron not only results in the formation of PSI-IsiA supercomplexes of various sizes, but

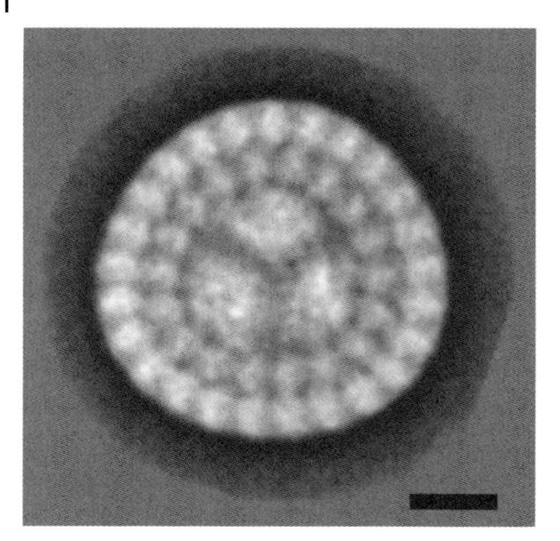

Figure 6.7 A projection map of the largest supercomplex of trimeric PSI and IsiA from *Thermosynechococcus vulcanus* found up till now by single particle electron microscopy. The space bar is 10 nm.

also of many supercomplexes without associated photosystems [87, 90]. In fact, after a very long time of growth in the absence of iron, IsiA appeared to be the only spectroscopically measurable pigment-protein complex. A spectroscopic study on isolated IsiA aggregates revealed a characteristic fluorescence emission band peaking at 687 nm at 5 K and very short fluorescence lifetimes at room temperature [91], suggesting a very pronounced non-photochemical quenching. The spectra and temperature dependence of this fluorescence were identical in the isolated aggregates and in the long-term iron-depleted cells, suggesting that the quenching is not an artefact induced by the isolation procedure. Recently, studies were done with *Synechocystis* PCC 6803 cells isolated in various stages of iron deficiency that showed that the typical 687 nm fluorescence of IsiA-only complexes occurs already at the earliest stages of iron deficiency [92]. Note that in isolated $(PSI)_3(IsiA)_{18}$ supercomplexes, efficient energy transfer occurs from IsiA to PSI; therefore these complexes do not give rise to a 687 nm fluorescence emission band [77]. Even at early stages of iron deficiency the temperature dependence of the IsiA emission was similar to that of isolated IsiA aggregates obtained from long-term iron-depleted cells, indicating that the photoprotective properties of IsiA already are significant at early stages of iron deficiency, in line with the predicted role of IsiA as photoprotector [85].

6.3.2
PSII-Pcb Supercomplexes

Based on measurements on the functional antenna size of PSII during iron depletion of *Synechocystis* PCC 6803, it was concluded that IsiA should also have a light-harvesting function for PSII [93]. Despite intensive search, however, PSII-

IsiA supercomplexes were never found, not even in low numbers. It is possible that the connection between PSII and IsiA (probably in the form of aggregates) is too weak to survive detergent solubilization.

A number of PSII-Pcb supercomplexes, however, have been isolated and characterized by EM (reviewed in [94]). A PSII-Pcb complex consisting of a PSII dimer and four Pcb proteins at each side of the dimer was found in several *Prochlorococcus* species [95], whereas a complex with five Pcb proteins at each side of the PSII dimer was found in *Prochloron didemni* [96]. The chlorophyll *d* containing organism *Acaryochloris marina* was found to have a megacomplex (a dimeric supercomplex) of two PSII dimers with a total of 16 Pcb subunits [97].

Acknowledgments

The authors acknowledge the support of the European Union by means of grants for the PSIC.O. Research Training Network (HPRN-CT-2002-00248) and the INTRO2 Marie Curie Research Training Network (MRT-CT-2003-505069).

References

1 Jordan, P., Fromme, P., Witt, H.T., Kuklas, O., Saenger, W. and Krauss, N. (2001) *Nature*, **411**, 909–17.

2 Amunts, A., Drory, O. and Nelson, N. (2007) *Nature*, **447**, 58–63.

3 Zouni, A., Witt, H.T., Kern, J., Fromme, P., Krauss, N., Saenger, W. and Orth, P. (2001) *Nature*, **409**, 739–43.

4 Ferreira, K.N., Iverson, T.M., Maghlaoui, K., Barber, J. and Iwata, S. (2004) *Science*, **303**, 1831–8.

5 Loll, B., Kern, J., Saenger, W., Zouni, A. and Biesadka, J. (2005) *Nature*, **438**, 1040–4.

6 Kamiya, N. and Shen, J.R. (2003) *Proceedings of the National Academy of Sciences of the United States of America*, **100**, 98–103.

7 Liu, Z., Yan, H., Wang, K., Kuang, T., Zhang, J., Gui, L. An, X. and Chang, W. (2004) *Nature*, **428**, 287–92.

8 Standfuss, J., Terwisscha van Scheltinga, A.C., Lamborghini, M. and Kühlbrandt, W. (2005) *The EMBO Journal*, **24**, 919–28.

9 Green, B.R., Anderson, J.M. and Parson, W.W. (2003) *Light-Harvesting Antennas in Photosynthesis* (eds B.R., Green and W.W., Parson), Kluwer Academic Publishers, Netherlands, pp. 1–28.

10 van Amerongen, H. and Dekker J.P. (2003) *Light-Harvesting Antennas in Photosynthesis* (eds B.R., Green and W.W., Parson), Kluwer Academic Publishers, Netherlands, pp. 219–51.

11 Pascal, A.A., Liu, Z., Broess, K., van Oort, B., van Amerongen, H., Wang, C., Horton, P., Robert, B., Chang, W. and Ruban, A. (2005) *Nature*, **436**, 134–7.

12 Li, X.-P., Björkman, O., Shih, C., Grossman, A.R., Rosenquist, M., Jansson, S. and Niyogi, K.K. (2000) *Nature*, **403**, 391–5.

13 Horton, P., Ruban, A.V. and Walters, R.G. (1996) *Annual Review of Plant Physiology and Plant Molecular Biology*, **47**, 655–84.

14 Adamska, I. (1997) *Physiologia Plantarum*, **100**, 794–805.

15 Yao, D., Kieselbach, T., Komenda, J., Promnares, K., Hernandez Prieto, M.A., Tichy, M., Vermaas, W. and Funk, C. (2007) *Journal of Biological Chemistry*, **282**, 267–76.

16 Boekema, E.J., Hankamer, B., Bald, D., Kruip, J., Nield, J., Boonstra, A.F., Barber, J. and Rögner, M. (1995) *Proceedings of the National Academy of Sciences of the United States of America*, **92**, 175–9.

17 Hankamer, B., Nield, J., Zheleva, D., Boekema, E.J., Jansson, S. and Barber, J.

(1997) *European Journal of Biochemistry*, **243**, 422–9.
18 van Roon, H., van Breemen, J.F.L., de Weerd, F.L., Dekker, J.P. and Boekema, E.J. (2000) *Photosynthesis Research*, **64**, 155–66.
19 Nield, J. and Barber, J. (2006) *Biochimica et Biophysica Acta*, **1757**, 353–61.
20 Nield, J., Kruse, O., Ruprecht, J., da Fonseca, P., Büchel, C. and Barber, J. (2000) *Journal of Biological Chemistry*, **275**, 27940–6.
21 Harrer, R. (2003) *Photosynthesis Research*, **75**, 249–58.
22 Morosinotto, T., Bassi, R., Frigerio, S., Finazzi, G., Morris, E. and Barber, J. (2006) *The FEBS Journal*, **273**, 4616–30.
23 Lunde, C., Jensen, P.E., Haldrup, A., Knoetzel, J. and Scheller, H.V. (2000) *Nature*, **408**, 613–15.
24 Swiatek, M., Kuras, R., Sokolenko, A., Higgs, D., Olive, J., Cinque, G., Müller, B., Eichacker, L.A., Stern, D.B., Bassi, R., Herrmann, R.G. and Wollman, F.-A. (2001) *Plant Cell*, **13**, 1347–67.
25 Boekema, E.J., van Roon, H. and Dekker, J.P. (1998) *FEBS Letters*, **424**, 95–9.
26 Boekema, E.J., van Roon, H., Calkoen, F., Bassi, R. and Dekker, J.P. (1999) *Biochemistry*, **38**, 2233–9.
27 Boekema, E.J., van Roon, H., van Breemen, J.F.L. and Dekker, J.P. (1999) *European Journal of Biochemistry*, **266**, 444–52.
28 Yakushevska, A.E., Jensen, P.E., Keegstra, W., van Roon, H., Scheller, H. V., Boekema, E.J. and Dekker, J.P. (2001) *European Journal of Biochemistry*, **268**, 6020–1.
29 Dekker, J.P. and Boekema, E.J. (2005) *Biochimica et Biophysica Acta*, **1706**, 12–19.
30 Dekker, J.P., van Roon, H. and Boekema, E.J. (1999) *FEBS Letters*, **449**, 211–14.
31 Kovács, L., Damkjær, J., Kereïche, S., Ilioaia, C., Ruban, A.V., Boekema, E.J., Jansson, S. and Horton, P. (2006) *Plant Cell*, **18**, 3106–20.
32 Boekema, E.J., van Breemen, J.F.L., van Roon, H. and Dekker, J.P. (2000) *Journal of Molecular Biology*, **301**, 1123–33.
33 Yakushevska, A.E., Keegstra, W., Boekema, E.J., Dekker, J.P., Andersson, J., Jansson, S., Ruban, A.V. and Horton, P. (2003) *Biochemistry*, **42**, 806–13.
34 Andersson, J., Walters, R.G., Horton, P. and Jansson, S. (2001) *Plant Cell*, **13**, 1193–204.
35 Ganeteg, U., Külheim, C., Andersson, J. and Jansson, S. (2004) *Plant Physiology*, **134**, 502–9.
36 Garab, G., Wells, S., Finzi, L. and Bustamante, C. (1988) *Biochemistry*, **27**, 5839–43.
37 Ruban, A.V., Wentworth, M., Yakushevska, A.E., Andersson, J., Lee, P.J., Keegstra, W., Dekker, J.P., Boekema, E.J., Jansson, S. and Horton, P. (2003) *Nature*, **421**, 648–52.
38 Ruban, A.V., Solovieva, S., Lee, P.J., Ilioaia, C., Wentworth, M., Ganeteg, U., Klimmek, F., Chow, W.S., Anderson, J.M., Jansson, S. and Horton, P. (2006) *Journal of Biological Chemistry*, **281**, 14981–90.
39 Mullineaux, C.W. (2005) *Trends in Plant Science*, **10**, 521–5.
40 Kereïche, S., Kouril, R., Oostergetel, G.T., Boekema, E.J., Doust, A.B., van der Weij – de Wit, C.D. and Dekker, J.P. (2008) Association of chlorophyll *a/c*2 complexes to photosystem I and photosystem II in the cryptophyte *Rhodomonas* CS24. *Biochimica et Biophysica Acta*, in press, doi:10.1016/i.66abio.2008.04.045.
41 Ben-Shem, A., Frolow, F. and Nelson, N. (2003) *Nature*, **426**, 630–5.
42 Nelson, N. and Yocum, C.F. (2006) *Annual Review of Plant Biology*, **57**, 521–65.
43 Melkozernov, A.N., Barber, J. and Blankenship, R.E. (2006) *Biochemistry*, **45**, 331–45.
44 Boekema, E.J., Jensen, P.E., van Breemen, J.F.L., van Roon, H., Scheller, H.V. and Dekker, J.P. (2001) *Biochemistry*, **40**, 1029–36.
45 Boekema, E.J., Dekker, J.P., van Heel, M.G., Rögner, M., Saenger, W., Witt, I. and Witt, H.T. (1987) *FEBS Letters*, **217**, 283–6.
46 Chitnis, V.P. and Chitnis, P.R. (1993) *FEBS Letters*, **336**, 330–4.
47 Ben-Shem, A., Frolow, F. and Nelson, N. (2004) *FEBS Letters*, **564**, 274–80.
48 Klimmek, F., Ganeteg, U., Ihalainen, J.A., van Roon, H., Jensen, P.E., Scheller, H.V., Dekker, J.P. and Jansson, S. (2005) *Biochemistry*, **44**, 3065–73.
49 Croce, R., Morosinotto, T., Castelletti, S., Breton, J. and Bassi, R. (2002) *Biochimica et Biophysica Acta*, **1556**, 29–40.

50 Morosinotto, T., Ballotari, M., Klimmek, F., Jansson, S. and Bassi, R. (2005) *Journal of Biological Chemistry*, **280**, 31050–8.

51 Klimmek, F., Sjödin, A., Noutsos, C., Leister, D. and Jansson, S. (2006) *Plant Physiology*, **140**, 793–804.

52 Lucinski, R., Schmid, V.H.R., Jansson, S. and Klimmek, F. (2006) *FEBS Letters*, **580**, 6485–8.

53 Ihalainen, J.A., Klimmek, F., Ganeteg, U., van Stokkum, I.H.M., van Grondelle, R., Jansson, S. and Dekker, J.P. (2005) *FEBS Letters*, **579**, 4787–91.

54 Germano, M., Yakushevska, A.E., Keegstra, W., van Gorkom, H.J., Dekker, J.P. and Boekema, E.J. (2002) *FEBS Letters*, **525**, 121–5.

55 Kargul, J., Nield, J. and Barber, J. (2003) *Journal of Biological Chemistry*, **278**, 16135–41.

56 Kargul, J., Turkina, M.V., Nield, J., Benson, S., Vener, A.V. and Barber, J. (2005) *The FEBS Journal*, **272**, 4797–806.

57 Boekema, E.J., Kouril, R., Dekker, J.P. and Jensen, P.E. (2006) *Photosystem I: The Light-Driven, Plastocyanin:Ferredoxin Oxidoreductase* (ed. J.H., Golbeck), Springer, The Netherlands, pp. 41–6.

58 Takahashi, Y., Yasui, T.-A., Stauber, E.J. and Hippler, M. (2004) *Biochemistry*, **43**, 7816–23.

59 Allen, J.F. (1992) *Biochimica et Biophysica Acta*, **1098**, 275–335.

60 Haldrup, A., Jensen, P.E., Lunde, C. and Scheller, H.V. (2001) *Trends in Plant Science*, **6**, 301–5.

61 Wollman, F.A. (2001) *The EMBO Journal*, **20**, 3623–30.

62 Zhang, S. and Scheller, H.V. (2004) *Journal of Biological Chemistry*, **279**, 3180–7.

63 Kouril, R., Zygadlo, A., Arteni, A.A., de Wit, C.D., Dekker, J.P., Jensen, P.E., Scheller, H.V. and Boekema, E.J. (2005) *Biochemistry*, **44**, 10935–40.

64 Dekker, J.P., Germano, M., van Roon, H. and Boekema, E.J. (2002) *Photosynthesis Research*, **72**, 203–10.

65 van der Weij – de Wit, C.D., Ihalainen, J.A., van Grondelle, R. and Dekker, J.P. (2007) *Photosynthesis Research*, **93**, 173–82.

66 Barzda, V., Gulbinas, V., Kananavicius, R., van Amerongen, H., van Grondelle, R. and Valkunas, L. (2001) *Biophysical Journal*, **80**, 2409–21.

67 Takahashi, H., Iwai, M., Takahashi, Y. and Minegawa, J. (2006) *Proceedings of the National Academy of Sciences of the United States of America*, **103**, 477–82.

68 Turkina, M.V., Kargul, J., Blanco-Rivero, A., Villarejo, A., Barber, J. and Vener, A.V. (2006) *Molecular and Cellular Proteomics*, **5**, 1412–25.

69 Bibby, T.S., Nield, J. and Barber, J. (2001) *Nature*, **412**, 743–5.

70 Bibby, T.S., Nield, J. and Barber, J. (2001) *Journal of Biological Chemistry*, **276**, 43246–52.

71 Boekema, E.J., Hifney, A., Yakushevska, A.E., Piotrowski, M., Keegstra, W., Berry, S., Michel, K.P., Pistorius, E.K. and Kruip, J. (2001) *Nature*, **412**, 745–8.

72 Bibby, T.S., Nield, J., Partensky, F. and Barber, J. (2001) *Nature*, **413**, 590.

73 Bumba, L., Prasil, O. and Vacha, F. (2005) *Biochimica et Biophysica Acta*, **1708**, 1–5.

74 Chen, M., Bibby, T.S., Nield, J., Larkum, A.W.D. and Barber, J. (2005) *Biochimica et Biophysica Acta*, **1708**, 367–74.

75 Nield, J., Morris, E.P., Bibby, T.S. and Barber, J. (2003) *Biochemistry*, **42**, 3180–8.

76 Barber, J., Nield, J., Duncan, J. and Bibby, T.S. (2006) *Photosystem I: the Light-Driven, Plastocyanin:Ferredoxin Oxidoreductase* (ed. J.H., Golbeck), Springer, The Netherlands, pp. 99–117.

77 Andrizhiyevskaya, E.G., Schwabe, T.M.E., Germano, M., D'Haene, S., Kruip, J., van Grondelle, R. and Dekker, J.P. (2002) *Biochimica et Biophysica Acta*, **1556**, 265–72.

78 Melkozernov, A.N., Bibby, T.S., Lin, S., Barber, J. and Blankenship, R.E. (2003) *Biochemistry*, **42**, 3893–903.

79 Andrizhiyevskaya, E.G., Frolov, D., van Grondelle, R. and Dekker, J.P. (2004) *Biochimica et Biophysica Acta*, **1656**, 104–13.

80 Riley, K.R., Zazubovich, V. and Jankowiak, R. (2006) *The Journal of Physical Chemistry B*, **110**, 22436–46.

81 Kou il, R., Yeremenko, N., D'Haene, S., Yakushevska, A.E., Keegstra, W., Matthijs, H.C.P., Dekker, J.P. and Boekema, E.J.

(2003) *Biochimica et Biophysica Acta*, **1607**, 1–4.

82 Burnap, R.L., Troyan, T. and Sherman, L.A. (1993) *Plant Physiology*, **103**, 893–902.

83 Jeanjean, R., Zuther, E., Yeremenko, N., Havaux, M., Matthijs, H.C.P. and Hagemann, M. (2003) *FEBS Letters*, **549**, 52–6.

84 Havaux, M., Guedeney, G., Hagemann, M., Yeremenko, N., Matthijs, H.C.P. and Jeanjean, R. (2005) *FEBS Letters*, **579**, 2289–93.

85 Sandström, S., Park, Y.I., Öquist, G. and Gustafsson, P. (2001) *Photochemistry and Photobiology*, **74**, 431–7.

86 Ivanov, A.G., Krol, M., Sveshnikov, D., Selstam, E., Sandström, S., Koochek, M., Park, Y.I., Vasil'ev, S., Bruce, D., Öquist, G. and Huner, N.P.A. (2006) *Plant Physiology*, **141**, 1436–45.

87 Yeremenko, N., Kouřil, R., Ihalainen, J.A., D'Haene, S., van Oosterwijk, N., Andrizhiyevskaya, E.G., Keegstra, W., Dekker, H.L., Hagemann, M., Boekema, E.J., Matthijs, H.C.P. and Dekker, J.P. (2004) *Biochemistry*, **43**, 10308–13.

88 Aspinwall, C.L., Duncan, J., Bibby, T., Mullineaux, C.W. and Barber, J. (2004) *FEBS Letters*, **574**, 126–30.

89 Kouřil, R., Yeremenko, N., D'Haene, S., Oostergetel, G.T., Matthijs, H.C.P., Dekker, J.P. and Boekema, E.J. (2005) *Biochimica et Biophysica Acta*, **1706**, 262–6.

90 Kouřil, R., Arteni, A.A., Lax, J., Yeremenko, N., D'Haene, S., Rögner, M., Matthijs, H.C.P., Dekker, J.P. and Boekema, E.J. (2005) *FEBS Letters*, **579**, 3253–7.

91 Ihalainen, J.A., D'Haene, S., Yeremenko, N., van Roon, H., Arteni, A.A., Boekema, E.J., van Grondelle, R., Matthijs, H.C.P. and Dekker, J.P. (2005) *Biochemistry*, **44**, 10846–53.

92 Van der Weij – de Wit, C.D., Ihalainen, J.A., van de Vijver, E., D'Haene, S., Matthijs, H.C.P., van Grondelle, R. and Dekker, J.P. (2007) Fluorescence quenching of IsiA in the early stage of iron-deficiency and at cryogenic temperatures. *Biochimica et Biophysica Acta*, **1767**, 1393–400.

93 Cadoret, J.-C., Demoulière, R., Lavaud, J., van Gorkom, H.J., Houmard, J. and Etienne, A.-L. (2004) *Biochimica et Biophysica Acta*, **1659**, 100–4.

94 Chen, M. and Bibby, T.S. (2005) *Photosynthesis Research*, **86**, 165–73.

95 Bibby, T.S., Mary, I., Nield, J., Partensky, F. and Barber, J. (2003) *Nature*, **424**, 1051–4.

96 Bibby, T.S., Nield, J., Chen, M., Larkum, A.W.D. and Barber, J. (2003) *Proceedings of the National Academy of Sciences of the United States of America*, **100**, 9050–4.

97 Chen, M., Bibby, T.S., Nield, J., Larkum, A.W.D. and Barber, J. (2005) *FEBS Letters*, **579**, 1306–10.

98 Macpherson, A.N. and Hiller, R.G. (2003) *Light-Harvesting Antennas in Photosynthesis*, (eds B.R., Green and W.W., Parson), Kluwer Academic Publishers, The Netherlands, pp. 323–52.

7
Cytochrome b_6f Complex, Colon Structure, Spectroscopy, and Function of Heme c_n: *n*-Side Electron and Proton Transfer Reactions

William A. Cramer, Danas Baniulis, Eiki Yamashita, Huamin Zhang, Anna I. Zatsman, and Mike P. Hendrich

7.1
Structure of the Cytochrome b_6f Complex; Comparison with the Cytochrome bc_1 Complex

The 217 kDa hetero-oligomeric cytochrome b_6f complex [1] provides the electronic connection between Photosystem II [2, 3] and Photosystem I [4, 5] (see also Chapters 2, 3 and 4). Electron transfer through the b_6f complex is coupled to proton translocation to the electrochemically positive side of the complex (the luminal side) and, with proton deposition from the water splitting reaction of PSII and the decrease of proton concentration on the stromal side by reduction of $NADP^+$ to NADPH, generation of a transmembrane proton electrochemical potential gradient, $\Delta\tilde{\mu}_{H}^{+}$, marked by the *p* (lumen) and *n* (stromal) sides of the complex in Figure 7.1a.

7.1.1
Principle Features of the Structure

At a resolution of 3.0–3.1 Å, the structure of the b_6f complex is almost identical in the 2.5–3.0 × 10^9 year old thermophilic cyanobacterium, *M. laminosus* and the green alga, *C. reinhardtii* that appeared ~10^9 years later [6]. The hetero-oligomeric b_6f complex consists of eight tightly bound subunits (Figure 7.1b), four large (MW = 17.5–32.3 kDa, cyt b, cyt f, subunit IV (suIV)), and four small (MW = 3.30–4.06 kDa, PetN, G, M. L) subunits, present in a 1 : 1 stoichiometry [7–9]. The dimeric complex contains 26 transmembrane helices, 13 in each monomer. The four large subunits, cyt *f*, cyt b_6, the Rieske [2Fe-2S] protein, and subunit IV (red, blue, yellow, and purple in Figure 7.1b) bind the redox prosthetic groups of the complex, four hemes (1 bound by cyt *f*, 3 by cyt b_6), and one [2Fe-2S] cluster. Of the three hemes bound by the cytochrome *b* polypeptide, two are classical bis-histidine coordinated, non-covalently bound *b*-type hemes; the third is a novel heme, c_n, covalently bound (*c*-type) to Cys35 on the stromal *n*-side of the transmembrane helix A (TMH-A) of

Photosynthetic Protein Complexes: A Structural Approach. Edited by P. Fromme

ISBN: 978-3-527-31730-1

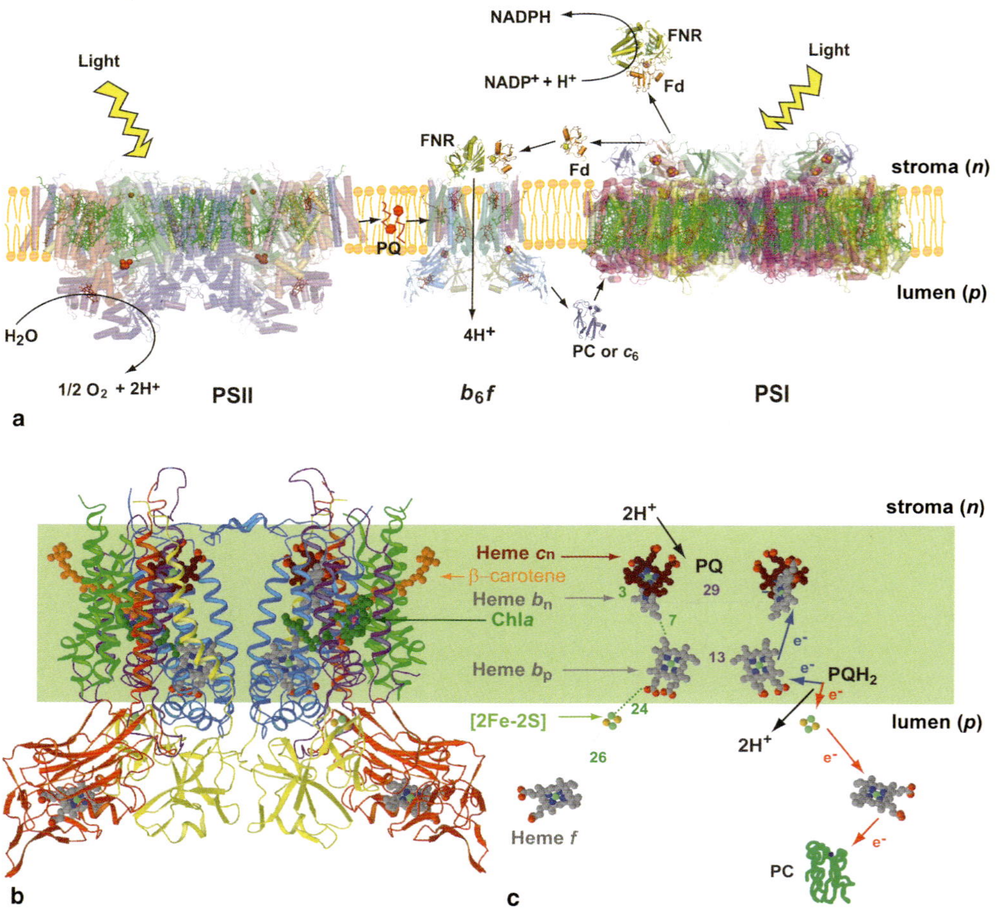

the cyt *b* polypeptide (Figure 7.2a). The binding sites of PQ/PQH_2 have been inferred from the binding sites of quinone analog inhibitors [8–10]. In addition, one chlorophyll *a* (green, Figure 7.1b) and one β-carotene (orange, Figure 7.1b) are present in each monomer [11–13]. The porphyrin ring of the Chl is inserted between the TMH-F and TMH-G, and the Chl phytyl chain is inserted through the portal in the "roof" of the inter-monomer quinone exchange cavity. The β-carotene is inserted through the petM and petG small subunits of the peripheral picket fence-like structure (green, Figure 7.1b) and extends to the E helix of suIV, a distance of closest approach of 14 Å to the Chl *a*. The picket fence-like structure of the four small subunits is a unique aspect of the structure, not only with respect to the bc_1 complex, but relative to all integral membrane proteins. It has been proposed that the picket fence and β-carotene toothpick may have an organizing function in assembly of the complex into the membrane [14].

Figure 7.1 **(a)** Electron transfer and proton pumping chain of oxygenic photosynthesis showing the pathway of electron transfer across the membrane from the luminal to the stromal side of Photosystem II reaction center (3.0 Å resolution; [3]) from water as the electron donor. Absorption of four photons and concomitant transfer of four electrons to the stromal *n*-side results in reduction of two plastoquinone (PQ) molecules, formation of one O_2 molecule from two waters, that is, $2H_2O \rightarrow O_2 + 4e^- + 4H^+$, and deposition of the $4H^+$ on the lumen side of the membrane. The reduced plastoquinone (PQH_2), which is hydrophobic, resides in a quinone "pool" (approximately 10–15 molecules per reaction center) in the center of the membrane bilayer, in which it can diffuse and enter the ~5000 Å^3 inter-monomer quinone exchange cavity. **(b)** Ribbon diagram of the native symmetric dimeric b_6f complex from the thermophilic cyanobacterium, *Mastigocladus laminosus*, solved to a resolution of 3.0 Å [6], which contains 8 polypeptide subunits and 13 transmembrane helices, and whose electron density shows 7 prosthetic groups (four hemes, one [2Fe-2S] cluster, 1 chlorophyll *a*, and 1 β-carotene), and which has a dimer molecular weight of 217 000. The "large" subunits are seen to bind the redox prosthetic groups (heme *f*, 2 hemes *b*, heme c_n, the [2Fe-2S] cluster, and the single chlorophyll *a* and β-carotene. The function of the Chl *a*, whose porphyrin ring plane is inserted in a lipid-like manner between the F and G helices of subunit IV, whose 20 carbon phytyl tail is inserted into the quinone exchange cavity through the small lumenal *p*-side portal in the quinone exchange cavity, is not known. A function of the β-carotene is presumably to quench the Chl *a* excited triplet state, although the two molecules are seen in the structure to be separated by 14 Å, much too long a distance to allow quenching by wave function overlap. The four small subunits (petG, L, M, and N) are uniquely arranged in a picket fence fashion on the periphery of each monomer of the complex. **(c)** Once in the inter-monomer cavity, PQH_2 ($E_{m7} \cong +90$ mV) is oxidized to the neutral and anionic plasto-semiquinone, $PQ^{\cdot}$ and $PQ^{\cdot -}$, respectively, by the [2Fe-2S] cluster. ($E_{m7} \cong +290$ mV) of the Rieske iron-sulfur protein, depositing two H^+ per electron transferred to the electrochemically positive (*p*) side of the membrane. The pathway of interheme *b* (b_p -> b_n) 1-electron transfer that results in stromal *n*-side reduction of plastoquinone to quinol (PQ -> PQH_2), concomitant with uptake of two H^+, is shown. From the binding of the quinone analogue inhibitor TDS or NQNO as a ligand to heme c_n [9] in the position distal to heme b_n, it was inferred that this site on the edge of the intermonomer quinone exchange cavity, is the physiological stromal *n*-side binding of the quinone, and thus the interface to the relatively mobile quinone in the cavity. The proximity (3.5 Å) of heme b_n to c_n and their strong interaction is indicated by EPR analysis [17, 89], and binding of the quinone analog inhibitors to an axial position of heme c_n. The presence of heme c_n and the coupled hemes b_n–c_n may avoid formation of plasto-semiquinone that could be formed by heme b_n reduction of O_2, which could then generate ROS. PQ reduction may occur through 2-electron reduction by the coupled hemes b_n-c_n (Figure 7.4).

7.1.2 Difference in Prosthetic Group Content of bc_1 and b_6f Complexes

The presence of heme c_n in the b_6f complex, originally observed by sensitive spectrophotometry [15, 16], and documented from the crystsal structures of the complex [7–9], implies the possibility of differences in mechanism in te b_6f complex relative to bc_1. The presence of the heme c_n and the existence in the oxygenic photosynthetic membrane of a ferredoxin-dependent cyclic electron transport pathway can certainly result in different reactions for stromal *n*-side quinone reduction. It may be possible that one pathway of ferredoxin-mediated reduction of the

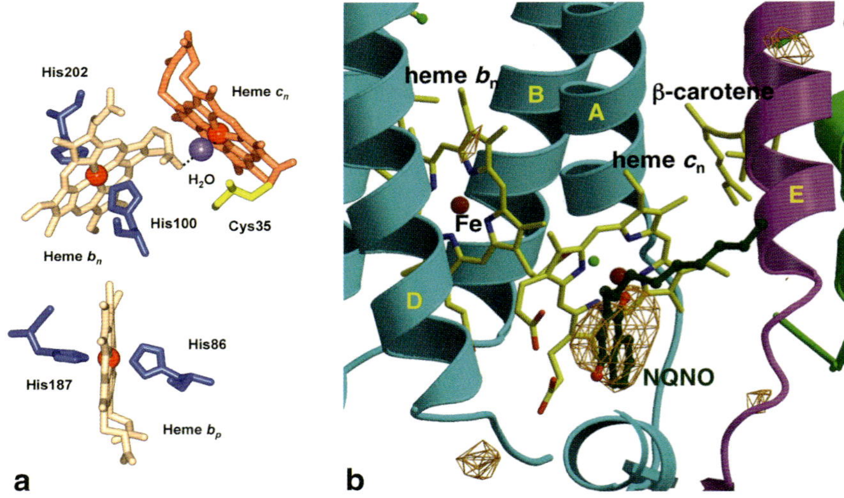

Figure 7.2 **(a)** Arrangement in the b_6f complex of the three hemes, b_p, b_n and c_n. The two *bis*-histidine coordinated *b* hemes are separated by an edge-edge distance of 7.5 Å normal to the membrane plane, and thus almost span the complex and the membrane bilayer. Heme c_n on the edge of the quinone exchange cavity, is seen to have an orientation approximately orthogonal to heme b_n, to be covalently bound to Cys35 in the A helix of the cytochrome *b* polypeptide, and to be ligated to a heme b_n propionate, 3.3 Å distant, by the O atom of the inter-heme water or OH^- molecule. **(b)** Fo–Fc difference map of stromal *n*-side binding site of NQNO in the b_6f complex. Fo–Fc difference map in the region of heme c_n on the stromal *n*-side of the b_6f complex shows the stromal *n*-side binding site of NQNO in a position that approaches that of an axial ligand of heme c_n. TMH and surface helices within loops are labeled as in [6]. Fo–Fc map is contoured at 4σ (published with permission of the *Journal of Molecular Biology*).

plastoquinone pool proceeds through hemes b_n and c_n (Table 7.1). Based on the crystal structure analysis of the b_6f complex, in which the distance between hemes b_n and c_n was found to be only 3.5 Å (Figures 7.1c and 7.2a), and analysis of the paramagnetic properties of the b_6f complex that show the electronic properties of the two hemes to be coupled [17], as described below, it is proposed that the n-side reduction of PQ proceeds through a two electron transfer (Table 7.1, Figure 7.3), thus avoiding the possibility of generating a semiquinone (SQ) and, thereby, reactive oxygen species (ROS) in a membrane that is oxygen-rich.

7.1.3
Symmetry and Asymmetry

(i) *Symmetry*: the structure of the dimeric b_6f complex is symmetric between the two monomers, with space groups $P6_122$ and I222, respectively, for the complexes from crystals in the native state and in the presence of the quinone-analog inhibitors, TDS and NQNO, of the cyanobacterium, *M. laminosus* [6] and for crystals grown in the presence of TDS for the b_6f complex from the green alga, *C. reinhardtii* [8];

Table 7.1 (A) Standard and (B, C) modified Q-cycle reactions in the cytochrome b_6f complex; modifications in B, C arises from participation of heme c_n; in (C), a modification entails stromal *n*-side electron donation by ferredoxin in a PSI-linked cyclic electron transport pathway.

A1. *p*-side quinol oxidation
PQH_2 + FeS (o) → PQ•- + FeS (r) + $2H^+$
PQ•- + b_p (o) → PQ + b_p (r)

A2. high potential chain
FeS (r) + cyt *f* (ox) → FeS (o) + cyt *f* (r)
cyt *f* (r) + Pc (ox) → cyt *f* (o) + Pc (r); in low Cu milieu, cyt c_6 substitutes for Pc
Pc (r) + P700 (ox) → Pc (o) + P700 (r)

A3. transmembrane *p*- to *n*-side electron transfer
heme b_p (red) + heme b_n (ox) → b_p (ox) + b_n (red)

A4. *n*-side 2 × 1 electron reduction of PQ (as in bc_1 complex)
b_n (r) + PQ (o) → b_n (o) + PQ•- heme
b_n (r) + PQ•- + $2H^+$ → b_n (o) + PQH_2 (b_n supplies second electron after it is reduced following second *p*-side turnover of PQH_2)

B. *n*-side 2 electron reduction of bound PQ using two electrons from *p*-side; intermonomer PQH_2
(i) b_n (r) /c_n (r) /PQ (o) + $2H^+$ → b_n (o)/c_n (o) /PQH_2
(b_n/c_n supply two electrons cooperatively to PQ after each heme is reduced by two lumenal *p*-side turnovers of PQH_2- not shown)
(ii) b_n (o)/c_n (o) /PQH_2 → b_n (o)/c_n (o) + intermonomer PQH_2

C. *n*-side 2 electron reduction of PQ by one electron from *p*-side, 1 from Fd (*n*-side)
(i) heme b_p (red) + heme b_n (ox) → b_p (ox) + b_n (red)
(ii) b_n (r) /c_n (o)/PQ (o) → b_n (o) /c_n (r)/PQ(o)
(iii) Fd (r) + b_n(o)/c_n(r)/PQ(o) → Fd(o) + b_n(r)/c_n (r)/PQ(o)
(iv) b_n (r)/c_n (r)/PQ(o) + $2H^+$ → b_n(o)/c_n(o)/PQH_2
(v) b_n (o)/c_n(o) /PQH_2 → b_n (o)/c_n(o) + intermonomer PQH_2

(ii) *Asymmetry*: The surface potentials in the *M. laminosus* complex that result from an asymmetric lumenal *p*-versus stromal *n*-side distribution of charged residues is −5.4 kT (in *M. laminosus*) and +4.6 kT, respectively, on the lumenal *p*- and stromal *n*-sides of the complex [1].

7.1.4 Questions about Structure–Function at the Outset

Among the mechanistic questions implicit in Figure 7.1a are: (i) the site(s) of entry/exit of reduced plastoquinol/oxidized plastoquinone (PQH_2/PQ) from/into the membrane lipid bilayer into/from the b_6f complex; (ii) the pathway of the movement of the snake-like PQH_2/PQ, a six carbon dimethyl-substituted benzene ring

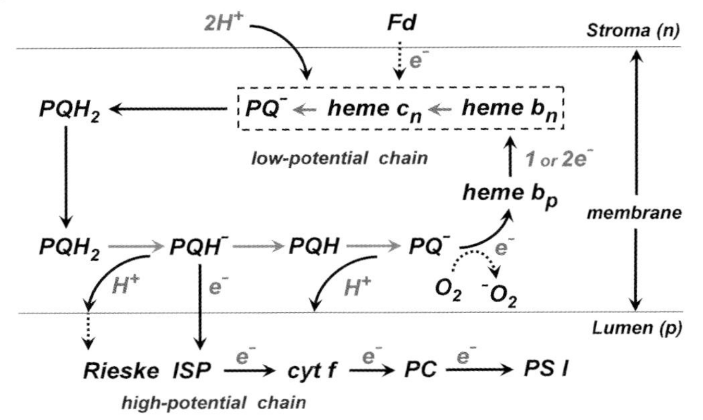

Figure 7.3 Modified "Q-cycle" model of electron transfer in the cytochrome b_6f complex of oxygenic photosynthesis. The modifications, relative to the standard model (Table 7.1A) used to describe the function of the cytochrome bc_1 complex in the mitochondrial respiratory chain and photosynthetic bacteria, involve (a) the presence of heme c_n in addition to the heme b_n that is present in both bc_1 and b_6f complexes, and the possibility of injection of one electron from ferredoxin that is reduced through the PSI-linked cyclic electron transport pathway.

that drags a 45 carbon tail made of nine isoprenoid units between its site of oxidation and reduction on the lumenal *p*-side and stromal *n*-side of the complex. The problem of diffusion of such quinones in the membrane bilayer has been discussed extensively [18], but not that of movement between redox sites through the interstices of a lipophilic oligomeric membrane protein. (iii) The pathway(s) for PSI-and ferredoxin-dependent cyclic electron transfer, as well as the regulation of cyclic relative to linear electron transfer, present major questions. A pathway involving electron feedback into the b_6f complex is shown (Figure 7.1a) in which electrons would be transferred via ferredoxin from the stromal side of PSI to the stromal *n*-side (heme b_n) of the b_6f complex [19], and thereby to bound and mobile PQ, allowing additional H^+ transfer and ATP synthesis without net reduction of $NADP^+$. It has been argued that heme b_n of the b_6f complex is not an intermediate in the cyclic pathway because of contrasting effects of antimycin A on inhibition of cyclic electron transfer and an absence of inhibition of flash-induced reduction of heme b_n [20]. However, realization of the short distance and tight coupling between hemes b_n and c_n [7–9, 17], and structure details of modes of action of quinone analog inhibitors, mean that the effect of antimycin A on the transient net reduction of heme b_n will depend on the detailed mode of action of this inhibitor.

7.1.5
Evolution: cyt b_6f vs bc_1

The similarity of the amino acid sequences, and particularly the distribution of hydrophobic residues and histidine heme ligands of the cyt b_6 and subunit IV core of the b_6f complex to the N- and C-terminal domains of the cytochrome *b* polypeptide in the cytochrome bc_1 complex of the mitochondrial respiratory chain and purple photosynthetic bacteria [21], implies a common evolutionary origin of this core [22]. Moving away from the intramembrane core of the structure in a direction either parallel or perpendicular to the plane of the membrane, the conservation of sequence and structure diminishes. The rubredoxin-like core of the Rieske [2Fe-2S] protein in the different *bc* complexes is more similar than the remainder of its soluble domain that is distal from the membrane [23]. It is striking that cytochromes *f* and c_1 have absolutely no significant sequence similarity beyond the signature Cys-X-Y-Cys-His sequence for the covalent binding of heme through two thioether cysteinyl bonds [24, 25]. Another unique feature of the b_6f complex is the presence of the heme c_n (see Section 7.3.4), which is also part of a b_6f-like complex in the gram-positive bacteria [26].

7.2 Electron and Proton Transfer Pathways

The arrangement of the redox prosthetic groups in the complex is shown in Figures 7.1c and 7.2a. Starting on the lumenal *p*-side of the complex, the edge–edge intramonomer distances (to the nearest Å) are: [2Fe-2S] cluster – cyt *f*, 26 Å; [2Fe-2S] – heme b_p, 24 Å; heme b_p – heme b_n, 7 Å; heme b_p – Chl *a*, 12 Å; Chl *a* – heme b_n, 6 Å; Chl *a* – β-carotene, 14 Å. The edge–edge intermonomer distances are: heme b_p – heme b_p, 13 Å; heme b_n – heme b_n, 29 Å. In the perspective of the structure of the complex (Figure 7.2a), the electron transfer scheme described in Figure 7.1c shows part of a Q-cycle model for electron and proton transfer through the b_6f complex. The model describes a bifurcated electron transfer scheme originating with the lumenal *p*-side oxidation of the quinol (PQH_2) which initially transfers one electron in the high potential chain downhill in redox potential to the [2Fe-2S] cluster ($E_m \cong +0.3$ V) while liberating 2 H^+ to the lumenal *p*-side aqueous phase. The electron is subsequently transferred to cytochrome *f* ($E_m \cong +0.35$ V), plastocyanin ($E_m \cong +0.38$ V) (Figures 7.1c and 7.4, Table 7.1), (see also Chapter 8 for details on the soluble lumenal electron carriers), and the primary donor of PSI (P700) ($E_m \cong +0.45$ V) (Table 7.1) (see also Chapters 2 and 3). Details of the interaction of the quinol with the Rieske [2Fe-2S] cluster at the lumenal *p*-side interface, and of proton-coupled electron transfer from the quinol to the cluster, are discussed for the cytochrome bc_1 complex in the *Rb. sphaeroides* photosynthetic bacterium using structure data for the bc_1 complex from the respiratory chain [27, 28].

7.2.1 The Q-Cycle and Modifications

The one electron transfer from PQH_2 to the [2Fe-2S] cluster can generate a lumenal *p*-side semiquinone radical (**$PQ^{\cdot-}$**) of more negative redox potential that reduces the heme b_p ($E_m \cong -50$ mV in the b_6f complex) and thereby initiates electron transfer across the complex through a low potential chain (heme b_p (r) + heme b_n (o) → b_p (o) + b_n (r); Table 7.1a). Questions about mechanism that relate to the stability of this semiquinone have been discussed recently [27, 29]. In the "Q-cycle" model, based to a major extent on data obtained for the bc_1 complex [30–34], two turnovers of the quinol (ubiquinol, UQH_2) on the lumenal *p*-side result in the transfer of two electrons through heme b_n to a stromal *n*-side quinone bound near heme b_n, as originally inferred from the effect of the quinone analog inhibitor, antimycin A, on the amplitude of flash-induced reduction of heme b_n [35] and the ubisemiquinone radical detected by EPR [36]. The position of the lumenal *p*- and stromal *n*-side redox components essential for the Q-cycle has been supported by crystal structure data that confirm the predicted positions of the stromal *n*-side ubiquinone relative to heme b_n, and the positions of stromal *n*- [37, 38] and lumenal *p*-side [38–42] quinone analog inhibitors. On the lumenal *p*- and stromal *n*-sides of the bc_1 complex, the quinone analog inhibitors, stigmatellin and antimycin A displace the lumenal *p*- and stromal *n*-side redox-active quinone as predicted by the Q-cycle model. In the b_6f complex, tridecyl-stigmatellin binds on the lumenal *p*-side in an analogous manner [8, 9], as well as binding as a ligand to heme c_n on the stromal *n*-side. The high affinity lumenal *p*-side inhibitor, DBMIB, is found at a lumenal *p*-side peripheral site 20 Å from the [2Fe-2S] cluster [11]. The phenomenon of "half of the sites reactivity of stigmatellin, seen in the bc_1 complex [43, 44], has not been seen in the b_6f complex. In b_6f, the stromal *n*-side inhibitor, NQNO, binds at a site quite different from that of antimycin in bc_1 [Section 1.4(i)].

7.2.2
Question of Obligatory Application of the Q-Cycle Model to the b_6f Complex

A significant part of the experimental evidence supporting the operation of the Q-cycle in the b_6f complex is different than that used for the bc_1 complex: (i) the overlap of the wavelength peaks of the difference spectra for the two *b* hemes (λ_m = 563–564 nm for both hemes) is much more extensive than in the bc_1 complex, which creates difficulties in the quantitation of the electron transport step between hemes b_p and b_n (Figure 7.1c); (ii) neither antimycin A nor any other quinone analog inhibitor is as potent and specific an inhibitor as is antimycin A for the bc_1 complex. On the basis of studies on the transmembrane electrical potential and electrochromic band shift associated with transmembrane electron and proton transfer in the Q-cycle (Figures 7.1c and 7.3), and the H^+/e^- = 3 stoichiometry of lumenal *p*-side proton transfer, operation of the Q-cycle has been inferred. Many of these measurements have been made under conditions of intermittent flashed light intensity and small $\Delta\tilde{\mu}_H^+$. The point is debated as to whether the Q-cycle operates ubiquitously under conditions of high as well as low $\Delta\tilde{\mu}_H^+$ [45, 46] obtained under conditions of low light intensity, and thereby small values of the $\Delta\tilde{\mu}_H^+$ [47]. This view of ubiquitous operation of the Q-cycle has led to the view that the b_6f complex is essentially a bc_1 complex, perhaps differing in details of fine structure, but with no major conceptual or mechanistic differences that alter the electron transfer pathways through which it generates the $\Delta\tilde{\mu}_H^+$ [48]. Thus, the scheme for coupled electron and proton transport shown in Table 7.1A1–A4 shows the standard Q-cycle model applied to the b_6f complex. The notable point, for purposes of this discussion, is that the stromal *n*-side reduction of PQ proceeds in two 1-electron steps via a plastosemiquinone. In contrast, Table 7.1B presents a Q-cycle model for stromal *n*-side PQ reduction accomplished by a concerted 2-electron transfer from the tightly coupled hemes b_n/c_n (see Sections 7.3.4, 7.3.5). In this model, the PQ binding site in a b_n/c_n/PQ complex is emphasized, as is release of PQH_2 from this site to the intermonomer space of the quinone-exchange cavity. The model shown in Table 7.1C emphasizes the possibility of one electron donation from ferredoxin (Fd) participating in PSI-linked cyclic electron transfer [19, 49–57], to the stromal b_n/c_n/PQ complex. One electron reduction of PQ by Fd (reduced) would be prevented by the 2-electron gate mechanism imposed by the coupled hemes b_n/c_n (Sections 7.3.4, 7.3.5, 7.3.6).

7.2.3 Production of Superoxide and Other ROS

Superoxide production has been clearly demonstrated in vitro for the mitochondrial cyt bc_1 complex. Superoxide production during ubiquinol oxidase activity of the bc_1 complex is observed under inhibited conditions where processing of electrons by the low-potential chain is hindered and oxidation of ubiquinone results in accumulation of the semiquinone (SQ) intermediate at the Q_p site [58–62]. Unstable SQ species at the Q_p site can reduce O_2, forming superoxide. Models of Q-cycle bypass reactions of mitochondrial bc_1 complex and possible mechanisms preventing O_2 reduction at the O_o site have been reviewed [48, 63]. To explain the significance of Q-cycle inhibition for the superoxide producing reaction, it has been suggested that this site may act as a sealed "reaction chamber" that effectively shields the unstable semiquinone intermediate under uninhibited conditions [59]. However, SQ can escape from the reaction chamber, or oxygen can diffuse into the chamber, when the low-potential electron transport chain is blocked. A similar mechanism resulting in superoxide production as a consequence of low-potential chain inhibition may exist at the Q_p site of the cyt b_6f complex. Semiquinone stabilization at Q_n site of the cyt bc_1 complex is believed to prevent reduction of O_2 [64].

Photosynthetic oxygen evolution creates local hyperoxic conditions, resulting in the production of reactive oxygen species (ROS) [65]. If lumenal *p*-side generation of semiquinone is the precursor to production of superoxide in the bc_1 complex, then it must be generated in the b_6f complex. This likelihood is indicated in Figure 7.3, although there is presently a lack of documentation (only [66]) for lumenal *p*-side superoxide production in the b_6f complex, and none for the stromal *n*-side. Structural differences at the Q_n site of the cyt b_6f complex, involving heme c_n located at the site homologous to that of Q_n in the bc_1 complex, suggest a different (2-electron) reaction for PQ reduction. Such a 2-electron reduction of PQ would provide a mechanism to prevent superoxide formation by cyt b_6f in spite of the hyperoxic conditions in oxygenic photosynthetic membranes.

7.3 Stromal *n*-Side Electron Transfer: Properties and Function of Heme c_n

7.3.1 Application of Novel EPR Analysis

The discovery of the new heme c_n in the crystal structure of cyt b_6f prompted reinvestigation of the EPR spectroscopy of the complex [17]. Many proteins contain several paramagnetic centers that may interact electronically, giving complicated EPR spectra with overlapping signals. For such complexes, the interpretation of EPR spectra is typically only qualitative. This holds true for the cyt b_6f complex where quantitative estimates of the cofactor concentrations from previous EPR studies, when attempted, resulted in unrealistically low concentrations [67]. Consequently, the low-field signals with $g > 4.3$ in all previous studies were attributed to impurity species, loss of an axial histidine of a low-spin heme, or to a mixture of isolated high-spin species [67–69]. However, since the assignment of specific resonance line positions was uncertain, these conclusions must be re-examined. Two advances in EPR spectroscopy have been developed over the past decade that provide information for interpretation of complicated EPR spectra, such as those of the cyt b_6f complex: (i) the use of an incident microwave radiation field, $\mathbf{B}_1$, polarized parallel to the static magnetic field, $\mathbf{B}$; and, (ii) the development of simulation software that allows quantitative assessment of signals arising from mono- or di-nuclear metal sites [70–72].

7.3.2 EPR Spectra of b_6f Complex

A typical X-band EPR spectrum of native b_6f complex from spinach thylakoids is shown (Figure 7.4a, trace *a*). The $g = 4.3$ signal is from an adventitious Fe (III) impurity. The signal at $g = 3.51$ is from the cyt *f* low-spin heme. The simulation shown in Figure 7.4a (trace *c*) indicates that approximately 60% of cyt *f* is reduced, with this amount varying between preparations. The other signals for this species are at $g = 1.70$ and 0.78; the latter value is predicted from the Griffith model [73], but is not observed due to a broad line shape. The signal at $g = 3.68$ is from low-spin heme b_p, and the simulation of this signal (Figure 7.4a, trace *b*) indicates a spin concentration comparable to that of the protein. In previous studies, this signal was assigned to both *b* hemes of the complex [67, 74]; however, this assignment must now be revised. The signal quantitation and presence of the spin interacting system, discussed below, both indicate that the $g = 3.68$ signal originates only from the b_p heme.

7.3.3 Low-Field Spectra

To address the assignments of the low-field signals ($g > 4.3$) of the b_6f complex, it is useful to first review basic spin characteristics of a high-spin Fe(III) heme.

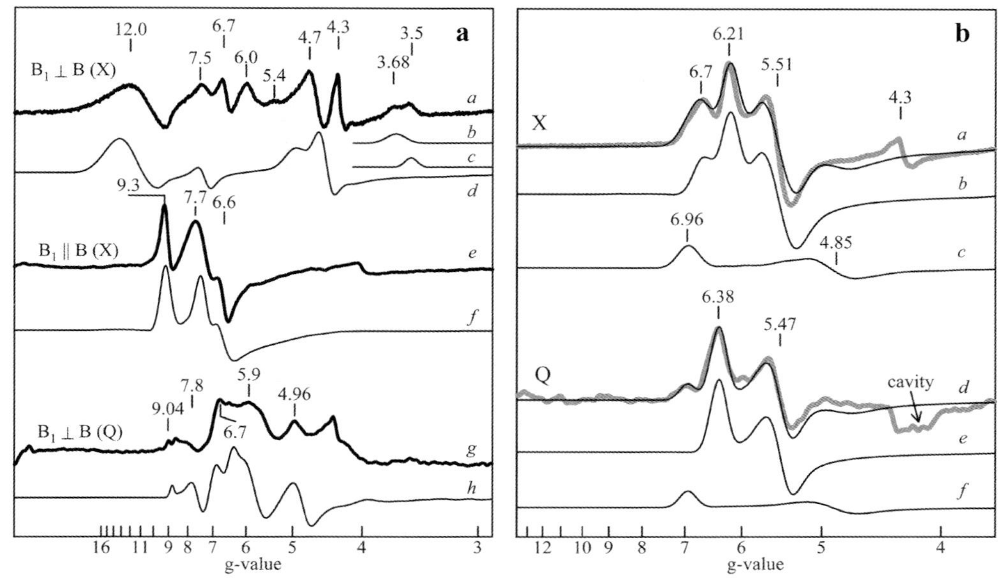

Figure 7.4 **(a)** Multifrequency, multimode EPR spectra (thick lines: ***a, e, g***) and simulations (thin lines: ***b, c, d, f, h***) of native cytochrome b_6f complex isolated from spinach thylakoid. (**b**) Simulation of a low-spin heme b_p signal ($\mathbf{g}$ = 0.63, 1.53, 3.68); (***c***) low-spin heme of cytochrome *f* ($\mathbf{g}$ = 0.78, 1.70, 3.51). Simulations ***d, f*** and ***h*** all use the same set of simulation parameters of an exchanged-coupled system: $S_1 = 5/2$, $S_2 = 1/2$, $\mathbf{g}_1 = 1.98$, $D_1 = 7\,\text{cm}^{-1}$, $E/D_1 = 0.033$, $\mathbf{g}_2$ = 0.4, 1.6, 3.7, $J_{iso} = 0.074\,\text{cm}^{-1}$, S_2 Euler rotation (0°, 90°, 90°). Experimental conditions, microwave mode as shown: (*a*) $\nu = 9.65$ GHz, $T = 13.3$ K; (*e*) $\nu = 9.35$ GHz, $T = 2$ K; (*g*) $\nu = 34.10$ GHz, $T = 8$ K. **(b)** Multifrequency, perpendicular mode EPR spectra (gray) and simulations (black) of cytochrome b_6f complex from spinach thylakoids after reaction with 10-fold excess of NQNO. Simulations ***a*** and ***d*** (overlaid on the data) are composed of two species: (i) (***b, e***) an exchanged-coupled system $S_1 = 5/2$, $S_2 = 1/2$, $\mathbf{g}_1 = 1.98$, $D_1 = 7\,\text{cm}^{-1}$, $E/D_1 = 0.02$, $\mathbf{g}_2$ = 0.9, 1.7, 3.5, $\mathbf{J}$ = 0.0035, 0.0085, 0.07 cm^{-1}, S_2 Euler rotation (0°, 90°, 90°), and (ii) (***c, f***) an isolated high-spin heme species with g = 6.96, 4.85 ($E/D = 0.045$). Experimental conditions: (*a*) $\nu = 9.65$ GHz, $T = 13.3$ K; (*d*) $\nu = 34.10$ GHz, $T = 8$ K.

The S = 5/2 spin manifold of a high-spin heme splits into three isolated doublets due to the spin-orbit interaction. The isolated or Kramers doublets ($\pm m_s$) for any half-integer spin system with $S \geq 3/2$ and large zero-field splitting ($D >> g\beta B$) can be described with an effective spin $S' = 1/2$ formalism. The quantization axis of the effective spin $\mathbf{S}'$ (or $\mathbf{S}' \cdot \mathbf{g}'$ if anisotropy is present) will always align with the direction of the static magnetic field, $\mathbf{B}$. A transition between the spin states of a Kramers doublet can only be stimulated by a component of the oscillating microwave magnetic field ($\mathbf{B}_1$) perpendicular to the spin quantization axis, giving the familiar $\Delta m_s = \pm 1$ selection rule. Consequently, EPR signals from such spin systems have maximum intensity for $\mathbf{B}_1 \perp \mathbf{B}$, and vanish for $\mathbf{B}_1 \parallel \mathbf{B}$ [71]. In addition, the resonance field for Kramers doublets is linearly related to the microwave frequency ($h\nu = g\beta B$), and thus the observed g-values are independent of microwave frequency. For a typical isolated high-spin Fe(III) heme with axial symmetry, the EPR signals originate from the lowest of the three Kramers doublets.

As Figure 7.4a (trace *e*) shows, the EPR spectrum of cyt b_6f for $\mathbf{B}_1 \parallel \mathbf{B}$ does not vanish. This implies that the standard treatment based on Kramers doublets is not valid and some paramagnetic heme species of cyt b_6f cannot be described as isolated. Thus, the assignments of the low-field signals must be reconsidered. While the observation of a parallel-mode signal is an indication of a novel species, the importance of any spectral assignment in a protein depends on a quantitative assessment of the signal relative to the protein concentration. Such an assessment can only come from simulations of signals, since the common double integration method of EPR spectra is only applicable for Kramers doublets. For simulation of spectra, the new structure information [7, 8] that shows heme b_n close to heme c_n is relevant, as it implies the possibility of spin interaction between the two hemes. The 5-coordinate heme c_n is expected to be in either an $S_1 = 5/2$ (oxidized) or 2 (reduced) configuration, and the bis-His coordinated heme b_n is expected to be in the low-spin ($S_2 = 1/2$) state. The simulations of the EPR spectra were calculated from diagonalization of the spin Hamiltonian:

$$H_s = J\mathbf{S}_1 \bullet \mathbf{S}_2 + \mathbf{S}_1 \bullet \mathbf{D}_{12} \bullet \mathbf{S}_2 + \beta \mathbf{B} \bullet (\mathbf{g}_1 \bullet \mathbf{S}_1 + \mathbf{g}_2 \bullet \mathbf{S}_2) + D_1 \left[S_{1z}^2 - \frac{S_1(S_1+1)}{3} \right] + E_1 (S_{1x}^2 - S_{1y}^2)$$

with standard parameter definitions [75–77]. The tensor $\mathbf{D}_{12}$ is the symmetric part of the full $\mathbf{J}$ tensor, containing anisotropic and symmetric off-diagonal contributions.

The simulated spectra of Figure 7.4a (traces *d, f*) are calculated for a high-spin heme c_n ($S_1 = 5/2$) exchange coupled to a low-spin heme b_n ($S_2 = 1/2$), using the same set of parameters given in the figure caption. The magnitude of the isotropic exchange term is small relative to the zero-field splitting of heme c_n ($J \ll D_1$). Thus, the exchange interaction splits the three doublets of the $S_1 = 5/2$ into three spin quartets. The temperature dependence of EPR signals indicates that they originate from the lowest of the three quartets. Although the simulations do not yet match all the features of the experimental spectra precisely, the predicted spin concentration from the simulation is close to that of the protein concentration. Furthermore, the same simulation parameters quantitatively predict both the perpendicular and parallel mode signals. This agreement depends on the spin states of both hemes, the zero-field parameters of heme c_n, the g-values of heme b_n, the relative orientation of the two hemes, and the exchange interaction between the hemes. Thus, the assignment of these signals to the heme b_n/c_n pair is unambiguous. Both hemes are ferric, with the heme b_n being in a low-spin and heme c_n in a high-spin configuration. There are several 5-coordinate Fe(III) heme complexes for which the axial ligand is either a hydroxide [78a] or water [78b]. The heme hydroxide ligand is characterized by a high-spin $S = 5/2$ state, while such hemes with water are found to be in an intermediate-spin $S = 3/2$ state. Because heme c_n is observed to be in a high-spin state, it is inferred that its coordination involves ligation by an axial hydroxide rather than a water molecule. The previous finding of a permanently reduced heme c_n in the isolated complex is, presumably, a consequence of photoreduction during the resonance Raman measurement [79].

The simulation fit requires that the g-tensor of low-spin heme b_n has one large g-value near 3.7 in a direction aligned with the in-plane g-values of high-spin heme c_n. A g-value greater than 3.2 is indicative of a crossed imidazole plane configuration for axial histidine ligands [80]. This agrees with the structure of the cyt b_6f complex, which shows a crossed imidazole plane configuration for the axial histidines of heme b_n [7, 8]. Furthermore, the alignment of g-values indicates that the heme b_n and c_n planes are perpendicular, also in agreement with the crystal structures.

Figure 7.4a (trace g) shows a Q-band EPR spectrum of the same protein state. All spectra are plotted on the same g-value scale ($g = h\nu/\beta B$) to allow identification of signals which have a nonlinear frequency-field relation. All major features of the X-band spectrum for $g > 4.3$ show a shift in g-value relative to the signals of the Q-band spectrum, which confirms that these signals are not from isolated high-spin hemes. A feature does exist near $g = 6$ in both spectra, but this is a minority high-spin species that is less then 3% of the protein concentration. The same simulation parameters used for the X-band spectra correctly predict the Q-band experimental spectrum as shown in Figure 7.4a (trace *h*). Further refinement of the simulation parameters is needed to achieve a better fit of both X- and Q-band data, since small contributions from $\mathbf{D}_{12}$, the magnetic dipole-dipole interaction, small non-perpendicular heme deviations, and possibly the antisymmetric exchange term, have not yet been fully considered. Nevertheless, the general agreement across multiple microwave frequencies and orientations is strong evidence for the validity of the current assignments.

7.3.4
Novel Properties of Heme c_n

Two properties of the heme b_n/c_n pair are novel: (i) heme c_n lacks an axial ligand from on amino acid side chain and (ii) the presence of an electronic exchange interaction via a bridging water molecule. All other naturally occurring hemes have at least one axial ligand provided by the protein. The other novel property pertains to heme-heme interaction. Electronic heme–heme interaction in proteins is a relatively new phenomenon and is now known for hydroxylamine oxidoreductase [72], cytochrome c_{554} [81], cytochrome *c* nitrite reductase [82], tetrathionate reductase [83], fumarate reductase [84], the split-Soret cytochrome *c* [85], and diheme cytochrome DHC2 [86]. The hemes in these proteins are generally observed in pairs, having approximately parallel alignment of heme planes, allowing the exchange interaction through direct π–π overlap of porphyrin orbitals [72]. Cytochrome b_6f thus represents a new type of heme-heme interaction motif where the orientation of the heme planes is perpendicular and the electronic exchange interaction is via a bridging water hydioxide molecule. The magnitude of the spin exchange is expected to be smaller for the heme b_n/c_n pair due to the longer bonding pathway between the iron ions. Indeed, the isotropic exchange interaction for heme b_n/c_n

($J = 0.07\,cm^{-1}$ [17]) is approximately 10-fold lower than for the heme pairs in the other proteins ($0.4 < J < 1.0\,cm^{-1}$) [72, 87]. In comparison to other inorganic metal complexes where the metals are bridged by a single atom ($J > 10\,cm^{-1}$), the magnitude of these exchange interactions is relatively small.

7.3.5
Plastoquinone as a Ligand of Heme c_n

The native crystal structure data for heme c_n does not show significant electron density at the Fe coordination site *trans* to the water molecule. However, the structure data for binding sites of the quinone analog inhibitors, NQNO (Figure 7.2b) and TDS, imply that this site, in the native state, is occupied by PQ [9]. Further information regarding the accessibility of this heme c_n site comes from small molecule and substrate analog additions. The addition of KCN at pH 8.5, or its hydrophobic analog butyl isocyanide to a native b_6f sample, had only minor effects on the EPR spectra, although BIC caused a shift in the position of the Soret band of the reduced comples [69]. With rare exception, cyanide will bind to high-spin ferric hemes to give a low-spin state and a significant change to the EPR signal. The absence of an effect on cyt b_6f indicates that the site is blocked or coordinated by a molecule, supporting the presence of a PQ in this site.

Binding of NQNO to heme c_n results in a significant EPR spectral change [7]. A consequence of the NQNO binding to heme c_n is a shift in the heme midpoint potential of approximately $-200\,mV$ [88]. The X- and Q-band spectra of cyt b_6f treated with a 10-fold excess of NQNO are shown (Figure 7.4b, traces *a*, *d*). The g-values of some of the signals clearly shift with the microwave frequency, indicating that spin interactions are still present. No significant parallel mode signal was observed in the presence of NQNO. The simulations overlaid on the data are composed of two species: a weakly interacting high-spin/low-spin heme pair, and a minority isolated high-spin species. The simulations of the individual species used for the construction of the sum are shown (Figure 7.4b). The minority species is a high-spin heme ($g = 6.96$, 4.85, $E/D = 0.045$), which corresponds to ~20% of the protein concentration. Only the $g = 6.96$ peak in the Q-band spectrum (Figure 7.4b, trace *d*) is resolved from the other signals; at X-band, this peak is a shoulder. At 20% or greater, this species would be noticeable in the spectrum of the native sample, but it is not observed. The spin-coupled signals of the native spectrum are nearly quantitatively converted to the signals of a different spin-interacting system upon addition of NQNO. Importantly, the in-plane symmetry parameter, E/D, for heme c_n changes from 0.033 to 0.02, and the strength of the exchange interaction in the presence of NQNO is significantly weaker than that of the native complex. The change in E/D indicates binding of NQNO to the heme c_n, with the exchange interaction still dominating over the weaker magnetic dipole-dipole interaction between the spins. The 10-fold decrease in the exchange interaction suggests a weakening of the H-bond between the bridging water hydroxide and propionate O-atom. Simulation of the signals with the inclusion of only a magnetic dipolar interaction and no electronic exchange terms gave insufficient splitting of the relevant features in the X-band spectrum. Therefore, the suggestion that the spin interaction in the presence of NQNO can be described with only a purely dipolar interaction term, and thus the loss of the bridging species [89], is not supported.

7.3.6
Functions and Unwanted Functions of Heme c_n

The above findings provide new insight into possible functions of heme c_n in the cyt b_6f complex. The presence of PSII creates an oxygenic environment that requires special protection against reaction with O_2 that is not necessary in the relatively low O_2 environment of cyt bc_1. The generation of semiquinone radicals that are reactive with O_2 would be more problematic in cyt b_6f with higher O_2 concentration. The unique coordination to heme c_n may provide important protective features due to its high reduction potential and direct ligation to the hydroquinone species. The high potential results in localization of an electron in heme c_n near the PQ acceptor. The heme-quinone ligation could function to anchor the quinone at the heme until the second electron arrives, at which time a fast 2-electron reduction of the quinone occurs, followed by release of hydroquinone: $Fe^{2+}_{cn}Fe^{2+}_{bn} + PQ + 2H+ \rightarrow Fe^{3+}_{cn}Fe^{3+}_{bn} + PQH_2$. Studies have shown that the quinone analog NQNO will bind at this site [9]. This direct ligation to an iron is a new mode of metal-quinone interaction not previously observed in nature. The heme $b_n - c_n$ diheme cluster and the metal-quinone binding could function together to ensure hydroquinone production without significant release of a plasto-semiquinone species.

A complication presented by this new mode of quinone reduction is the possibility of O_2 reduction to superoxide or peroxide at the same heme c_n site. How does the cyt b_6f complex provide specificity for reduction of quinone over O_2? As stated above, cyanide does not bind at heme c_n, which is presumably due to the occupation of the site by PQ. However, PQ only transiently occupies the site during normal Q-cycle activity, and thus O_2 access could be allowed when PQ is not present. In this regard, the novel features of heme c_n coordination may be important. The probability of superoxide formation after the binding of O_2 to the reduced heme c_n site may be decreased due to the relatively positive $E_m \sim +100$ mV potential [88]) of heme c_n. Regarding possible H_2O_2 production, the diheme site in cyt b_6f could allow 2-electron reduction of O_2 to H_2O_2. To minimize this reaction, Q-cycle reactions might have to be synchronized so that the local concentration of PQ near heme c_n is sufficiently high such that heme c_n will tend to be ligated by PQ.

7.4
Stromal *n*-Side Proton Uptake Pathway

The conclusions from crystal structures of b_6f with the quinone analog inhibitors, NQNO and TDS [9], and the EPR data for b_6f with NQNO referred to above, imply that the n-side PQ binding site is, or overlaps with, an axial ligand position of heme c_n, where it faces the stromal *n*-side of the membrane. The site is located in a hydrophobic α-helical bundle formed by the cyt b_6 and subunit IV polypeptides and is buried ~4 Å from the protein surface. Therefore a proton uptake pathway is required to provide protons for the reduction of plastoquinone (Table 7.1). Possible stromal *n*-side proton uptake pathways could be assessed by analysis of the residue composition around identified quinone analog binding sites in the three-dimensional structures of the b_6f complex and comparative structure analysis using functionally related proteins.

The cyt *b* subunit of the bc_1 complex and cyt b_6, together with subunit IV in the b_6f complex, belongs to the cytochrome b/b_6 protein family [21] and consists of a core of electron transferring groups that support transmembrane proton translocation. A molecular fit of the polypeptide backbones reveals a high degree of protein fold conservation among the three proteins. The purple bacterial reaction center (RC) also mediates proton uptake to a quinone reduction site buried in a TMH bundle. Several conserved protein sequence and structural features have been suggested for the RC and cytochrome *b* family proteins [90]. Proton uptake pathways have been proposed and confirmed by site directed mutagenesis for the bc_1 and RC complexes. These pathways include several water molecules stabilized by interactions with neighboring residues. The present resolution of available crystal structures of the b_6f complex do not allow identification of water molecules; however, aspects of residue conservation, protein folding, and ligand binding can be considered to assess similarities with known proton uptake pathways.

Two distinct proton uptake pathways were described for the yeast bc_1 complex [64, 91]. The two pathways are located on either side of the quinone substrate and can directly provide protons to the two carbonyl group reduction sites. A water molecule, Wat36, stabilized by His202 of cyt *b*, may be the primary proton donor on one side of the ubiquinone bound at the *n*-site; the second proton is donated either by Wat31 or Asp229 of the cyt *b* polypeptide. The E/R-pathway that leads to His202 includes six water molecules that are connected by hydrogen bonds and stabilized by interactions with neighboring residues Asp208, Thr213, and Arg218 of cytochrome *b*, and Glu52 of Qcr7p. The cardiolipin (CL)/K-pathway is located on the side of ubiquinone facing Asp229 of cyt b. In addition to water molecule (Wat31) located at the quinone reduction site, three other water molecules (Wat141, 257, and 415), and the Lys228 of cyt *b* mediate proton transfer from the CL phosphodiester group that provides the entry point for proton uptake at the membrane surface. It has been suggested that a high surface charge density of anionic phospholipids acts as a buffer that concentrates protons and passes them directly to a proton-translocating pore [64]. A cluster of protonatable residues: Lys228, Lys296, Lys288, Lys289, and His85 (cyt *b*, cyt *f*, and Qcr7p), may act as a proton buffer system [91].

For proton uptake coupled to the reduction of the Q_B ubiquinone in the RC, the region of the protein near the Q_B reduction site contains a number of acidic and polar residues that have been shown to be involved in a branched pathway for transfer of two protons [92, 93]. The entry point of this proton-uptake pathway is near His126, His128, and Asp124 of the H-subunit. From the histidines, the pathway leads through Asp210, Asp213 of the L-, and Asp17 of the M subunit, and their coordinated water molecule network, and diverge into two sides of the Q_B reduction site. Ser223 and Glu212 of L are thus involved in the two proton transfer routes.

The presence of an additional heme (c_n) at the PQ reduction site in the b_6f complex does not significantly alter the position of transmembrane helices or folding of the amino acid chain backbone, compared to the UQ binding site in cyt *b* of the bc_1 complex. This suggests that the two types of complexes may retain similar proton uptake pathways to the quinone reduction site. However, several important differences exist between the structures of the bc_1 and b_6f complexes that preclude the presence of identical proton uptake pathways in the b_6f complex. The most obvious differences are:

(i) The heme c_n position that overlaps the UQ antimycin binding site in bc_1 shift the proposed PQ reduction site that is approximately 2 Å (for NQNO) closer to the stromal *n*-side surface and the inter-dimer cavity. As a result, a position analogous to the proton transfer mediating acidic residue Asp229 in CL/K pathway of cytochrome *b* is occupied by hydrophobic residue Phe40 of subunit IV that is more likely to be important for heme c_n or substrate stabilization. Overall, the N-terminal amino acid sequence of subunit IV is not conserved compared to the corresponding region of cyt *b* (bc_1), a conspicuous difference being the absence of Lys228, which is crucial for proton uptake in the CL/K-pathway.

(ii) It was proposed that the acid-anion of the cardiolipin (CL) may have a role as a proton sink for oxidative phosphorylation in eubacteria [94]. Although cardiolipin type lipids are common in bacteria, and provide an important element for proton uptake pathway in complex bc_1, at the present level of resolution, 3.0 Å, such lipids have not been identified in crystal structures of the b_6f complex in *M. laminosus* [7] or *C. reinhardtii* [8]. Instead, the cavity analogous to cardiolipin binding site is partially filled by the petM and petN subunits. Considering the absence of a residue analogous to Lys228, the non-homologous replacement for Asp229, and the absence of CL, it is unlikely that an H^+ uptake pathway analogous to the CL/K-pathway exists in the cyt b_6f complex.

(iii) Further comparison of the sequences and structures of the cyt *b* (bc_1) and cyt b_6-subunit IV identify replacements for the residues shown to be important in the E/R-pathway of complex bc_1. His202 is replaced with Arg207 in the cytochrome B_6 sequence. The significance of the replacement of the residue in the b_6f complex has been addressed by site-directed mutagenesis [95], suggesting that Arg207 may be important in modulating substrate binding affinity, but it is not likely to be involved in the H^+ transfer pathway to PQ.

(iv) In addition, a loop analogous to the cyt *b* (bc_1) fold containing Arg218 and Thr213 does not exist in the b_6f complex. This structure is partially replaced by folding of the C-terminus of cyt b_6 and the N-terminus of subunit IV. These differences argue against the presence of a conserved E/R H^+ uptake pathway.

Despite the absence of sequence similarity between the proton-channel regions of the cyt bc_1 and cyt b_6f complexes, more than one H^+ uptake channel and several features related to the location of the entrance of the H^+-uptake pathways may be preserved in the stromal *n*-side surface of the b_6f complex. The stromal *n*-side surface charge distribution on the cyt b_6f complex is similar to that on cyt bc_1 and is predominantly basic. However, the part of the stromal *n*-side surface facing the monomer-monomer interface in the b_6f complex contains a larger number of acidic residues. These are located near the N-terminus of cyt b_6 (Asp6, Glu10, Glu13, Asp19, and Asp20), and suIV (Asp8, Asp11, Glu29, Asp35), the C-terminus of cyt *f* (Glu280, Glu286), and the N-terminus of ISP (Asp9, Asp12). Many of the listed residues are located within a radius of 10 Å from their center. It was demonstrated that charged residue patches on the surface of the protein molecule, together with anionic phospholipids, may act as a buffer that concentrates protons and passes them directly to a proton-translocating pore [91]. An involvement of negatively charged residues in a proton-collecting antenna on a surface of other proton pumps was suggested for purple bacteria RC and subunit III of cytochrome *c* oxidase [96]. An analogous acidic residue patch does not exist on the corresponding surface of cyt bc_1, possibly because the surface is involved in interaction with the core subunit, Cor1p, which has an acidic surface at the *n*-side membrane-stromal interface. The CL/K-pathway leads protons from the cardiolipin binding side on the opposite side of the monomer.

The two closest acidic residues to the proposed PQ reduction site (NQNO/TDS binding site) are Asp20 of cyt b_6 and Asp35 of suIV, which are at a distance of approximately 6 Å, from the putative PQ site, facing it from two different directions at a right-angle (Figure 7.5). This Asp35 is positioned from the same direction, relative to the inferred position of the PQ bound to heme c_n, as the CL/K-pathway in the bc_1 complex. The residue is at the same level as the NQNO binding site and ~4 Å closer to the stromal *n*-side surface compared to Lys228 or Asp229 of cyt bc_1. The latter are the two key charged residues of the CL/K-pathway that are closest to the UQ reduction site, but which do not have an analog in the b_6f complex. Another acidic residue close (3.4 Å) to Asp35 is Glu29 of suIV. Since no interaction corresponding to the binding of the Cor1p subunit is present in the cyt b_6f complex, the H^+ uptake pathway may take a more direct route leading to the stromal *n*-side surface H^+-collecting antenna that involves acidic residues at the intersection of the two monomers.

A different possibility is present on the opposite side of the PQ binding site. Here, residues corresponding to Arg218, Asn208, and Thr213 at the E/R-pathway entrance of bc_1 complex may be substituted by His29 of cyt b_6 and His24 of suIV that are placed in a similar position at the surface of the molecule (Figure 7.5). The two histidines may represent an entrance/transition from the proton buffer

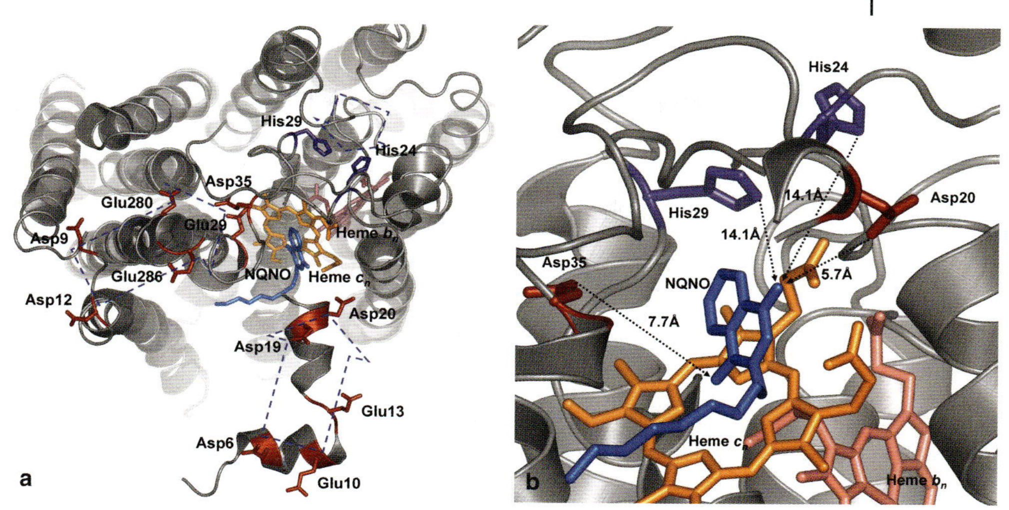

Figure 7.5 **(a)** H^+ uptake from the stromal *n*-side bulk aqueous phase to the heme c_n through possible pathways that enriched in carboxylate residues. The b_6f complex oriented to display the stromal *n*-side surface of the structure. Arrows point at the entrance points (residues Asp20, Asp35 and His29/His24) of the proposed proton uptake pathways and demonstrate localization of the proton collecting antenna on the protein surface. **(b)** Termination of the H^+ uptake pathway at plastoquinone bound as axial ligand of heme c_n, modeled as the NQNO quinone analog inhibitor, whose binding at his site has been determined [6]. This view is perpendicular to the plane of the phospholipid bilayer, looking from the other monomer in the b_6f complex. Some of the cyt b_6 residues, Ser23-Tyr25 (panel A); Met1-Ala18 and Thr22-Ser23 (panel B), are removed to reveal the PQ reduction site.

to a directed proton transport pathway leading to the PQ reduction site through the array of hydrogen bonded water molecules that resembles the six water molecule network proposed for the E/R-pathway in the cyt bc_1 complex.

The basic stromal *n*-side region of the complex may facilitate a proposed ferredoxin docking site [1, 7, 97]. Ferredoxin has an acidic isoelectric point and also a negative surface potential near its FeS cluster and C-terminus. A similar interaction is found between the basic surface of cyt *b* (bc_1) and the acidic surface of Qcr7p, an extrinsic subunit of the cyt bc_1 complex. Glu52 of the Qcr7p coordinates Wat176 and Wat351 at the entrance of the E/R-pathway in the bc_1 complex. Acidic residues at the ferredoxin surface may constitute a part of proton-conducting pathway or a part of proton-collecting antenna leading protons to His29/His24 gate. Similarly, two histidine residues provide a gateway from the surface proton-collection antenna to a coordinated water molecule proton-uptake pathway in the RC [93, 96]. The binding of ferredoxin may also be facilitated by binding of the extremely basic C-terminal half of the peptide PGR5 which has been found to be involved in cyclic electron flow around PhotosystemI [98].

Acknowledgments

The studies described in this review have been supported by NIH grant GM-38323 (WAC) and a grant-in-aid [B] to Young Scientists from the Japanese Ministry of Education, Culture, Sports, and Technology (EY). The higher resolution crystal structure analysis from our laboratory described herein was carried out at beam line SBC-19ID at the Advanced Photon Source, Argonne National Lab (DOE 1031-109-ENG-389).

References

1 Cramer, W.A., Zhang, H., Yan, J., Kurisu, G. and Smith, J.L. (2006) Transmembrane traffic in the cytochrome b_6f complex. *Annual Review of Biochemistry*, **75**, 769–90.

2 Ferreira, K.N., Iverson, T.M., Maghlaoui, K., Barber, J. and Iwata, S. (2004) Architecture of the photosynthetic oxygen-evolving center. *Science*, **303**, 1831–8.

3 Loll, B., Kern, J., Saenger, W., Zouni, A. and Biesiadka, J. (2005) Towards complete cofactor arrangement in the 3.0 A resolution structure of photosystem II. *Nature*, **438**, 1040–4.

4 Amunts, A., Drory, O. and Nelson, N. (2007) The structure of a plant photosystem I supercomplex at 3.4 A resolution. *Nature*, **447**, 58–63.

5 Jordan, P., Fromme, P., Witt, H.T., Klukas, O., Saenger, W. and Krauss, N. (2001) Three-dimensional structure of cyanobacterial photosystem I at 2.5 A resolution. *Nature*, **411**, 909–17.

6 Knoll, A.H. (2003) *Life on a New Planet: The First Three Billion Years of Evolution on Earth*, Princeton University Press, Princeton, NJ.

7 Kurisu, G., Zhang, H., Smith, J.L. and Cramer, W.A. (2003) Structure of the cytochrome b_6f complex of oxygenic photosynthesis: tuning the cavity. *Science*, **302**, 1009–14.

8 Stroebel, D., Choquet, Y., Popot, J.L. and Picot, D. (2003) An atypical haem in the cytochrome b_6f complex. *Nature*, **426**, 413–18.

9 Yamashita, E., Zhang, H. and Cramer, W.A. (2007) Structure of the cytochrome b_6f complex: quinone analogue inhibitors as ligands of heme c_n. *Journal of Molecular Biology*, **370**, 39–52.

10 Yan, J., Kurisu, G. and Cramer, W.A. (2006) Intraprotein transfer of the quinone analogue inhibitor 2,5-dibromo-3-methyl-6-isopropyl-p-benzoquinone in the cytochrome b_{6f} complex. *Proceedings of the National Academy of Sciences of the United States of America*, **103**, 69–74.

11 Huang, D., Everly, R.M., Cheng, R.H., Heymann, J.B., Schagger, H., Sled, V., Ohnishi, T., Baker, T.S. and Cramer, W.A. (1994) Characterization of the chloroplast cytochrome b_6f complex as a structural and functional dimer. *Biochemistry*, **33**, 4401–9.

12 Pierre, Y., Breyton, C., Lemoine, Y., Robert, B., Vernotte, C. and Popot, J.L. (1997) On the presence and role of a molecule of chlorophyll *a* in the cytochrome b_6f complex. *Journal of Biological Chemistry*, **272**, 21901–8.

13 Zhang, H., Huang, D. and Cramer, W.A. (1999) Stoichiometrically bound beta-carotene in the cytochrome b_6f complex of oxygenic photosynthesis protects against oxygen damage. *Journal of Biological Chemistry*, **274**, 1581–7.

14 Cramer, W.A., Yan, J., Zhang, H., Kurisu, G. and Smith, J.L. (2005) Structure of the cytochrome b_6f complex: new prosthetic groups, Q-space, and the “hors d'oeuvres hypothesis” for assembly of the complex. *Photosynthesis Research*, **85**, 133–43.

15 Joliot, P. and Joliot, A. (1988) The low-potential electron-transfer chain in the cytochrome *b/f* complex. *Biochimica et Biophysica Acta*, **933**, 319–33.

16 Lavergne, J. (1983) Membrane potential-dependent reduction of cytochrome b_6 in an algal mutant lacking photosystem I centers. *Biochimica et Biophysica Acta*, **725**, 25–33.

17 Zatsman, A.I., Zhang, H., Gunderson, W.A., Cramer, W.A. and Hendrich, M.P. (2006) Heme-heme interactions in the cytochrome b_6f complex: EPR spectroscopy and correlation with structure. *Journal of the American Chemical Society*, **128**, 14246–7.

18 Whitmarsh, J. (1986) Mobile electron carriers in Thylakoids, *in Encyclopedia of Plant Physiology: New Series: Photosynthesis III* (eds L.A. Staehelin and T.J. Avenson), Springer-Verlag, Berlin, pp. 508–27.

19 Joliot, P. and Joliot, A. (2006) Cyclic electron flow in C3 plants. *Biochimica et Biophysica Acta*, **1757**, 362–8.

20 Moss, D.A. and Bendall, D.S. (1984) Cyclic electron transport in chloroplasts: the Q-cycle and the site of action of antimycin. *Biochimica et Biophysica Acta*, **767**, 389–95.

21 Widger, W.R., Cramer, W.A., Herrmann, R.G. and Trebst, A. (1984) Sequence homology and structural similarity between cytochrome *b* of mitochondrial complex III and the chloroplast b_6f complex: position of the cytochrome *b* hemes in the membrane. *Proceedings of the National Academy of Sciences of the United States of America*, **81**, 674–8.

22 Schutz, M., Brugna, M., Lebrun, E., Baymann, F., Huber, R., Stetter, K.O., Hauska, G., Toci, R., Lemesle-Meunier, D., Tron, P., Schmidt, C. and Nitschke, W. (2000) Early evolution of cytochrome *bc* complexes. *Journal of Molecular Biology*, **300**, 663–75.

23 Carrell, C.J., Zhang, H., Cramer, W.A. and Smith, J.L. (1997) Biological identity and diversity in photosynthesis and respiration: structure of the lumen-side domain of the chloroplast Rieske protein. *Structure*, **5**, 1613–25.

24 Martinez, S.E., Huang, D., Ponomarev, M., Cramer, W.A. and Smith, J.L. (1996) The heme redox center of chloroplast cytochrome *f* is linked to a buried five-water chain. *Protein Science*, **5**, 1081–92.

25 Soriano, G.M., Smith, J.L. and Cramer, W.A. (2001) Cytochrome *f*, in *Handbook of Metalloproteins* (eds A. Messerschmidt, R. Huber, K. Wieghardt and T. Poulos), Wiley, London, pp. 172–81.

26 Yu, J., Hederstedt, L. and Piggot, P.J. (1995) The cytochrome *bc* complex (menaquinone:cytochrome *c* reductase) in *Bacillus subtilis* has a nontraditional subunit organization. *Journal of Bacteriology*, **177**, 6751–60.

27 Crofts, A.R., Lhee, S., Crofts, S.B., Cheng, J. and Rose, S. (2006) Proton pumping in the bc_1 complex: a new gating mechanism that prevents short circuits. *Biochimica et Biophysica Acta*, **1757**, 1019–34.

28 Crofts, A.R. (2004) The cytochrome bc_1 complex: function in the context of structure. *Annual Review of Physiology*, **66**, 689–733.

29 Osyczka, A., Moser, C.C., Daldal, F. and Dutton, P.L. (2004) Reversible redox energy coupling in electron transfer chains. *Nature*, **427**, 607–12.

30 Berry, E.A., Guergova-Kuras, M., Huang, L.S. and Crofts, A.R. (2000) Structure and function of cytochrome *bc* complexes. *Annual Review of Biochemistry*, **69**, 1005–75.

31 Darrouzet, E., Moser, C.C., Dutton, P.L. and Daldal, F. (2001) Large scale domain movement in cytochrome bc_1: a new device for electron transfer in proteins. *Trends in Biochemical Sciences*, **26**, 445–51.

32 Mitchell, P. (1975) The protonmotive Q-cycle: a general formulation. *FEBS Letters*, **59**, 137–9.

33 Mulkidjanian, A.Y. (2005) Ubiquinol oxidation in the cytochrome bc_1 complex: reaction mechanism and prevention of short-circuiting. *Biochimica et Biophysica Acta*, **1709**, 5–34.

34 Rich, P.R. (2004) The quinone chemistry of *bc* complexes. *Biochimica et Biophysica Acta*, **1658**, 165–71.

35 Glaser, E.G., Meinhardt, S.W. and Crofts, A.R. (1984) Reduction of cytochrome *b*-561 through the antimycin-sensitive site of the ubiquinol-cytochrome c_2 oxidoreductase complex of Rhodopseudomonas sphaeroides. *FEBS Letters*, **178**, 336–42.

36 de Vries, S., Albracht, S.P., Berden, J.A. and Slater, E.C. (1981) A new species of bound ubisemiquinone anion in QH_2: cytochrome *c* oxidoreductase. *Journal of Biological Chemistry*, **256**, 11996–8.

37 Gao, X., Wen, X., Esser, L., Quinn, B., Yu, L., Yu, C.A. and Xia, D. (2003) Structural basis for the quinone reduction in the bc_1 complex: a comparative analysis of crystal structures of mitochondrial cytochrome bc_1 with bound substrate and inhibitors at the Q_i site. *Biochemistry*, **42**, 9067–80.

37 Huang, L.S., Cobessi, D., Tung, E.Y. and Berry, E.A. (2005) Binding of the respiratory chain inhibitor antimycin to the mitochondrial bc_1 complex: a new crystal structure reveals an altered intramolecular hydrogen-bonding pattern. *Journal of Molecular Biology*, **351**, 573–97.

39 Esser, L., Quinn, B., Li, Y.F., Zhang, M., Elberry, M., Yu, L., Yu, C.A. and Xia, D. (2004) Crystallographic studies of quinol oxidation site inhibitors: a modified classification of inhibitors for the cytochrome bc_1 complex. *Journal of Molecular Biology*, **341**, 281–302.

40 Gao, X., Wen, X., Yu, C., Esser, L., Tsao, S., Quinn, B., Zhang, L., Yu, L. and Xia, D. (2002) The crystal structure of mitochondrial cytochrome bc_1 in complex with famoxadone: the role of aromatic-aromatic interaction in inhibition. *Biochemistry*, **41**, 11692–702.

41 Hunte, C., Koepke, J., Lange, C., Rossmanith, T. and Michel, H. (2000) Structure at 2.3 Å resolution of the cytochrome bc_1 complex from the yeast *Saccharomyces cerevisiae* co-crystallized with an antibody Fv fragment. *Structure*, **8**, 669–84.

42 Kim, H., Xia, D., Yu, C.A., Xia, J.Z., Kachurin, A.M., Zhang, L., Yu, L. and Deisenhofer, J. (1998) Inhibitor binding changes domain mobility in the iron-sulfur protein of the mitochondrial bc_1 complex from bovine heart. *Proceedings of the National Academy of Sciences of the United States of America*, **95**, 8026–33.

43 Covian, R. and Trumpower, B.L. (2006) Regulatory interactions between ubiquinol oxidation and ubiquinone reduction sites in the dimeric cytochrome bc_1 complex. *Journal of Biological Chemistry*, **281**, 30925–32.

44 Gutierrez-Cirlos, E.B. and Trumpower, B.L. (2002) Inhibitory analogs of ubiquinol act anti-cooperatively on the yeast cytochrome bc_1 complex. Evidence for an alternating, half-of-the-sites mechanism of ubiquinol oxidation. *Journal of Biological Chemistry*, **277**, 1195–202.

45 Rich, P.R. (1988) A critical examination of the supposed variable proton stoichiometry of the chloroplast cytochrome *bf* complex. *Biochimica et Biophysica Acta*, **932**, 33–42.

46 Sacksteder, C.A., Kanazawa, A., Jacoby, M.E. and Kramer, D.M. (2000) The proton to electron stoichiometry of steady-state photosynthesis in living plants: a proton-pumping Q cycle is continuously engaged. *Proceedings of the National Academy of Sciences of the United States of America*, **97**, 14283–8.

47 Graan, T. and Ort, D.R. (1983) Initial events in the regulation of electron transfer in chloroplasts. The role of the membrane potential. *Journal of Biological Chemistry*, **258**, 2831–6.

48 Cape, J.L., Bowman, M.K. and Kramer, D.M. (2006) Understanding the cytochrome *bc* complexes by what they don't do. The Q-cycle at 30. *Trends in Plant Science*, **11**, 46–55.

49 Bendall, D.S. and Manasse, R. (1995) Cyclic photophosphorylation and electron transport. *Biochimica et Biophysica Acta*, **1229**, 23–38.

50 Breyton, C., Nandha, B., Johnson, G.N., Joliot, P. and Finazzi, G. (2006) Redox modulation of cyclic electron flow around photosystem I in C3 plants. *Biochemistry*, **45**, 13465–75.

51 Joliot, P. and Joliot, A. (2005) Quantification of cyclic and linear flows in plants. *Proceedings of the National Academy of Sciences of the United States of America*, **102**, 4913–18.

52 Joliot, P. and Joliot, A. (2002) Cyclic electron transfer in plant leaf. *Proceedings of the National Academy of Sciences of the United States of America*, **99**, 10209–14.

53 Kramer, D.M., Avenson, T.J. and Edwards, G.E. (2004) Dynamic flexibility in the light reactions of photosynthesis governed by both electron and proton transfer reactions. *Trends in Plant Science*, **9**, 349–57.

54 Munekage, Y., Hashimoto, M., Miyake, C., Tomizawa, K., Endo, T., Tasaka, M. and Shikanai, T. (2004) Cyclic electron flow around photosystem I is essential for photosynthesis. *Nature*, **429**, 579–82.

55 Nelson, N. and Yocum, C.F. (2006) Structure and function of photosystems I and II. *Annual Review of Plant Biology*, **57**, 521–65.

56 Shikanai, T. (2007) Cyclic electron transport around photosystem I: genetic approaches. *Annual Review of Plant Biology*, **58**, 199–217.

57 Zhang, H., Whitelegge, J.P. and Cramer, W.A. (2001) Ferredoxin: $NADP^+$ oxidoreductase is a subunit of the chloroplast cytochrome b_6f complex. *Journal of Biological Chemistry*, **276**, 38159–65.

58 Cape, J.L., Strahan, J.R., Lenaeus, M.J., Yuknis, B.A., Le, T.T., Shepherd, J.N., Bowman, M.K. and Kramer, D.M. (2005) The respiratory substrate rhodoquinol induces Q-cycle bypass reactions in the yeast cytochrome bc_1 complex: mechanistic and physiological implications. *Journal of Biological Chemistry*, **280**, 34654–60.

59 Forquer, I., Covian, R., Bowman, M.K., Trumpower, B.L. and Kramer, D.M. (2006) Similar transition states mediate the Q-cycle and superoxide production by the cytochrome bc_1 complex. *Journal of Biological Chemistry*, **281**, 38459–65.

60 Muller, F., Crofts, A.R. and Kramer, D.M. (2002) Multiple Q-cycle bypass reactions at the Q_o site of the cytochrome *bc*1 complex. *Biochemistry*, **41**, 7866–74.

61 Muller, F.L., Roberts, A.G., Bowman, M.K. and Kramer, D.M. (2003) Architecture of the Q_o site of the cytochrome *bc*1 complex probed by superoxide production. *Biochemistry*, **42**, 6493–9.

62 Sun, J. and Trumpower, B.L. (2003) Superoxide anion generation by the cytochrome bc_1 complex. *Archives of Biochemistry and Biophysics*, **419**, 198–206.

63 Kramer, D.M., Roberts, A.G., Muller, F., Cape, J. and Bowman, M.K. (2004) Q-cycle bypass reactions at the Q_o site of the cytochrome bc_1 (and related) complexes. *Methods in Enzymology*, **382**, 21–45.

64 Hunte, C., Palsdottir, H. and Trumpower, B.L. (2003) Protonmotive pathways and mechanisms in the cytochrome bc_1 complex. *FEBS Letters*, **545**, 39–46.

65 Apel, K. and Hirt, H. (2004) Reactive oxygen species: metabolism, oxidative stress, and signal transduction. *Annual Review of Plant Biology*, **55**, 373–99.

66 Ouyang, Y., Grebe, R., Guo, L.W., Horn, D., Nelson, M., Wang, Q., Whitmarsh, J., Finazzi, G., Facciotti, C., Wollman, F.A., de Vitry, C. and Kallas, T. (2004) Mutational analysis of the Rieske iron-sulfur protein and the cytochrome *bf* complex quinone reductase site. *Cellular and Molecular Biology Letters*, **9**, 38–42.

67 Salerno, J., McGill, J.W. and Gerstle, G.C. (1983) The electron paramagnetic resonance spectra of partially purified cytochrome b_6f complex from spinach. *FEBS Letters*, **162**, 257–61.

68 Nitschke, W. and Hauska, G. (1987) On the nature of the $g = 6$ EPR signal in isolated cytochrome b_6f complex from spinach chloroplasts. *Biochimica et Biophysica Acta*, **892**, 314–19.

69 Zhang, H., Primak, A., Bowman, M.K., Kramer, D.M. and Cramer, W.A. (2004) Characterization of the high-spin heme *x* in the cytochrome b_6f complex of oxygenic photosynthesis. *Biochemistry*, **43**, 16329–36.

70 Golombek, A.P. and Hendrich, M.P. (2003) Quantitative analysis of dinuclear manganese(II) EPR spectra. *Journal of Magnetic Resonance*, **165**, 33–48.

71 Hendrich, M.P. and Debrunner, P.G. (1989) Integer-spin electron paramagnetic resonance of iron proteins. *Biophysical Journal*, **56**, 489–506.

72 Hendrich, M.P., Petasis, D., Arciero, D.M. and Hooper, A.B. (2001) Correlations of structure and electronic properties from EPR spectroscopy of hydroxylamine oxidoreductase. *Journal of the American Chemical Society*, **123**, 2997–3005.

73 Griffith, J.S. (1961) *The Theory of Transition-Metal Ions*, Cambridge University Press, London.

74 Schunemann, V., Trautwein, A.X., Illerhaus, J. and Haehnel, W. (1999) Mossbauer and electron paramagnetic resonance studies of the cytochrome *bf* complex. *Biochemistry*, **38**, 8981–91.

75 Abragam, A. and Bleaney, B. (1970) *Electron Paramagnetic Resonance of*

Transition Ions, Clarendon Press, Oxford.

76 Bencini, A. and Gatteschi, D. (1989) *Electron Paramagnetic Resonance of Exchange Coupled Systems*, Springer-Verlag, Berlin.

77 Palmer, G. (2000) *Electron Paramagnetic Resonance of Heme Proteins*, University Science Books, Sausalito, CA, pp. 121–85.

78 (a) Yeh, C.-Y., Chang, C.J. and Nocera, D.G. (2001) Hangman Porphyrins for the Assembly of a Model Heme Water Channel. *Journal of the American Chemical Society*, **123**, 1514–4.
(b) Cheng, B., Safo, M.K., Orosz, R.D., Reed, C.A., Debrunner, P.G. and Scheidt, W.R. (1994) Synthesis, Structure, and Characterization of Five-Coordinate Aquo (octaethylporphinato)iron(III) Perchlorate. *Inorganic Chemistry*, **33**, 1319–24.

79 de Vitry, C., Desbois, A., Redeker, V., Zito, F. and Wollman, F.A. (2004) Biochemical and spectroscopic characterization of the covalent binding of heme to cytochrome b_6. *Biochemistry*, **43**, 3956–68.

80 Shokhirev, N.V. (1998) Co- and counterrotation of magnetic axes and axial ligands in low-spin ferriheme systems. *Journal of the American Chemical Society*, **120**, 981–90.

81 Andersson, K.K., Lipscomb, J.D., Valentine, M., Munck, E. and Hooper, A.B. (1986) Tetraheme cytochrome c-554 from *Nitrosomonas europaea*. Heme-heme interactions and ligand binding. *Journal of Biological Chemistry*, **261**, 1126–38.

82 Einsle, O., Stach, P., Messerschmidt, A., Simon, J., Kroger, A., Huber, R. and Kroneck, P.M. (2000) Cytochrome *c* nitrite reductase from *Wolinella succinogenes*. Structure at 1.6 Å resolution, inhibitor binding, and heme-packing motifs. *Journal of Biological Chemistry*, **275**, 39608–16.

83 Mowat, C.G., Rothery, E., Miles, C.S., McIver, L., Doherty, M.K., Drewette, K., Taylor, P., Walkinshaw, M.D., Chapman, S.K. and Reid, G.A. (2004) Octaheme tetrathionate reductase is a respiratory enzyme with novel heme ligation. *Nature Structural and Molecular Biology*, **11**, 1023–4.

84 Bamford, V., Dobbin, P.S., Richardson, D.J. and Hemmings, A.M. (1999) Open conformation of a flavocytochrome c_3 fumarate reductase. *Nature Structural and Molecular Biology*, **6**, 1104–7.

85 Matias, P.M., Morais, J., Coelho, A.V., Meijers, R., Gonzalez, A., Thompson, A.W., Sieker, L., Legall, J. and Carrondo, M.A. (1997) A preliminary analysis of the three-dimensional structure of dimeric di-haem split-Soret cytochrome *c* from *Desulfovibrio desulfuricans* ATCC 27774 at 2.5-Ã resolution using the MAD phasing method: a novel cytochrome fold with a stacked-haem arrangement. *Journal of Biological Inorganic Chemistry*, **2**, 507–14.

86 Heitmann, D. and Einsle, O. (2005) Structural and biochemical characterization of DHC2, a novel diheme cytochrome *c* from *Geobacter sulfurreducens*. *Biochemistry*, **44**, 12411–19.

87 Upadhyay, A.K., Petasis, D.T., Arciero, D.M., Hooper, A.B. and Hendrich, M.P. (2003) Spectroscopic characterization and assignment of reduction potentials in the tetraheme cytochrome c_{554} from *Nitrosomonas europaea*. *Journal of the American Chemical Society*, **125**, 1738–47.

88 Alric, J., Pierre, Y., Picot, D., Lavergne, J. and Rappaport, F. (2005) Spectral and redox characterization of the heme c_i of the cytochrome b_6f complex. *Proceedings of the National Academy of Sciences of the United States of America*, **102**, 15860–5.

89 Baymann, F., Giusti, F., Picot, D. and Nitschke, W. (2007) The c_i/b_H moiety in the b_6f complex studied by EPR: a pair of strongly interacting hemes. *Proceedings of the National Academy of Sciences of the United States of America*, **104**, 519–24.

90 Xiong, J. and Bauer, C.E. (2002) A cytochrome *b* origin of photosynthetic reaction centers: an evolutionary link between respiration and photosynthesis. *Journal of Molecular Biology*, **322**, 1025–37.

91 Lange, C., Nett, J.H., Trumpower, B.L. and Hunte, C. (2001) Specific roles of protein-phospholipid interactions in the yeast cytochrome bc_1 complex structure. *The EMBO Journal*, **20**, 6591–600.

92 Lancaster, C.R. (2003) The role of electrostatics in proton-conducting membrane protein complexes. *FEBS Letters*, **545**, 52–60.

93 Paddock, M.L., Feher, G. and Okamura, M.Y. (2003) Proton transfer pathways and mechanism in bacterial reaction centers. *FEBS Letters*, **555**, 45–50.

94 Haines, T.H. and Dencher, N.A. (2002) Cardiolipin: a proton trap for oxidative phosphorylation. *FEBS Letters*, **528**, 35–9.

95 Nelson, M.E., Finazzi, G., Wang, Q.J., Middleton-Zarka, K.A., Whitmarsh, J. and Kallas, T. (2005) Cytochrome b_6 arginine 214 of *Synechococcus* sp. PCC 7002, a key residue for quinone-reductase site function and turnover of the cytochrome *bf* complex. *Journal of Biological Chemistry*, **280**, 10395–402.

96 Adelroth, P. and Brzezinski, P. (2004) Surface-mediated proton-transfer reactions in membrane-bound proteins. *Biochimica et Biophysica Acta*, **1655**, 102–15.

97 Cramer, W.A. and Zhang, H. (2006) Consequences of the structure of the cytochrome b_6f complex for its charge transfer pathways. *Biochimica et Biophysica Acta*, **1757**, 339–45.

98 Munekage, Y., Hojo, M., Meurer, J., Endo, T., Tasaka, M. and Shikanai, T. (2002) PGR5 is involved in cyclic electron flow around photosystem I and is essential for photoprotection in *Arabidopsis*. *Cell*, **110**, 361–71.

8
Plastocyanin and Cytochrome c_6: the Soluble Electron Carriers between the Cytochrome b_6f Complex and Photosystem I

Antonio Díaz-Quintana, Manuel Hervás, José A. Navarro, and Miguel A. De la Rosa

8.1
Introduction

In oxygenic photosynthesis, the transport of electrons from the cytochrome (Cyt) b_6f complex to Photosystem I (PSI) is performed by either plastocyanin (Pc) or Cyt c_6, two small soluble metalloproteins located inside the thylakoidal lumen of chloroplasts and cyanobacteria (Figure 8.1). (see also Chapter 7 on the cytochrome b_6f complex and Chapters 2 and 3 on Photosystem I). Whereas Pc is the only carrier in plants, both Pc and Cyt c_6 are synthesized by most cyanobacteria and eukaryotic algae [1, 2]. Actually, the two proteins can replace each other and play the same physiological role in most primitive oxygen-evolving photosynthetic organisms, but the synthesis of either one is controlled by copper availability within each organism [3].

Despite the fact that their overall structures are unrelated, the surface properties of the interaction regions of the two proteins in any given organism are sufficiently similar, with regard to recognition and electron transfer, for them to act interchangeably. Actually, the global charge and the distribution of hydrophobic and charged residues of both electron carriers are crucial in assessing their respective affinities and interactions with their two membrane-embedded partners, Cyt b_6f and PSI. Thus, Pc and Cyt c_6 can be presented as an excellent case study of biological evolution, which is not only convergent – two different structures playing the same role – but also parallel, two proteins varying in a concerted way from one organism to another [4].

Extensive work has been carried out to understand the functions of these two metalloproteins and how they depend on their structures [1, 5, 6]. Despite this intense work, several unanswered questions remain. Why do photosynthetic organisms need two different proteins for the same function? Why do they use two different metals, namely iron and copper? To what extent do the relative availabilities of metals act as constraints during evolution of the proteins? How does the chemical composition of the earth's atmosphere, specifically reducing or oxidizing, play into these constraints? Does the reaction mechanism of the two

Photosynthetic Protein Complexes: A Structural Approach. Edited by P. Fromme

ISBN: 978-3-527-31730-1

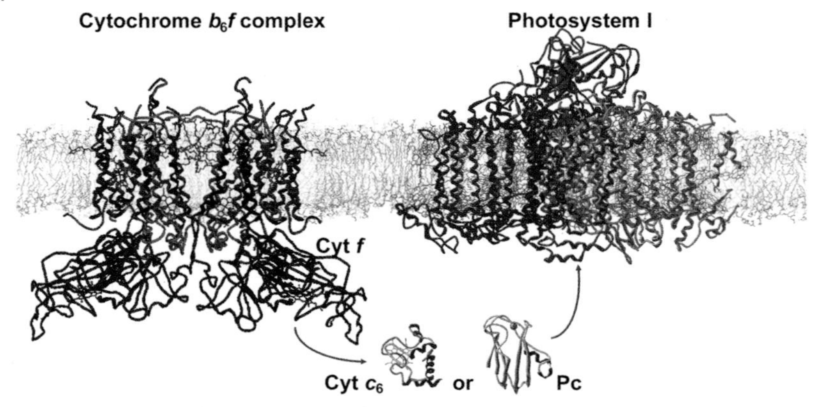

Figure 8.1 Scheme showing the role of Pc and Cyt c_6 as soluble electron-carrying proteins in the photosynthetic electron transfer chain. The two molecules are located inside the thylakoidal lumen, so acting as a shuttle of electrons from Cyt *f*, at the Cyt b_6f complex, to photooxidized P700 in Photosystem I.

electron-carrying proteins correlate with the changes in the interaction areas with their membrane partners? In this chapter, these and other questions are addressed by reviewing and summarizing, from an evolutionary point of view, the most recent experimental data in the literature on the structural and functional features of Pc and Cyt c_6.

8.2
An Evolutionary Proposal

It is now widely accepted that the evolution of earth's atmosphere is linked tightly to the evolution of its biota, with microorganisms determining its basic composition since the origin of life. This must be attributed to fundamental connections between genome evolution and the geochemical environment. Such a strong interplay of geochemical change and biological evolution raises questions about how new biochemical capabilities arise and spread in response to environmental changes in biogeochemical cycling [7]. Oxygenic photosynthesis, in particular, had to be initiated only once by a primitive cyanobacterium for the process to evolve. Microorganisms thus created the breathable, O_2-rich air that we enjoy today; but they also caused the formation of dioxygen in an anaerobic earth [8]. Consequently, cyanobacteria altered the solubility of metals – in particular, iron and copper – on a global scale, as a by-product of dioxygen-evolving photosynthesis.

In a reducing environment, the predominance of relatively soluble ferrous iron readily permits cellular acquisition; during early evolution this will have encouraged widespread recruitment of iron as a cofactor in biological redox chemistry. By contrast, under oxidizing conditions ferric iron is poorly soluble [9]. Today, all multi-cellular and, essentially, all single-cell organisms require iron for growth, despite the biological availability of iron being extremely limited by the insolubility of iron hydroxide. This is the reason why microbes synthesize low-molecular weight chelating agents, called siderophores, to bind and solubilize iron. Such ferric siderophore complexes are then transported into the bacteria by specific receptor proteins. In fact, competition for iron between a host and a bacterium is an important factor in determining the course of a bacterial infection. Because of that, different organisms utilize structurally varied siderophores to also competitively bind iron and gain selective growth advantages. Such competition even occurs in mammals where dietary iron is absorbed and bound to transferrin, the iron transport protein, and is then stored by the protein ferritin in cells.

Copper, on the other hand, first existed as cuprous sulfide – which is very insoluble in aqueous media – in the early and middle Precambrian period when the stationary oxygen pressure in the atmosphere was quite low; thus, copper might have been unavailable to organisms. Copper became Cu(II) upon the rise of atmospheric oxygen pressure. This probably occurred in the middle of the Proteozoic Era, when the first eukaryotic organisms appeared on earth. The change in the oxidation state of copper, combined with the new environment, made copper more accessible to more organisms. Copper may thus be considered an indicator element for the atmospheric evolutionary switch from anoxygenic to oxygenic, and for the evolution of higher organisms [10].

Iron and copper play an important role in the living world. From a brief consideration of their chemistry and biochemistry, one can conclude that the early chemistry of life used water soluble ferrous iron but had only limited access to copper as the latter was present in the form of water-insoluble Cu(I). The rise of atmospheric O_2 enabled the oxidation of iron, which led to a loss of bioavailability as insoluble Fe(III). Conversely, the oxidation of insoluble Cu(I) led to soluble Cu(II) [11].

Based on these findings, it has been proposed that Cyt c_6 was first incorporated in nature at a time when the reducing character of earth's atmosphere made iron more available than copper [8]. As the atmospheric molecular oxygen concentration was rising because of photosynthetic activity (see Chapters 4 and 5 for structure of Photosystem II and the oxygen-evolving complex), the relative bioavailabilities of iron and copper were going down and up, respectively, and Cyt c_6 was replaced with Pc [2]. In plants, copper is usually not a limiting element so Cyt c_6 would have disappeared, whereas Pc would have become a constitutively synthesized protein.

In fact, the absence of Cyt c_6 in plants was a widely accepted paradigm for many years, until a number of plant genomes became sequenced. In 2002, a modified Cyt c_6 – the so-called Cyt c_P [2], or Cyt c_{6A} [12] – was discovered in some plants [13], and it was proposed that such a cytochrome could replace Pc in *Arabidopsis* [14]. However, such a conclusion was challenged from two different approaches. On the one hand, the structural and functional analysis of the *Arabidopsis* Cyt c_P, compared with plant Pc and algal Cyt c_6, demonstrated that the plant heme protein is not an effective donor to its own PSI. Actually, the physicochemical parameters and surface electrostatic potential of Cyt c_P and Pc are so different in plants that the former cannot replace the latter [15]. On the other hand, inactivation of the two Pc genes of *Arabidopsis* by stable frame-shift mutations resulted in plants unable to grow photoautotrophically, even when the Cyt c_6-like protein was over-expressed at the same time [16].

8.3
General Features of Pc and Cyt c_6 Structures

Pc and Cyt c_6 are well-characterized proteins, both at the structural and functional levels, showing that their main structural features are well conserved among prokaryotic and eukaryotic photosynthetic organisms [2, 4]. Cyt c_6 is a typical class-I *c*-type cytochrome containing four α-helices and a heme group, in which the iron atom exhibits a typical histidine-methionine axial coordination [17–19]. The edge of pyrrole ring C and the ring D propionic group are solvent accessible, thus establishing a tentative electron transfer pathway to and from the heme iron atom [17, 20–23]. On the other hand, Pc consists of a single polypeptide chain forming a β-barrel with eight β strands and a small α-helix, along with a type-I blue copper center. The metal atom is coordinated by two histidines, one methionine and one cysteine, in a distorted tetrahedral coordination geometry [24–26]. His87 (plant residue numbering) is the only solvent-exposed Cu-ligand, thus making this group the most probable redox port of the protein (Figure 8.2).

Despite their different structures, both proteins share a number of physicochemical properties: their molecular masses are circa 10 kDa; their midpoint redox potential values are circa +350 mV at pH 7, in accordance with their role as electron shuttle between Cyt b_6f and PSI; and their isoelectric points are similar within the same organism, but can vary in parallel from one organism to another. In fact, Pc and Cyt c_6 show a global negative electrostatic charge in eukaryotes, but they are almost neutral or positively charged in cyanobacteria [4, 29, 30]. Such a concerted similarity and diversity in the isoelectric point is a remarkable aspect of the comparative evolutionary analysis of these two proteins.

The coincidence in their redox potential value is another notable similarity between Pc and Cyt c_6. As already pointed out, Pc is a member of the small type-I blue copper proteins, or cupredoxins. Other members of this class include azurins, pseudoazurins, and amicyanins – all having typical midpoint redox potential values around +300 mV. It is the protein-constrained trigonal pyramidal geometry, along with the copper-ligand distances of the redox center, which stabilizes the oxidized form over the reduced species [31–33]. In its turn, Cyt c_6 exhibits a midpoint redox potential value higher than most other soluble *c*-type cytochromes with Met-His coordination. As the geometry of the heme cofactor in Cyt c_6 resembles those in other class-I *c*-type cytochromes [17, 18], the variation in redox potential may result from other structural features in Cyt c_6, namely its hydrogen bonding network involving the iron axial ligands, the relative decrease in heme surface exposure

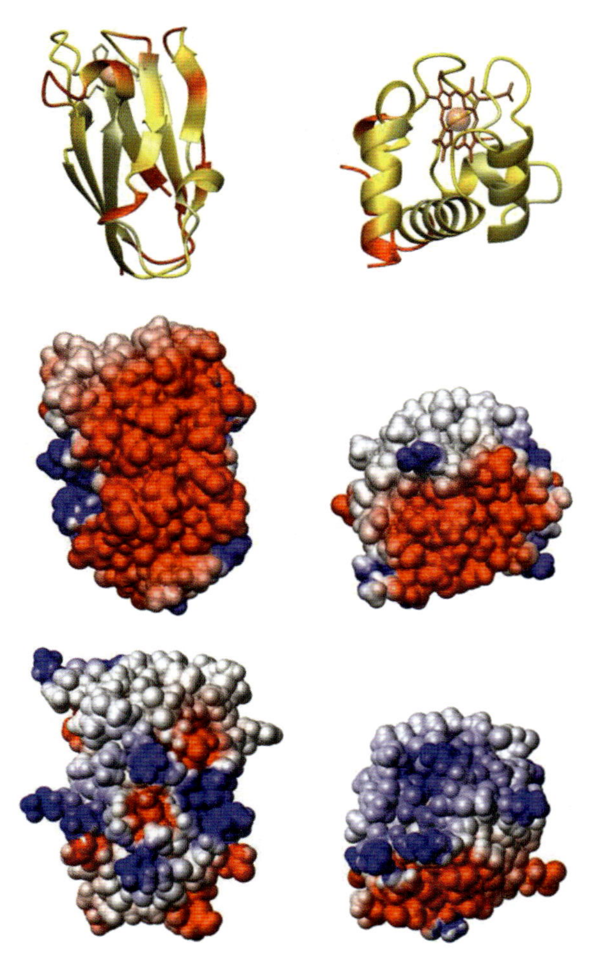

Figure 8.2 Structural comparison of Pc (left) and Cyt c_6 (right). Robertson diagrams (upper) and van der Waals representations of the surface electrostatic potential calculated at 50 mM ionic strength (middle and lower). Pc and Cyt c_6 are from either *Chlamydomonas* (upper and middle) or *Nostoc* (lower). The 3D structures used are: Pc (PBD entry, 2 plt) and Cyt c_6 (PDB entry, 1 cyj) from *Chlamydomonas*, Pc (PDB entry, 1 gim) and Cyt c_6 (homology model) [27] from *Nostoc*. Both proteins are oriented in a similar way to display their only arginine residue centered at the front and site 1 at the top. Graphics were drawn with MOLMOL [28]. In the Robertson diagrams, colors range from light yellow to red according to the b-factor of the Cα atoms. In the van der Waals representations, colors range from −3 (red) to +3 (blue).

and increased exposure of propionate D, and the small number of buried water molecules in the vicinity of the heme group [18, 19].

In this context, it is worth noting that the redox potential value of Cyt c_P, phylogenetically related to Cyt c_6, is circa 200 mV lower [2, 15, 34]. The recently solved structure of Cyt c_P (or Cyt c_{6A}) from *Arabidopsis* [12, 35] shows an overall folding that is very similar to that of Cyt c_6, with the peculiar extra loop of Cyt c_P lying at the protein surface far from the heme pocket. Such a loop is not responsible for the change in redox potential value of Cyt c_P as the loop deletion has no effect on this parameter [34]. A detailed structural comparison between Cyt c_P and Cyt c_6 reveals that the lower redox potential value of Cyt c_P may specifically arise from the lower polarity of residues packing the heme cofactor in positions 31 (alanine in Cyt c_P versus histidine or lysine in Cyt c_6) and 52 (valine or isoleucine versus glutamine) [12, 35].

The specific distribution of electrostatic charges on the surface of Pc and Cyt c_6 has a fundamental role in driving their attractive movements and in forming the transient complexes with Cyt b_6f and PSI [29] (see also Chapters 2, 3 and 7). Two well-defined areas were first identified at the surface of plant Pc: a hydrophobic patch around His87 (the so-called North pole, site 1), and a negatively charged patch around the highly conserved Tyr83 group (the so-called East face, site 2) [24, 36]. A striking exception to this general model is fern Pc, in which the acidic region drifts to the north of the molecule and spreads to the hydrophobic area, affecting its reactivity towards PSI [37, 38].

Site-directed mutagenesis studies of Pc from eukaryotes have provided relevant information on the role of specific residues at sites 1 and 2 [5, 39, 40]. It is now well established that certain residues at the hydrophobic patch are essential for the surface complementarity between Pc and PSI, thus allowing the formation of an active Pc–PSI complex [40–43]. Further studies with several cyanobacterial Pc molecules showed similar results [44, 45]. With regard to the interaction between Pc and the soluble domain of Cyt *f*, NMR data [20, 46–49] and kinetic analyses [50–52] invariably show that site 1 of Pc is involved in complex formation in both prokaryotic and eukaryotic organisms.

Whereas the hydrophobic patch is well conserved in all Pc molecules, the negative groups of the acidic area of eukaryotic Pc are substituted by neutral, or positively charged, residues in cyanobacteria (Figure 8.2). This explains the observed variations in isoelectric point [4, 29]. The presence of basic groups around site 1 in cyanobacteria (namely Lys33 and Arg88), which are not found in plants, seems to compensate the diversity of site 2 in these organisms. Such positive residues make the dipole moment of the molecule align along its main axis, so as to orient the encounter complex and allow Pc to establish proper hydrophobic interactions with its partners. On the other hand, the variations in site 2 of cyanobacterial Pc affect the long-range electrostatic interactions with its partners and determine the differences observed among organisms in the reaction mechanism (see Section 8.4.2). In plants, however, the dipole vector tilts to become almost perpendicular to the main axis. This yields a clear splitting of the binding site in electrostatic and hydrophobic interaction areas, which can indeed be related to the presence of a lysine-rich helix at the luminal domain of eukaryotic PsaF [53, 54].

There is a striking analogy between the surface properties of Pc and Cyt c_6. Actually, the surface region surrounding the accessible part of the heme group in Cyt c_6 resembles the hydrophobic patch of Pc, whereas the negatively charged area in eukaryotic Cyt c_6 resembles that at the east site of Pc [17, 55, 56]. In addition, the surface charge configuration of cyanobacterial Cyt c_6 changes from one organism to another, in parallel to the changes observed at the surface of Pc (Figure 8.2) [29, 30]. Moreover, cyanobacterial Pc and Cyt c_6 possess just one arginine group, each similarly located between their hydrophobic and electrostatic areas. In *Nostoc*, this single arginine residue (Arg88 in Pc, Arg67 in Cyt c_6) is required for the efficient oxidation of Cyt *f* [51] and the reduction of PSI [45]. Such similarities in the surface properties of Pc and Cyt c_6 within the same organism, along with their parallel variations among organisms, suggest that the two proteins possess regions that are functionally equivalent [4, 17, 55], as discussed in the following sections.

8.4 Reaction Mechanisms

8.4.1 Kinetic Models

The functional interaction of Pc and Cyt c_6 with their electron acceptor PSI has been intensively studied in a variety of photosynthetic organisms ranging from cyanobacteria to green algae and plants [1, 2, 5, 30]. Such a comparative analysis of PSI reduction by its two donor proteins has shed new light on the mechanism of electron transfer and its evolution. In contrast, the kinetic data concerning the interaction with the Cyt *f* subunit of the Cyt b_6f complex is limited to Pc from a few organisms; actually, no information with Cyt c_6 is available because of technical difficulties due to the similarity of Cyt c_6 and Cyt *f* spectral properties.

To facilitate the analysis of experimental data, three different kinetic models for the interaction of Pc and Cyt c_6 with PSI have been proposed [29, 57]. These models can be extrapolated not only to that of the interaction of the two soluble proteins with Cyt *f*, but also to other transient protein-protein interactions. A diagram of these kinetic models is shown in Figure 8.3. The type-I model applies when the membrane partner interacts with Pc or Cyt c_6 by following a collisional mechanism wherein the transient complex between the two reaction partners is not kinetically detectable. Type II involves the formation of a transient complex with a lifetime long enough to be calculated from kinetic data. And the type-III model–which does correspond to the full description of the whole reaction [5]–applies when it is possible to observe an additional step ascribed to the rearrangement of redox partners within the transient complex prior to electron transfer. The different kinetic behaviors are determined by the step acting as the bottleneck in each case; in other words, the kinetic model applied in every case will depend on the relative values of the equilibrium constants for complex association (K_A) and complex rearrangement (K_R), as well as the rate constant for electron transfer itself (k_{et}) (Figure 8.3). In the particular case of PSI reduction by Pc or Cyt c_6, monophasic

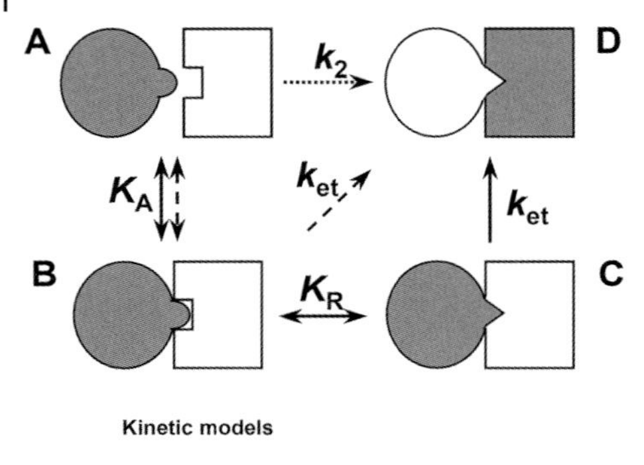

Kinetic models

Type I: A ⋯→ D

Type II: A ↔ B - - → D

Type III: A ↔ B ↔ C → D

Figure 8.3 Kinetic models proposed for the reaction mechanism of electron transfer within transient protein–protein complexes, including those of Pc and Cyt c_6 with its two membrane partners. The reduced and oxidized molecules are depicted in dark and white, respectively. See text for explanation.

kinetics are observed for type-I and II mechanisms, whereas the type-III model is always characterized by a biphasic kinetic profile [2, 29, 57].

The evolution from type I to types II and III does involve an improvement in the efficiency of the operating mechanism. In fact, the kinetics of PSI reduction by Pc and Cyt c_6 fit the more sophisticated type-III kinetic model in most green algae and plants, but predominantly apply to the two others in cyanobacteria [57–59]. No matter the source of proteins (cyanobacteria, algae, or plants), the reduction of Pc by Cyt *f* always follows monophasic kinetics–in some cases under limited time resolution. Thus, the mechanism of Cyt *f* oxidation by Pc cannot be easily assigned to a specific kinetic model [5, 51, 52] since the only parameters that can be calculated are the bimolecular rate constant and, eventually, the apparent electron transfer rate constant (see Section 8.4.2).

8.4.2
Nature of the Interaction Forces

The available kinetic data suggest that the relative weight of electrostatic forces in driving the interaction of Pc and Cyt c_6 with PSI has increased along evolution, from prokaryotes to eukaryotes. However, there is not a clear evolution pattern within the wide group of cyanobacteria, which live in many distinct environments. They may have adapted the interaction as a function of many different factors, namely temperature, ionic strength, composition of the growing media, or changes in the environmental conditions [2].

Extensive studies of the effect of ionic strength on the interaction of Pc and Cyt c_6 with PSI have provided crucial evidence on the nature of the forces involved in the reaction mechanism and its evolution. In most prokaryotes, the rate constant changes monotonically – either increasing or decreasing – with increasing ionic strength [27, 60]; no change is observed in others [61] (see below). It depends on the absence or presence of electrostatic interactions: hydrophobic interactions are not affected by changes in ionic strength, while electrostatic interactions decrease with increases in ionic strength. Eukaryotic systems show a bell-shape dependence of the PSI reduction rate constant with respect to ionic strength [57]. This indicates that, at very low ionic strength, the electrostatic forces are so strong that the transient Pc–PSI and Cyt c_6–PSI complexes are frozen in a rather rigid orientation that is not proper for electron transfer. At high ionic strength, in contrast, the electrostatic attraction is so weak that the partners do not recognize and bind each other. So, the optimal configuration becomes accessible only at mild ionic strength. This bell-shape profile only appears when kinetic data fit the type-III model, that is when the rearrangement becomes the rate-limiting step.

The rate constant of electron transfer from Cyt *f* to Pc, on the other hand, typically decreases monotonically with increasing ionic strength, although the extent of the effect depends upon the organism source of the reaction partners [51, 52, 62, 63].

Although electrostatics does affect the binding equilibrium, its long-range nature makes it drive the reaction partners toward many different unproductive conformations, with almost equal energies, rather than toward a single, productive conformation [6, 64, 65]. Only the state that is further stabilized by hydrophobic forces becomes both active and populated enough to be kinetically detected. Such a description disagrees with the former interpretation of biphasic ionic strength dependences of PSI reduction rates [57], according to which electrostatics leads to a restrained set of conformations rather than to a scattered ensemble. The kinetic data also indicate that the active orientation predominates in the rearrangement equilibrium under standard conditions [57].

Global charges of proteins are important for the establishment of the encounter complex through long-range interactions, but it is the surface charge distribution of the binding molecules which determines the specificity of the interaction. In fact, there is a close correlation between the rate of Cyt *f* oxidation by Pc mutants and their net charges [66]. The rate of PSI reduction by Pc mutants shows a similar trend [44], but a better correlation is found if the calculations are based on the electrostatic potential at the interface area – which depends on the surface charge distribution around the non-polar core – rather than on the global charge of the interacting proteins [67]. In addition, amino acid substitutions in Pc at different positions but involving the same net charge modification differently affect the kinetics of Cyt *f* oxidation [68]. Finally, theoretical Brownian dynamics calculations indicate that the electrostatics primarily restrain the conformational space of the complex [69, 70].

Assuming that the surface distribution of electrostatic charges affects the reactivity of the reaction partners, could it be a key parameter determining the evolution

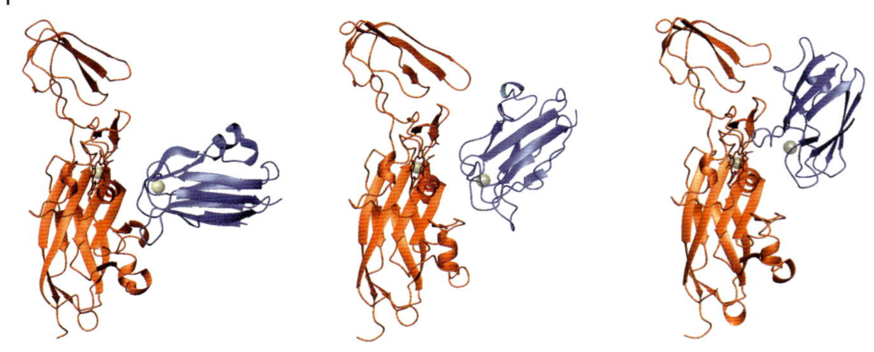

Figure 8.4 Structural comparison of cyanobacterial and plant Pc–Cyt *f* complexes. From left to right, the 3D structures used are from *Phormidium* [47], *Nostoc* [72] (PDB entry, 1tu2) and plant [65] (PDB entry, 1tkw). Cyt *f* and Pc are in red and blue, respectively. The heme porphyrin ring is depicted in sticks, the copper and iron atoms in spheres.

towards distinct reaction mechanisms? As pointed out above, cyanobacteria show an ample diversity in the mechanisms of PSI reduction that is closely related to the diversity in the electrostatic properties of the two electron donors, Pc and Cyt c_6. In fact, cyanobacteria exhibit a great variation at site 2 and adjacent regions of Pc, but they all conserve a number of specific electrostatic charges around site 1 that are absent in plants. Most of the mutations of such specific residues in cyanobacterial Pc have drastic effects on the reaction rates, but there are some exceptions. Arg88, for instance, is essential for Cyt *f* oxidation and PSI reduction in *Nostoc*, where attractive binding electrostatics predominates [45, 51, 52]. Such is not the case in *Phormidium*, where the bold, long-range electrostatics do not play any major role for Cyt *f* oxidation and PSI reduction, as inferred from both kinetic [63, 71] and NMR data [47]. Thus Pc from *Nostoc* and *Phormidium* may adopt a different orientation upon binding to PSI, as already found in the complexes of Pc with Cyt *f* in these two organisms (see Figure 8.4, and Section 8.6).

As expected, the changes in ionic groups near site 1 of Pc hinder hydrophobic interactions; however, the changes at remote regions do not. The effect of point mutations in Pc on the apparent activation parameters of PSI reduction has been widely studied [59, 71, 73, 74]. Noteworthy, mutations at or near site 1 show large enthalpy-entropy compensation effects, with a small compensation temperature, and a significant drop in the activation entropy [71, 74]. In addition, the enthalpy-entropy compensation effect corresponding to most mutations at site 2 of Pc [73] is not significant enough, according to Sharp's criteria [75]. The mutations near site 1 impair short-range electrostatic and hydrophobic interactions and lead to a poorly defined encounter complex, similar to that between myoglobin and Cyt b_5 [76], thus yielding a highly negative value for the activation entropy of electron transfer.

8.5
Evolution of Pc and Cyt c_6

8.5.1
Evolution of the Reaction Mechanisms

The comparative analysis of PSI reduction by Pc and Cyt c_6 in a wide range of photosynthetic organisms indicates that the kinetic mechanism of each donor protein has evolved, from type I to type II, to reach the maximum efficiency with type III [1, 2]. In fact, type I and II models are typical of cyanobacteria [57, 60, 77], and type III predominates in eukaryotic organisms and in cyanobacteria with positively charged donors [27, 57]. In organisms following the more elaborated type-III mechanism, the electrostatic attractions drive the formation of a stable intermediate complex, wherein the soluble carrier is oriented to allow the establishment of the hydrophobic interactions required for efficient electron transfer upon complex reorganization [57]. This pattern demands the complex to be highly dynamic, in agreement with spectroscopic data displaying a rather small correlation time of Pc bound to plant PSI [78]. The reorganization step, which is characteristic of this type of mechanism, occurs close to equilibrium and would thus appear in the evolution when electrostatic forces gain relevance in the interaction [57]. This may guarantee the final dissociation of the complex after electron transfer [50].

It has been proposed that Cyt c_6 optimized its interaction with PSI much earlier than Pc, as its reaction mechanism in most cyanobacteria ranges from type I to type II, and even to type III. Pc, in turn, is always less efficient than Cyt c_6, and only reacts according to type I or type II in these organisms [30, 79]. In eukaryotic algae, however, the experimental observation is just the opposite, with Pc and Cyt c_6 following the type-III kinetic model but with Pc being even more reactive than Cyt c_6. This suggests that Pc has also evolved to interact with PSI by following the most efficient way. Finally, in plants, Cyt c_6 has lost its original function and is constitutively replaced by the copper protein as donor to PSI [2].

In eukaryotes, the further optimization attained with the type-III mechanism could be related to the spatial separation of the surface patches involving distinct types of interactions, as well as to the appearance of the positively charged luminal domain of the PsaF subunit in PSI. Interestingly, the double substitution of Asp44 and Asp47 by arginine in *Synechocystis* Pc accelerates in vitro PSI reduction, with the rate constant showing a bell-shape dependence on ionic strength [44]. In nature, the absence of mutations that enhance the positive electrostatic potential at site 2 may be due to the requirement for a fast physiological turnover of soluble electron carriers in the functional cycle [50, 80].

8.5.2
Increase in Specificity and Efficiency

Experimental data show that the specificity and efficiency of the interaction of Pc and Cyt c_6 with their membrane partners have increased during evolution, with the most relevant forces involved in the process being predominantly hydrophobic at the beginning but slowly shifting to be both hydrophobic and electrostatic [2]. In cyanobacteria, the PSI binding site for the donor proteins is poorly defined [81–83] and PsaF does not appear to play any significant role [84, 85] (see also Chapter 2). It seems that the function of the hydrophobic patch of Pc and Cyt c_6 becomes well established in cyanobacteria, but the function of the charged patch, which may even be absent, is not yet well defined. The crucial evolutionary steps in the reaction mechanisms would thus involve concerted modifications in parallel both in the donor protein and in PSI. On the one hand, the donor proteins from eukaryotes acquire a well-defined negative patch for electrostatic interaction with PSI. On the other hand, the PsaF subunit of eukaryotic PSI develops a lysine-rich positive patch to form a specific binding pocket for Pc and Cyt c_6 [59, 86, 87] (see also Chapter 3).

8.5.3 Role of the Soluble Proteins in Driving the Interactions

The effect of ionic strength on the cross-reactions between cyanobacterial Cyt c_6–either slightly acidic (*Synechocystis*), neutral (*Arthrospira*), or basic (*Nostoc*)–and PSI isolated from any of these cyanobacteria [61] indicates that it is the heme protein which determines the differences observed in the rate constants (Figure 8.5). In any case, the highest efficiency is always attained with the native systems, that is, with Cyt c_6 and PSI isolated from the same organism. It thus seems that the specific structural features acquired by Cyt c_6 along evolution in each cyanobacterial species are responsible for the different kinetic properties observed in the interactions with PSI, whereas the docking site in PSI does not change significantly from one cyanobacterium to another [29, 61].

In addition, recent NMR analyses of the interaction between Pc and Cyt *f* from different cyanobacteria suggest that the global charges of the proteins play a major role in determining the affinity constant and that the binding mode is mainly controlled by the soluble electron carrier [49, 51]. The predominant role of Pc over Cyt *f* in driving the redox interactions between them is likewise demonstrated by a recent kinetic study using two sets of Pc and Cyt *f* variants that were both specifically mutated at electrostatically charged residues [51, 52].

8.6 Structural Analysis of the Transient Complexes

In the last years, many efforts have been made to analyze and solve the structures of the transient complexes formed by Pc and Cyt c_6 with their membrane partners [6]. The complex between Pc and Cyt *f*, in particular, has been one of the most extensively studied and the structures of several Pc-Cyt *f* complexes, either from prokaryotic or eukaryotic organisms, have already been solved [20, 47, 65, 72]. Plant complexes at the PDB [88] include those formed by turnip Cyt *f* with either spinach

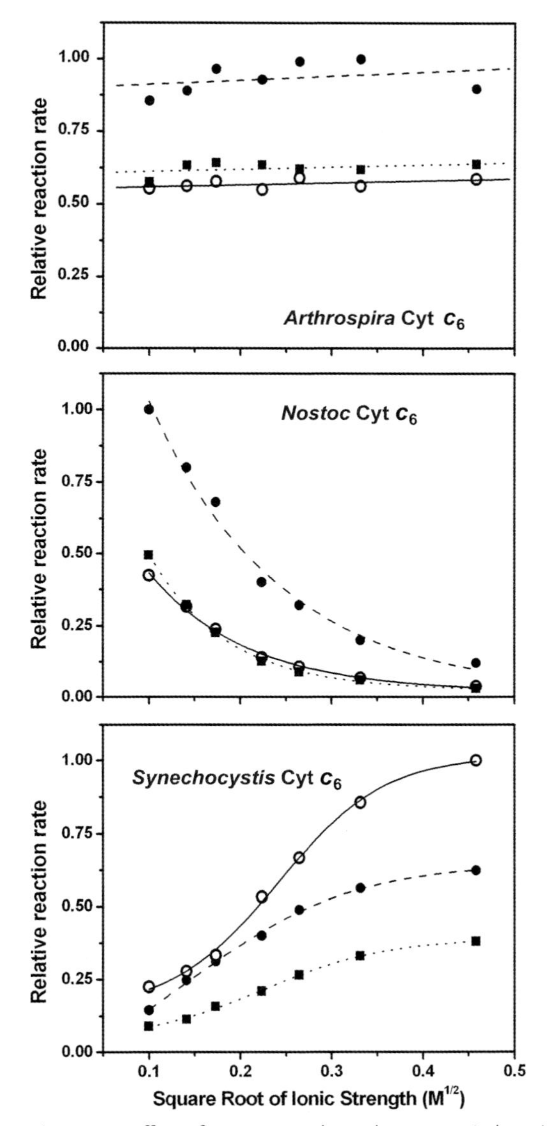

Figure 8.5 Effect of ionic strength on the cross-reactions between PSI and Cyt c_6 isolated from three different cyanobacteria. Relative values for the second-order rate constant (k_2) of the reduction of PSI from *Arthrospira* (squares), *Nostoc* (black circles) and *Synechocystis* (open circles) are plotted versus the square root of ionic strength. Original experimental data are from Hervás and coworkers [61].

Pc [20] (PDB entry, 1tkw) or poplar Pc [65] (PDB entry, 2pcf), whereas the cyanobacterial complexes have been mainly analyzed using Pc and Cyt *f* from the same organism, namely *Phormidium* [47] and *Nostoc* [72] (PDB entry, 1tu2).

As a common structural feature, the Pc–Cyt *f* structures show similar, modest interface areas (circa 600–850 $Å^2$ per protein). In all cases, site 1 of Pc and, in particular, the Cu-ligand His87 lie near the iron-coordinating Tyr1 of Cyt *f*, thus providing an efficient electron transfer pathway from the heme iron of Cyt *f* to the copper atom of Pc via Tyr1 and His87 [47, 65, 72]. Several other charged residues from both proteins are located at the rim of the interface. Interestingly, the key residue Arg88 of Pc (see Section 8.4.2) is not placed at the interaction area, but somewhat further (9 Å) from the closest charged residue of Cyt *f* (Glu165).

However, significant differences can be found not only between plant and cyanobacterial complexes, but also between those from cyanobacteria such as *Nostoc* and *Phormidium*. In fact, Pc binds to Cyt *f* in a head-on conformation in *Phormidium*, with the hydrophobic patch at site 1 accounting for the whole recognition patch in Pc. Conversely, the binding mode in plants and *Nostoc* fits a side-on orientation, wherein site 2 of Pc is also part of the interface area. The chemical shift perturbations of Pc amides at the *Phormidium* Pc-Cyt *f* complex do not depend on ionic strength [64], in agreement with kinetic data [63], whereas the plant and *Nostoc* complexes show a significant electrostatic dependence [51, 52, 68]. Thus, it seems that the stronger the electrostatics, the larger the shift from head-on to side-on configurations.

The *Phormidium* Pc-Cyt *f* complex shows a weak electrostatics and poor definition of lower energy solutions [47]. This may reflect a low convergence of calculations due to a small amount of experimental restraints because of the small association constant of the complex. However, Brownian dynamics calculations predict a lower homogeneity for the physiological complex between Cyt *f* and Pc from *Phormidium* than for the non-physiological complexes between *Phormidium* Cyt *f* and Pc from several other cyanobacteria [89]. This finding strengthens the idea that electrostatics restrain the conformational space of the transient complexes.

Unfortunately, there is currently no structural information available on the interaction area of Pc upon binding to PSI. However, the comparative kinetic analysis of PSI reduction and Cyt *f* oxidation by site-directed mutants of Pc clearly demonstrates that the copper protein uses roughly the same surface region for the interaction with the two membrane partners [45, 51, 52]. As can be seen in Figure 8.6, the replacement of specific residues in *Nostoc* Pc yields a similar effect on the efficiency of the copper protein either to donate electrons to PSI or to accept them from Cyt *f*, with most of the point mutations; the only exception is D54K making Pc decrease its reactivity.

With regard to the transient complexes formed by Cyt c_6 with its two reaction partners, NMR data is available on both of them. At first glance, the data concerning the complexes between Cyt c_6 and Cyt *f* from cyanobacteria [21] demonstrate the relevance of net charges in the binding equilibrium. Recently, a low-resolution model of the native Cyt c_6–Cyt *f* complex from *Nostoc* [23] has been reported. It is

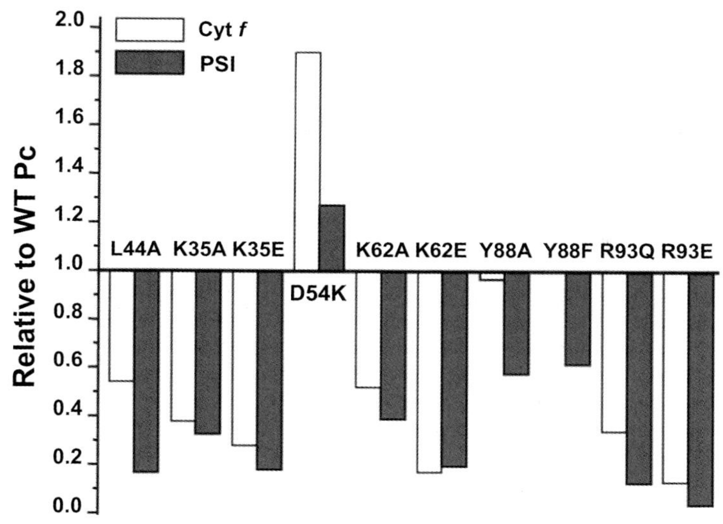

Figure 8.6 Effect of site-directed mutations in Pc on its reactivity towards Cyt *f* and PSI in *Nostoc*. Original experimental data are from Molina-Heredia and coworkers [45] and Albarrán and coworkers [51].

worth noting that *Nostoc* Cyt c_6 shows a similar, but not identical, orientation relative to Cyt *f* from *Nostoc* [23] and *Phormidium* [21]. In fact, Arg64 of *Nostoc* Cyt c_6, which is not involved in the non-physiological interaction with *Phormidium* Cyt *f*, exhibits a strong chemical shift perturbation in the physiological complex with *Nostoc* Cyt *f*. In the NMR-based docking model of the *Nostoc* Cyt c_6–Cyt *f* complex, the side-chain of Arg64 spans from the ridge to the core of the complex interface and accounts for circa 17% of the interaction area. Conversely, two lysine residues of *Nostoc* Cyt c_6 show significant perturbations in the non-physiological complex, but not in the physiological one. This illustrates how charge configuration can be crucial for the arrangement of the functional complex.

Although the 3D structure of the Cyt c_6–Cyt *f* complex from eukaryotes is not available, some theoretical models based on Brownian dynamics calculations have been reported for the *Chlamydomonas* complex [70]. The resulting modeled structures are different from those previously reported in cyanobacteria (see above), but the differences are the same as those found between cyanobacterial and plant Pc–Cyt *f* complexes [23]. Thus, it seems that Cyt c_6 approaches the small domain of Cyt *f* in eukaryotes, as does Pc, with electrostatics gaining predominance.

The interaction between Cyt c_6 and PSI from *Nostoc*, which follows a type-III kinetic model [57], has also been studied by NMR spectroscopy [22]. In this case, the signals of Cyt c_6 amides at the interface area exhibit a slow exchange regime within the chemical shift time scale, but most peaks corresponding to PSI-bound Cyt c_6 species broaden beyond detection limits because of the large photosystem size. Also, a specific line broadening, due to exchange, appears in some amide

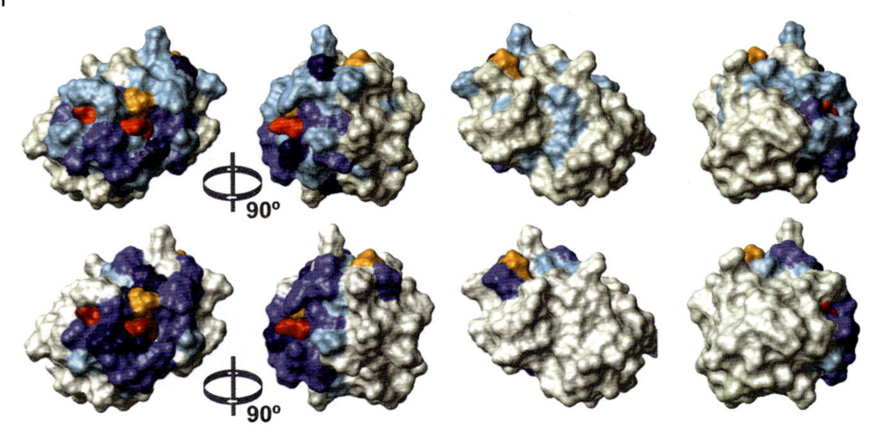

Figure 8.7 Identification of surface residues of Cyt c_6 responsible for the interaction with PSI (upper) and Cyt *f* (lower), as determined by NMR spectroscopy. The NMR perturbation map of Cyt c_6 amide signals is determined upon titration with either PSI monomers [22] or truncated Cyt *f* [23]. Ranging from lower to larger perturbations, the residues are colored in cyan, light blue, and dark blue. The heme group, prolines and unaltered residues are in red, orange, and white, respectively.

signals upon binding to PSI monomers; but most of them disappear, however, when PSI trimers are used instead. In spite of such technical difficulties, the interaction surface of Cyt c_6 in the Cyt c_6–PSI complex has been determined; it is in agreement with previous kinetic studies using site-directed mutants of the heme protein [27].

As can be seen in Figure 8.7, the specific residues of Cyt c_6 displaying any signal perturbation upon interaction with either Cyt *f* or PSI are practically the same. These findings demonstrate that Cyt c_6 does use the same functional area for the interaction with its two membrane partners, as Pc does. A precise comparative analysis of functional regions between Pc and Cyt c_6 is, however, more difficult. Nevertheless, one can conclude that Cyt c_6 and Pc possess, in principle, equivalent surface regions to interact with both Cyt *f* and PSI.

8.7 Concluding Remarks

The comparative structural and functional analyses of Pc and Cyt c_6 isolated from different organisms, ranging from cyanobacteria to green algae and plants, reveal that the two proteins possess two functional equivalent areas at the surface level – one hydrophobic (site 1), the other electrostatically charged (site 2) – for the interaction with both Cyt *f* and PSI. The electrostatic patch should be responsible for driving the formation of the transient complex with the corresponding membrane partner, and the hydrophobic patch should serve to establish the electron transfer pathway. In spite of the intrinsic structural differences between Pc and Cyt c_6, their two functional areas have evolved in a parallel way, thus giving rise to similarities such that each donor protein can replace the other with similar kinetic efficiency.

Lastly, the redox interaction between the soluble proteins, Pc and Cyt c_6, and their membrane-embedded partners, Cyt b_6f and PSI, has been optimized during evolution to improve the specificity and efficiency. In fact, the reaction mechanism would have evolved from a simple collision, type I, to a more sophisticated model involving not only formation of a transient complex, type II, but even a further rearrangement of the two partners to facilitate the final electron transfer step, type III.

Acknowledgments

Experimental research work at the authors' laboratory was supported by grants from the Spanish Ministry of Education and Science, the European Commission, and the Andalusian Government.

References

1 Hervás, M., Navarro, J.A. and De la Rosa, M.A. (2003) *Accounts of Chemical Research*, **36**, 798–805.
2 De la Rosa, M.A., Molina-Heredia, F.P., Hervás, M. and Navarro, J.A. (2006) Photosystem I: the ligth driven plastocyanin:ferredoxin oxidoreductase, in *Advances in Photosynthesis and Respiration Series* (ed. J.H. Golbeck), Springer, Dordrecht, The Netherlands, pp. 683–96.
3 Wood, P.M. (1978) *European Journal of Biochemistry*, **87**, 9–19.
4 Navarro, J.A., Hervás, M. and De la Rosa, M.A. (1997) *Journal of Biological Inorganic Chemistry*, **2**, 11–22.
5 Hope, A.B. (2000) *Biochimica et Biophysica Acta*, **1456**, 5–26.
6 Prudencio, M. and Ubbink, M. (2004) *Journal of Molecular Recognition*, **17**, 524–39.
7 Macalady, J. and Bansfield, J.F. (2003) *Earth and Planetary Science Letters*, **209**, 1–17.
8 Williams, R.J.P. and Fraústo Da Silva, J.J.R. (1997) *The Natural Selection of the Chemical Elements*, Oxford University Press, Oxford, UK.
9 Cavet, J.S., Borrelly, G.P.M. and Robinson, N.J. (2003) *FEMS Microbiology Reviews*, **27**, 165–81.
10 Ochiai, I. (1983) *Biosystems*, **16**, 81–6.
11 Crichton, R.R. and Pierre, J.L. (2001) *Biometals*, **14**, 99–112.
12 Marcaida, J., Schlarb-Ridley, B.G., Worrall, J.A.R., Wastl, J., Evans, T.J., Bendall, D.S., Luisi, B.F. and Howe, C.J. (2006) *Journal of Molecular Biology*, **360**, 968–77.
13 Wastl, J., Bendall, D.S. and Howe, C.J. (2002) *Trends in Plant Science*, **7**, 244–5.
14 Gupta, R., He, Z. and Luan, S. (2002) *Nature*, **30**, 567–71.
15 Molina-Heredia, F.P., Wastl, J., Navarro, J.A., Bendall, D.S., Hervás, M., Howe, C.J. and De la Rosa, M.A. (2003) *Nature*, **424**, 33–4.
16 Weigel, M., Varotto, C., Pesaresi, P., Finazzi, G., Rappaport, F., Salamini, F. and Leister, D. (2003) *Journal of Biological Chemistry*, **15**, 31286–9.
17 Frazão, C., Soares, C.M., Carrondo, M.A., Pohl, E., Dauter, Z., Wilson, K.S., Hervás, M., Navarro, J.A., De la Rosa, M.A. and Sheldrick, G.M. (1995) *Structure*, **3**, 1159–69.

18 Kerfeld, C.A., Anwar, H.P., Interrante, R., Merchant, S. and Yeates, T.O. (1995) *Journal of Molecular Biology*, **250**, 627–47.

19 Sawaya, M.R., Krogmann, D.W., Serag, A., Ho, K.K., Yeates, T.O. and Kerfeld, C.A. (2001) *Biochemistry*, **40**, 9215–25.

20 Ubbink, M., Ejdebäck, M., Karlsson, B.G. and Bendall, D.S. (1998) *Structure*, **6**, 323–35.

21 Crowley, P.B., Díaz-Quintana, A., Molina-Heredia, F.P., Nieto, P., Sutter, M., Haehnel, W., De la Rosa, M.A. and Ubbink, M. (2002) *Journal of Biological Chemistry*, **277**, 48685–9.

22 Díaz-Moreno, I., Díaz-Quintana, A., Molina-Heredia, F.P., Nieto, P.M. Hansson, Ö., De la Rosa, M.A. and Karlsson, B.G., (2005) *Journal of Biological Chemistry*, **280**, 7925–31.

23 Díaz-Moreno, I., Díaz-Quintana, A., Ubbink, M. and De la Rosa, M.A. (2005) *FEBS Letters*, **579**, 2891–6.

24 Guss, J.M. and Freeman, H.C. (1983) *Journal of Molecular Biology*, **169**, 521–63.

25 Guss, J.M., Harrowell, P.R., Murata, M., Norris, V.A. and Freeman, H.C. (1986) *Journal of Molecular Biology*, **192**, 361–87.

26 Inoue, T., Sugawara, H., Hamanaka, S., Tsukui, H., Suzuki, E., Kohzuma, T. and Kai, Y. (1999) *Biochemistry*, **38**, 6063–9.

27 Molina-Heredia, F.P., Díaz-Quintana, A., Hervás, M., Navarro, J.A. and De la Rosa, M.A. (1999) *Journal of Biological Chemistry*, **274**, 33565–70.

28 Koradi, R., Billeter, M. and Wüthrich, W. (1996) *Journal of Molecular Graphics*, **14**, 51–5.

29 De la Rosa, M.A., Navarro, J.A., Díaz-Quintana, A., De la Cerda, B., Molina-Heredia, F.P., Balme, A., Murdoch, P.S., Díaz-Moreno, I., Durán, R.V. and Hervás, M. (2002) *Bioelectrochemistry*, **55**, 41–5.

30 Díaz-Quintana, A., Navarro, J.A., Hervás, M., Molina-Heredia, F.P., De la Cerda, B. and De la Rosa, M.A. (2003) *Photosynthesis Research*, **75**, 97–110.

31 Redinbo, M.R., Yeates, T.O. and Merchant, S. (1994) *Journal of Bioenergetics and Biomembranes*, **26**, 49–66.

32 Gray, H.B., Malmström, B.G. and Williams, R.J.P. (2000) *Journal of Biological Inorganic Chemistry*, **5**, 551–9.

33 Paraskevopoulos, K., Sundararajan, M., Surendran, R., Hough, M.A., Eady, R.R., Hillier, I.H. and Hasnain, S.S. (2006) *Dalton Transactions*, **25**, 3067–76.

34 Wastl, J., Molina-Heredia, F.P., Hervás, M., Navarro, J.A., De la Rosa, M.A., Bendall, D.S. and Howe, C.J. (2004) *Biochimica et Biophysica Acta*, **1657**, 115–20.

35 Chida, H., Yokoyama, T., Kawai, F., Nakazawa, A., Akazaki, H., Takayama, Y., Hirano, T., Suruga, K., Satoh, T., Yamada, S., Kawachi, R., Unzai, S., Nishio, T., Park, S.-Y. and Oku, T. (2006) *FEBS Letters*, **580**, 3763–8.

36 Sigfridsson, K. (1998) *Photosynthesis Research*, **57**, 1–28.

37 Kohzuma, T., Inoue, T., Yoshizaki, F., Sasakawa, Y., Onodera, K., Nagatomo, S., Kitagawa, T., Uzawa, S., Isobe, Y., Sugimura, Y., Gotowda, M. and Kai, Y. (1999) *Journal of Biological Chemistry*, **274**, 11817–23.

38 Navarro, J.A., Lowe, C., Amons, R., Kohzuma, T., Canters, G., De la Rosa, M.A. Ubbink, M. and Hervás, M. (2004) *European Journal of Biochemistry*, **271**, 3449–56.

39 Nordling, M., Sigfridsson, K., Young, S., Lundberg, L.N. and Hansson, Ö. (1991) *FEBS Letters*, **291**, 327–30.

40 Sigfridsson, K., Young, S. and Hansson, Ö. (1996) *Biochemistry*, **35**, 1249–57.

41 Haehnel, W., Jansen, T., Gause, K., Klösgen, R.B., Stahl, B., Michl, D., Huvermann, B., Karas, M. and Herrmann, R.G. (1994) *The EMBO Journal*, **13**, 1028–38.

42 Hippler, M., Reichert, J., Sutter, M., Zak, E., Altschmied, L., Schröer, U., Herrmann, R.G. and Haehnel, W. (1996) *The EMBO Journal*, **15**, 6374–84.

43 Sigfridsson, K., Young, S. and Hansson, Ö. (1997) *European Journal of Biochemistry*, **245**, 805–12.

44 De la Cerda, B., Navarro, J.A., Hervás, M. and De la Rosa, M.A. (1997) *Biochemistry*, **36**, 10125–30.

45 Molina-Heredia, F.P., Hervás, M., Navarro, J.A. and De la Rosa, M.A. (2001) *Journal of Biological Chemistry*, **276**, 601–5.

46 Ejdebäck, M., Bergkvist, A., Karlsson, B.G. and Ubbink, M. (2000) *Biochemistry*, **39**, 5022–7.

47 Crowley, P.B., Otting, G., Schlarb-Ridley, B., Canters, G. and Ubbink, M. (2001) *Journal of the American Chemical Society*, **123**, 10444–53.

48 Bergkvist, A., Ejdebäck, M., Ubbink, M. and Karlsson, B.G. (2001) *Protein Science*, **10**, 2623–6.

49 Díaz-Moreno, I., Díaz-Quintana, A., De la Rosa, M.A. Crowley, P.B. and Ubbink, M. (2005) *Biochemistry*, **44**, 3176–83.

50 Illerhaus, J., Altschmied, L., Reichert, J., Zak, E., Herrmann, R.G. and Haehnel, W. (2000) *Journal of Biological Chemistry*, **275**, 17590–5.

51 Albarrán, C., Navarro, J.A., Molina-Heredia, F.P., Murdoch, P.S., De la Rosa, M.A. and Hervás, M. (2005) *Biochemistry*, **44**, 11601–7.

52 Albarrán, C., Navarro, J.A., De la Rosa, M.A. and Hervás, M. (2007) *Biochemistry*, **46**, 997–1003.

53 Ben-Shem, A., Frolow, F. and Nelson, N. (2003) *Nature*, **426**, 630–5.

54 Amunts, A., Drory, O. and Nelson, N. (2007) *Nature*, **447**, 58–63.

55 Ullmann, M., Hauswald, M., Jensen, A., Kostic, N.M. and Knapp, E.-W. (1997) *Biochemistry*, **36**, 16187–96.

56 Dikiy, A., Carpentier, W., Vandenberghe, I., Borsari, M., Safarov, N., Dikaya, E., Van Beeumen, J. and Ciurli, S. (2002) *Biochemistry*, **41**, 14689–99.

57 Hervás, M., Navarro, J.A., Díaz, A., Bottin, H. and De la Rosa, M.A. (1995) *Biochemistry*, **34**, 11321–6.

58 Sigfridsson, K., Hansson, Ö., Karlsson, B.G., Baltzer, L., Nordling, M., Lundberg, L.G. (1995) *Biochimica et Biophysica Acta*, **1228**, 28–36.

59 Hippler, M., Drepper, F., Farah, J. and Rochaix, J.D. (1997) *Biochemistry*, **36**, 6343–9.

60 Hervás, M., Ortega, J.M., Navarro, J.A., De la Rosa, M.A. and Bottin, H. (1994) *Biochimica et Biophysica Acta*, **1184**, 235–41.

61 Hervás, M., Díaz-Quintana, A., Kerfeld, C.A., Krogmann, D.W., De la Rosa, M.A. and Navarro, J.A. (2005) *Photosynthesis Research*, **83**, 329–33.

62 Soriano, G.M., Ponamarev, M.V., Piskorowski, R.A. and Cramer, W.A. (1998) *Biochemistry*, **37**, 15120–8.

63 Schlarb-Ridley, B.G., Bendall, D.S. and Howe, C.J. (2002) *Biochemistry*, **41**, 3279–85.

64 Crowley, P.B. and Ubbink, M. (2003) *Accounts of Chemical Research*, **36**, 723–30.

65 Lange, C., Cornvik, T., Díaz-Moreno, I. and Ubbink, M. (2005) *Biochimica et Biophysica Acta*, **1707**, 179–88.

66 Kannt, A., Young, S. and Bendall, D.S. (1996) *Biochimica et Biophysica Acta*, **1277**, 115–26.

67 Watkins, J.A., Cusanovich, M.A., Meyer, T.E. and Tollin, G. (1994) *Protein Science*, **3**, 2104–14.

68 Gong, X.S., Wen, J.Q. and Gray, J.C. (2000) *European Journal of Biochemistry*, **267**, 1732–42.

69 De Rienzo, F., Gabdouline, R.R., De Benedetti, P.G. and Wade, R. (2001) *Biophysical Journal*, **81**, 3090–104.

70 Haddadian, E.J. and Gross, E.L. (2006) *Biophysical Journal*, **90**, 566–77.

71 Schlarb-Ridley, B.G., Navarro, J.A., Spencer, M., Bendall, D.S., Hervás, M., Howe, C.J. and De la Rosa, M.A. (2002) *European Journal of Biochemistry*, **269**, 5893–902.

72 Díaz-Moreno, I., Díaz-Quintana, A., De la Rosa, M.A. and Ubbink, M. (2005) *Journal of Biological Chemistry*, **280**, 18908–15.

73 De la Cerda, B., Díaz-Quintana, A., Navarro, J.A., Hervás, M. and De la Rosa, M.A. (1999) *Journal of Biological Chemistry*, **274**, 13292–7.

74 Díaz-Quintana, A., De la Cerda, B., Hervás, M., Navarro, J.A. and De la Rosa, M.A. (2002) *Photosynthesis Research*, **72**, 223–30.

75 Sharp, K. (2001) *Protein Science*, **10**, 661–7.

76 Worrall, J.A.R., Liu, Y.J., Crowley, P.B., Nocek, J.M., Hoffman, B.M. and Ubbink, M. (2002) *Biochemistry*, **41**, 11721–30.

77 Navarro, J.A., Myshkin, E., De la Rosa, M.A., Bullerjahn, G.S. and Hervás, M. (2001) *Journal of Biological Chemistry*, **276**, 37501–5.

78 Danielsen, E., Scheller, H.V., Bauer, R., Hemmingsen, L., Bjerrum, M.J. and Hansson, Ö. (1999) *Biochemistry*, **38**, 11531–40.

79 Durán, R.V., Hervás, M., De la Rosa, M.A. and Navarro, J.A. (2005) *Biochemical and Biophysical Research Communications*, **334**, 170–5.

80 Durán, R.V., Hervás, M., De la Cerda, B., De la Rosa, M.A. and Navarro, J.A. (2006) *Biochemistry*, **45**, 1054–60.

81 Sun, J., Xu, W., Hervás, M., Navarro, J.A., De la Rosa, M.A. and Chitnis, P.R. (1999) *Journal of Biological Chemistry*, **274**, 19048–54.

82 Fromme, P., Jordan, P. and Krauss, N. (2001) *Biochimica et Biophysica Acta*, **1507**, 5–31.

83 Navarro, J.A., Hervás, M., Sun, J., De la Cerda, B., Chitnis, P.R. and De la Rosa, M.A. (2001) *Photosynthesis Research*, **65**, 63–8.

84 Chitnis, P.R., Purvis, D. and Nelson, N. (1991) *Journal of Biological Chemistry*, **266**, 20146–51.

85 Xu, Q., Chitnis, V.P., Yu, L. and Chitnis, P.R. (1994) *Journal of Biological Chemistry*, **269**, 3205–11.

86 Haehnel, W., Hesse, V. and Pröpper, A. (1980) *FEBS Letters*, **111**, 79–82.

87 Sommer, F., Drepper, F. and Hippler, M. (2002) *Journal of Biological Chemistry*, **277**, 6573–81.

88 Berman, H.M., Westbrook, J., Feng, Z., Gilliland, G., Bhat, T.N., Weissig, H., Shindyalov, I.N. and Bourne, P.E. (2000) *Nucleic Acids Research*, **28**, 235–42.

89 Gross, E.L. and Rosenberg, I. (2006) *Biophysical Journal*, **90**, 366–80.

9
The Structure of the H^+-ATP Synthase from Chloroplasts

Bettina Böttcher and Peter Gräber

9.1
Introduction

Proton translocating ATPases (H^+-ATP synthase, F-type ATPase, F_0F_1, EC 3.6.1.34) occur in the cytoplasmic membranes of bacteria (e.g. *E.coli*, EF_0F_1), in the thylakoid membranes of chloroplasts (CF_0F_1), and in the inner mitochondrial membranes (MF_0F_1). The protein complexes of the electron transport chains are also located in these energy converting membranes. During photosynthesis and respiration, the electron transport is coupled with a transmembrane proton transport that leads to the generation of a transmembrane eletrochemical potential difference of protons, $\Delta\tilde{\mu}_{H^+}$. The H^+-ATP synthases couple the $\Delta\tilde{\mu}_{H^+}$-driven backflux of protons across the membrane with the synthesis of ATP from ADP and inorganic phosphate [1]. Hydrolysis of ATP provides the free energy for almost all energy requiring processes in cells; therefore, H^+-ATP synthases are central enzymes of energy metabolism in cells. For example, the mitochondrial H^+-ATP synthases in humans produce between (50–70) kg ATP per day.

H^+-ATP synthases from different sources have a similar bipartite overall structure. A hydrophilic F_1-part extrudes from the membrane into the aqueous phase. It contains the nucleotide and phosphate binding sites and catalyzes the hydrolysis/synthesis of the β-γ-phosphate bond in ATP. A hydrophobic F_0-part is integrated into the membrane. It contains the proton binding sites and catalyzes the proton transport through the membrane. The F_1- and F_0-part are connected by a central and a peripheral stalk to which subunits from both parts contribute (Figure 9.1).

H^+-ATP synthases are multi-protein complexes. Although they are similar in shape, their subunit composition depends on the source of the enzyme. The H^+-ATP synthase from *Escherichia coli*, EF_0F_1 is the simplest enzyme, with eight different subunits. The most complex one is the enzyme from mitochondria, MF_0F_1, with at least 13 different subunits. The H^+-ATP synthase from chloroplasts, CF_0F_1, has nine different subunits (see Table 9.1 for subunit compositions of CF_0F_1, EF_0F_1, and MF_0F_1). CF_1 consists of five different subunits which are

Photosynthetic Protein Complexes: A Structural Approach. Edited by P. Fromme

ISBN: 978-3-527-31730-1

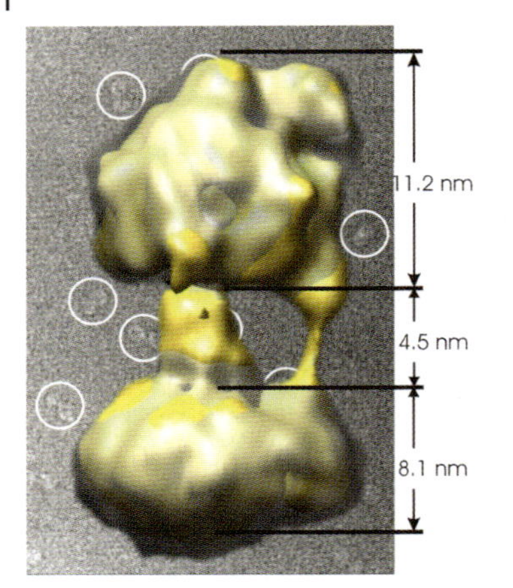

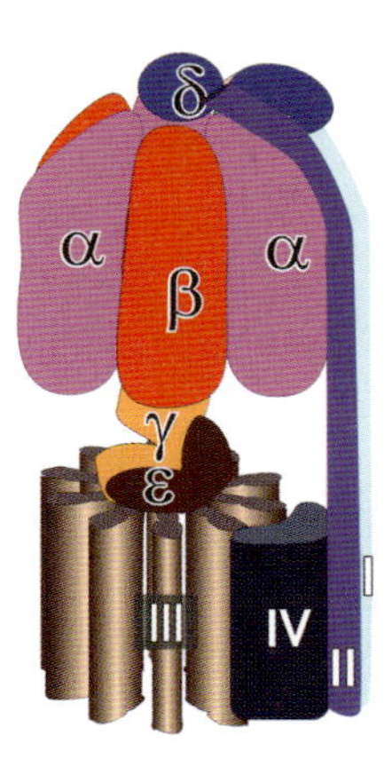

Figure 9.1 Left: Background electron micrograph of CF_0F_1 in vitrified buffer. Some enzymes are marked with white circles. Image analysis was carried out with 10000 particles and a three-dimensional map was calculated. A surface representation of this map is shown is shown in the foreground. Right: Cartoon of the subunit organization in CF_0F_1. Subunits designated with Greek letters form F_1, subunits labeled with Roman numbers form F_0.

designated by Greek letters (α, β, γ, δ, ε) [58]. They have the stoichiometry $\alpha_3\beta_3\gamma\delta\varepsilon$ [59, 60]. The four subunits of CF_0 are numbered by Roman numerals (I, II, III, IV) [61, 62] and have the subunit stoichiometry I II III_{14} IV [53, 56]. The nomenclature of the subunits in the different organisms is inconsistent. This is especially true for the F_0 subunits. Subunit IV is the largest subunit of CF_0 and is homologous to subunit a in *E. coli*. Subunits I and II form the second stalk and have homologies to subunit b from *E. coli*. Subunit III_{14} forms the rotor of the H^+-ATP synthase and corresponds to the c-ring in *E. coli*. In Table 9.1 homologous subunits are collected in the same rows.

The enzyme mechanism of H^+-ATP synthases from different sources is similar. According to the binding change mechanism [64, 66], the three catalytic nucleotide binding sites, located mainly on the β-subunits, adopt three different conformations. Each conformation allows a specific catalytic step: substrate binding, formation, and splitting of the β-γ–phosphate bond and product release. The three different conformations are caused by different interactions of each β-subunit with the γ-subunit. Rotation of the γ-subunit changes simultaneously its interaction with all three β-subunits. This gives rise to a strong cooperativity of the three catalytic sites. The high resolution structure of F_1 from mitochondria [71] shows directly that the γ-subunit interacts differently with the three β-subunits (see 9.2).

The mechanism of coupling of proton translocation and rotation is described by the following model [67, 68]: Subunits III form a ring in the membrane (see 9.3). They can be protonated from the internal aqueous phase through an entrance channel formed by subunits III and IV. The uncharged protonated amino acid

Table 9.1 Stoichiometry of subunits and structural data of F-type H^+ATPases and their subcomplexes.

E. coli		Chloroplasts		Mitochondria	
Subunits (Stoichiometry)	**Structural data Method, pdb-id, reference**	**Subunits (Stoichiometry)**	**Structural data Method, reference**	**Subunits (Stoichiometry)**	**Structural data Method, reference**
α (3)	–	α (3)	–	α (3)	–
β (3)	–	β (3)	–	β (3)	–
γ (1)	–	γ (1)	–	γ (1)	–
δ (1)	NMR N-term-domain **1ABV**, **2A7U** [2, 3]	δ (1)	–	OSCP (1)	NMR, 1BO5 [4]
ε (1)	NMR, 1BSH,1BSN [5, 6] X-ray, 1AQT[7]	ε (1)	–	δ (1)	–
–		–	–	ε (1)	–
–		–	–	IF1 (1)	X-ray, 1GMJ [8]
EF_1	EM [9–12]	CF_1	EM [13–15]	MF_1	EM [16–18]
a (1)	NMR, **1C17** [19]	IV (1)	–	a (1)	–
b (2)	NMR **1B9U** [20] **1L2P** [21]	I (1)	–	b (1)	–
		II (1)	–		
c (11)	NMR, 1C0V; 1A91[22] 1C99[19]	III (14)	–	c (10)	–
–	–	–	–	d (1)	–
–	–	–	–	F_6 (1)	NMR, 1VZS [23]
–	–	–	–	e (1)	–
–	–	–	–	f (1)	–
–	–	–	–	g (1)	–
EF_0	AFM [24–26]	CF_0	–	MF_0	–
EF_0F_1	EM [27–29]	CF_0F_1	AFM[30], EM [31–36]	MF_0F_1	EM [35, 37–40]
Sub-complexes					
–	–	$(\alpha\beta)_3$	X-ray 1FX0 [41] 1KMH [42]	–	–
$(\alpha\beta)_3\gamma\varepsilon$	X-ray, 1D8S [43]	–	–	$(\alpha\beta)_3\gamma\delta\varepsilon$	X-ray 2HLD [44] 1E79 [45]
$\gamma\varepsilon$	X-ray 1FS0 [46] 1JNV [47]	–	–	–	–
–	–	–	–	$((\alpha\beta)_3\gamma IF1)_2$	X-ray, 1QHH [48] EM [49]
				$(\alpha\beta)_3\gamma\delta\varepsilon c_{10}$	X-ray, 1QO1 [50]
c11(tartaricus)	X-ray **1YCE** [51] AFM/EM [52]	III_{14}	AFM [53, 54] EM [55]	–	
ac_{12}	NMR/ modelling **1C17** [19]	IV $(III)_{14}$	AFM [56]	–	
–	–	–	–	bdF_6	X-ray 2CLY [57]

Homologous subunits of the different enzymes are given in the same rows.
Subunit stoichiometries of the F_0-parts are estimated.
IF_1 is the inhibitor protein of MF_0F_1.
EM: electron microscopy; NMR: nuclear magnetic resonance; X-ray: X-ray diffraction.
Structures used for modelling of CF_0F_1 in Figures 9.2– 9.4 are shown bold.

IIIglu61 can move from the hydrophilic entrance channel into the hydrophobic membrane interior. This movement of the ring shifts simultaneously a protonated IIIglu61 into an asymmetrically located, hydrophilic channel with an exit to the outer aqueous phase where the proton is released. This mechanism explains the generation of a rotational movement by proton translocation through the membrane and the transmembrane Δ pH determines the direction of rotation. Since the III-ring interacts non-covalently with γ and ε, these subunits rotate together with the III-ring. During proton transport coupled ATP hydrolysis the direction of rotation is opposite to that during ATP synthesis as shown with single molecule spectroscopy [75, 76]. In the F_1 part chemical energy is used to generate rotational movement of the γ subunit, in the F_0 part osmotic energy is used to generate rotational movement of the III-ring. In the holoenzyme both parts are connected and form a chemiosmotic machine. Since the magnitude of the enzyme is in the nm-range, it is called a "nanomotor."

In this review, collected structural data for H^+-ATP synthases and their subunits (see Table 9.1 for overview) are presented to generate a pseudo atomic model of CF_0F_1. Most of the direct structural knowledge on CF_0F_1 comes from electron microscopy (EM), atomic force microscopy (AFM), and the crystallization and structure resolution with x-ray crystallography. The only available atomic model of a CF_0F_1-subcomplex represents the $\alpha_3\beta_3$-complex [41, 42]. In order to overcome this lack of high-resolution structural information, homology models for most of the subunits (γ,δ,ε, I, II, III, IV) were generated with MODELLER [69] using known structures of orthologous subunits as references (Table 9.1, bold). For the modeling, the sequences of target and reference were aligned according to the PFAM data-base [70]. The 3D-map of CF_0F_1 (Figure 9.1 left, [32]) derived from electron microscopy served as a scaffold for integrating the diverse structural information into a pseudo-atomic model of CF_0F_1.

9.2
The Structure of CF_1

The hydrophilic CF_1-part can be separated from the membrane integrated CF_0-part. First insights into the structure of CF_1 came from electron microscopy and image analysis of negatively stained particles that show a pseudo hexagonal arrangement of $\alpha_3\beta_3$ with $\gamma\delta\varepsilon$ located asymmetrically in the centre [13, 14]. These projection maps still lacked three-dimensional information. Only later localization of subunit δ of the *E. coli* ATP synthase by immuno electron microscopy revealed its location at the top, outside of the $\alpha_3\beta_3$-complex [65]; whereas, γ and ε form the central stalk that is anchored inside the $\alpha_3\beta_3$-hexagon by two long helices of γ [45, 71]. In the pseudo-hexagonal view of $\alpha_3\beta_3$, the centers of mass of the α-subunits extend further from the two central helices of γ than those of the β-subunits, which also enables distinguishing between α and β subunits in low-resolution maps.

The three $\alpha\beta$-pairs in $\alpha_3\beta_3\gamma$ can adopt different conformations. In the mitochondrial complex, the orientation of the central stalk relative to the $\alpha\beta$-pairs deter-

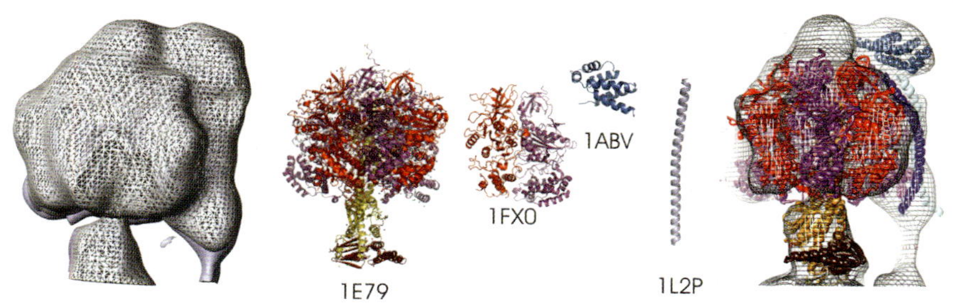

Figure 9.2 Superposition of the CF_1-part of the three dimensional map of CF_0F_1 derived from electron microscopy (grey) with models of the subunits. The α-subunits are shown in magenta, β in red, γ in yellow, ε in brown, δ in light blue, I and II in dark blue. The homology models of γ and ε are calculated from the $\alpha_3\beta_3\gamma\delta$ template (center, 1E79, mitochondrial δ is homologous to chloroplast ε). The αβ-pairs are derived from the chloroplast $\alpha_3\beta_3$(center, 1FXO). The homology model of the N-terminal region of δ was based on the structure of δ from *E. coli* (center, 1ABV), the models of subunit I and II are both based on the structure of the dimerization region of the b-subunits from *E. coli* (center, 1L2P).

mines the conformational state of the catalytic nucleotide binding sites. These sites are located on the β-subunits at the interface to the adjacent α subunits. In the mitochondrial $\alpha_3\beta_3\gamma\delta\varepsilon$-complex, one of the catalytic binding sites is empty (E), one has an ADP (D) loosely bound, and the third has a tightly bound non-hydrolysable ATP analogue(AMPPNP) (T).

This is different in the $\alpha_3\beta_3\gamma\varepsilon$ complex of chloroplasts [41, 42], where the crystallographic threefold axis superimposes with the pseudo-sixfold axis of the complex, making the αβ-pairs indistinguishable and the electron density for γ and ε uninterpretable. However, the αβ pair is well resolved and shows a similar conformation as the αβ-pair of the mitochondrial complex in the D or T conformation, but is incompatible with an αβ-pair in the E state. The significance of this difference to the mitochondrial enzyme is still unclear.

In order to obtain a pseudo-atomic model for CF_1, the structure of the mitochondrial $(\alpha\beta)_3\gamma\delta\varepsilon$ (1E79, [45]) was used as a template (Figure 9.2 center). The three copies of the αβ-pairs of CF_1 (1FX0, [41], Figure 9.2 centre) and the modeled γ and ε subunits were matched to the template. The newly generated sub-complex was placed into the upper half of the 3D-map of CF_0F_1 (Figure 9.2, right). The model accounted for most of the density in CF_1, except for a region at the top of $\alpha_3\beta_3$ and at the periphery of one of the α-subunits. Into the region at the top, the modeled N-terminal domain of the δ-subunit was placed adjacent to an α-subunit as determined by NMR-spectroscopy. This left room for the C-terminal domain of δ at the central top of $\alpha_3\beta_3$, in agreement with the localization by immuno-electron microscopy in EF_0F_1 [65]. Models of the dimerization region of subunits I and II were fitted into the unaccounted density at the periphery of the α-subunit to which δ is bound. The positioning of subunits I and II in close proximity to the α-subunit is in agreement with cross-linking data between the α- and the b-subunits in *E.*

coli [72]. Figure 9.2, right, shows a superposition of the 3D-map derived from electron cryo-microscopy (grey) with $\alpha_3\beta_3$ from chloroplasts and the homology models of γ, ε, δ and I + II (in color).

9.3
The Structure of CF_0

CF_0 consists of four different subunits, I, II, III, and IV. Subunit IV (homology to *E. coli a*) was initially missed by SDS-PAGE as a component of CF_0 [62] because it stains only weakly by Coomassie Blue. Later, subunit IV was identified in silver stained gels as a genuine component of CF_0F_1 [61].

Subunit III (homology to *E. coli c*) exists in multiple copies and is the major component of CF_0. It forms a stable complex that remains intact even during SDS gel electrophoresis and requires heating in SDS-buffer for dissociation into monomers [55]. The subunit-III complex has an oblate shape with a membrane spanning length of 6.1 nm and a diameter of 6.2 nm and a stoichiometry of III_{12} [55] was suggested. Later, investigations of the subunit III-complex with atomic force microscopy (AFM) allowed the identification of the individual subunits in the complex resulting in 14 monomers per complex [53]. The stoichiometry of the subunits in the ring seems to be fixed by the shape of the subunits and the contacts to their nearest neighbors as revealed by incomplete complexes that maintain their diameter [54]. Comparison of AFM images of CF_0, $III_{14}IV$ and III_{14} show an additional density inside the III-ring in sub-complexes that contain subunit IV [56]. This lead to the assumption that subunit IV is located in the centre of the III-ring. However, this notion is in contradiction to the current structural and functional understanding of CF_0. An alternative explanation for the presence of the additional density is the variable amounts of lipids that plug the ring similarly as observed for the c-ring of *I. tartaricus* [73].

For modeling the III_{14}-ring, the c-ring of *I. tartaricus* (1YCE, [51]) was used as a template (Figure 9.3 top, centre). This ring consists of only 11 subunits. Each subunit is a helical hairpin. The N-terminal helices form an unusually tightly packed inner ring. The C-terminal helices fill the grooves between the N-terminal helices and form an outer ring. Sodium ions bind between C-terminal and N-terminal helices in the center of the membrane, where the diameter of the ring is smallest. Since *I. tartaricus* is a sodium-translocating ATP synthase, it is likely that these sodium ions identify the potential proton-binding sites in H^+-ATP synthases.

With 11 subunits, the c-ring of *I. tartaricus* is significantly smaller in diameter than the III_{14}-ring in chloroplasts. For modeling of the III-ring, the inter-subunit packing of the homology models of subunit III was kept similar as in *I. tartaricus,* but the stoichiometry of subunits and the diameter of the ring were adjusted according to the AFM data ([53, 74], Figure 9.3 top, left). The modeled ring matched the surface topology measured by AFM (Figure 9.3 top, right).

Currently, no experimental data on the structure of subunit IV (subunit a) is available. For placing the subunit IV relative to the III-ring, the model of the *E.*

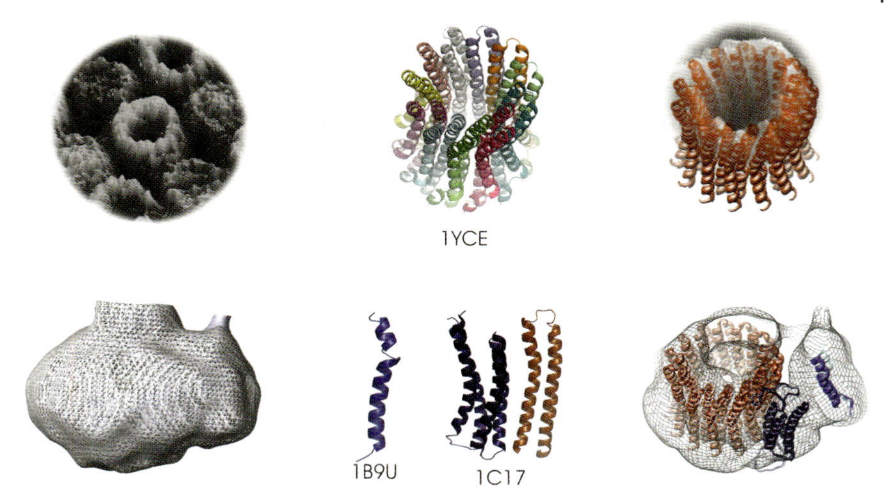

Figure 9.3 Top: Superposition of the structure of the subunit III_{14}-complex from atomic force microscopy (grey, left) with the homology model of subunit III_{14} (light brown, right), based on the structure of III_{14}-complex from *Ilyobacter tartaricus* (center, 1YCE). Bottom: Superposition of the CF_0-part of the three-dimensional map of CF_0F_1 from electron microscopy (grey) with homology models. The subunit III_{14}–complex is shown in light brown. The N-terminal domains of subunit I and II are based on the structure of this region of the b-subunit from *E. coli* (1A9U, centre), blue. The model of subunit IV is based on the structure of the a-subunit from *E. coli* (1C17, centre) blue.

coli ac_{12}-complex served as a template, based on computational structure prediction (1C17, [19], Figure 9.3, bottom, centre). The resulting $III_{14}IV$-model was placed as a whole into the CF_0-part of the 3D-map of CF_0F_1 (Figure 9.3 bottom, left). The CF_0-part in the 3D-map is considerably larger than the modeled $III_{14}IV$, which makes the placement tentative. The reason for the discrepancy in size is the presence of a detergent micelle that shields the hydrophobic regions of CF_0, which are usually integrated into the thylakoid membranes. The outline of the III-ring can be recognized in the 3D-map inside the detergent micelle. The III-ring protrudes at the bottom from the detergent micelle (Figure 9.3 bottom, left) and is closed by an unidentified density, which is similar to the lipids that plug the c-ring of *I. tartaricus* [73]. Adjacent to the III-ring, underneath the peripheral stalk, a free volume is detected that is not accounted for by the detergent micelle (Figure 9.3 bottom, right). The free volume is a region with electron density in the 3D-map from electron cryo microscopy, in which no protein is found after placing the homology model. This volume is separated from the III-ring by a gap, which is much wider than the distance between the III-ring and subunit IV in the model. Therefore, the $III_{14}IV$ model was placed with subunit IV adjacent to the gap. The membrane parts of subunits I and II were modeled according to the N-terminal region of the *E. coli* subunit b ([20], Figure 9.3, bottom, centre) and were positioned into the free volume directly underneath the peripheral stalk (Figure 9.3, bottom right).

9.4 The Structure of CF_0F_1

The homology model of CF_0F_1 (Figure 9.4 center) is obtained by placing the models of CF_1 and CF_0 (Figures 9.2 and 9.3) into the 3D-map of CF_0F_1 (Figure 9.4, left). Despite the wealth of structural information on orthologous subunits, some domains and fragments of subunits could not be modeled. These are the C-terminal region of the δ-subunit (Figure 9.4, green columns at the top of F_1), two helices of subunit IV (Figure 9.4, green columns in front of the rotor), and the region between the membrane integrated parts of subunits I and II and the CF_1 attached regions of subunits I and II (Figure 9.4, green columns in the peripheral stalk). According to EPR-measurements of Spin-Spin interactions after spin-labeling of subunits b in the stalk region, the b-subunits in EF_0F_1 are spaced by 2.9 nm in this region [63], which is far more than expected for a coiled coil structure. Therefore, the two green columns, representing the missing helical regions of I and II, were placed well apart, but inside, the scaffold provided by the image reconstruction (Figure 9.4, right).

The superposition of the pseudo atomic model of CF_0F_1 with the EM map shows that the model accounts for all density except the detergent micelle that surrounds the region that is usually membrane integrated, and the plug that closes the III-ring from the lumenal side. Thus, the combination of the low resolution electron microscopy data with the homology models of subunits and subunit complexes gives a structural model of CF_0F_1 with a higher resolution

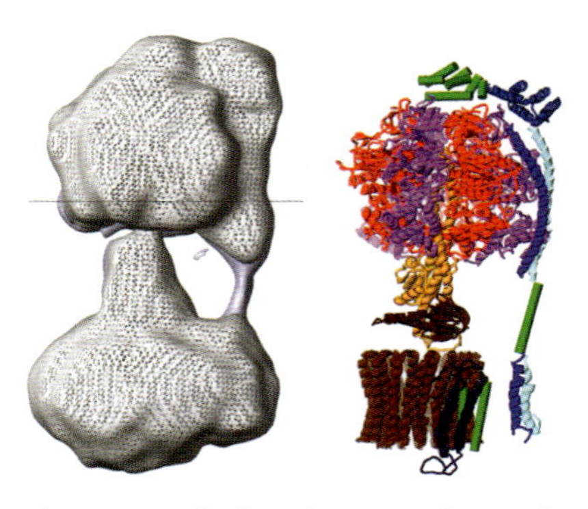

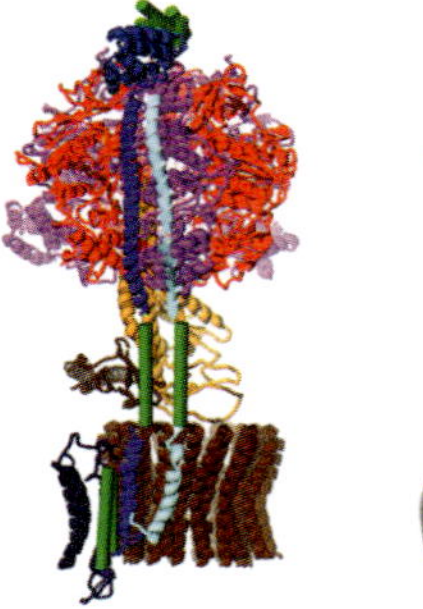

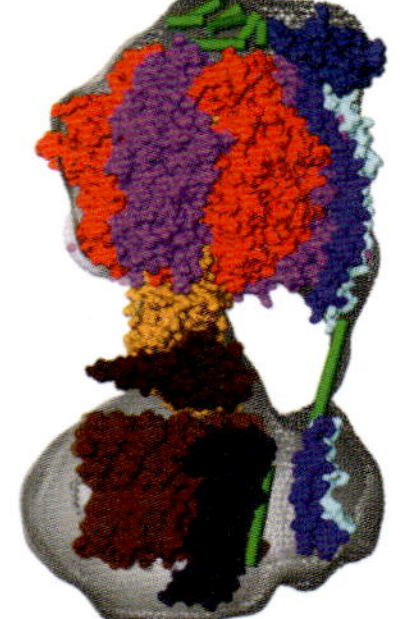

Figure 9.4 Left: Three-dimensional map of CF_0F_1 (grey). The line indicates the slice which is represented in Figure 9.5. Center: The homology models shown in Figures 9.2 and 9.3 are combined to a pseudo-atomic model of CF_0F_1. Regions which could not be modeled are filled with green columns. These are the C-terminal domain of subunit δ, the tether domain of subunits I and II between CF_0 and CF_1 and two helices of subunit IV. The pseudo-atomic model of CF_0F_1 is shown in two different side views, rotated by 90°. The rotation axis is perpendicular the membrane (color code as in Figures 9.2 and 9.3). Right: Superposition of the 3D-map from electron microscopy with the pseudo atomic model of CF_0F_1. Unaccounted density in CF_0 can be mainly attributed to the detergent micelle and a plug at the lumenal side of the III-ring.

which will be useful in designing future experiments on structure function relations.

9.5 Structure and Function of CF_0F_1

H^+-ATP synthases have three catalytic nucleotide binding sites on the β-subunits and three non-catalytic nucleotide binding sites on the α-subunits; they use a remarkable mechanism to synthesize, or to hydrolyze, ATP ("binding change mechanism" [64, 66]). According to this mechanism, the catalytic sites work cooperatively. At a given time, each site has a different conformation. For example, site 1 is in an open conformation and can bind substrates; site 2 is in a closed conformation, it does not exchange substrates with the aqueous phase and there is equilibrium between bound products (ATP) and bound substrates (ADP and phosphate) at this site. Site 3 is partly open and contains loosely bound products. These different conformations of the biochemically identical β-subunits are generated by the different interactions with the γ-subunit located asymmetrically in the centre of the $\alpha_3\beta_3$-complex. If the γ-subunit rotates in the F_1-part, the interactions with all three β-subunits change and, therefore, the conformations of all catalytic sites change simultaneously. At each catalytic site the sequence of the three conformation leads to the synthesis of one ATP per site, when the γ-subunit rotates by 360°.

The coupling between proton transport and chemical rotation is effected as follows [67, 68]. The deprotonated amino acid Glu61 of subunit III is located in a hydrophilic access channel, provided by subunit IV, in the middle of the membrane. Protonation of this amino acid via subunit IV drives the rotation of the III_{14} ring thereby moving the protonated (uncharged) subunit III from the access channel into the hydrophobic membrane core. Since subunit III forms a ring, the neighbouring unprotonated subunit III simultaneously moves into the access channel and can be protonated in the following step. Deprotonation occurs when the protonated subunit III has moved so far that it reaches the second hydrophilic channel in subunit IV, which has an exit to the other side of the membrane. Protons are released to the aqueous phase on the other side of the membrane. Since subunit III is in contact with subunits γ and ε, this leads to a rotation of the γ ε III_{14}-complex (rotor) relative to the $\alpha_3\beta_3$ δ I II IV-complex (stator). During ATPsynthesis, the proton transport through the enzyme leads to a clockwise movement of the rotor, when viewed from F_0 to F_1, and to synthesis of ATP. During ATP hydrolysis, the sequential binding and hydrolysis of ATP at the three catalytic sites lead to a movement of the rotor in the opposite direction and a concomitant proton translocation through the enzyme. The opposite rotation direction during ATPsynthesis and ATP hydrolysis has been shown recently with single molecule fluorescence spectroscopy [75, 76].

In F_1, subcomplexes that lack subunit δ and subunits I and II, the three αβ-pairs are intrinsically identical. Only the different interactions between γ, ε, and each

$\alpha\beta$-pair give rise to the different conformations. In all X-ray structures reported up to now, this situation is realized. It is essential for the binding change mechanism that all catalytic sites are intrinsically identical; that is, rotation of the central stalk by 120° leads to interactions with the next $\alpha\beta$-pair which are energetically equal to those with the previous $\alpha\beta$-pair. In CF_0F_1 (see Figure 9.1), the situation is different. The presence of subunit δ, and subunits I and II that form the peripheral stalk, gives rise to an asymmetry and, possibly, the binding of I, II, and δ, changes the properties of one α-subunit. If the α-subunit tagged by subunits I and II has a conformation which differs from the two other $\alpha\beta$-pairs, it may have different interactions with the γ-subunit. There are three possibilities for the interaction of the central stalk with the tagged α-subunit: either the central stalk is bound stronger, or weaker, or its binding is unchanged, as compared to the non-tagged α-subunits. Based on this consideration, it is expected to find in an enzyme ensemble one orientation of the central stalk relative to the peripheral stalk in case of stronger interaction, two orientations in case of weaker interaction, and the three orientations in case of identical interaction.

The image reconstruction of vitrified CF_0F_1 shows only one conformation of the enzyme; the homology model of the $\alpha_3\beta_3\gamma\varepsilon$-complex can be fitted optimally only in one position to the central stalk observed in the 3-D map from electron microscopy [32]. This shows that, in the presence of AMPPNP, the H^+-ATP synthase is found in a unique resting position. Notice that, in this state, the conformation of one α-subunit is changed by interaction with the peripheral stalk, leading to an increase in stability of one position of the central stalk within the $\alpha_3\beta_3$-barrel.

Figure 9.5 shows a schematic view of the subunit arrangement looking on to the membrane plane from CF_0 to the CF_1-part. A 0.3 nm slice from the electron density map at CF_0F_1 (grey) is shown, taken from the electron density map at the position marked by a line in Figure 9.4 left. The α-subunits are indicated by magenta circles, the β-subunits by red circles. The α-subunit with the highest density is the one tagged by the peripheral stalk. The model of the central stalks fits optimally only in one position into the $\alpha\beta$-hexagon. In the $\alpha_3\beta_3\gamma$-complex, without the peripheral stalk, the different interactions of the γ-subunit with the $\alpha\beta$-pairs have been used to define the conformations of the different α- and β-subunits (E, D and T) [71]. This nomenclature is adopted for CF_0F_1 and, on this basis, the different conformations can be assigned with respect to the peripheral stalk (see Figure 9.5). The α-subunit tagged by the peripheral stalk has the conformation T.

The activity of CF_0F_1 is strongly regulated by a redox reaction and by a ΔpH dependent activation [77–79]. The redox regulation is due to the redox state of a -S–S-group in the γ subunit which is either in the reduced or oxidized state [80–82]. In the oxidized state, no ATP hydrolysis is observed and, at the same ΔpH, the rate of ATP synthesis is higher in the reduced state than in the oxidized state. The redox regulation seems to be a peculiarity of the photosynthetic H^+-ATP synthases. These enzymes must have a very strict control of activity since they are exposed to both light and dark circles and they have to avoid ATP hydrolysis in the dark period.

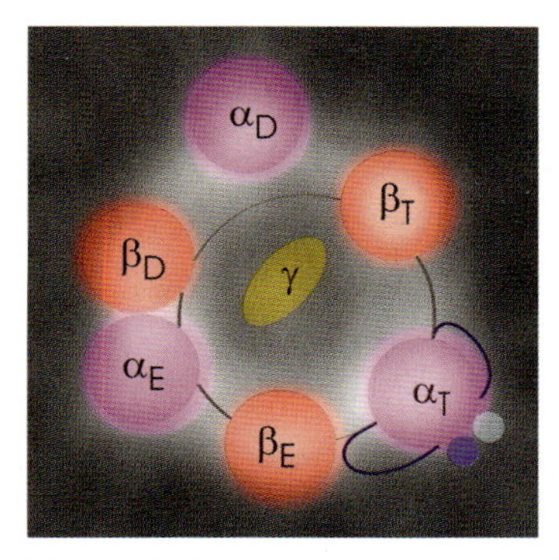

Figure 9.5 Orientation of the peripheral stalk (subunits I and II) and the central stalk (subunit γ) relative to the $\alpha_3\beta_3$-complex in the resting (inactive) state of CF_0F_1. A slice of 0.3 nm thickness from the electron density map of CF_0F_1 as indicated in Figure 9.4 left (line) is shown in a view from CF_0 to the CF_1-part. Magenta circles indicate α-subunits, red circles show β-subunits. Subunits I and II are shown in blue, the γ-subunit is shown in yellow. E, D and T refer to the conformational states of the catalytic nucleotide binding sites. The shape at the III ring is indicated by a brown line and subunit IV is outlined in blue.

Therefore, the redox-state of the H^+-ATP synthase is coupled via ferredoxin and thioredoxin to the redox state of $NADP^+$/NADPH, which changes strongly at light–dark transitions.

CF_0F_1 (in the oxidized and the reduced state) catalyzes neither ATP synthesis nor ATP hydrolysis, when the membrane is not energized. This is expected for the energy requiring ATP synthesis, but it is a surprising observation for energy releasing ATP hydrolysis. Obviously, in the absence of membrane energization, CF_0F_1 is in an inactive state in both redox states. Energization of the membrane by ΔpH or an electric potential difference leads to a conformational change. The enzyme releases a previously tightly bound ADP and it is turned into the metastable active state. Only in this state is the the enzyme able to carry out catalysis [78].

The structure of CF_0F_1 analyzed in the presence of AMPPNP represents the inactive ("resting") state of the enzyme [32]. In the inactive state, there is one preferential position of the central stalk relative to the peripheral stalk. That is, interaction with one of the αβ-pairs leads to a conformational state with the lowest energy. On the other hand, the binding change mechanism proposes a sequence of identical interactions of the central stalk with each of the three αβ-pairs and, consequently, during catalysis, three different orientations of the central stalk, relative to the peripheral stalk, should have the same energy. Based on this consideration, it is concluded that the activation process leads to a structural change between α and the peripheral stalk. In this structure, identical interactions of all three αβ-pairs with the central stalk are possible. Only in this state ("active state") the enzyme is able to catalyze reversibly proton transport coupled ATP synthesis and ATP hydrolysis.

The mechanism of coupling of proton transport with the movement of the III-ring and the $\gamma\varepsilon$-complex in combination with the binding change mechanism predicts the number of protons which are translocated through CF_0F_1 per ATP (H^+/ATP ratio). After rotation of the $\gamma\varepsilon III_{14}$ complex by 360° the enzyme has produced 3 ATP (one at each catalytic site) and translocated 14 H^+, that is the H^+/ATP ratio is given by the subunit stoichiometry of subunit III to β.

$$\frac{H^+}{ATP} = \frac{\text{number of III-subunits}}{\text{number of }\beta\text{-subunits}} = \frac{14}{3} = 4.7$$

However, it has been shown for CF_0F_1 that the thermodynamic H^+/ATP ratio is four [83, 84]. This difference may result from the activation. For estimation of H^+/ATP from subunit stoichiometries, it is assumed that at the start and the end of the 360° rotation, the enzyme is in the same state. If the enzyme is in the inactive state before rotation starts, it must be first converted into the metastable active state. Also, this process requires protons, and it might be speculated that 2 protons are necessary for activation of the enzyme and that 12 protons are used for ATP synthesis.

References

1 Mitchell, P. (1961) Coupling of phosphorylation to electron and hydrogen transfer by a chemi-osmotic type of mechanism. *Nature*, **191**, 144–8.
2 Wilkens, S., Borchardt, D., Weber, J. and Senior, A.E. (2005) Structural characterization of the interaction of the delta and alpha subunits of the Escherichia coli F1F0-ATP synthase by NMR spectroscopy. *Biochemistry*, **44**, 11786–94.
3 Wilkens, S., Dunn, S.D., Chandler, J., Dahlquist, F.W. and Capaldi, R.A. (1997) Solution structure of the N-terminal domain of the delta subunit of the E-coli ATPsynthase. *Nature Structural Biology*, **4**, 198–201.
4 Carbajo, R.J., Kellas, F.A., Runswick, M.J., Montgomery, M.G., Walker, J.E. and Neuhaus, D. (2005) Structure of the F1-binding domain of the stator of bovine F1Fo-ATPase and how it binds an alpha-subunit. *Journal of Molecular Biology*, **351**, 824–38.
5 Wilkens, S., Dahlquist, F.W., McIntosh, L.P., Donaldson, L.W. and Capaldi, R.A. (1995) Structural features of the epsilon subunit of the Escherichia coli ATP synthase determined by NMR spectroscopy. *Nature Structural Biology*, **2**, 961–7.
6 Wilkens, S. and Capaldi, R.A. (1998) Solution structure of the epsilon subunit of the F1-ATPase from Escherichia coli and interactions of this subunit with beta subunits in the complex. *The Journal of Biological Chemistry*, **273**, 26645–51.
7 Uhlin, U., Cox, G.B. and Guss, J.M. (1997) Crystal structure of the epsilon subunit of the proton-translocating ATP synthase from Escherichia coli. *Structure*, **5**, 1219–30.
8 Cabezon, E., Runswick, M.J., Leslie, A.G. and Walker, J.E. (2001) The structure of bovine IF1, the regulatory subunit of mitochondrial F-ATPase. *The EMBO Journal*, **20**, 6990–6.
9 Gogol, E.P., Lücken, U., Bork, T. and Capaldi, R.A. (1989) Molecular architecture of Escherichia coli F1 adenosinetriphosphatase. *Biochemistry*, **28**, 4709–16.

10 Gogol, E.P., Aggeler, R., Sagermann, M. and Capaldi, R.A. (1989) Cryoelectron microscopy of Escherichia coli F1 adenosinetriphosphatase decorated with monoclonal antibodies to individual subunits of the complex. *Biochemistry*, **28**, 4717–24.

11 Gogol, E.P., Johnston, E., Aggeler, R. and Capaldi, R.A. (1990) Ligand-dependent structural variations in Escherichia coli F1 ATPase revealed by cryoelectron microscopy. *Proceedings of the National Academy of Sciences of the United States of America*, **87**, 9585–9.

12 Wilkens, S. and Capaldi, R.A. (1994) Asymmetry and structural changes in ECF1 examined by cryoelectronmicroscopy. *Biological Chemistry Hoppe-Seyler*, **375**, 43–51.

13 Boekema, E.J., Van Heel, M. and Gräber, P. (1988) Structure of the ATP synthase from chloroplasts studied by electron-microscopy and image-processing. *Biochimica et Biophysica Acta*, **933**, 365–71.

14 Boekema, E.J., Xiao, J.P. and Mccarty, R.E. (1990) Structure of the ATP synthase from chloroplasts studied by electron-microscopy–localization of the small subunits. *Biochimica et Biophysica Acta*, **1020**, 49–56.

15 Boekema, E.J. and Böttcher, B. (1992) The structure of ATP synthase from chloroplasts–conformational-changes of Cf1 studied by electron-microscopy. *Biochimica et Biophysica Acta*, **1098**, 131–43.

16 Boekema, E.J., Berden, J.A. and van Heel, M.G. (1986) Structure of mitochondrial F1-ATPase studied by electron microscopy and image processing. *Biochimica et Biophysica Acta*, **851**, 353–60.

17 Akey, C.W., Crepeau, R.H., Dunn, S.D., McCarty, R.E. and Edelstein, S.J. (1983) Electron microscopy and single molecule averaging of subunit-deficient F1-ATPases from Escherichia coli and spinach chloroplasts. *The EMBO Journal*, **2**, 1409–15.

18 Tsuprun, V.L., Mesyanzhinova, I.V., Kozlov, I.A. and Orlova, E.V. (1984) Electron microscopy of beef heart mitochondrial F1-ATPase. *FEBS Letters*, **167**, 285–90.

19 Rastogi, V.K. and Girvin, M.E. (1999) Structural changes linked to proton translocation by subunit c of the ATP synthase. *Nature*, **402**, 263–8.

20 Dmitriev, O., Jones, P.C., Jiang, W. and Fillingame, R.H. (1999) Structure of the membrane domain of subunit b of the Escherichia coli F0F1 ATP synthase. *The Journal of Biological Chemistry*, **274**, 15598–604.

21 Del Rizzo, P.A., Bi, Y., Dunn, S.D. and Shilton, B.H. (2002) The "second stalk" of Escherichia coli ATP synthase: structure of the isolated dimerization domain. *Biochemistry*, **41**, 6875–84.

22 Girvin, M.E., Rastogi, V.K., Abildgaard, F., Markley, J.L. and Fillingame, R.H. (1998) Solution structure of the transmembrane H+-transporting subunit c of the F1F0 ATP synthase. *Biochemistry*, **37**, 8817–24.

23 Carbajo, R.J., Silvester, J.A., Runswick, M.J., Walker, J.E. and Neuhaus, D. (2004) Solution structure of subunit F(6) from the peripheral stalk region of ATP synthase from bovine heart mitochondria. *Journal of Molecular Biology*, **342**, 593–603.

24 Takeyasu, K., Omote, H., Nettikadan, S., Tokumasu, F., Iwamoto-Kihara, A. and Futai, M. (1996) Molecular imaging of Escherichia coli F0F1-ATPase in reconstituted membranes using atomic force microscopy. *FEBS Letters*, **392**, 110–13.

25 Singh, S., Turina, P., Bustamante, C.J., Keller, D.J. and Capaldi, R. (1996) Topographical structure of membrane-bound Escherichia coli F1F0 ATP synthase in aqueous buffer. *FEBS Letters*, **397**, 30–4.

26 Birkenhäger, R., Hoppert, M., Deckers-Hebestreit, G., Mayer, F. and Altendorf, K. (1995) The F_0 complex of the Escherichia coli ATP synthase. Investigation by electron spectroscopic imaging and immunoelectron microscopy. *European Journal of Biochemistry*, **230**, 58–67.

27 Böttcher, B., Bertsche, I., Reuter, R. and Gräber, P. (2000) Direct visualization of conformational changes in $EF0F_1$ by electron microscopy. *Journal of Molecular Biology*, **296**, 449–57.

28 Gogol, E.P., Lücken, U. and Capaldi, R.A. (1987) The stalk connecting the F1 and F0 domains of ATP synthase visualized by electron microscopy of unstained specimens. *FEBS Letters*, **219**, 274–8.

29 Wilkens, S. and Capaldi, R.A. (1998) Electron microscopic evidence of two stalks linking the F1 and F0 parts of the Escherichia coli ATP synthase. *Biochimica et Biophysica Acta*, **1365**, 93–7.

30 Neff, D., Tripathi, S., Middendorf, K., Stahlberg, H., Butt, H.J., Bamberg, E. and Dencher, N.A. (1997) Chloroplast F0F1 ATP synthase imaged by atomic force microscopy. *Journal of Structural Biology*, **119**, 139–48.

31 Böttcher, B., Gräber, P., Boekema, E.J. and Lücken, U. (1995) Electron cryomicroscopy of 2-Dimensional crystals of the H^+-ATPase from chloroplasts. *FEBS Letters*, **373**, 262–4.

32 Mellwig, C. and Böttcher, B. (2003) A unique resting position of the ATP-synthase from chloroplasts. *Journal of Biological Chemistry*, **278**, 18544–9.

33 Böttcher, B., Lücken, U. and Gräber, P. (1995) The structure of the H^+-ATPase from chloroplasts by electron cryomicroscopy. *Biochemical Society Transactions*, **23**, 780–5.

34 Böttcher, B., Schwarz, L. and Gräber, P. (1998) Direct indication for the existence of a double stalk in CF0F1. *Journal of Molecular Biology*, **281**, 757–62.

35 Boekema, E.J., Schmidt, G., Gräber, P. and Berden, J.A. (1988) Structure of the Atp-synthase from chloroplasts and mitochondria studied by electron-microscopy. *Zeitschrift fur Naturforschung. C, Journal of Biosciences*, **43**, 219–25.

36 Mellwig, C. and Böttcher, B. (2001) Dealing with particles in different conformational states by electron microscopy and image processing. *Journal of Structural Biology*, **133**, 214–20.

37 Soper, J.W., Decker, G.L. and Pedersen, P.L. (1979) Mitochondrial ATPase complex–dispersed, cytochrome-deficient, oligomycin-sensitive preparation from Rat-Liver containing molecules with a tripartite structural arrangement. *Journal of Biological Chemistry*, **254**, 1170–6.

38 Mörschel, E. and Staehelin, L.A. (1983) Reconstitution of cytochrome-F/B6 and CF_0F_1 ATP synthetase complexes into phospholipid and galactolipid liposomes. *Journal of Cell Biology*, **97**, 301–10.

39 Karrasch, S. and Walker, J.E. (1999) Novel features in the structure of bovine ATP synthase. *Journal of Molecular Biology*, **290**, 379–84.

40 Rubinstein, J.L., Walker, J.E. and Henderson, R. (2003) Structure of the mitochondrial ATP synthase by electron cryomicroscopy. *The EMBO Journal*, **22**, 6182–92.

41 Groth, G. and Pohl, E. (2001) The structure of the chloroplast F1-ATPase at 3.2 Å resolution. *The Journal of Biological Chemistry*, **276**, 1345–52.

42 Groth, G. (2002) Structure of spinach chloroplast F1-ATPase complexed with the phytopathogenic inhibitor tentoxin. *Proceedings of the National Academy of Sciences of the United States of America*, **99**, 3464–8.

43 Hausrath, A.C., Gruber, G., Matthews, B.W. and Capaldi, R.A. (1999) Structural features of the gamma subunit of the Escherichia coli F_1 ATPase revealed by a 4.4-A resolution map obtained by x-ray crystallography. *Proceedings of the National Academy of Sciences of the United States of America*, **96**, 13697–702.

44 Kabaleeswaran, V., Puri, N., Walker, J.E., Leslie, A.G. and Mueller, D.M. (2006) Novel features of the rotary catalytic mechanism revealed in the structure of yeast F1 ATPase. *The EMBO Journal*, **25**, 5433–42.

45 Gibbons, C., Montgomery, M.G., Leslie, A.G. and Walker, J.E. (2000) The structure of the central stalk in bovine F_1 ATPase at 2.4 A resolution. *Nature Structural Biology*, **7**, 1055–61.

46 Rodgers, A.J. and Wilce, M.C. (2000) Structure of the gamma-epsilon complex of ATP synthase. *Nature Structural Biology*, **7**, 1051–4.

47 Hausrath, A.C., Capaldi, R.A. and Matthews, B.W. (2001) The conformation of the epsilon- and gamma-subunits within the Escherichia coli F_1 ATPase. *The Journal of Biological Chemistry*, **276**, 47227–32.

48 Cabezon, E., Montgomery, M.G., Leslie, A.G. and Walker, J.E. (2003) The structure of bovine F1-ATPase in complex with its regulatory protein IF1. *Nature Structural Biology*, **10**, 744–50.

49 Cabezon, E., Arechaga, I., Jonathan, P., Butler, G. and Walker, J.E. (2000) Dimerization of bovine F1-ATPase by binding the inhibitor protein, IF1. *The Journal of Biological Chemistry*, **275**, 28353–5.

50 Stock, D., Leslie, A.G. and Walker, J.E. (1999) Molecular architecture of the rotary motor in ATP synthase. *Science*, **286**, 1700–5.

51 Meier, T., Polzer, P., Diederichs, K., Welte, W. and Dimroth, P. (2005) Structure of the rotor ring of F-Type Na^+-ATPase from Ilyobacter tartaricus. *Science*, **308**, 659–62.

52 Stahlberg, H., Müller, D.J., Suda, K., Fotiadis, D., Engel, A., Meier, T., Matthey, U. and Dimroth, P. (2001) Bacterial Na^+-ATP synthase has an undecameric rotor. *EMBO Reports*, **2**, 229–33.

53 Seelert, H., Poetsch, A., Dencher, N.A., Engel, A., Stahlberg, H. and Müller, D.J. (2000) Proton-powered turbine of a plant motor. *Nature*, **405**, 418–19.

54 Müller, D.J., Dencher, N.A., Meier, T., Dimroth, P., Suda, K., Stahlberg, H., Engel, A., Seelert, H. and Matthey, U. (2001) ATP synthase: constrained stoichiometry of the transmembrane rotor. *FEBS Letters*, **504**, 219–22.

55 Fromme, P., Boekema, E.J. and Gräber, P. (1987) Isolation and characterization of a supramolecular complex of subunit-Iii of the ATP synthase from chloroplasts. *Zeitschrift fur Naturforschung. C, Journal of Biosciences*, **42**, 1239–45.

56 Seelert, H., Dencher, N.A. and Müller, D.J. (2003) Fourteen protomers compose the oligomer III of the proton-rotor in spinach chloroplast ATP synthase. *Journal of Molecular Biology*, **333**, 337–44.

57 Dickson, V.K., Silvester, J.A., Fearnley, I.M., Leslie, A.G. and Walker, J.E. (2006) On the structure of the stator of the mitochondrial ATP synthase. *The EMBO Journal*, **25**, 2911–18.

58 Nelson, N., Nelson, H. and Racker, E. (1972) Partial resolution of the enzymes catalyzing photophosphorylation. XII. Purification and properties of an inhibitor isolated from chloroplast coupling factor 1. *The Journal of Biological Chemistry*, **247**, 7657–62.

59 Moroney, J.V., Lopresti, L., McEwen, B.F., McCarty, R.E. and Hammes, G.G. (1983) The Mr-value of chloroplast coupling factor 1. *FEBS Letters*, **158**, 58–62.

60 Süss, K.-H. and Schmidt, O. (1982) Evidence for an α_3, β_3, γ, δ, I, II, ε, III_5 subunit stoichiometry of chloroplast ATP synthetase complex (CF_1-CF_0). *FEBS Letters*, **144**, 213–18.

61 Fromme, P., Gräber, P. and Salnikow, J. (1987) Isolation and identification of a 4th subunit in the membrane part of the chloroplast ATP-Synthase. *FEBS Letters*, **218**, 27–30.

62 Pick, U. and Racker, E. (1979) Purification and reconstitution of the N,N'-dicyclohexylcarbodiimide-sensitive ATPase complex from spinach chloroplasts. *The Journal of Biological Chemistry*, **254**, 2793–9.

63 Steigmiller, S., Börsch, M., Gräber, P. and Huber, M. (2005) Distances between the b-subunits in the tether domain of F0F1-ATP synthase from E. coli. *Biochimica et Biophysica Acta*, **1708**, 143–53.

64 Boyer, P.D. (1993) The binding change mechanism for ATP synthase–some probabilities and possibilities. *Biochimica et Biophysica Acta*, **1140**, 215–50.

65 Wilkens, S., Zhou, J., Nakayama, R., Dunn, S.D. and Capaldi, R.A. (2000) Localization of the delta subunit in the Escherichia coli F(1)F(0)-ATPsynthase by immuno electron microscopy: the delta subunit binds on top of the F(1). *Journal of Molecular Biology*, **295**, 387–91.

66 Boyer, P.D. (1998) ATP synthase–past and future. *Biochimica et Biophysica Acta*, **1365**, 3–9.

67 Junge, W. (2004) Protons, proteins and ATP. *Photosynthesis Research*, **80**, 197–221.

68 Junge, W., Lill, H. and Engelbrecht, S. (1997) ATP synthase: an electrochemical transducer with rotatory mechanics. *Trends in Biochemical Sciences*, **22**, 420–3.

69 Marti-Renom, M.A., Stuart, A.C., Fiser, A., Sanchez, R., Melo, F. and Sali, A. (2000)

Comparative protein structure modeling of genes and genomes. *Annual Review of Biophysics and Biomolecular Structure*, **29**, 291–325.

70 Finn, R.D., Mistry, J., Schuster-Bockler, B., Griffiths-Jones, S., Hollich, V., Lassmann, T., Moxon, S., Marshall, M., Khanna, A., Durbin, R., Eddy, S.R., Sonnhammer, E.L. and Bateman, A. (2006) Pfam: clans, web tools and services. *Nucleic Acids Research*, **34**, D247–51.

71 Abrahams, J.P., Leslie, A.G.W., Lutter, R. and Walker, J.E. (1994) Structure at 2.8-angstrom resolution of F1-Atpase from bovine heart-mitochondria. *Nature*, **370**, 621–8.

72 McLachlin, D.T., Coveny, A.M., Clark, S.M. and Dunn, S.D. (2000) Site-directed cross-linking of b to the alpha, beta, and a subunits of the Escherichia coli ATP synthase. *The Journal of Biological Chemistry*, **275**, 17571–7.

73 Meier, T., Matthey, U., Henzen, F., Dimroth, P. and Müller, D.J. (2001) The central plug in the reconstituted undecameric c cylinder of a bacterial ATP synthase consists of phospholipids. *FEBS Letters*, **505**, 353–6.

74 Eisfeld, J. (2003) *Physikalische Chemie II*, University of Freiburg, Freiburg.

75 Diez, M., Zimmermann, B., Borsch, M., Konig, M., Schweinberger, E., Steigmiller, S., Reuter, R., Felekyan, S., Kudryavtsev, V., Seidel, C.A. and Graber, P. (2004) Proton-powered subunit rotation in single membrane-bound F0F1-ATP synthase. *Nature Structural and Molecular Biology*, **11**, 135–41.

76 Zimmermann, B., Diez, M., Zarrabi, N., Graber, P. and Borsch, M. (2005) Movements of the epsilon-subunit during catalysis and activation in single membrane-bound H(+)-ATP synthase. *The EMBO Journal*, **24**, 2053–63.

77 Kramer, D.M., Wise, R.R., Frederick, J.R., Alm, D.M., Hesketh, J.D., Ort, D.R. and Crofts, A.R. (1990) Regulation of coupling factor in field-grown sunflower–a redox model relating coupling factor activity to the activities of other thioredoxin-dependent chloroplast enzymes. *Photosynthesis Research*, **26**, 213–22.

78 Gräber, P., Schlodder, E. and Witt, H.T. (1977) Conformational change of the chloroplast ATPase induced by a transmembrane electric field and its correlation to phosphorylation. *Biochimica et Biophysica Acta*, **461**, 426–40.

79 Junesch, U. and Gräber, P. (1987) Influence of the redox state and the activation of the chloroplast ATP-synthase on proton-transport coupled ATP-synthesis/hydrolysis. *Biochimica et Biophysica Acta*, **893**, 275–88.

80 Nalin, C.M. and McCarty, R.E. (1984) Role of a disulfide bond in the gamma subunit in activation of the ATPase of chloroplast coupling factor 1. *The Journal of Biological Chemistry*, **259**, 7275–80.

81 Groth, G. and Strotmann, H. (1999) *Physiologia Plantarum*, **106**, 142–8.

82 Richter, M.L., Samra, H.S., He, F., Giessel, A.J. and Kuczera, K.K. (2005) Coupling proton movement to ATP synthesis in the chloroplast ATP synthase. *Journal of Bioenergetics and Biomembranes*, **37**, 467–73.

83 Van Walraven, H.S., Strotmann, H., Schwarz, O. and Rumberg, B. (1996) The H+/ATP coupling ratio of the ATP synthase from thiol-modulated chloroplasts and two cyanobacterial strains is four. *FEBS Letters*, **379**, 309–13.

84 Turina, P., Samoray, D. and Gräber, P. (2003) H^+/ATP ratio of proton transport-coupled ATP synthesis and hydrolysis catalysed by $CF0F_1$-liposomes. *The EMBO Journal*, **22**, 418–26.

10
Structure of the Light-Harvesting Complex II

Zhenfeng Liu and Wenrui Chang

10.1
Introduction

In photosynthesis, the light-harvesting antennae are very efficient at absorbing solar energy and transferring the excitation energy toward the reaction center, the place where the input energy is utilized to trigger the charge separation of a special pair of chlorophyll molecules. Light-harvesting complexes are associated at the periphery of Photosystem II of higher plants and consist of four homologous members, namely LHC-II, CP29, CP26, and CP24 (for structure of the photosystem and its supercomplexes see also Chapters 4 and 6). Among them, the LHC-II is the most abundant in chloroplasts and is known as the major light-harvesting complex II; the other three have been termed minor light-harvesting complexes II, or chlorophyll-binding proteins (CPs), because of their low abundance. The LHC-II exists as a homo- or hetero-trimer formed by the products of three highly homologous nuclear genes, *Lhcb1, Lhcb2,* and *Lhcb3,* in unequal stoichiometries. The molecular weight of the apoprotein varies from 24 000 to 29 000 daltons. According to early biochemical data, each LHC-II monomer binds 13–15 Chl *a* and Chl *b* molecules [1], 3–4 carotenoids [2, 3], and 1 tightly bound phospholipid [4].

LHC-II has been found to be a multifunctional complex in the thylakoid membrane. Firstly, it serves as efficient light-harvesting machinery, powering the photosynthetic reactions. Under high-light conditions, when the input energy becomes excessive, LHC-II is able to switch from the efficient light-harvesting state to a dissipative state and safely dissipate the excess energy as heat. Through this act, it provides a photoprotective mechanism for the plants [5–7]. LHC-II also has a role in regulating the distribution of excitation energy to photosystems II and I through the reversible phosphorylation of the its N-terminus [8]. In addition, the stacking of thylakoid membranes in chloroplast grana is largely related to the close interactions between LHC-IIs in two adjacent layers of the membrane [9].

In 1994, the structure of pea LHC-II was determined by electron crystallography at 3.4 Å resolution, parallel to the membrane plane, and at approximately 4.9 Å

Photosynthetic Protein Complexes: A Structural Approach. Edited by P. Fromme

ISBN: 978-3-527-31730-1

perpendicular to this plane [10]. Some characteristic features of the LHC-II structure, including three long transmembrane α-helices (helix A, B, C) and a short amphiphilic helix (helix D) at the luminal surface, were revealed for the first time in this model. Twelve chlorophylls and two carotenoids have been positioned with rough coordinates and orientations. Using the electron crystallographic model, the chlorophyll identities and orientations in LHC-II have been intensively studied through site-directed mutagenesis, in vitro reconstitution, and spectroscopic studies [11–13]. Some significant differences in the chlorophyll assignment stem from different research groups being present during the investigation process [14]. For a better understanding of the basic functional mechanism of LHC-II, it is very important to solve the complete structure of LHC-II at a resolution that allows for an unambiguous determination of the orientation of the transition dipole moments of chlorophylls as well as their identities (Chl *a* or Chl *b*).

Toward this end, the X-ray crystallographic structure of spinach LHC-II has been determined at 2.72 Å resolution [15]. With a high-quality dataset of structure factor amplitude and an accurate phase set, the electron densities for the oxygen atom of the C7-formyl group of Chl *b*, as well as the side chains and head groups of each individual chlorophyll molecules, were observed for the first time, leading to the final precise assignment of chlorophyll identities and orientations. Because of this, many new structural features of LHC-II have been unraveled at atomic detail. About one year after the reporting of the assignments, the X-ray structure of pea LHC-II was solved at 2.5 Å resolution from a different crystal form, by the Kühlbrandt group [16]. Superposition of the two structures yielded extremely small root mean square deviation of α-carbon atoms at 0.35 Å, indicating the high degree of similarity between the two species.

10.2
Crystal Packing: A Novel Type of Membrane Protein Crystal

The spinach LHC-II was crystallized in a novel crystal form that is very different from the type-I and type-II membrane protein crystals originally described by Michel [17]. Firstly, 20 LHC-II trimers, together with numerous lipid and detergent molecules, organize into a highly symmetric icosahedral proteoliposome vesicle with inner and outer diameters of about 160 and 261 Å, respectively (Figure 10.1a). The orientation and position of LHC-IIs in the vesicle conform to the 532 point group symmetry of a $T = 1$ icosahedron. Every two adjacent LHC-II trimers, in the spherical shell of the vesicle, are bridged by two molecules of digalactosyl diacylglycerol (DGDG) at the interface by way of hydrogen bonds and hydrophobic interactions. The proteoliposome vesicles have extensive hydrophilic surfaces formed by the stromal surfaces of LHC-IIs and lipid head groups. The rigid and uniform features of such vesicles with densely packed LHC-IIs enable them to stack in the 3-D lattice and form highly ordered 3-D crystals that isotropically diffract X-rays to around 2.5 Å resolution. The contacts between two proteoliposomes

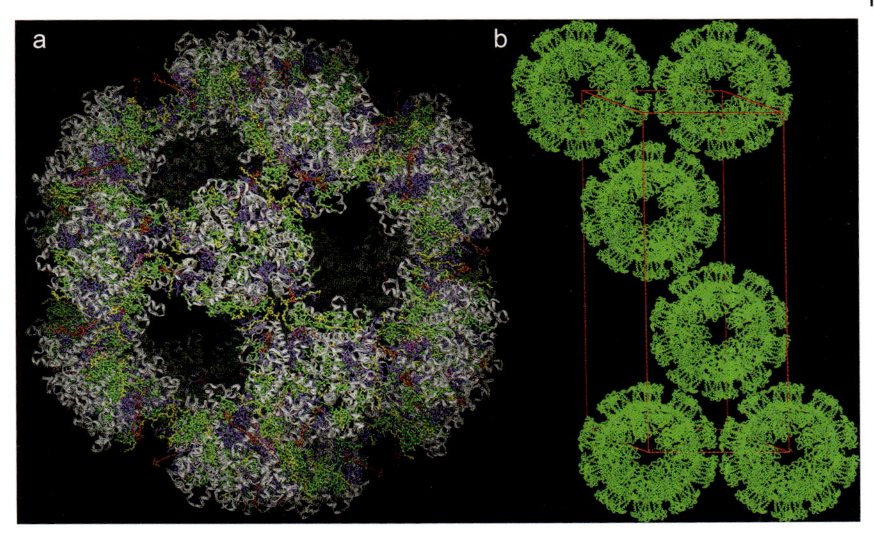

Figure 10.1 A novel type of membrane protein crystal composed of proteoliposome vesicles. **(a)** The structure of a proteoliposome vesicle with 20 LHC-II trimers assembled in an icosahedral symmetry on the shell. The hollow region around C_5 axes is supposed to be filled by lipid molecules, most of which are disordered and do not show up in the electron density map. Proteins are sketched as silver ribbons, while cofactors are shown as colored stick models. **(b)** Packing of the vesicles in a hexagonal unit cell. Cofactors are omitted for clarity.

in the crystal lattice are completely contributed by the polar interactions between the hydrophilic stromal surfaces of LHC-IIs (Figure 10.1b). The hydrophobic intramembranous surfaces of LHC-II trimers are covered by the presumed lipid bilayer of the vesicular shell and do not participate in the crystal packing.

This novel type of 3-D crystal is called a type-III membrane-protein crystal. It was reported that bacteriorhodopsin was also able to form a similar type of crystal under appropriate conditions [18]. However, the crystal was poorly ordered and only diffracted X-rays to 30 Å. More recently, pea LHC-II has also been crystallized in this type, but in a different space group, which yielded a low resolution model at 9.5 Å [19]. The study on spinach LHC-II crystal reveals the detailed organizational pattern of this unique type of crystal and proves that it is possible to obtain highly ordered type-III membrane protein crystals for X-ray crystallographic structure determination. From a functional view, it has been shown that the chlorophyll fluorescence of LHC-II will be quenched if it is reconstituted into a liposome containing DGDG [20]. Since the LHC-II in a type-III crystal does exist in the DGDG liposome, similar phenomena are expected to be present in the crystal sample, an occurrence which may be related to a photoprotective mechanism in vivo. This was proven to be true in the follow-up spectroscopic studies [7] and will be explored in more detail in Section 10.6.2 of this chapter.

10.3
LHC-II Trimer and Monomer

10.3.1
LHC-II Trimer

The LHC-IIs predominantly form trimers both in vivo and in vitro. The trimeric state was indicated to be functionally significant for the LHC-II in enhancing protein stability, enabling efficient light-harvesting, and optimizing the control of energy dissipation [21]. The LHC-II trimer assumes a trigonal-prism shape with a height of 50 Å and an edge of 69 Å (Figure 10.2a and b). Both surfaces on the stromal and luminal sides are flat and almost in parallel with the membrane plane. Using the positions of lipid (DGDG) and detergent (NG) head groups, one can estimate the approximate location of LHC-II trimer in the thylakoid membrane (Figure 10.2b). Most parts of the complex are immersed within the membrane, while small portions on the stromal and luminal sides protrude from the membrane by about 7–10 and 5 Å, respectively. All kinds of cofactors, including chlorophylls, carotenoids, and lipids, surround the polypeptide scaffolds and are embedded within the intramembrane regions.

The trimerization domain includes the N-terminal region, the C-terminus from one monomer, the stromal side of helix B, some hydrophobic residues of helix C from the neighboring monomer, and also the pigments and lipids bound to these parts of polypeptides (Figure 10.2c). The N-terminal region (from Ser 14 to Asp 54) at the stromal side forms a cap over the trimerization interface and provides binding sites for a Chl *b* (Chl *b* 601), a phosphatidylglycerol (PG), and the head group of a lutein (Lut 621). The phosphatic head group of PG is linked to the polypeptide through a hydrogen bond with Tyr 44 and a strong ionic bond with the nearby side chain of Lys 182. It also serves as the axial ligand of a Chl *a* (Chl *a* 611) that is coordinated on the other side of PG. Both fatty acid chains of PG were modeled as palmitic acid chains with 16 carbon atoms and exhibit completely different conformations. The 1-fatty acid chain bends over before chain 2 and runs parallel to the membrane plane in the middle of the thylakoid membrane. The other chain, at position 2, bears extended conformation, stretching straight toward the luminal surface and across the membrane. It is nearly in parallel with the phytyl chain of Chl *b* 601 and the polyene chain of a xanthophyll (Violaxanthin, Vio) at the monomer-monomer interface (Figure 10.2c). The extensive hydrophobic interactions among these three cofactors provide strong forces to bind neighboring monomers together. It is very likely that the chain 2 corresponds to the trans-Δ^3 hexadecenoic acid existing in the PG associated with oligomeric LHC-II [22]. This special PG has been proposed as essential for the oligomerization of LHC-II [23]. It was also found that Chl *b* 601 is coordinated to the backbone carbonyl of Tyr 24, which is located at the N-terminal part of the polypeptide that was unresolved in the old structure. It was reported that removing the first 49 or 51 amino acid residues of the polypeptide by proteolytic cleavage would result in loss of PG and complete dissociation of the trimer

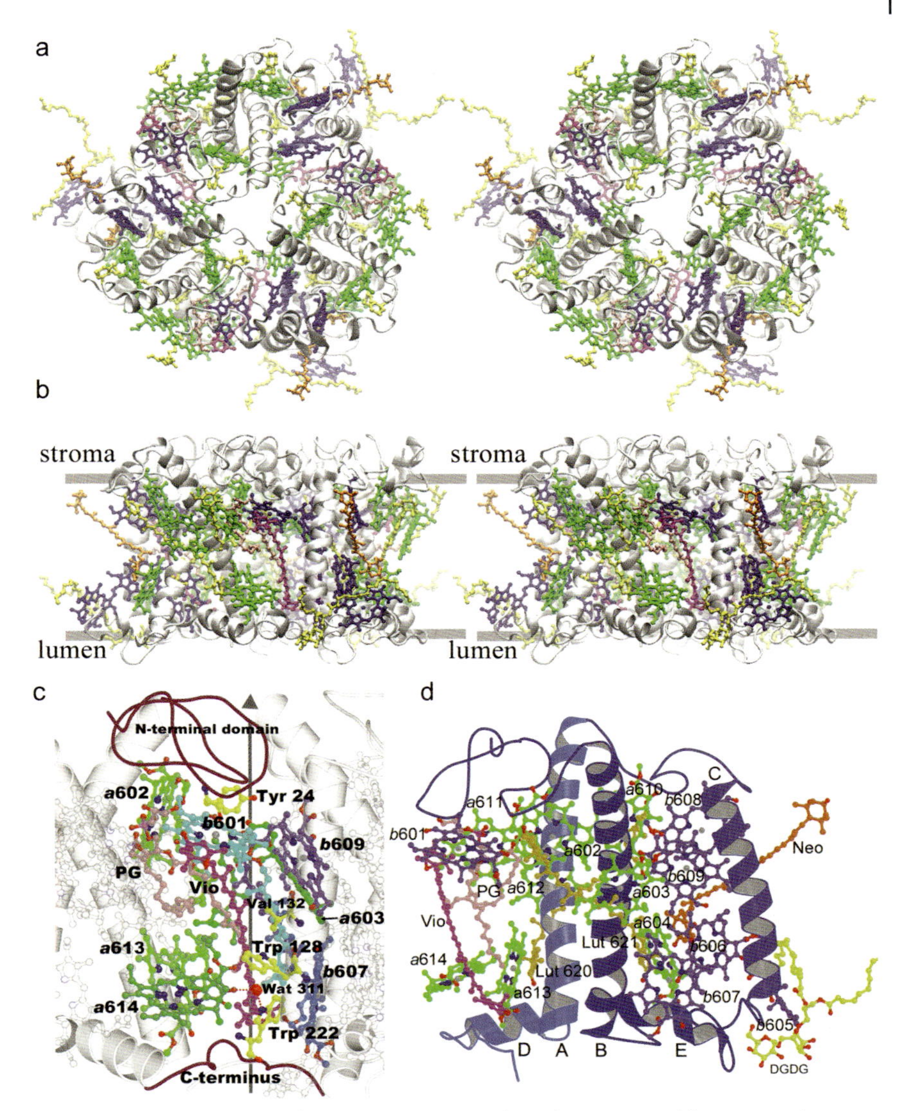

Figure 10.2 LHC-II trimer and monomer. (**a** and **b**) Stereographs of LHC-II trimer viewed along the membrane normal from the stromal side and in parallel with membrane plane, respectively. **(c)** Side view of the trimerization interface. Adapted from [15]. **(d)** Side view of LHC-II monomer. Color code: yellow–lutein, DGDG, NG, and amino acid residues; green–Chl *a;* cyan and blue–Chl *b;* magenta–Vio; pink–PG; red–water; orange–Neo. Polypeptide C traces are in silver, maroon, and blue ribbons. Cofactors are in ball-and-stick model. The vertical line is the local C_3 axis of LHC-II trimer. Phytyl chains of chlorophylls are omitted for clarity.

into monomers [4]. This kind of cleavage will lead to the removal of the physical shelter over the trimerization interface, decrease the binding strength of PG by taking away its hydrogen bond partner (Tyr 44), and also destroy the stable PG–Chl–Vio triplet complex structure via the dissociation of Chl *b* 601 from the interface. Hydrolysis of the PG in LHC-II by phospholipase A2 [4] will have a similar effect in breaking down the trimer through destabilizing the trimerization interface. The phytyl chains of Chl *b* 607 and 609 from the neighboring monomer line up antiparallelly with the PG–Chl–Vio triplet hydrophobic chains along the monomer-monomer interface; thus, they extend the hydrophobic interactions toward the center of LHC-II trimer. The phytyl chains of Chl *a* 602 (from one monomer) and 603 (from the neighboring monomer) cross over the five-chain sheet in a nearly perpendicular direction, further strengthening the linkage between two adjacent monomers. There are six Chl *a* molecules (Chl *a* 602 and 603 from each monomer) constituting the core of the LHC-II trimer. Their phytyl chains extend from the stromal surface into the trimeric core to form extensive hydrophobic interactions and block the potential central hole connecting lumen and stroma like a plug. It is foreseeable that the leakage of such a hole in the middle of the LHC-II trimer would destroy the selective permeability of the thylakoid membrane in case that these central chlorophylls were removed from the complex.

Trp16 and/or Tyr17, as well as Arg21 in the WYXXXR motif, are found to be essential for the formation of LHC-II trimers [24]. In the structure, it was found that these residues are important in stabilizing the local structure of the N-proximal domain, rather than being directly involved in the trimerization. The bulky side chains of Trp 46, Trp 128, and Trp 222 are all important elements contributing to trimerization. The absence of Trp 222 from the C-terminus destabilizes the LHC-II trimer [25]. Its bulky side chain inserts between the polyene chain of the Xanc and the chlorines ring of Chl *b* 607 from adjacent monomers, tightening the connection between monomers (Figure 10.2c). Direct protein-protein interactions between monomers are discovered merely on both stromal and luminal surfaces, either by hydrogen bonds or by van de Waals contacts. The whole trimerization domain extends laterally from the periphery of the trimer to the core region around the C_3 axis. Hydrophobic interactions dominate the associations between monomers within the LHC-II trimer. This explains the previous observation that treating the LHC-II preparation with a high concentration of non-ionic detergent would dissociate the LHC-II trimer and the resulting monomers would reassociate into trimers upon dilution of the detergent [4].

A cup-shaped cavity around the central C_3 axis of the LHC-II trimer, walled by hydrophobic residues, can be observed from the luminal side. It opens up to the lumen and extends deep into the complex core where the three closely associated phytyl chains of Chl *a* 603 molecules form the bottom of the cavity. There are some residual weak electron densities inside the cavity that might be attributed to detergents or lipids introduced during the purification and crystallization processes. To be careful and precise, and because of their obscure appearance, the densities were not modeled. In the pea LHC-II structure reported by the Kühlbrandt group, these features, within the same cavity, have been fit with partial models of DGDG molecules [16]. In the spinach LHC-II structure, one DGDG molecule per monomer was found at the trimer-trimer interface, with well-defined electron densities on the head group and part of the hydrophobic tails. These DGDG molecules have critical roles in holding adjacent LHC-II trimers together in the proteoliposome.

10.3.2 LHC-II Monomer

The high-quality electron density map made it possible to continuously trace the polypeptides of all 10 monomers in 1 asymmetric unit from Ser 14 to Gly 231. The first 13 amino acid residues at the N-terminus contain a phosphorylation site; this site is predicted to be able to change its conformation upon phosphorylation [8]. Phosphorylation and dephosphorylation of the N-terminus are involved in the migration of LHC-II between Photosystem II (see Chapter 4) and Photosystem I (see Chapters 2, 4 and 6), thus adjusting the distribution of excitation energy between the two photosystems. The polypeptide heterogeneity may also account for the weak appearance of the electron density in the N-terminal region. The crystals that were used to determine the structure contain two highly homologous polypeptides, Lhcb1 and Lhcb2 (amino acid sequence of spinach Lhcb2 is not available yet). The AA sequences between Lhcb1 and Lhcb2 from barley share 82% identity [26], while those from *Arabidopsis* [27] are 66.8% identical. The differences between the two polypeptides are confined to the N-terminal region. Therefore, we determined that the polypeptide structure is statistically an averaged structure of Lhcb1 and Lhcb2.

The apoprotein structures of the 10 LHC-II monomers in an asymmetric unit are almost identical, with a root mean square deviation of α carbon (C_α) atoms ranging from 0.12 to 0.23 Å between every two monomers. The apoprotein of each monomer adopts a simple fold with three long transmembrane helices (B, C, and A), two short amphipathic helices (E and D) at the luminal surface, and loops connecting the helices (Figure 10.2d). The helices B and A cross over in the middle of the monomer with a mutual angle of 59.4° and form a left-handed supercoil. There is a pseudo-C_2 axis running through the supercoil and perpendicular to the membrane plane that relates helices B and E to helices A and D. The helix C flanks the central supercoil and disrupts the local pseudo-C2 symmetry. The secondary structure of spinach LHC-II is very similar to that of pea LHC-II [10, 16]. This is in good agreement with the fact that 89% of the 232 amino acid residues are conserved between the two species. Nevertheless, there are some minor deviations in the residue range, length, turns, and orientation between helices of the two species. Helix E is a newly identified 3_{10} helix located in the BC loop region at the luminal surface with an amphipathic nature. It is inclined from the membrane plane by an angle of about 30°. In the following EC loop, the polypeptide folds into two short antiparallel strands that are stabilized by an inter-strand ionic pair (Asp 111-His 120) and some hydrogen bonds.

Two strong inter-helix ionic pairs (Glu 65-Arg 185, Arg 70-Glu 180) stabilize the central supercoil. Both pairs are further linked to the N-proximal domain and A-C loop by forming hydrogen bonds to the backbone carbonyl of Gly 42 and Leu 155 through the guanidinium groups of Arg 185 and Arg 70, respectively. One inner-helix ionic pair (Glu139-Arg142) is found on Helix C. Another strong ionic interaction occurred between Lys182 and the phosphatic head group of PG. Each of these four ionic pairs coordinates a chlorophyll molecule in LHC-II. Six additional strong ionic pairs (Arg 21-Asp 43, Lys 60-Glu 63, Asp 111-His 120, Glu 175-Lys 179, Lys 203-Asp 211, and Lys 203-Glu 207) are distributed in various part of the monomer. The formation of these ionic bonds (10/monomer) within the complex will contribute to the thermostability of LHC-II.

The N-terminal region (Ser 14-Asp 54) and the AC loop (Gly 145-Asp 169) are located at the stromal surface, whereas the BE loop (Asn88-Val96), helix E, EC loop (Ser 106-Ser 123), helix D, and the C-terminus (Asp 215-Gly 231) are exposed at the luminal surface. The N-terminal region is part of the trimerization domain as discussed above. The head group of Lut 621 interacts with Trp 46 and a short β turn (Asp47-Gly50) in this region. The AC loop contains the binding site (β turn Asp162-Gly165) for Lut 620 and interacts with the head groups of two chlorophylls (Chl *b* 608 and Chl *a* 610) through hydrogen bonds. The EC loop provides the binding site for the 3-hydroxy head group of neoxanthin (Neo) via the hydroxyl group of Tyr 112. The other newly identified chlorophyll (Chl *b* 605) adopts the backbone carbonyl of Val 119 in the EC loop as its axial ligand. Two β turns lead the EC loop to bend over and form two short antiparallel strands. The C-terminal region extends to the bottom of the adjacent monomer and stabilizes the trimeric state. It is involved in the binding of violaxanthin through the bulky hydrophobic side chain of Trp 222. In summary, the N-terminal, C-terminal, and the loop regions are not only necessary for the structureal integrity of the LHC-II, but also are indispensable for the proper functioning of the complex.

In the spinach LHC-II structure, 14 chlorophylls, 4 carotenoids, 1 PG, 1 DGDG, 1 detergent molecule (nonyl-β-D-glucoside–NG), and about 70 water molecules have been located within each LHC-II monomer. The polypeptide provides a simple but effective scaffold to position all of the cofactors precisely in the 3-D space so that they can function properly in the light harvesting and the regulatory processes. The cofactors are attached to the protein scaffold through direct and indirect interactions with the polypeptide (Figure 10.2d). The density of pigments (chlorophyll and carotenoid) in LHC-II is so high that they accounts for 38% of the total mass of the whole complex [28]. This is remarkable because, if these pigments were not properly oriented and positioned in the complex to avoid strong interactions between them, the chlorophyll fluorescence would be quenched due to the concentration quenching effect [29], and efficient light harvesting would not be possible. On the other side, close interactions between pigments at specific sites, to form an excited state chlorophyll dimer or carotenoid–chlorophyll heterodimer, hypothetically induced by reversible protein conformational change, may provide a mechanism to dissipate excess excitation energy [30].

10.4
Chlorophylls: Identity, Binding and Interactions

10.4.1
Identification of Chlorophylls

The only chemical difference between Chl *a* and Chl *b* is at the C7 head group, namely, methyl for Chl *a* and formyl for Chl *b*. Provided that hydrogen atoms are not visible in X-ray structures above 1 Å resolution, the difference is just one oxygen atom that will only be distinguishable at sufficient resolution. Theoretically, a good diffraction dataset, with high quality phases at resolution higher than 2.8 Å, will yield clear electron density maps that show carbonyl features. Using the $2F_o$-F_c and F_o-F_c electron density maps at 2.72 Å resolution, eight chlorophylls were assigned as Chl *a* and six as Chl *b* (Table 10.1). The resulting Chl *a*/*b* molar ratio of 1.33 is consistent with the value determined by biochemical analysis [1, 2]. In the latest pea LHC-II structure, the same identification results have been obtained [16].

10.4.2
Chlorophyll Binding Sites

Chlorophylls of the same type, either Chl *a* or Chl *b*, occupy the same binding sites in the 10 noncrystallographic symmetry-related LHC-II monomers. No observation was made of the mixed binding of Chl *a* and Chl *b* in the structure. This does not exclude the possibility that some chlorophyll binding sites in LHC-II may accept either Chl *a* or Chl *b* under conditions of in vitro reconstitution, as suggested in a previous report [11]. The selective forces of some sites may not be strong enough to distinguish between Chl *a* and Chl *b* in the artificial refolding system; this may lead to the appearance of mixed binding sites. Our observation in the structure does not support the proposal of mixed chlorophyll binding in the native LHC-II. Except for Chl *b* 601, the other five Chl *b* are selectively bound in the complex with their C7-formyl group forming hydrogen bonds with corresponding partners, such as the side chains of amino acid residues, the backbone amide of polypeptide, or a water molecule that coordinates Chl *b* 607 (Table 10.1). Although the chlorophyll at site 601 does not have a hydrogen bond partner for its C7-formyl group, strong positive difference electron density (at the $3 \times \sigma$ level) still shows up at the C7 position, unambiguously identifying this Chl as a Chl *b*. The force selecting Chl *b*, instead of Chl *a*, at this site may come from the central ligand, the backbone carbonyl of Tyr 24, as the only two chlorophyll molecules bound to backbone carbonyl groups are both Chl *b*. It was shown that LHC-II monomers can fold in vitro with Chl *b*, carotenoids, and traces of Chl *a* under specific conditions and that Chl *a* molecules and carotenoids alone do not support the folding of LHC-II [31, 32]. Considering the finding in the structure, it can be inferred that the incorporation of Chl *b* molecules in LHC-II stabilizes the complex structure through forming selective interactions with the polypeptide scaffold. As

Table 10.1 Chlorophyll identities, ligands and their interactions with local environments.[a]

Chlorophylls	***b* 601**	***a* 602**	***a* 603**	***a* 604**	***b* 605**	***b* 606**	***b* 607**	***b* 608**	***b* 609**	***a* 610**	***a* 611**	***a* 612**	***a* 613**	***a* 614**
Central ligands	Tyr 24	Glu 65	His 68	Wat 309	Val 119	Wat 310	Wat 308	Wat 302	Glu 139	Glu 180	PG	Asn 183	Gln 197	His 212
H bond partners of C7-formyl (Chl *b*)	–	–	–	–	Gln 122 N, Ser 123 N	Wat 308	Gln 131 NE2	Leu 148 N	Gln 131 NE2	–	–	–	–	–
H bond partner of C131-keto	–	Tyr 44 N, Trp 46 N	Wat	Leu 113 N	–	–	–	Arg 70 NH1	His 68 ND1	Gly 158 N	–	–	–	–
Nomenclature 2[b]	none	a4	a5	a6	none	b6	a7	b1	b5	a1	b2	a2	a3	b3
Nomenclature 3[c]	Chl 9	Chl 4	Chl 5	Chl 6	Chl 14	Chl 13	Chl 10	Chl 11	Chl 12	Chl 1	Chl 7	Chl 2	Chl 3	Chl 8

a) Table adapted from [15].
b) Nomenclature of chlorophylls in the electron crystallographic structure of pea LHC-II [10].
c) Nomenclature of chlorophylls in the X-ray crystallographic structure of pea LHC-II (PDB ID code: 2bhw) [16].

for the selective binding of Chl *a*, it was noticed that the environment surrounding the C7-methyl groups of Chl *a* molecules is mostly nonpolar. The binding of Chl *b* to these Chl *a* binding sites will be weaker than Chl *a* as a result of the hydrophobic repulsion or steric hindrance against Chl *b*.

The side chains of seven amino acid residues, two backbone carbonyl groups, four water molecules, and the phosphodiester group of PG serve as the central ligands of chlorophylls (Table 10.1). The coordination of Chl *a* 611 to the head group of PG was surprising, even though this coordination mode has once been discovered in the crystal structure of Photosystem I [33] (see Chapter 2). In LHC-II, four chlorophylls (three Chl *b* and one Chl *a*) are coordinated by water molecules. These chlorophylls are actually dihydrates. One water molecule occupies the fifth coordination site of the central Mg; a second simultaneously forms hydrogen bonds with the coordinated water and other ligands, such as the $C13^2$-carbonyl group of the methyl ester from the same chlorophyll, polypeptide backbone, or side chains, or the hydroxyl group of lutein. Gln 131 was previously assigned as the fifth ligand of Chl *b* 606 [10]. In the new structures, a water molecule (Wat 310) was found to serve as the axial ligand of this Chl *b*; the side chain carbonyl group of Gln 131 is hydrogen-bonded to Wat 310. Gln131 is involved in the binding of three Chl *b* molecules via its amide side chain. Two additional hydrogen bonds are formed via its NH_2 interacting with the C7-formyls of Chl *b* 607 and 609. Moreover, the C7-formyl of Chl *b* 606 is hydrogen-bonded to the coordinated water of Chl *b* 607. These interactions bring three Chl *b* molecules into close proximity, resulting in the clustering of Chl *b* molecules in this region. It was found that mutation of Gln 131 in LHC-II to Glu results in a decreased Chl *b* content [11, 34]. An adverse effect was observed when the corresponding site (Glu 166) in CP29 was substituted by Gln [34]. From a structural view, the authors were changing the available number of hydrogen bonds from the binding site to be formed with Chl *b* formyl, leading to the variation of Chl *b* content.

10.4.3
Interactions between Chlorophylls

In LHC-II, the chlorophyll density is so high that close interactions between chlorophylls are inevitable. These interactions will not only change the spectroscopic features of chlorophylls, but also promote the excitation energy transfer between chlorophylls and, possibly, lead to the dissipation of excitation energy. In order to detect potential strong interactions between chlorophylls within LHC-II, attempts were made to calculate the excitonic coupling strengths between every two chlorophylls by applying the point-dipole approximation formula [14]. The coupling strengths between chlorophylls inside a monomer with absolute values greater than $50\,cm^{-1}$ are listed in Table 10.2. The strongest couplings between chlorophylls from adjacent monomers in a trimer and between chlorophylls from two adjacent trimers are also included. Strong excitonic Chl *a*/*a*, *a*/*b* couplings have previously been proposed on the basis of some spectroscopic observations [35–37]. The results

Table 10.2 Strong excitonic coupling between chlorophylls in LHC-II[a].

Chlorophyll pairs	Excitonic coupling strength V (cm^{-1})	Cosβ[b]	R_{center}[c] (Å)	R_{π}[d] (Å)
Intra-monomeric couples				
Chl *a* 611–612	144.9	–0.79	9.76	3.77
Chl *b* 606–Chl *a* 604	123.4	0.80	8.05	4.42
Chl *a* 603–Chl *b* 609	106.7	–0.88	9.74	3.97
Chl *b* 608–Chl *a* 610	69.2	–0.54	11.57	5.00
Chl *b* 607–606	63.6	0.88	9.46	3.67
Chl *a* 613–614	–62.5	0.21	9.30	3.96
Chl *b* 601–Chl *a* 602	54.1	–0.69	12.79	5.51
Inter-monomeric couple within a trimer				
Chl *b* 609–601	42.0	0.10	11.79	4.95
Inter-trimeric couple within an icosahedron				
Chl *b* 605–Chl *a* 614	62.0	–0.61	12.16	5.40

a) Table adapted from [15].
b) The angle between two transition dipoles.
c) The distance between the molecular centers of two chlorophylls. The coordinates of molecular centers of chlorophylls are calculated by arithmetically averaging the coordinates of the four pyrrole nitrogen atoms.
d) Closest distance between two chlorin conjugated π-systems..

of the calculations, based on structural information, provides supporting evidence for the existence of at least one pair of Chl *a*/*a* dimer and two pairs of Chl *a*/*b* dimers that are coupled at strength of over 100 cm^{-1}. The coupling between the Chl *a*/*a* dimer, *a*611-*a*612 is the strongest within the monomer. The absorption of Chl *a* 612 was observed to be red-shifted to about 680 nm; this was proposed to be the lowest Q_y energy level in the complex [12, 34]. The strong excitonic coupling is, most likely, responsible for the red shift. As this pair is located at the periphery of the trimer, it is proposed to be the terminal fluorescence emitter. They may transmit energy to neighboring LHCs, or to the core antenna proteins CP43 and CP47 of Photosystem II, when the LHC-II is coupled to the PSII complex [12, 34]. The next two strongly coupled pairs are Chl *a* 603-Chl *b* 609 and Chl *a* 604-Chl *b* 606. Even though the mutual orientation and distance are optimal for the coupling between Chl *b* 606–607, the coupling strength of this Chl *b* dimer is relatively small because the dipole strength of Chl *b* is much weaker than that of Chl *a* [14]. This pair might account for the sub-picosecond Chl *b*-Chl *b* transfer observed in LHC-II [38–40]. The couplings between every two chlorophylls from two different layers (stromal and lumenal) in the complex are all found to be relatively weak with strength smaller than 20 cm^{-1}. The inter-monomer and inter-trimer chlorophyll coupling is also weak as a result of poor mutual orientations and large distances between chlorophylls.

10.5
Carotenoids

Four carotenoid binding sites per monomer have been located in the LHC-II structure. The central two carotenoids, with all-trans configurations, are bound in the grooves on both sides of the supercoil (Helices A and B) to form a cross brace. They were assigned as lutein molecules (Figure 10.2d). The β rings of both luteins are oriented towards the luminal surface, while the ε rings point to the stromal surface. The polyene chains of Lut 620 and 621 are inclined, with respect to the membrane normal, by angles of about 59° and 62°, respectively. It was noted that four internal homologous peptide segments exist in LHC-II that are responsible for the binding of lutein head groups [41]. Both ring-shaped end groups of the two luteins form van der Waals contacts and hydrogen bonds with these four segments located on both ends of transmembrane Helices A and B. Their polyene chains are firmly fixed in two elongated, narrow hydrophobic cavities on both sides of the supercoil, providing a strong and rigid linkage between Helices A and B. Lutein was found to be indispensable for the proper folding of LHC-II into stable complexes in vitro [42–45]. Resonance Raman spectroscopy revealed that these two luteins in LHC-II trimer assume different configurations and that trimerization leads to the twisting of one lutein [46, 47]. Attempts were made to superimpose Lut 621 on Lut 620, based on the overlap of the central part of polyene chains (C13, C14, C15, C33, C34, and C35, r.m.s.d. = 0.05 Å), and resulted in the finding that Lut 621 appears to be more twisted than Lut 620 (Figure 10.3). In the case that Lut 621 assumed the same configuration as Lut 620, there would be close contacts between Chl a 603′ (from the adjacent monomer) and the ε ring, and between the indole side chain of Trp 97 and the β ring. The van der Waals repulsion forces from Chl a 603′ and Trp 97 are responsible for the twisting of Lut 621 on both ends. The end close to ε ring is more twisted than the other end, suggesting that the repulsion force from Chl *a* 603′ is stronger than the effect of Trp 97. These two groups are missing in the environment surrounding Lut 620. Therefore, Lut 620 has a relaxed configuration that allows a better overlap between the π electron orbitals of its polyene chain and the tetrapyrrole plane of Chl *a* 612. Upon monomerization of LHC-II, the repulsion force from Chl *a* 603′ will be removed and the configuration of Lut 621 will come back to a relaxed state as observed in the Raman spectra [46, 47]. The functional significance of Lut 621 twisting upon trimerization of LHC-II is still unclear and requires further investigation.

The third carotenoid, with a shape like a bent-over hook, is located in the Chl *b*-rich region around Helix C and was assigned as 9′-*cis* neoxanthin (Figure 10.2d). More than 90% of thylakoid neoxanthin is bound to LHC-II [48, 49]. The polyene chain is inclined from the membrane normal at an angle of about 58°, which is consistent with the value of about 57 ± 1.5° derived from linear dichrosim spectra [48]. The epoxycyclohexane ring of Neo is attached to the hydroxyl of Tyr 112 via its C3′-hydroxyl and is positioned right above the tetrapyrrole macrocycle of Chl *a* 604. The polyene chain of Neo is bound in a hydrophobic cleft walled by the side chains of Leu 134, Met 135, Val 138 from helix C and Trp 71 from helix B, as well

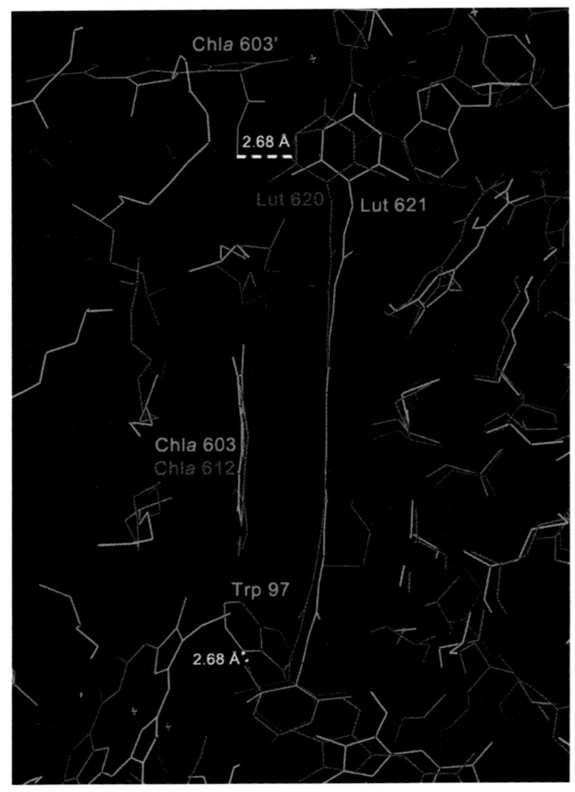

Figure 10.3 Two different luteins in LHC-II. Lut 621 are superimposed on Lut 620 by fitting the central part of polyene chain. The surrounding groups are all rotated by the same operation.

as chlorin rings and phytyl chains of Chl *b* 606 and 608. This binding site has been shown to be highly selective for neoxanthin [42, 43]. The cyclohexane ring of neoxanthin on the other end sticks outward into the lipidic region; the high temperature factor on this end indicates that it has a flexible nature. Since neoxanthin is exposed at the periphery of the trimer, its configuration will be more or less influenced by the surrounding environment, such as the detergent micelles in the solution, or the lipid bilayer in the thylakoid membrane. Oligomerization of LHC-II trimer will lead to distortion of the neoxanthin molecule, as compared to the free trimer in detergent solution [46]; the neoxanthin structure will be even more twisted when the LHC-II trimers are incorporated into the crystal [7].

It was found in previous biochemical studies that LHC-II contains a fourth carotenoid binding site which will loosely bind violaxanthin, a xanthophyll-cycle carotenoid (Xanc), at the periphery of trimer [2, 3]. More details about this carotenoid will be discussed in Section 10.6.2.

10.6
Mechanism of Excitation Energy Transfer and the Role of LCH-II in Non-photochemical Quenching

10.6.1
Mechanism of Excitation Energy Transfer

With the chromophore identities being revealed in the structure, a close look at the pigment arrangement pattern in LHC-II can be taken and attempts to correlate the structural feature with the mechanism of excitation energy transfer within the complex can be made. The chlorophylls in LHC-II are vertically distributed into two layers within the membrane, each layer lying close to the stromal and luminal surfaces, respectively (Figure 10.4). Within a monomer, the layer close to the

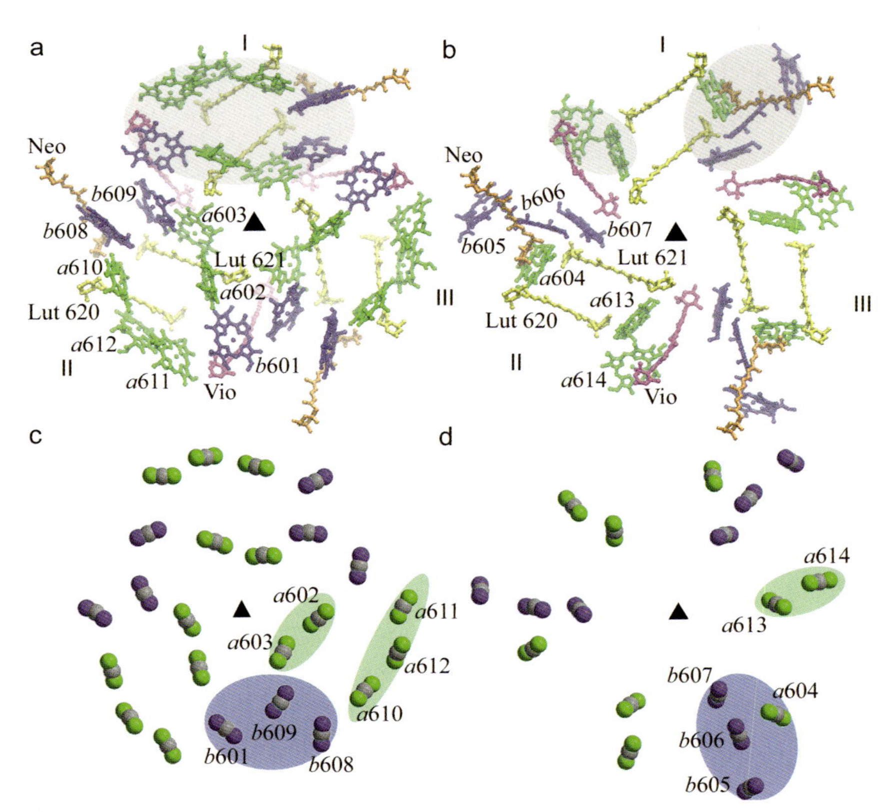

Figure 10.4 Pigment arrangement pattern in LHC-II. **(a, b)** Distribution of pigments at the levels close to stromal and luminal surfaces of LHC-II trimer, respectively. **(c, d)** The pattern of chlorophyll Q_y transition dipole moment arrangement at stromal and luminal layers, respectively. Chlorophylls are simplified as three-atom models with the central magnesium and two nitrogen atoms. The connecting line between the two nitrogen atoms defines the approximate orientation of the Q_y transition dipole.

stromal surface contains eight chlorophylls (five Chl *a* and three Chl *b*), which more or less evenly surround the central helix A and B to form an elliptical ring (Figure 10.4a). Each of these chlorophylls has a symmetric mate related by the internal pseudo-C_2 axis. The average center-to-center distance between two adjacent chlorophylls within the ring is about 11.26 Å, with a maximum of 12.79 Å and a minimum of 9.74 Å. The remaining six chlorophylls (three Chl *a* and three Chl *b*) are arranged in the layer close to luminal surface. They are arranged into two separate clusters with four chlorophylls (three Chl *b* and one Chl *a*) and a Chl *a*-Chl *a* dimer, respectively (Figure 10.4b). Among them, Chl *b* 606 and Chl *a* 604 are associated with the smallest center-to-center distance (8.05 Å) in LHC-II. The maximum distance between two adjacent chlorophylls inside every cluster is 12.62 Å. The minimum distance between two intra-monomeric clusters inside the luminal layer is about 21.02 Å, whereas that between two inter-monomeric clusters is smaller (16.58 Å). The shortest distance between the two chlorophyll layers is about 13.89 Å (Chl *b* 609–606 distance).

The head groups of all chlorophylls have been accurately positioned in the electron density map. With this structural information, we can approximate the Q_y transition dipole moments of chlorophylls along the molecular y-axis. The distribution patterns of Q_y transition dipole moments in two chlorophyll layers have different characteristics (Figure 10.4c and d). The Q_y transition moments in the stromal layer lies nearly in the membrane plane, with a small average inclination angle of 16.4 ° (the absolute value was used in averaging); those in the luminal layer are inclined with a larger average angle of 45.2 °.

Surprisingly, all Chl *b* in LHC-II are enriched in the region around helix C and at the monomer-monomer interface within the trimer. Five of them are from one monomer, and one (Chl *b* 601) is from the adjacent monomer. Chl *b* 601(II) and 609(I) (distance, 11.79 Å) are the closest associated couple of chlorophylls between adjacent monomers within a trimeric LHC-II, indicating that this Chl *b*-rich region is of critical importance for the fast energy equilibration inside a functional trimer.

At the trimeric level, 24 chlorophylls from the stromal layer are organized into 2 irregular circular rings (Figure 10.4a). The inner ring located in the core region of a trimer is composed of six Chl *a* molecules which are thought to play an important role in the inter-monomeric energry transfer [50]. The remaining nine Chl *a* and nine Chl *b* form the outer ring and are arranged in a mosaic pattern with three Chl *b* alternating with three Chl *a*. Such an even pigment arrangement pattern would facilitate the efficient absorption of incident light energy from all directions in a broad spectral region and also the transfer of the excitation energy to the nearest exit, the putative terminal fluorescence emitter Chl *a* 612, in few steps and at high rates. Apparently, energy transfers between luminal clusters are much less efficient than energy transfers within the stromal layer–they are separated by larger distances. It was proposed that these luminal chlorophyll clusters might serve as upstream energy collectors, absorbing energy and transmitting it to the stromal chlorophylls in a relatively independent way. The energy absorbed by the stromal chlorophylls is quickly focused on Chl *a*

612/611 and is further transmitted to the neighboring LHCs or the reaction centers.

The understanding of the energy transfer mechanism in LHC-II has moved one step forward through a series of follow-up studies combining spectroscopic analyses and theoretical modeling based on the high-resolution structure [51–53]. The aims of these studies are to determine the site energies for individual chlorophylls in LHC-II, the only unknown parameters essential for obtaining an accurate picture of the energy transfer in LHC-II after the chlorophyll orientation and identities have been determined in the structure. Two research groups have come up with different results of site energies by using different approaches [51, 53]. With these newly determined data, in combination with the structural data, the authors were able to obtain a satisfactory fitting of various spectra, calculate energy transfer rates between chlorophylls, and map the transfer pathways within the complex [51–53]. Considering excitonic interactions between chlorophylls, the Grondelle group divided the 14 chlorophylls into 5 groups, a602-a603, a610-a612-a611, b601′-b608-b609, b605-b606-b607-a604, and a613-a614 (Figure 10.4c and d) [53]. Followed by the excitation of Chl *b* at 650 nm, the energy is expected to flow downhill toward Chl *a* at a lower energy level. The excitation energy relaxes very quickly, 100 fs, within the b601-b608-b609 cluster at the stromal side, and transfers the energy to the stromal-side Chl *a* clusters (a610-a612-a611 and a602-a603) at a sub-picosecond rate [52]. Equilibration within the luminal b605-b606-b607-a604 cluster takes place in a timescale of sub-picoseconds, and the excitation populates at b605 and a604, bottleneck sites in the b-to-a and a-to-a transfers, respectively. This means that energy transfer from b605 and a604 to the other sites is relatively slow (in 1–4 ps and 40–50 ps respectively). The intra-cluster relaxation in the Chl *a* region is ultra-fast (90–300 fs), whereas the energy transfer between Chl *a* clusters takes a little longer, between 250 and 600 fs. When final equilibrium is reached, the excitation will be mostly focused on the a610-a611–612 cluster at the lowest energy level [52]. The location of this cluster at the periphery of LHC-II trimer is favorable for the further transfer of excitation energy to the neighboring complexes.

In addition to chlorophylls, carotenoids (lutein and neoxanthin) can also function as efficient light-harvesting antennae. They absorb light in the blue-green spectral region as a complement to Chl *a*/*b* absorbing in the red region, thus broadening the absorption spectrum of the whole complex. It was discovered that singlet excitation energy of luteins is transferred exclusively to Chl *a* molecules, and not to Chl *b* [54], whereas neoxanthin was found to transfer its energy mostly toward Chl *b* [54, 55]. The total Car-to-Chl transfer efficiency was estimated to be about 80% [54]. The rate of singlet excitation energy transfer between carotenoids and chlorophylls is correlated with the mutual orientation between them, the center-to-center inter-molecular distance, and the closest distance between two conjugated parts. It was found that six Chl *a* are in favorable orientations and distances, with respect to two Lut, for efficient singlet energy transfer from Lut to Chl *a* (Lut620→Chl *a* 610, 612, 613; Lut 621→ Chl *a* 602, 603, 604). The structural data also demonstrates that efficient energy transfer from Neo to Chl *b* 606 and 608 is very likely. The six possible Lut-to-Chl *a* singlet energy transfer pathways and two potential Neo-to-Chl *b* transfer pathways provide a structural basis for interpreting the spectrally observed carotenoid-to-chlorophyll transfer [49, 54–56]. It was proposed that the fourth carotenoid is not directly involved in the light harvesting function of LHC-II [3]. Instead, it may function in the photoprotective mechanism of LHC-II, as will be discussed in the following Section 10.6.2.

10.6.2
The Role of LHC-II in Non-photochemical Quenching

In the light reaction of photosynthesis, the consumption of excitation energy to trigger charge separation in the reaction center will lead to a decrease of chlorophyll fluorescence quantum yield, a phenomenon termed photochemical quenching. When the light intensity exceeds the photosynthetic capacity, the excess excitation energy has to be dissipated non-radiatively as heat so that its potential damaging effects can be avoided. The non-radiative dissipation (NRD) can be detected as non-photochemical quenching of chlorophyll fluorescence (NPQ), a regulatory mechanism consisting of feedback de-excitation quenching (qE) and photoinhibition (qI) [30, 57]. qE responds rapidly to the light intensity change and is regulated by the transmembrane pH gradient (ΔpH) across the thylakoid membrane. Under high-light conditions, the continuous operation of photosynthetic light reaction will result in the acidification of thylakoid lumen and the low pH in the lumen will activate violaxanthin de-epoxidase (VDE) to catalyze conversion of vioxanthin to zeaxanthin (Zea) via intermediate antheraxanthin (Ant). The other enzyme, zeaxanthin epoxidase, will convert zeaxanthin back to vioxanthin under light-limiting conditions. The whole cycle is known as the xanthophyll cycle. The light intensity-dependant operation of the xanthophyll cycle is correlated with the formation and relaxation of qE [58, 59]. The appearance of Zea and/or Ant is associated with the formation of qE. An absorbance change at 535 nm (ΔA_{535}) is closely correlated with qE; it was shown that ΔA_{535} arises from the red shift of zeaxanthin due to the alteration of pigment configuration [60]. It was proposed that the protonation of qE-related PSII proteins under low lumen pH will cause conformational changes within these proteins, activate zeaxanthin binding sites, lead to the emergence of ΔA_{535}, and eventually, contribute to NPQ [58, 61]. A 22 kDa protein, PsbS, has been found to be necessary for qE in vivo [62]. Although the mechanism of NPQ has been intensively investigated for many years, the fundamental questions concerning qE – the identity and location of the quencher(s) and the way in which they dissipate the excess energy – remain controversial and require more data for an improved understanding.

Here, structural evidence is provided that supports the proposal that LHC-II is involved in NPQ, and that LHC-II may contain the quenching sites that are probably composed of closely interacting pigments. The first evidence comes from the finding that LHC-II contains a binding site for violaxanthin, a doubly epoxidated member of the xanthophyll. Carotenoid composition analysis results indicate that the major component of xanthophyll in the LHC-II preparation used for crystallization is violaxanthin (Table 10.3). The fourth carotenoid binding site is located at

Table 10.3 Carotenoid composition of the spinach leaves and LHC-II preparations used for crystallization.

Sample	%/MR	Lut	Neo	Vio	Ant + Zea	Xanc
Spinach Leaves	%	55.0 ± 0.9	20.0 ± 0.5	24.3 ± 1.3	<0.83 ± 0.05	25.1
LHC-II	%	67.9 ± 1.3	22.0 ± 0.9	10.1 ± 0.3	0	10.1
	MR	2.8	0.84	0.39	0	0.39

Carotenoid composition was determined by reverse phase HPLC according to [63]. Carotenoids were identified by their absorption spectra, their retention times compared with standards, and by Electrospray Ionization Mass Spectrometry (ESI-MS). They were quantified using published extinction coefficients [64]. %, carotenoid content as a percentage of total; MR, estimated molar ratio normalized to yield a sum of 4; Lut, lutein; Neo, neoxanthin; Vio, violaxanthin; Ant, antheraxanthin; Zea, zeaxanthin; Xanc, Vio + Ant + Zea.

the monomer-monomer interface. The polyene chain of this carotenoid has an all-trans configuration and forms a small angle (34°) with the membrane normal. A hydrophobic pocket is formed at the interface by several chlorophylls, hydrophobic residues from the polypeptide, and the PG. Part of the polyene chain together with one of its end groups is accommodated inside this pocket. The opposite end group sticks outside the binding pocket and faces the chlorin plane of Chl *b* 601 at the stromal side. The two terminal hydroxyl groups at stromal and lumenal sides are hydrogen bonded to the glycerol hydroxyl of PG and 17-propionic carbonyl of Chl *b* 607, respectively. The two ring-shaped end groups of this carotenoid exhibit distinct shapes of electron densities. At the position between C-5 and C-6, the one facing the lumen is flat, whereas the one facing the stroma is bulgy (Figure 10.5a and b). The densities for the polyene chain part are extremely well defined (Figure 10.5c). Initial thoughts could be to model this carotenoid as antheraxanthin, but there is only a trace amount of antheraxanthin in the sample. Additionally, because negative difference densities (at the $-3 \times \sigma$ level) will show up on the epoxy group at some sites, it is also contradictory if it is assigned as violaxanthin at full occupancy. When the carotenoid composition data was normalized with four carotenoids per monomer, the result turns out to be 2.8 luteins and 0.39 violaxanthin. This indicates that there may be some lutein mixed with violaxanthin at this site, which may account for the ambiguous appearance of the end group densities. Although we did not detect any zeaxanthin in the sample, we still expect zeaxanthin to bind at the same site with higher affinity since zeaxanthin is more hydrophobic than violaxanthin. This site is expected to be more accessible to the lipidic environment than the two intrinsic lutein sites and less selective than the Neo binding site. It would be able to accommodate three members of the xanthophyll cycle carotenoids and lutein in the LHC-II complex. This structural assignment result is consistent with early biochemical data showing that violaxanthin bound at the peripheral site could easily be lost during purification [2, 3].

It was suggested that zeaxanthin can act as direct quenchers of excess excitation by accepting singlet excitation energy transferred from chlorophyll and dissipate

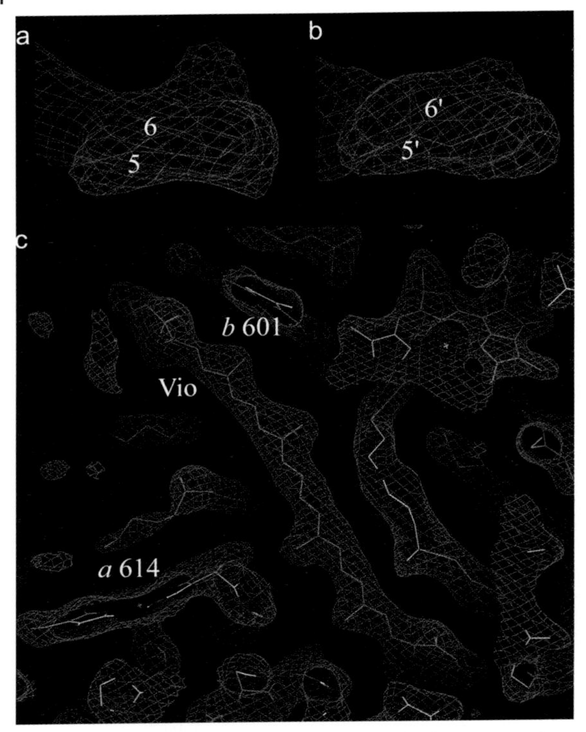

Figure 10.5 $2F_o$-F_c electron density of the fourth carotenoid in LHC-II (1.5 × σ level). **(a)** flat end ring at luminal side. **(b),** bulgy end-ring at stromal side. **(c)** well-defined polyene chain. Interpretation of the density is shown by the refined models fit in the density.

it as heat [65, 66]. In the case that zeaxanthin binds at this site, there are four chlorophylls (Chl *b* 601, Chl *b* 607, Chl *a* 613, and Chl *a* 611) located nearby that have favorable orientations for efficient singlet excitation transfer from these chlorophylls to zeaxanthin. Besides the direct quenching role of zeaxanthin, it may serve as an allosteric regulator whose binding would promote a structural change in LHC-II, leading to the formation of chlorophyll excimers that could potentially be powerful quenchers [30, 67, 68].

A follow-up study using fluorescence lifetime imaging microscopic and spectroscopic techniques on the crystal sample indicates that the fluorescence of LHC-II in the crystals was highly quenched [7]. This is surprising since there is no zeaxanthin present in the sample. There must be some quenchers other than zeaxanthin within the crystal that are responsible for the observed phenomena. Besides the shortening of fluorescence lifetime, a prominent red shift of fluorescence emission spectrum is detected in the crystal sample, indicating that the chlorophyll configuration undergoes some changes that may be attributed to the formation of strongly coupled chlorophyll dimers. These dimers could potentially be the quenching sites in LHC-II. A Resonance Raman spectroscopic study revealed that the neoxanthin molecule is twisted in the crystal, as compared to the free trimer [7]. As observed in the structure, neoxanthin is in close van der Waals interaction with Chl *b* 606 and Chl *b* 608. An evident change in the Chl *b* Raman spectra suggests that the carbonyl of at least one Chl *b* is involved in a hydrogen bond in the crystal; this interaction is not present in the free trimer. By relating the Raman signal to the structure, Pascal et al. [7] assigned this hydrogen bond to the one between Wat 308 (central ligand of Chl b 607) and the formyl group of Chl *b* 606. As mentioned above, the side chain of Gln 131 binds the central ligand (Wat310) of Chl *b* 606 and the C7-formyl of Chl *b* 607 at the same time. Consequently, Chl *b*606 and Chl *b*607 are brought closely together, allowing both π-electron orbitals to overlap extensively. It was hypothesized that this pair of Chl *b* may be a potential quenching site [7]. The other possible quenching site is located in the terminal emitter that constitutes Chl *a* 610-611-612 and Lut 620, as proposed previously [69].

In order to address the question of how the conformational changes would occur in LHC-II to form the quenching sites mentioned earlier, an LHC-II structure in an unquenched, or fully fluorescent state is essential. Coincidently, the pea LHC-II structure was believed to be in a quenched state, too [16]. Attempts were made to superimpose the two structures by fitting the C_α coordinates (from Ser14 to Gly231). No significant differences, either in the protein or pigment region, are discernable. The two structures overlap almost perfectly, except for the floppy regions, such as lipid tails and chlorophyll phytyl tails. The 0.4 Å displacement of Chl *a* 612 to Chl *a* 611, reported by Standfuss and coworkers (2005), could be the error introduced in the superimposition or arising from the structure refinement. It is evident that both structures are in a quenched state, despite the fact that the pH for crystallization is very different. The pH for crystallizing spinach LHC-II is higher than 7.5, while the pea LHC-II was crystallized at pH 5.2–5.6. In the case of LHC-II, low pH alone does not seem to be sufficient to induce structural changes within the complex. It might be necessary to combine low pH with the binding of Zea in order to trigger conformational changes in LHC-II. Perhaps the current observed quenching in LHC-II is primarily triggered by the lateral pressure present in the highly curved lipid bilayer of the liposome and by the close interactions between LHC-II trimers on the proteoliposome vesicles. The contacts between LHC-II trimers may have a smaller role since the trimer-trimer contacts are minimal at the interface, involving only two pairs of chlorophylls and two DGDG molecules. The lateral pressure arising from the curvature of the lipid bilayer could have a major role in triggering the conformational changes in LHC-II and, as a consequence, fluorescence quenching. It was discovered that the force arising from the curvature stress induced by the asymmetric incorporation of lysophosphatidylcholine (LPC) into the proteoliposomes is strong enough to trigger the transition of the mechanosensitive channel of large conductance from a closed state to a fully open state [70]. Similarly, in the LHC-II-proteoliposome, the additive deoxy-bigchap and the detergent nonyl-glucoside may serve the same role to increase the curvature stress in the proteoliposome. This hypothesis only explains the quenching observed in the crystal; this kind of curvature stress is unlikely to be present in the thylakoid membrane. To account for the NPQ in vivo, it is suggested that the protein-protein interactions, perhaps between LHC-II and PsbS, may be the driving force of conformational changes in LHC-II.

10.7 Summary

In summary, the current available LHC-II structures represent their quenched state. A xanthophyll-cycle carotenoid binding site has been located in LHC-II, which is occupied by a mixture of violaxanthin and lutein. The potential quenching sites are tentatively attributed to two chlorophyll domains, namely the Chl *b* 606–607 dimer and the Chl *a* 610-611-612-Lut 620. The curvature stress in the proteoliposomes may be responsible for the induction of conformational changes within LHC-II that lead to the formation of these quenching sites. Future work on the structural biology of LHC-II should be focused on solving the LHC-II structure in the unquenched state. This is indispensable for a full understanding of how the conformational changes occur during the transition. Site-directed mutagenesis work, in combination with spectroscopy, can be used to produce and identify specific LHC-II mutants deficient in fluorescence quenching. The results will be helpful in the evaluation of the current NPQ model based on the structures.

Acknowledgements

This work was financially supported by grants from the National Key Research Development Project of China (G1998010110), the National Natural Science Foundation of China (30170202 and 30530210), the Knowledge Innovation Project of the Chinese Academy of Sciences (KJCX2-SW-N06), and the President Special Foundation of the Chinese Academy of Sciences.

References

1 Peter, G.F. and Thornber, J.P. (1991) Biochemical composition and organization of higher plant photosystem II light-harvesting pigment-proteins. *Journal of Biological Chemistry*, **266**, 16745–54.

2 Ruban, A.V., Lee, P.J., Wentworth, M., Young, A.J. and Horton, P. (1999) Determination of the stoichiometry and strength of binding of xanthophylls to the photosystem II light harvesting complexes. *Journal of Biological Chemistry*, **274**, 10458–65.

3 Caffarri, S., Croce, R., Breton, J. and Bassi, R. (2001) The major antenna complex of photosystem II has a xanthophyll binding site not involved in light harvesting. *Journal of Biological Chemistry*, **276**, 35924–33.

4 Nußberger, S., Dörr, K., Wang, D.N. and Kühlbrandt, W. (1993) Lipid-protein interactions in crystals of plant light-harvesting complex. *Journal of Molecular Biology*, **234**, 347–56.

5 Horton, P., Ruban, A.V. and Walters, R.G. (1994) Regulation of Light Harvesting in Green Plants (Indication by Nonphotochemical quenching of chlorophyll fluorescence). *Plant Physiology*, **106**, 415–20.

6 Elrad, D., Niyogi, K.K. and Grossman, A.R. (2002) A major light-harvesting polypeptide of photosystem II functions in thermal dissipation. *The Plant Cell*, **14**, 1801–16.

7 Pascal, A.A., Liu, Z., Broess, K., van Oort, B., van Amerongen, H., Wang, C., Horton, P., Robert, B., Chang, W. and Ruban, A. (2005) Molecular basis of photoprotection and control of photosynthetic light-harvesting. *Nature*, **436**, 134–7.

8 Nilsson, A., Stys, D., Drakenberg, T., Spangfort, M.D., Forsen, S. and Allen, J.F. (1997) Phosphorylation controls the three-dimensional structure of plant light harvesting complex II. *Journal of Biological Chemistry*, **272**, 18350–7.

9 Allen, J.F. and Forsberg, J. (2001) Molecular recognition in thylakoid structure and function. *Trends in Plant Science*, **6**, 317–26.

10 Kühlbrandt, W., Wang, D.N. and Fujiyoshi, Y. (1994) Atomic model of plant light-harvesting complex by electron crystallography. *Nature*, **367**, 614–21.

11 Remelli, R., Varotto, C., Sandona, D., Croce, R. and Bassi, R. (1999) Chlorophyll binding to monomeric light-harvesting complex. A mutation analysis of chromophore-binding residues. *Journal of Biological Chemistry*, **274**, 33510–21.

12 Rogl, H. and Kühlbrandt, W. (1999) Mutant trimers of light-harvesting complex II exhibit altered pigment content and spectroscopic features. *Biochemistry*, **38**, 16214–22.

13 Yang, C., Kosemund, K., Cornet, C. and Paulsen, H. (1999) Exchange of pigment-binding amino acids in light-harvesting chlorophyll a/b protein. *Biochemistry*, **38**, 16205–13.

14 Amerongen, H.V. and Grondelle, R.V. (2001) Understanding the energy transfer function of LHCII, the major light-harvesting complex of green plants. *The Journal of Physical Chemistry B*, **105**, 604–17.

15 Liu, Z., Yan, H., Wang, K., Kuang, T., Zhang, J., Gui, L., An, X. and Chang, W. (2004) Crystal structure of spinach major light-harvesting complex at 2.72 A resolution. *Nature*, **428**, 287–92.

16 Standfuss, J., Terwisscha van Scheltinga, A.C., Lamborghini, M. and Kühlbrandt, W. (2005) Mechanisms of photoprotection and nonphotochemical quenching in pea light-harvesting complex at 2.5 A resolution. *EMBO Journal*, **24**, 919–28.

17 Michel, H. (1983) Crystallization of membrane proteins. *Trends in Biochemical Sciences*, **8**, 56–9.

18 Takeda, K., Sato, H., Hino, T., Kono, M., Fukuda, K., Sakurai, I., Okada, T. and Kouyama, T. (1998) A novel three-dimensional crystal of bacteriorhodopsin obtained by successive fusion of the vesicular assemblies. *Journal of Molecular Biology*, **283**, 463–74.

19 Hino, T., Kanamori, E., Shen, J.R. and Kouyama, T. (2004) An icosahedral assembly of the light-harvesting chlorophyll a/b protein complex from pea chloroplast thylakoid membranes. *Acta crystallographica. Section D, Biological Crystallography*, **60**, 803–9.

20 Moya, I., Silvestri, M., Vallon, O., Cinque, G. and Bassi, R. (2001) Time-resolved fluorescence analysis of the photosystem II antenna proteins in detergent micelles and liposomes. *Biochemistry*, **40**, 12552–61.

21 Wentworth, M., Ruban, A.V. and Horton, P. (2004) The functional significance of the monomeric and trimeric states of the photosystem II light harvesting complexes. *Biochemistry*, **43**, 501–9.

22 Remy, R., Tremolieres, A., Duval, J.C., Ambard-Bretteville, F. and Dubacq, J.P. (1982) Study of the supramolecular organization of light-harvesting chlorophyll protein (LHCP): conversion of the oligomeric form into the monomeric one by phospholipase A2 and reconstitution with liposomes. *FEBS Letters*, **137**, 271–5.

23 Krupa, Z., Williams, J.P., Khan, M.U. and Hunter, N.P.A. (1992) The role of acyl lipids in reconstitution of lipid-depleted light-harvesting complex II from cold-hardened and nonhardened rye. *Plant Physiology*, **100**, 931–8.

24 Hobe, S., Förster, R., Klingler, J. and Paulsen, H. (1995) N-proximal sequence motif in light-harvesting chlorophyll a/b-binding protein is essential for the trimerization of light-harvesting chlorophyll a/b complex. *Biochemistry*, **34**, 10224–8.

25 Kuttkat, A., Hartmann, A., Hobe, S. and Paulsen, H. (1996) The C-terminal domain of light-harvesting chlorophyll-a/b-binding protein is involved in the stabilisation of trimeric light-harvesting complex. *European Journal of Biochemistry*, **242**, 288–92.

26 Caffarri, S., Croce, R., Cattivelli, L. and Bassi, R. (2004) A look within LHCII: differential analysis of the Lhcb1-3 complexes building the major trimeric antenna complex of higher-plant photosynthesis. *Biochemistry*, **43**, 9467–76.

27 Standfuss, J. and Kühlbrandt, W. (2004) The three isoforms of the light-harvesting complex II: spectroscopic features, trimer formation, and functional roles. *Journal of Biological Chemistry*, **279**, 36884–91.

28 Noy, D., Moser, C.C. and Dutton, P.L. (2006) Design and engineering of photosynthetic light-harvesting and electron transfer using length, time, and energy scales. *Biochimica et Biophysica Acta*, **1757**, 90–105.

29 Beddard, G.S., Carlin, S.E. and Porter, G. (1976) Concentration quenching of chlorophyll fluorescence in bilayer lipid vesicles and liposomes. *Chemical Physics Letters*, **43**, 27–32.

30 Horton, P., Wentworth, M. and Ruban, A. (2005) Control of the light harvesting function of chloroplast membranes: the LHCII-aggregation model for non-photochemical quenching. *FEBS Letters*, **579**, 4201–6.

31 Paulsen, H. (1997) Pigment ligation to proteins of the photosynthetic apparatus in higher plants. *Physiologia Plantarum*, **100**, 760–8.

32 Kleima, F.J., Hobe, S., Calkoen, F., Urbanus, M.L., Peterman, E.J., van Grondelle, R., Paulsen, H. and van Amerongen, H. (1999) Decreasing the chlorophyll a/b ratio in reconstituted LHCII: structural and functional consequences. *Biochemistry*, **38**, 6587–96.

33 Jordan, P., Fromme, P., Witt, H.T., Klukas, O., Saenger, W. and Krauß, N. (2001) Three-dimensional structure of cyanobacterial photosystem I at 2.5 A resolution. *Nature*, **411**, 909–17.

34 Bassi, R., Croce, R., Cugini, D. and Sandonà, D. (1999) Mutational analysis of a higher plant antenna protein provides identification of chromophores bound into multiple sites. *Proceedings of the National Academy of Sciences of the United States of America*, **96**, 10056–61.

35 Leupold, D., Teuchner, K., Ehlert, J., Irrgang, K.-D., Renger, G. and Lokstein, H. (2002) Two-photon excited fluorescence from higher electronic states of chlorophylls in photosynthetic antenna complexes: a new approach to detect strong excitonic chlorophyll a/b coupling. *Biophysical Journal*, **82**, 1580–5.

36 Schubert, A., Beenken, W.J.D., Stiel, H., Voigt, B., Leupold, D. and Lokstein, H. (2002) Excitonic coupling of chlorophylls in the plant light-harvesting complex LHC-II. *Biophysical Journal*, **82**, 1030–9.

37 Krikunova, M., Voigt, B. and Lokstein, H. (2002) Direct evidence for excitonically coupled chlorophylls a and b in LHC II of higher plants by nonlinear polarization spectroscopy in the frequency domain. *Biochimica et Biophysica Acta*, **1556**, 1–5.

38 Agarwal, R., Krueger, B.P., Scholes, G.D., Yang, M., Yom, J., Mets, L. and Fleming, G.R. (2000) Ultrafast energy transfer in LHC-II revealed by three-pulse photon echo peak shift measurements. *The Journal of Physical Chemistry B*, **104**, 2908–18.

39 Salverda, J.M., Vengris, M., Krueger, B.P., Scholes, G.D., Czarnoleski, A.R., Novoderezhkin, V., Amerongen, H.V. and Grondelle, R.V. (2003) Energy transfer in light-harvesting complexes LHCII and CP29 of spinach studied with three pulse echo peak shift and transient grating. *Biophysical Journal*, **84**, 450–65.

40 Novoderezhkin, V., Salverda, J.M., Amerongen, H.V. and Grondelle, R.V. (2003) Exciton modeling of energy-transfer dynamics in the LHCII complex of higher plants: a redfield theory approach. *The Journal of Physical Chemistry B*, **107**, 1893–912.

41 Bassi, R., Sandonà, D. and Croce, R. (1997) Novel aspects of chlorophyll a/b-binding proteins. *Physiologia Plantarum*, **100**, 769–79.

42 Croce, R., Weiss, S. and Bassi, R. (1999) Carotenoid-binding sites of the major light-harvesting complex II of higher plants. *Journal of Biological Chemistry*, **274**, 29613–23.

43 Hobe, S., Niemeier, H., Bender, A. and Paulsen, H. (2000) Carotenoid binding sites in LHCIIb. Relative affinities towards major xanthophylls of higher plants. *European Journal of Biochemistry*, **267**, 616–24.

44 Paulsen, H., Finkenzeller, B. and Kuhlein, N. (1993) Pigments induce folding of light-harvesting chlorophyll a/b-binding protein. *European Journal of Biochemistry*, **215**, 809–16.

45 Plumley, F.G. and Schmidt, G.W. (1987) Reconstitution of chlorophyll a/b light-harvesting complexes: xanthophyll-dependent assembly and energy transfer. *Proceedings of the National Academy of Sciences of the United States of America*, **84**, 146–50.

46 Ruban, A.V., Pascal, A.A. and Robert, B. (2000) Xanthophylls of the major photosynthetic light-harvesting complex of plants: identification, conformation and dynamics. *FEBS Letters*, **477**, 181–5.

47 Ruban, A.V., Pascal, A.A., Robert, B. and Horton, P. (2001) Configuration and dynamics of xanthophylls in light-harvesting antennae of higher plants. Spectroscopic analysis of isolated light-harvesting complex of photosystem II and thylakoid membranes. *Journal of Biological Chemistry*, **276**, 24862–70.

48 Croce, R., Remelli, R., Varotto, C., Breton, J. and Bassi, R. (1999) The neoxanthin binding site of the major light harvesting complex (LHCII) from higher plants. *FEBS Letters*, **456**, 1–6.

49 Peterman, E.J.G., Gradinaru, C.C., Calkoen, F., Borst, J.C., Grondelle, R.V. and Amerongen, H.V. (1997) Xanthophylls in light-harvesting complex II of higher plants: light harvesting and triplet quenching. *Biochemistry*, **36**, 12208–15.

50 Gradinaru, C.C., Ozdemir, S., Gulen, D., Stokkum, I.H.M.V., Grondelle, R.V. and Amerongen, H.V. (1998) The flow of excitation energy in LHCII monomers: implications for the structural model of the major plant antenna. *Biophysical Journal*, **75**, 3064–77.

51 Linnanto, J., Martiskainen, J., Lehtovuori, V., Ihalainen, J., Kananavicius, R., Barbato, R. and Korppi-Tommola, J. (2006) Excitation energy transfer in the LHC-II trimer: a model based on the new 2.72 A structure. *Photosynthesis Research*, **87**, 267–79.

52 van Grondelle, R. and Novoderezhkin, V.I. (2006) Energy transfer in photosynthesis: experimental insights and quantitative models. *Physical Chemistry Chemical Physics*, **8**, 793–807.

53 Novoderezhkin, V.I., Palacios, M.A., van Amerongen, H. and van Grondelle, R. (2005) Excitation dynamics in the LHCII complex of higher plants: modeling based on the 2.72 Angstrom crystal structure. *The Journal of Physical Chemistry. B, Materials, Surfaces, Interfaces & Biophysical*, **109**, 10493–504.

54 Gradinaru, C.C., Stokkum, I.H.M.V., Pascal, A.A., Grondelle, R.V. and Amerongen, H.V. (2000) Identifying the pathways of energy transfer between carotenoids and chlorophylls in LHCII and CP29. A multicolor, femtosecond pump-probe study. *The Journal of Physical Chemistry B*, **104**, 9330–42.

55 Croce, R., Müller, M.G., Bassi, R. and Holzwarth, A.R. (2001) Carotenoid-to-chlorophyll energy transfer in recombinant major light-harvesting complex (LHC-II) of higher plants. I. femtosecond transient absorption measurements. *Biophysical Journal*, **80**, 901–15.

56 Connelly, J.P., Müller, M.G., Bassi, R., Croce, R. and Holzwarth, A.R. (1997) Femtosecond transient absorption study of carotenoid to chlorophyll energy transfer in the light-harvesting complex II of photosystem II. *Biochemistry*, **36**, 281–7.

57 Holt, N.E., Fleming, G.R. and Niyogi, K.K. (2004) Toward an understanding of the mechanism of nonphotochemical quenching in green plants. *Biochemistry*, **43**, 8281–9.

58 Gilmore, A.M. (1997) Minireview. Mechanistic aspects of xanthophyll cycle dependent photoprotection in higher plant chloroplasts and leaves. *Physiologia Plantarum*, **99**, 197–209.

59 Demmig-Adams, B. and Adams, W.W. (1992) Photoprotection and other responses of plants to high light stress. *Annual Review of Plant Physiology and Plant Molecular Biology*, **43**, 599–626.

60 Ruban, A.V., Pascal, A.A., Robert, B. and Horton, P. (2002) Activation of zeaxanthin is an obligatory event in the regulation of photosynthetic light harvesting. *Journal of Biological Chemistry*, **277**, 7785–9.

61 Horton, P., Ruban, A.V. and Walters, R.G. (1996) Regulation of light harvesting in green plants. *Annual Review of Plant Physiology and Plant Molecular Biology*, **47**, 655–84.

62 Li, X.P., Björkman, O., Shih, C., Grossman, A.R., Rosenquist, M., Jansson, S. and Niyogi, K.K. (2000) A pigment-binding protein essential for regulation of photosynthetic light harvesting. *Nature*, **403**, 391–5.

63 Müller-Moulé, P., Conklin, P.L. and Niyogi, K.K. (2002) Ascorbate deficiency can limit violaxanthin de-epoxidase activity in vivo. *Plant Physiology*, **128**, 970–7.

64 Davies, B.H. (1976) Carotenoids, in *Chemistry and Biochemistry of Plant Pigments*, 2nd edn (ed. T.W. Goodwin), Academic Press, London, pp. 38–165.

65 Frank, H.A., Cua, A., Chynwat, V., Young, A., Gosztola, D. and Wasielewski, M.R. (1994) Photosynthesis of carotenoids associated with xanthophyll cycle in photosynthesis. *Photosynthesis Research*, **41**, 389–95.

66 Ma, Y.Z., Holt, N.E., Li, X.P., Niyogi, K.K. and Fleming, G.R. (2003) Evidence for direct carotenoid involvement in the regulation of photosynthetic light harvesting. *Proceedings of the National Academy of Sciences of the United States of America*, **100**, 4377–82.

67 Horton, P., Ruban, A.V., Rees, D., Pascal, A.A., Noctor, G. and Young, A.J. (1991) Control of the light-harvesting function of chloroplast membranes by aggregation of the LHCII chlorophyll-protein complex. *FEBS Letters*, **292**, 1–4.

68 Crofts, A.R. and Yerkes, C.T. (1994) A molecular mechanism for qE-quenching. *FEBS Letters*, **352**, 265–70.

69 Wentworth, M., Ruban, A.V. and Horton, P. (2003) Thermodynamic investigation into the mechanism of the chlorophyll fluorescence quenching in isolated photosystem II light-harvesting complexes. *Journal of Biological Chemistry*, **278**, 21845–50.

70 Perozo, E., Cortes, D.M., Sompornpisut, P., Kloda, A. and Martinac, B. (2002) Open channel structure of MscL and the gating mechanism of mechanosensitive channels. *Nature*, **418**, 942–8.

11
Structure of the Phycobilisome Antennae in Cyanobacteria and Red Algae

Noam Adir

11.1
Introduction

11.1.1
Photosynthetic Antennas

All photosynthetic organisms contain a two-component photochemical apparatus comprised of an antenna complex and a reaction center [1, 2]. One can thus assume that useful biological conversion of light-energy requires the presence of a high enough input of energy by the antennas to sustain efficient electron transfer by the reaction centers. In the most primary sense, antenna complex proteins have been designed by evolution to perform three tasks: bind high densities of pigments (chlorophylls, bilins and carotenoids); create a functional funnel that efficiently transfers the absorbed energy into the reaction center; and self-assemble into units that allow the performance of the former tasks. These antenna structures must function correctly while preventing potentially deleterious side effects that could occur due to the proximity of excited molecules with other proteins. A variety of very different geometric solutions has evolved to attain these very same goals, to a much greater extent than the differences between reaction centers [3–5]. The four major antenna forms are: the circular transmembrane light-harvesting complexes (LH1 and LH2) of purple non-sulfur bacteria [6]; the compact transmembrane light-harvesting complexes (LHC-I, LHC-II) of prochlorophytes, green algae, and plants [7, 8]; the membrane associated phycobilisome of cyanobacteria and red algae [9, 10]; and, the organelle-like chlorosome of green sulfur and green filamentous bacteria (*Chlorobi* and *Chloroflexi*) [11]. As a result of the exceptionally different morphological nature of these antenna systems, each one is described in a separate chapter. This chapter describes in depth the knowledge obtained from the determination of near-atomic resolution crystal structures of isolated components of the phycobillisome. Models of the entire complex based on these structures will be presented along with a description of how the phycobillisome performs its tasks.

Photosynthetic Protein Complexes: A Structural Approach. Edited by P. Fromme

ISBN: 978-3-527-31730-1

11.1.2
Photosynthetic Organisms that Utilize the Phycobilisome

The phycobilisome serves as the major light-harvesting antenna complex of cyanobacteria, rhodophyta (eukaryotic red algae), and cryptophyta. These organisms collectively make up earth's largest producers of oxygen and absorbers of CO_2. They can be found in environments ranging from oceans to hot-springs (up to ~70 °C) to dry deserts and to the Arctic [12]. In all environments they can use light throughout the visible spectrum due to the coverage provided by the phycobilisome and reaction center pigments. The phycobilisome pigments absorb strongly in the region between 550–660 nm, which complements the spectral region covered by chlorophyll *a* in the blue (soret band) and red regions. Red algae can grow in much deeper water due to the addition of a larger compliment of phycoerythrin (PE), the blue absorbing phycobiliprotein (PBP).

The study of the phycobilisome and its component protein subunits began almost 150 years ago (for a detailed description of the early period of phycobilisome research see [13]). However, the real first structural studies that led to insights on phycobilisome function began in the 1960s and 1970s in the research groups of E. Gantt, A. Glazer, D. Bryant, H. Zuber, and their coworkers [14–20]. These original structural studies coupled the visualization by electron microscopy (EM) of the phycobilisome along the thylakoid membranes, with biochemical and genetic studies that associated the phycobilisome with both the bilin-bearing PBPs and the colorless linker proteins. This was followed by additional EM studies on isolated phycobilisomes using negatively stained samples from different organisms. The intricate structure revealed in these micrographs showed that the phycobilisome is assembled by a large number of seemingly similarly sized disks. The fragility of the complex during sample fixation had a notable affect on its shape and integrity, as can be seen by the lack of uniformity between complexes. However, it became quite clear that the phycobilisome contains two main substructures – a set of rod like stacks of disks surrounding a core of two to five close-packed disks [21]. The number of disks in the core was found to be species dependent [22]; but, in all cases, the rods appeared to radiate out from the core. The fashion and architecture of rod attachment to the core is of major functional importance and will be the focus of the final section (Section 11.3 of this review.

11.2
Basic Structural Characteristics of the Phycobilisome Component Proteins

All modern PBPs (those that exist in present day organisms), can be traced back to a common founder gene product – an ancestral PBP [23]. This ancestral gene can be associated even further back to the globin family of proteins [24]; thus, one can chart the development of the phycobilisome through comparisons of protein sequences and cofactor changes. The primary unit of all PBPs is a heterodimer

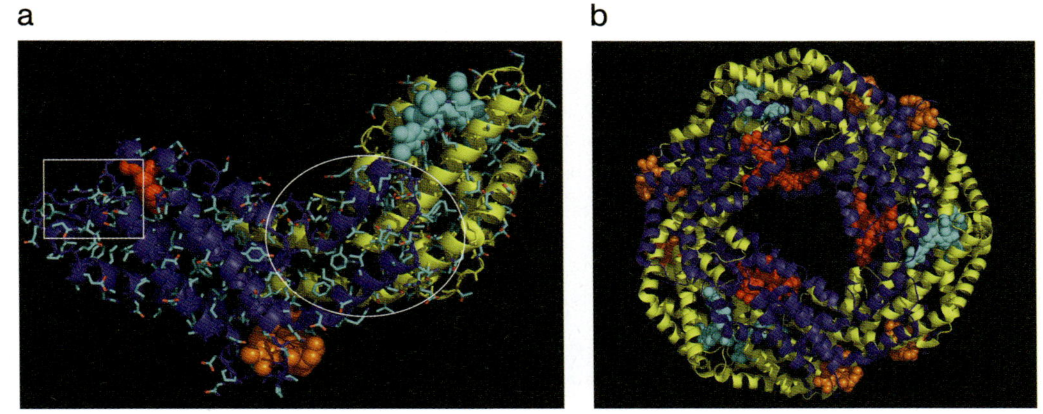

Figure 11.1 **(a)** ($\alpha\beta$) phycocyanin monomer. The monomer formation surface formed by the X-Y helices of both subunits is identified by the white oval. The white square identifies the position of two conserved phycobiliprotein modifications: the βThr^{77} residue (with non standard peptide dehydral angles) and βAsn^{72} which is methylated. Amino acid residues are depicted in stick representation and colored according to the usual CPK color code. **(b)** $(\alpha\beta)_6$ hexameric form of phycocyanin. In both panels the α and β subunits are depicted in yellow and blue cartoon helices, respectively. The PCB chromophores are shown in sphere representation: αPCB^{84} in cyan, βPCB^{84} in red and βPCB^{155} in orange. Figures 11.1, 11.2 and 11.4 were prepared using Pymol (pymol.sourceforge.net).

composed of two homologous subunits, α and β, known as the ($\alpha\beta$) monomer, shown in Figure 11.1a. There are four major forms of PBPs, classified by the number and type of bilin cofactor to which they are associated. These are allophycocyanin (APC, λ_{max} = 652 nm), phycocyanin (PC, λ_{max} = 620 nm), phycoerythrin (PE, λ_{max} = 560 nm), and the infrequently found phycoerythrocyanin (PEC, λ_{max} = 575 nm). The molecular weights of the α and β subunits are 16–20 kDa, depending on the type of PBP. The major structural facet of all PBP subunits is a compact globular structure made up of six α-helices (A, B, E, F, F′, and H) that are denoted according to their similarity to other members of the globin family (Figure 11.1a). All cofactors are covalently bound to these core helices or to additional loops that have arisen by insertion into the gene sequence. Two additional α-helices (X and Y) extend out from the core and serve as the assembly interface of the monomer and of additional levels of assembly (Figure 11.1a, white circle).

All PBPs, except for a special class of soluble PE found only in cryptomonads (see Section 11.2.3.5), further assemble into $(\alpha\beta)_3$ trimers. Further levels of assembly of PC, PEC, and PE occur rapidly: $(\alpha\beta)_3$ trimers can associate into $(\alpha\beta)_6$ hexamers and these then associate further into extended structures called **rods**. Figure 11.1a and b show a representative PC monomer and hexamer (respectively) from the *Thermosynechococcus vulcanus* crystal structure at the highest resolution determined to date (1.4 Å, deposition into the PDB in progress). The APC trimers associate in a somewhat different manner, forming cylinders containing four

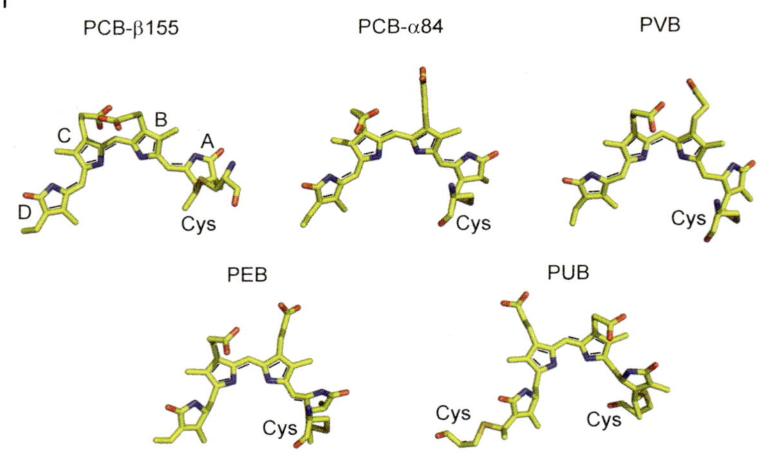

Figure 11.2 Three-dimensional structures of bilin chromophores thio-ether linked to cysteine residues. The coordinates of each bilin structure were carved out of the crystal structures: phycocyanoblins (PCB) from *T. vulcanus* PC–[1KTP], phycoviolobilin (PVB) from *M. laminosus* PEC–[1C7L], phycoerythrobilin (PEB) and doubly linked phycourobilin (PUB) from *G. monilis* PE–[1B8D]. The positions of the conjugated double bonds systems are denoted by black lines. Note that identical bilins can have different conformations (see two PCBs) and that the orientations of the propionic acid groups is highly variable.

trimers. Two to five of these cylinders pack into what is called the **core** of the phycobilisome.

All of the cofactors of the phycobilisome belong to the bilin family (Figure 11.2). These molecules are linear tetrapyrroles, which differ by the number and position of the conjugated double bonds that serve to tune the general region of light absorption. Both APC and cyanobacterial PCs contain only phycocyanobilin (PCB) cofactors, which have eight conjugated double bonds, while rhodophyte PC and all PECs and PEs can contain phycoviolobilins (PVB, seven conjugated double bonds), phycoerythrobilins (PEB, six conjugated double bonds), and phycoeurobilins (PUB, five conjugated double bonds). Some PVBs, PEBs, and PUBs can be covalently linked to two cysteine residues via a thio-ether linkage, while all PCBs are singly linked. The molecular structures shown in Figure 11.2 have been carved out of the PDP coordinates, as determined by X-ray crystallography of the isolated PBPs (two-dimensional representations of the bilins can be seen in (Grossman, Schaefer, Chiang and Collier, 1993)). In this way, the bent and twisted nature of the cofactors can be appreciated. It can be clearly seen that the degree of linearity, the positions of the propionic acids, and the relative orientation of the bilins with respect to the cysteines (to which each bilin is covalently bound) differ. These structural modifications are of course caused by the polypeptide environment surrounding each bilin. It is apparent that, by use of the open chain structure, the proteins can apply a second level of tuning by bending and twisting the molecules. Such tuning results in a broader range of absorption spectra and, potentially, serves to induce directed energy transfer from the outside of the phycobilisome down into the core and finally into the reaction center (see Section 11.4.2).

11.2.1
Isolation of Phycobilisomes and Phycobiliproteins

One of the major questions that must be addressed when describing information obtained by structural methods is the state of the molecule to be analyzed. Unfortunately, the state of the molecule in vivo is rarely exactly the same as it is in its isolated form prior to the manipulations required for the performance of measurements (such as crystallization, concentration, or flash-freezing). Upon performance of such manipulations, the state may change again. The resulting structures are very often spectacular in the amount of information obtained; thus, in many cases, the question of the relevance of the structure to the true state of the molecule is ignored. The relevance of crystal structures of membrane bound LHCs has been debated [3], and indeed the results obtained at high-resolution by X-ray crystallography do not always agree with lower-resolution electron microscopy (performed on less traumatized samples). In the case of phycobilisome, this has certainly been a major problem throughout the history of its structural investigation. To date, crystallography has provided structural information on isolated PBPs [9] that may, or may not be entirely useful in the building of larger phycobilisome structures. In many of the published reports of PBP crystallization, the level of organization (see Section 11.2) obtained during isolation is not noted. While this does not negate the importance of the structural results on the level of the monomeric protein, it does limit the applicability of the structure to the understanding of the function of the entire phycobilisome.

PBPs are readily isolated from a variety of organisms. The amount of protein in each cell is quite high, and the covalent nature of the bilin binding to the protein makes monitoring the purification of the protein quite simple. Upon cell disruption, large amounts of PBP, which can be easily purified by standard biochemical methods, are typically released. Experience shows that in low ionic strength buffers, much of the isolated PBPs remain in $(\alpha\beta)_3$ trimeric form. In almost all reported crystal forms, these trimers, or perhaps in some cases monomers, reform during the crystallization process into $(\alpha\beta)_6$ hexameric form, or even further into infinitely long rods.

Intact phycobilisomes are rarely found in the supernatant of disrupted cells. This may indicate that the phycobilisome are rather loosely held together. Most of the published EM studies of thin sections of cyanobacteria show series of similarly sized complexes lining the thylakoid membrane [25]. Each phycobilisome may actually be surrounded by densely packed cytosolic proteins, perhaps preserving complex integrity. Upon disruption and dilution, these proteins (bound to the phycobilisome in a specific manner or not) may dissociate from the phycobilisome, allowing the complex to fragment. In order to obtain intact phycobilisome, cell disruption must be performed in the presence of high concentrations of phosphate buffer (0.5–0.9 M); this buffer appears to mimic, in some fashion, the conditions that preserve the complex [19]. To date, isolated phycobilisomes have only served for EM studies (negatively-stained and flash-frozen cryoprotected samples [26]). Crystallization of such an immense complex is certainly a difficult prospect, but the availability of even a low-resolution structure would be extremely helpful, as it could serve as a template for the building of a true structure based model.

11.2.2
Crystallization of Phycobilisomes

Following their isolation, PBPs are typically readily crystallized. Crystals have been obtained from both polymer (e.g. polyethylene glycols) and salt (e.g. ammonium sulphate) precipitating reagents [9]. Many of the crystals obtained belong to trigonal or hexagonal space groups, including rhombohedral space groups in hexagonal habit, as a result of the threefold symmetry of the basic $(\alpha\beta)_3$ trimer. However, a number of additional space groups have been obtained including monoclininc and orthorhombic. In these cases, the asymmetric unit contains multiple trimers or hexamers which are, in some cases, in non-physiologically related packing arrangements [27].

The ease with which PBPs, especially PC, are crystallized is most likely due to their propensity to self-assemble in vivo. While the presence of chaperones in the self assembly process has been suggested [28], it has not been experimentally verified. If the same chemical properties come into play during both self assembly and crystallization, this would indicate that the extended crystal structures, beyond that of the asymmetric unit, are informative as far as understanding the actual in vivo substructures that make up the phycobilisome. It is interesting to note that the interaction surfaces of the different hierarchal levels of phycobilisome assembly (α and β subunits to monomer, monomers to trimers, trimers to hexamers, etc.) are quite flat and without prominent invaginations. The details of the interactions between PBPs at different association levels will be discussed in Section 11.3.

11.2.3
X-Ray Structures of PBPs

To date, a total of 23 PBP crystal structures exist in the Protein Data Bank (www.rcsb.org). They include three APC structures (PDB codes [1ALL], [1B33] and [1KN1]), seven unique (from different organisms) PC structures ([1CPC], [1F99], [1GH0], [1HA7], [1I7Y], [1JBO], [1KTP], [1ON7], [1PHN], and [2BV8]), one unique PEC structure ([2C7J], [2C7K], and [2C7L]), four PE structures from the phycobilisome containing organisms ([1B8D], [1EYX], [1LIA], and [2G9M]), and one unique cryptophyte PE structure ([1QGW], [1XF6] and [1XG0]). Only a single linker protein has been visualized by crystallographic methods within the [1B33] APC structure [27]. In the following sections the details of each PBP class are described. The structures are of varying diffraction quality, with the resolution of diffraction ranging between 1.45 Å to 3 Å. The crytophyte, PE_{545}, structures are unique in their superior quality. The PE_{545} crystals diffracted to better than 1 Å, enabling true atomic resolution and determination of anisotropic B-factors. Most PBP structures contain the entire polypeptide chain (due to the fact that the entire chain participates in the formation of tertiary/quaternary interactions), all cofactors, and various (but almost always significant) numbers of solvent molecules [9]. Refinement of the structures has typically converged at good R_{crys}/R_{free} values with relatively low average B-factors. On the basis of these crystallographic parameters, it is safe to assume that these structures represent very close approximations of the actual protein structures found in vivo. However, the general lack of linker proteins within these structures may indicate that significant conformational changes may occur within the trimeric/hexameric internal cavities.

In the following sections, each of the various components of the phycobilisome are described in detail. The amino acid residues mentioned in the text will be denoted according to the sequence numbering of the C-phycocyanin structures, as proposed by Huber and coworkers [29, 30]. For an extensive study of PBP sequence alignment see [23].

11.2.3.1 **Allophycocyanin**

APC is the only type of PBP found in phycobilisome cores. While only about 25–30% sequence identity exists between APC, PC, and PE from the same species [23], the three-dimensional structures are highly similar; typically there is pair-wise r.m.s.d. of ~1.3 Å between equivalent α-carbons. Each APC monomer covalently binds two phycobillisome cofactors to Cys^{84} of each subunit. The molecular structures and protein environment of the βPCB^{84} are almost identical to those found in PC, while the environment of the αPCB^{84} cofactors is quite different. It has been suggested that the significant shift in the λ_{max} of APC versus PC (652 nm and 620 nm respectively) in the total $(\alpha\beta)_3$ APC absorption spectra is a result of this change in the αPCB^{84} environment [31]. However no direct proof has yet been obtained that this is indeed the case. It has previously been noted that the conserved βTyr^{62} residue strongly interacts with the αPCB^{84} during trimer assembly [32]. The phenolic half penetrates in between the two PCB propionic acids, potentially modifying its absorption properties. Interestingly, the small linker protein visualized in the [1B33] structure [27] also contains a tyrosine residue at (Tyr^{33}) which upon assembly penetrates in a similar fashion into one of the βPCB^{84} cofactors. Thus four out of six APC cofactors interact in a similar fashion with Tyr residues. This could explain the increase in the 652 absorption peak and concomitant decrease in the 620 nm shoulder in the APC trimer + linker spectrum in comparison with linker-less APC trimer [27]. No such Tyr-PCB interaction exists in PC, PEC or PE subunits.

The phycobilisome core can contain two to five cylinders (each cylinder containing two to four trimers) depending on the species. Two cylinders associate with the outer surface of the thylakoid membrane (basal cylinders); the remaining cylinders are stacked as a second layer in closed packed fashion (Figure 11.3). Second level cylinders typically contain only APC monomers (the gene products of the *apcA* and *abcB* genes) and a small core linker. The two basal cylinders contain

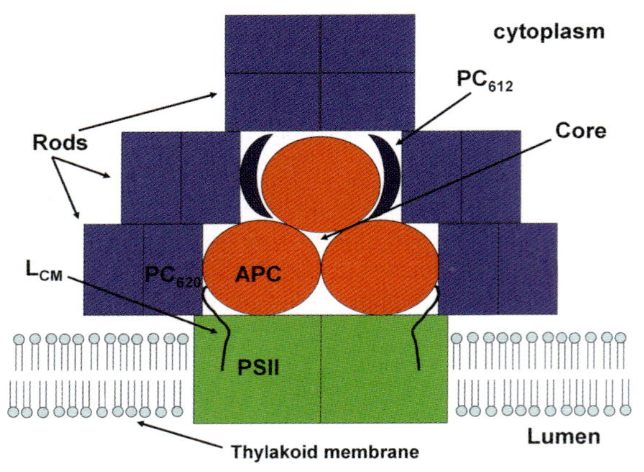

Figure 11.3 Model of a tricylindrical core phycobilisome consistent with crystallographic and biochemical studies. A Photosystem II (PSII) dimer (green) is situated in the thylakoid membrane. The phycobilisome core containing allophycocyanin (APC, red circles) is attached to PSII via the L_{CM} subunit on the cytoplasmic side of the membrane. Six rods are attached as three doublets surrounding the core. The rods always contain phycocyanin (PC) hexamers (shown as blue rectangles, but actually round disks shown side-on). In some species, additional hexamers of phycoerythrocyanin (PEC) or phycoerythrin attach to the rod distal end. Linker proteins found in the internal spaces of the core cylinders and rods are not depicted.

both regular $(\alpha\beta)_3$ APC and trimers that contain modified APC monomers: either a modified α subunit, called α^B or a modified β subunit, called β^{18} (encoded by the *apcD* and *apcF* genes respectively). The β^{18} is paired with the L_{CM} core-membrane linker protein, encoded by the *apcE* gene [22]. These minor constituents all have a single PCB cofactor that absorbs further to the red than the bulk APC PCBs, probably to complete the overlap with the PSII chlorophyll *a* pigments and assist in efficient energy transfer. Of the subunits that physically link the entire phycobilisome to the reaction center of Photosystem II (PSII) (see Chapter 4) within the membrane, L_{CM} is the largest–between 70–128 kDa, depending on the number of core cylinders. It has been suggested that the C-termini of the L_{CM} subunits penetrate into PSII by way of a cavity formed between the D1/D2 subunits and the peripheral small PSII subunits (psbJ, psbK, and cytochrome b_{559}; [33, 34]). The recently determined PSII crystal structures reveal a dimeric structure with the longest dimension, ~205 Å, closely matching the combined diameters of two APC cylinders, ~220 Å. Figure 11.3 shows schematically the position of the L_{CM} anchor penetrating into PSII. APC crystal structures and to date have been obtained only for regular $(\alpha\beta)_3$ APC trimers and not for any of the minor APC forms.

One of the major differences in the APC subunit is the formation in crystal of a more loosely associated hexamer, arising from slight sequence variations in the APC versus PC or PE [31, 32]. It is not clear what the source of this looseness might be; however, the somewhat lower rigidity could facilitate the perpendicular association between the APC core and the rods. Evidence that the native form of

APC, within the context of the complete phycobilisome, is a trimer has been suggested [35]; whether or not a hexameric nature is actually necessary for APC function is unclear. Concentrated APC can, however, form long rod-like crystalline needles [18], showing that APC also has the ability to form long stacks. In the [1B33] APC structure [27], APC is found in trimeric form. Two trimers make up the asymmetric unit; they are associated in a perpendicular orientation that is clearly non-physiological. This mode of association is not similar to the fashion of rod-core assembly as visualized in EM studies, and may be due to the presence of the internal linker protein which may disturb the association of the trimers into hexamers in crystal. In the other two APC structures, the linker is absent and the association into hexamers occurs, although the hexamers are not equivalent. In the [1ALL] *S. platensis* structure, the hexamer is formed via interactions between β subunits [31]; while, in the [1KN1] *P. yezoensis* structure, the association is via the α subunits [32] and is similar to that found for PC hexamers. Since each core cylinder is made up of four APC trimers, it is possible that these two crystal structures mimic two different trimer-trimer association modes. Clearly, crystallization of a larger complex (either a core subcomplex or the entire phycobilisome) will help in solving this problem.

Isolated APC, in either monomeric or trimer form, has served as a model system to study the effects of absorption and energy transfer [36–39]. This system is especially attractive because of the large red-shift that occurs in the absorption spectra following trimerization. A detailed review of the photophysical experiments performed on APC was recently published by MacColl [40]. Further details on possible modes of energy transfer in the phycobilisome are detailed in Section 11.4.2.

11.2.3.2 **Phycocyanin**

Phycocyanin (PC) is found in all phycobilisome and is the major rod component. Typically isolated in trimeric form, this PBP has an absorption maxima at 620 nm, as a result of the overlap of its three cofactors, α^{84}, β^{84}, and β^{155} (Figure 11.1a), which absorb between 600–630 nm [41]. In the presence of bound linkers, the λ_{max} is red-shifted by up to 20 nm [42]. The overall molecular dimensions of the trimeric discs are the same as found for APC; however, an additional loop made up of nine residues (residues 146–154) is inserted into the β subunit sequence between helices G and H. Such loops protrude from the outer circumference of the trimeric/hexameric disk (Figure 11.1b) and partially shield ring A of the βPCB^{155}, which is covalently linked to βCys^{155}, from the solvent. It has been proposed that cofactors on the outside circumference of the disks enable efficient cross-rod energy transfer. The extra loop may prevent movement of ring A (Figure 11.2), and perhaps ring B, but allow movement of rings C and D upon further complex assembly. Both α and β subunits in the monomer have thio-linked PCB chromophores at symmetry related positions, Cys^{84}, located between helices E, F′, and G. The chemical surroundings of these two PCBs is, however, very different upon formation of the higher-order $(\alpha\beta)_3$ trimer. The α PCB^{84} is almost totally buried and hidden from the surrounding solvent (Figure 11.1b), unlike the βPCB^{84} which extends out into the trimeric/hexameric disc interior.

PC crystal structures have been determined from seven species of cyanobacteria, (four mesophiles and three thermophiles) and three rhodophytes, (two mesophiles and one thermophile). Two of the early PC structures have not been deposited in the PDB; however, the published reports are quite detailed and informative [29, 30, 43]. The structures obtained are at high resolution and provide an excellent molecular description of the structural characteristics required for efficient energy transfer, preserved during evolution. These structures also provide insights into the fine-tuning required by species that have evolved in different surroundings, and clues on the process of phycobilisome assembly. The crystal structures, [1I7Y], [1KTP], and [1JBO], of two thermophilic cyanobacterial species show a significant change in the relative position of pyrolle ring D of the βPCB^{155}. The altered conformation of this ring was first identified in the [1I7Y] *T. vulcanus* structure determined at room temperature [44], and then further confirmed at cryogenic temperature [45]. This alteration in ring orientation was proposed to occur at a critical site with respect to the PC assembly process. This position may stabilize the $(\alpha\beta)$ monomer interaction interface as well as the $(\alpha\beta)_6$ hexamer formation interface. In both structures, solvent molecules are located in almost identical positions, about equidistant from αAsp^{28}, βAsn^{35}, and αArg^{33} from the lower $(\alpha\beta)_3$ trimer. The presence of bound water in these positions at the typical growth temperatures of 55–60 °C is probably transient, but could add a significant amount of stabilization energy to both the $(\alpha\beta)$ monomer and the $(\alpha\beta)_6$ hexamer. Additional solvent molecules were located in close contact, 2.6–3.4 Å, with each of the three cofactors. Those solvent molecules associated with βPCB^{84} and βPCB^{155} interact with the molecular edge adjacent to the bulk solvent, while those associated with αPCB^{84} are positioned in the trimer association interface. These waters may have a role similar to that of the surrounding protein residues in stabilizing the PCB configuration to obtain optimal functionality.

The two internal pyrolle rings, B and C, of all bilin chromophores contain propionic acids whose functions are not clear (Figure 11.2). One possibility is that they may be required to help keep the bilins in a less folded configuration, thereby modifying their absorption spectra. Interactions between the propionic acids and polar residues have been previously identified for the PCBs in a number of crystal structures; these residues are highly conserved. In the low temperature *T. vulcanus* structure ([1KTP]), all three chromophores bind solvent molecules in the gap between the two propionic carboxyl groups [45]. When a superposition of the three cofactors is performed, these solvent molecules spatially coincide. It can be thus suggested that these solvent molecules are important for the positioning of the propionic side chains in specific conformations. Indeed, molecular dynamics simulations of the propionic acid positions in the absence, or presence, of the intervening solvent molecules show that in their absence, the propionic acids both repel each other and are attracted to adjacent positively charged residues. This induces a greater degree of separation and strain on the PCB structure. Examination of other PC structures, [1CPC] from *F. diplosiphon,* [1JBO] from *T. elongatus,* and so on, show solvent molecules in similar positions, indicating that this requirement is not a characteristic trait limited to PC from thermophilic species. By obtaining their proper position, the propionic acid groups may maintain the two central pyrolle rings on almost the same plane, and thus may be important for efficient energy transfer. The propionic acids may also interact with linker proteins, as was visualized in the *M. laminosus* [1B33] APC structure [27].

As indicated above, isolated PC maximal absorption is at 620 nm in its $(\alpha\beta)_3$ trimeric form (PC_{620}). PC monomer absorption is blue-shifted to 614 nm [41] while the absorption of trimers isolated in the presence of linkers is red-shifted (629–638 nm). These observations reaffirm the complexity of clarifying the mechanism of energy transfer in the phycobilisome using isolated components. However, spectroscopic measurements made on isolated components, in conjunction with high-resolution studies, continue to serve as the major source of information on phycobilisome function. In the course of isolation of the phycobilisome and APC from *T. vulcanus*, a minor fraction of PC was isolated that absorbed maximally at 612 nm while in $(\alpha\beta)_3$ trimeric form [46]. This form of PC, PC_{612}, was characterized crystallographically ([1ON7]) and appeared to have a number of unique structural traits. The difference in absorption was proposed to be the result of the lack of methylation of βAsn^{72} (see Section 11.2.4.2). This residue has been found to be methylated in all PBPs [47, 48], and to affect the absorption characteristics of the conserved βPCB^{84} (see Section 11.2.4.2). Although crystallized under conditions similar to that of *T. vulcanus* PC_{620}, the PC_{612} did not form $(\alpha\beta)_6$ hexamers. This flexibility could be advantageous in the formation of the interaction surface between the end of the rod and the circumference of the APC core cylinders. The presence of a trimer at the end of two of the rods could serve as a "filler" ring to make up for the differential in position between the lower base core cylinders and the upper cylinder (Figure 11.3). This hypothesis is strengthened by the fact that the PC_{612} was only isolated in conjunction with APC cores, and thus it is probably not a small population of unmethylated PC. The lack of methylation could allow the rod linker protein, present at the end of the rod closest to the core, to interact strongly with the β^{84} cofactor, affording a red-shift towards 630–640 nm. This, in turn, would serve as a functional bridge between the bulk PC_{620} in the rods and the APC in the cores.

Along with the detailed three-dimensional description of the PC structures, one of the most important goals of all of the crystallographic studies of PC was to describe the possible pathways of energy transfer between adjacent cofactors. Since, in many cases, the crystallized protein was either in the $(\alpha\beta)_3$ or $(\alpha\beta)_6$ forms, the clarification of more extensive energy transfer pathways requires the assumption that the further assembly of the trimeric rings into hexamers and rods during the crystallization process resembles the organization of the phycobilisome in vivo. This assumption has been a mainstay of all structural studies, despite the fact that all PC structures lack linker components. The relevance of the crystal structures to the phycobilisome structure will be detailed in Sections 11.3 and 11.4.

11.2.3.3 **Phycoerythrocyanin**

PEC is a relatively minor PBP component, found in only some cyanobacterial species [41, 49, 50]. It is similar to PC, with a PVB chromophore bound to the α-subunit instead of PCB. This results in an absorption shift to the blue, positioning this PBP between PE and PC in the energy transfer funnel. While first crystallographically determined in 1990 [51], the PEC structure from *M. laminosus* was not deposited in the PDB. However, this same protein was recently studied crystallographically under different conditions of data collection – room-temperature versus cryogenic temperature, and monochromatic versus polychromatic Laue irradiation, yielding a molecular view of this protein [49]. The major motivation for this study is that the PVB chromophore goes through a reversible, photochemically induced *Z*/*E* isomerization, involving a rotation of pyrolle ring D, with respect to the rest of the chromophore. The crystals obtained in this study allowed a more precise determination of the chromophore structures within the protein environment, an important aspect for the understanding of the PEC functional absorption and transmission properties. The authors also report the first successful use of the Laue method, which can potentially be used for time-resolved crystallographic measurements. Since, to date, all structures are static, it would be interesting to see if any protein dynamics occur on the time-scale of energy transfer within the complex.

The *M. laminosus* PEC described in this study crystallized in the same space group as the [1ON7] structure of unmethylated PC from *T. vulcanus* (Section 11.2.3.2). In both cases, the $(\alpha\beta)_3$ trimers used for crystallization did not associate further into $(\alpha\beta)_6$ hexmers. Both structures show elevated B-factors for the helix-turn-helix motif adjacent and interacting with the α-subunit chromophore, PCB in PC and PVB in PEC. The elevated B-factors reveal a high degree of flexibility in the motif, which may afford the small proportion of the PVB found to undergo isomerization in the trimeric form. Isomerization could modify the functional properties of the PEC by loosening the trimer-trimer interactions in the hexamer.

11.2.3.4 **Phycoerythrin**

PE exists as an additional PBP only in some cyanobacterial species, but exists in all red algae. The basic PE structure is a $(\alpha\beta)_6$ hexamer in cyanobacteria and a $(\alpha\beta)_6\gamma$ in red algae: the unique chromophore-bearing 30 kDa γ subunit is found inserted within the hexameric disk of red-algal PE. The first PE structures to be determined were those of *Porphyridium sordidum* bPE [52] and *Porphyridium cruentum* bPE – organisms that differ in the number of PE/PUB chromophors bound to the (αβ) monomer and to the γ subunit. Unfortunately, the coordinates of these structures were not deposited in the PDB. A number of years later, four additional PE structures were determined and their coordinates deposited in the PDB. These structures are all red algal PE that contain both singly and doubly linked cofactors, which has a noticeable effect on their conformation and absorption. All proteins prior to crystallization had the internal γ subunit, except for the bPE structure; the presence of the protein in crystal was identified by SDS-PAGE of solubilized crystals. However, except for a very small number of residues, the γ subunit structure could not be visualized in the electron density. In the [1B8D] structure, three residues of the γ subunit were built into electron density within the $(\alpha\beta)_6$ disk [53], thus providing evidence for the general position of this subunit. The identified residues were located in the vicinity of the PE chromophores. It was proposed that the reason for the absence of the γ subunit in the electron density was due to the threefold crystallographic averaging applied during structure determination [52, 53], although heterogeneity in the position of this subunit cannot be discounted. Another possibility is that crystal packing forces on the PE disks change the affinity toward the γ-subunit to a degree that allows it to bind in a less specific manner, thereby lowering its apparent occupancy in the crystal structure.

The absorption of PE is blue-shifted compared to PC; some organisms take advantage of this difference to change their absorption cross-section by a mechanism known as complimentary chromatic adaptation (CCA) [54]. In this process, cells that contain phycobilisome rods with PC grown under red light modify the expression of PBP encoding genes, as a result of a shift to green light. The newly expressed genes encode for variants of PC, PE, and for the appropriate linker proteins. PE now becomes the terminal subunit on the newly assembled phycobilisome. A number of excellent reviews describe the mechanism of CCA in detail [41, 54, 55]. In normal sunlight, the phycobilisome contains intermediate levels of PC and PE, indicating that cells can respond correctly to the actual light quality, affording maximal energy transfer. Changes in light intensity, with or without additional changes in the light quality, have also been known to have an effect on both the number of phycobilisome and the length of the phycobilisome rods. These observations show that the phycobilisome is a dynamic complex, as far as composition, and requires the presence of additional accessory proteins in order to obtain its functional structure.

11.2.3.5 **Crytophyte Phycoerythrin**

Cryptophytes are fresh-water, or marine unicellular, organisms with a unique photosystem arrangement, different from cyanobacteria, red algae, green algae, and so on. The crytophytes contain plastid organelles with a red algae origin that have lost the ability to form phycobilisomes. Instead, proteins similar to PE β-subunits, in complex with a unique non-PBP α-subunit, form a $\alpha\alpha\beta\beta$ tetramer that binds chromophores and is translocated into the membrane lumen. Within this space, they are apparently soluble, perhaps interacting with the reaction center via an additional protein. Since these proteins are not phycobilisome forming, they are outside the scope of this review. However, it is important to note that this protein class has afforded crystals that diffract to the highest resolution (0.97 Å) of any photosynthetic antenna protein, thus allowing a true atomic resolution structure of an energy-harvesting protein [56, 57].

11.2.3.6 **Linker Proteins**

First identified biochemically in the 1970s [58, 59], the linker proteins (LPs) that lack bound chromophores have remained an enigma, both in functional and structural terms. Using electrophoretic methods on purified phycobilisomes, the number and molecular weight of these proteins could be estimated. More precise analysis of electron micrographs showed a protein component within the core cylinders and rod disks, promoting the idea that the role of these proteins is to serve as physical and stabilizing links within the phycobilisome. With the advantage of genomics, the ability to predict the entire linker complement of a number of phycobilisome containing organisms became possible, and the genes identified could be associated with the proteins identified (by molecular weight) biochemically. In accordance with the putative linking role, four different classes of LPs were suggested: (i) core linkers (L_C); (ii) rod linkers (L_R) that can be either large, ~25–30 kDa, for linking hexamers to hexamers, or small, ~8 kDa, that cap or terminate the rods; (iii) linkers involved in rod-core attachment (L_{RC}); and, (iv) the special APC subunit, which links the entire assembly to the membrane, called L_{CM}. This last LP is the product of gene fusion between a gene encoding for an APC α-subunit and a number of repetitions of genes encoding for LPs. The size of the L_{CM} is proportional to the size of the core, between two to five cylinders. In red algae, the PE hexamers contain a γ subunit that serves as both linker and additional chromophore-binding subunit. The γ subunits are dissimilar in sequence from both other linkers and the PE α and β subunits, and are a relatively new addition to the phycobilisome. These γ proteins are encoded by nuclear genes, unlike the other phycobilisome genes which are found in the plastid genome.

With the crystallographic visualizations of the PC and PEC trimeric/hexameric structures, it became clear that if the linkers are present in the disk cavity, they will come into contact with the chromophores that jut out into this solvent accessible space (for instance, the βPCB^{84} in PC, PEC, and APC). Since the chromophore environment has a very strong effect on its spectral characteristics, the presence of different LPs could potentially modify the functionality of these similar, or even identical, chromophores. By specifically altering the environment of certain chromophores, the LPs could induce directed energy flow down the rods and into the core, and from there into the reaction center. Thus, a second role for LPs was proposed. Indeed, biophysical measurements made on isolated PBPs show a significant shift in their absorption spectra in the presence or absence of LPs [41]. No phycobilisome has yet been isolated that does not contain LPs; but, mutagenesis that causes the lack of the LPs structural presence results in the loss of the phycobilisome.

Over 20 crystal structures of PBPs have now been determined, yet only a single crystal structure has an identifiable linker protein – the [1B33] APC structure [27]. In this structure, in which the APC trimers associate in a perpendicular fashion, the linker is seen completely embedded within the trimer ring. The linker contains a three strand β-sheet with two short α-helices, and makes contact with only two of the three APC monomers. Most of the contact surface is with residues and rings II and IV of the bilin on the APC β-subunits, with concomitant structural changes to both. The presence of the linker has a significant effect on the entire trimer, inducing a significant conformational change that restrains the natural curvature of the trimer, making it somewhat more flat. The asymmetric unit of this APC-linker form included two trimers; the positions of the linkers within each trimer were not identical, suggesting that the linker binding may be somewhat flexible. This flexibility may be compounded by the additional heterogeneity in the APC cylinders due to the presence of the minor forms of APC, as described above. It should be noted, however, that the ability to visualize the linker in the crystals negates to a certain extent the reasoning behind that lack of visualization of the γ subunits of PE (see Section 11.2.3.4). It was proposed that this subunit, which serves as both linker and additional chromophore binding protein, does not appear in the electron density maps because of threefold averaging around the hexamer ring. However, if the linkers bind asymmetrically within the cavity, it would be expected that the unit cell would be larger, and the crystals would belong to a space group that would allow for the presence of this protein. While it is certainly possible that binding of all linkers is not identical, rather being bound in symmetrical fashion aligned along the threefold axis, it is also possible that in the 1B33 structure, the linker was stabilized in a non-physiological position. Only the successful crystallization of other PBP-LP complexes will resolve this question.

11.2.4 Phycobiliprotein Post-translational Modifications

Following translation, the PBPs assemble into the subunits that form the phycobilisome. During this process, they undergo a number of structural and chemical modifications required for proper functionality. One of the non-covalent changes that occurs is the stabilization of the almost invariant Thrβ^{77} residue (Figure 11.1a, white square), in a configuration of dihedral angles that belong to a typically non-permitted region of the Ramachandran plot [51]. This is due to constraints that occur during the trimerization process that brings this residue into close contact with the α84 bilin, thereby shielding it from the solvent and shifting its absorption to the red. This configuration could also, potentially, shield the α84 bilin from the effects of linker binding, unlike the β84 bilin which is accessible to the linker in the trimer cavity.

11.2.4.1 Bilin Lyases

The phycobilisome is unique in that all chromophores are covalently linked to the polypeptide components. Thus, the most important of all post-translational modifications to the PBPs is the creation of thio-ether linkages between conserved Cys residues and the appropriate bilins. In the cases of PEC and PE, this requires the linkage of different bilin types to the same protein, indicating that the enzyme that catalyzes this bond formation has specificity for both substrates. Lyases have been studied for a number of systems [60–64]. Attachment of bilins to PC and PE α-subunits requires a two-protein complex consisting of the products of the genes *cpcE/F*. Very recently, a second set of genes (particularly *cpcS* and *cpcT or cpeS and cpeT)* that encode lyases that catalyze the attachment of bilins to the β subunits were identified [63, 64]. The lyases bind the chromophore and substrate apoprotein and may modulate the chromophore configuration prior to transfer to the PBP [61]. While there are cases where the PBP contains its own lyase activity [65] and autocatalytically creates the thio-ether linkage, other PBPs are totally dependent on the presence of lyase components. The recent expression of functional α-PC and α-PE subunits in *E. coli* was totally dependent on the coexpression of the *cpcE/F* and *pecE/F* lyase genes, respectively [66, 67].

11.2.4.2 Methylation of βAsn^{72}

Glazer and coworkers [47] first identified this post-translational modification of βAsn^{72} in *Anabaena variabilis* APC, today properly named *Nostoc* PCC7937. The asparagine residue was found to be modified by the addition of a methyl group to the γ-nitrogen of the side chain [47] to form N-methyl asparagine (NMA). The enzyme responsible for this modification was isolated, [48] and was shown to perform the methylation the βAsn^{72} residue in vitro, indicating that this modification may occur after trimer assembly in vivo. NMA residues were identified in PC, PE, and PEC and were also identified crystallographically by reinterpretation of the electron density maps of early PC structures [52, 68]. Utilization of the three-dimensional structures visualized the NMA position, showing that it is in van der Waals contact with ring II of the β^{155} bilin chromophore, close to the propionic acid, thereby modifying the polar environment of this chromophore which, in turn, could alter its energy absorption and transfer properties (Figure 11.1a, white square). When compared to PC isolated from methylase-lacking mutant strains, methylated PC is more efficient in energy transfer from PC to APC. Consistent with this idea, the linker visualized with APC [27] does interact with the NMA residue.

The existence of Asn residues at this position is highly conserved, but there are species in which not all of the PBPs contain this residue [23]. This indicates that, for certain cases, perhaps due to changes in the linker protein, methylation is not required, or is deleterious. It is possible to envision that isolation of the β84 bilin from the solvent could be achieved by a combination of natural amino-acid substitutions coupled with the linker protein. Indeed, one could envision that the evolutionary development of a methylation system might only be required if it served to control some aspect of function, thus suggesting that under other conditions, non-methylated βAsn^{72} might be required. As mentioned in Section 11.2.3.2, a non-methylated βAsn^{72} PC might serve as a functional bridge between some of the bulk PC rods and the APC in the tricylindrical core (Figure 11.3). In this case, the absence of NMA and the presence of the proper linker protein, could modify the absorption of the PC further to the red, efficiently transferring energy from the rods to the core. It is possible that this fraction is unique to this thermophilic fresh-water species – such a blue-shifted fraction of PC was isolated from the eukaryotic thermophilic cyanobacteria *S. lividus* [69].

11.2.4.3 Phosphorylation and Glycosylation of Linker Proteins

The presence of phosphorylation of cyanobacterial thylakoid proteins has been experimentally investigated on a number of systems, with somewhat contradictory results [70, 71]. The latter study indicated that dephosphorylation was required prior to complex disassembly (see Section 11.3.2) and that phosphorylation of the linker proteins could have a stabilizing influence on the entire structure. However, the experimental evidence for this modification lacks further corroboration. A similar state of uncertainty exists for the glycosylation of linker proteins, with one positive report [72] and one negative [73]. Obviously, further studies of the linker proteins, including crystal structures, will help clarify these matters.

11.3
Self Assembly and Disassembly of the Phycobilisome

An intriguing aspect of phycobilisome function is its ability to assemble into its extremely large, multi subunit structure. With the structural information obtained by X-ray crystallography of the isolated subunits described in Section 11.2, this propensity becomes even more remarkable as a consequence of the high degree of homology between the different subunits on all structural levels.

11.3.1
Phycobilisome Assembly

Assembly of protein subunits is typically considered to be an ordered sequence of events. In the case of PBPs, each newly synthesized α and β apo-subunit must be rapidly converted into (αβ) heterodimers, the basic monomeric PBP unit, perhaps, with attached chromophores. As shown by Anderson and coworkers, lack of either bilin binding or conversion to the heterodimer results in rapid protein degradation [28, 74–76]. This indicates that these non assembled proteins are very quickly recognized as incorrectly folded by the cellular proteolytic machinery, requiring their immediate removal. One of the potential results of lack of removal of misfolded protein subunits may be the formation of associations between subunits of different components, such as APC with PC, and so on. This could occur as a result of the similarity in the conserved residues identified as important for assembly.

Using the basic PC monomer as an example and using the standard PC sequence numbering, it is seen that all of the interactions between the two subunits are between α-helices X, Y (Figure 11.1a, white circle), and A (residues 1–42) and the E-F′ helix-turn-helix structure (residues 92–110) between both subunits in a symmetrical fashion. The surface electrostatic potential has the ability to attract, or repulse, at a greater distance than other chemical forces, and probably plays a role in the initial positioning of the subunits. The entire surfaces of both α and β subunits on the side that forms the dimer is almost completely flat, thus potential residues that will position the two subunits in the correct orientation must be found. Helices on the surface that do not interact have both positive and negative patches; however, a major positive patch can be identified on the E helices of both subunits. These patches could potentially repulse further overlap of the two subunits during association. In the αβ interaction surface, a few prominent electrostatic interactions form: between αMet^{1}-βAsp^{3}-αArg^{30}, αArg^{93}-βAsp^{13}, and βArg^{93}-αAsp^{13}-βArg^{110}. These electrostatic locks are invariable and can be found in APC, PEC, and PE structures as well. There must be additional locking molecules that impart specificity and prevent the formation of mixed PBPs. In the case of PC, such a specificity element is αArg^{42}-βAsp^{25}; while in APC, using the *M.*

laminosus [1B33] structure as the structural example, the specificity element is formed by αLys^{26}-βGlu^{35}. Both of these positive residues, αArg^{42} in PC and αLys^{26} in APC, jut out significantly from the main body of the protein. An additional specificity element occurs in PC from cyanobacterial thermophiles, between αAsp^{28} and the N atom of ring D of βPCB^{155}-βAsn^{35}.

These electrostatic interactions are, however, not sufficient to explain the stability of the PBP monomers, whose separation into subunits requires more than just the disruption of electrostatic interactions. Isolation of the subunits from higher aggregates typically requires the use of chaotropic agents (i.e. urea or KSCN) to induce polypeptide denaturation and separation [77, 78]. The ability of urea to disrupt the structures of polypeptides has been described as a result of direct interaction with backbone contacts (including non-polar interactions) which lead to secondary and tertiary structure destabilization [79]. Thus, along with polar interactions, it is clear that there are important hydrophobic interactions which stabilize the monomer formation interface. The major hydrophobic locks are due to Phe and Tyr residues, whose side chains extend out from one subunit and become buried in the second. Positions 18 and 97 are typically either Tyr or Phe residues and serve as additional common locks on all PBPs (both subunits). Additional aromatic residues further strengthen the monomer interaction in a more PBP specific manner, such as: βPhe^{5}, conserved in PC, PEC, and PE, absent in APC; αPhe^{31}, conserved in PC and PEC, but not in APC or PE; or, βTyr^{30}/βPhe^{31}, conserved in APC and PE, but not in PC or PEC. The proximity of such a large number of aromatic residues in each subunit prior to monomer formation may exert forces which keep the subunits in a partially unfolded conformation that is then recognized by either the other subunit, followed by rapid monomer formation, or by the proteolytic system, in cases of the presence of one excess subunit. Figure 11.4 shows the interaction surface of the monomer, with the unique locking residues (in *T. vulcanus* PC) encircled.

Further evidence of the importance of the hydrophobicity in the stabilization of the monomer interaction surface was obtained by the isolation of α-subunits of PEC from intact phycobilisomes [80]. Isolation required the application of rather harsh conditions, 0.3% formic acid which induces destabilization and denaturation, followed by isocratic hydrophobic interaction chromatography. The isolated subunit undergoes substantial aggregation, which could be alleviated by the proteolytic removal of the N-terminal helices X and Y. The initiation of aggregation was found to be caused by the formation of α_2 homodimers, reinforcing the need for the subunit specificity elements to prevent such non-physiological association. Removal of the helices prevented aggregation, but was found to have a destabilizing effect on the tertiary structure of the remaining helices.

Following monomer assembly, the process of trimer formation follows a similar pattern, including both polar and hydrophobic interactions. The buried monomer-monomer interaction surface that occurs during trimer formation is only about 30% of the α and β subunit interaction surface in the monomer. However, this interaction is quite strong, resisting decomposition at even very low protein con-

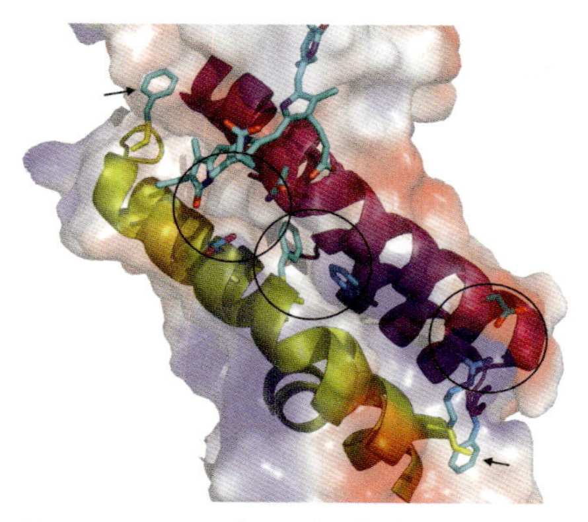

Figure 11.4 Specific residue locks in the PC αβ interaction interface. The subunits are in cartoon representation (α – yellow, β – purple). Sequence specific interactions (not conserved) mentioned in the text are encircled. Arrows show Phe^{18} conserved residues (on both subunits) that demarcate the interface.

centrations. In thermophilic organisms, additional polar locking residues can be found at both the monomer and trimer interfaces. These additional stabilizing contacts can be considered to be of intermediate strength, with distances of 4–8 Å. Thus protein structure thermostability requires only small sequence modifications, which do not change the other characteristics of the PBPs, such as their absorption spectra.

The next steps in phycobilisome assembly are hexamer, rod, and core formation, again utilizing similar principles. Most of the PC, PEC, and PE crystal structures contain such extended structures indicating that assembly can occur in the absence of LPs in vitro, at least in the presence of high PBP concentrations and dehydrating crystallization precipitants. In vitro reconstitution experiments performed on a number of systems indicated the importance of LPs in nucleating, extending, and terminating the rod structures. It thus appears that the major structural facets needed for assembly exist within the PBPs themselves. The role of the LPs may be to modify, control, and stabilize these contacts in the phycobilisome.

Figure 11.3 shows a model of the phycobilisome that is consistent with many of the structural details outlined in this and other, reviews. It should be pointed out that other models exist, with the aesthetically pleasing form showing the rods radiating out at equivalent angles from the core. Recent cryo-electron microscopy results [26] show doublets of rods at right-angles surrounding the core, in a fashion similar to that shown in Figure 11.3. These measurements show that all of the rods surround the core on the same plane, excluding the possibility that the rods are in a staggered formation. A description of the phycobilisome architecture that accounts for the actual sizes of the rods and cores can be found in [10, 46].

11.3.2
Phycobilisome Disassembly

Cyanobacteria can be exposed to changes in their typical growth conditions that modify both the number and characteristics of the phycobilisome. As already noted, the expression of the genes encoding for phycobilisome components can be altered by external environmental changes within a short period of time. However, more long-term changes that require changes in the phycobilisome makeup may occur. One such change is the response of cyanobacteria to the lack of specific nutrients, especially nitrogen and sulfur. Under these conditions, the cyanobacterial cells change color from blue-green to yellow-green in a process known as chlorosis, or bleaching [81]. Using spectroscopy, and other techniques, it was found that the bleaching occurs within a few hours of the onset of nutrient starvation and was a result of the loss of the phycobilisome in an ordered fashion [55]. It has been proposed that the disassembly of the phycobilisome can have two beneficial roles. During starvation the cellular metabolism decreases, while the rate of photosynthesis may be unchanged. This imbalance could lead to excessive absorption of excitation energy, which then leads to the production of harmful radical species [82]. Under extreme conditions this may lead to cell death and must be avoided. Since the rate of photosynthesis is directly coupled to the rate of energy absorption, uncoupling and degradation of the major antenna will be highly beneficial in avoiding overexcitation. A second role for phycobilisome degradation is to serve as an internal nutrient reservoir. The phycobilisome can supply the cell with amino acid residues for the synthesis of protein systems required for high affinity nutrient uptake. Additionally, by further degradation, it can supply building blocks for the synthesis of other metabolites or as a source of energy. Collier and Grossman used a genetic screen to identify mutants unable to degrade phycobilisomes during nutrient starvation in order to identify the molecular pathway of phycobilisome degradation [83]. The non-bleaching A, *nblA*, gene was thus identified and has been found to be present in all cyanobacteria. Nitrogen is the major nutrient whose lack can lead to a 50-fold increase in the amount of NblA protein [83]. In some cyanobacteria, such as *Synechococcus* sp. PCC 7942, sulfur and phosphorus limitation can also induce the *nblA* response [83]; while, in other species, sulfur limitation does not influence the expression of *nblA* [84]. During starvation, wild-type cells, but not *nblA* mutants, degrade two rod-linker proteins, L_R 33 and L_R 34.5, indicating a possible functional connection between the linkers and the NblA protein [84]. This observation supports previous data showing that wild-type cells degrade the phycobilisome by first decreasing the length of the rods [28, 85]. It was shown that in the cyanobacterium *Tolypothrix* PCC 7601, the NblA protein has affinity for both PC and PE subunits, but not for APC or for PBPs from other cyanobacterial species [86].

Beside the *nblA*, there are four other genes that have been identified as being related to phycobilisome degradation: *nblR*, *nblS*, *nbl,B*, and *nblC*. The NblR protein controls *nblA* expression and is critical for survival under stress conditions [87]. It has been suggested that *nblR* is controlled by *nblS*, a sensor histidine kinase that has a PAS domain [88]. The NblB protein is necessary for phycobilisome degradation during starvation; but, unlike the NblA protein, the NblB levels are similar prior to, or during, nutrient starvation [89]. This protein is partially homologous with PCB lyases, the enzyme involved in chromophore attachment to the PC apoprotein. The most recent addition to the arsenal of proteins involved in the disassembly of the phycobilisome is the nblC protein, which is also required for *nblA* expression [90].

The size of proteins belonging to the NblA family is quite small, ranging from 54 to 65 residues with molecular masses of about 7–7.5 kDa. The sequence identity between NblA proteins is relatively poor, ~30%, which is somewhat surprising since the phycobilisome components on which the NblA interacts are highly homologous, >70%. The first crystal structure ([1OJH] of a member of the NblA protein family (from *Anabena* sp. PCC 7120) was recently determined by X-ray crystallography [91]; this has been followed up with two additional NblA members (from *T. vulcanus* ([2Q8V] and [2QDO])and *S. elongatus* sp. 7942;[92]). All structures are extremely similar, exhibiting a helix-turn-helix motif that dimerizes into a flat four-helical bundle. Based on direct measurements between PC and the NblA (both wild-type and mutant forms), Bienert and coworkers suggested a mode of activity whereby the NblA dimer interacts with PC monomers via its amino and carboxyl termini [91]; however no proposal as to the method by which the NblA mechanically disassembles the phycobilisome is presented by the authors. Alternative interaction modes are certainly possible, especially between similar helix-turn-helix motifs found in all PBPs, but await further investigation to prove their validity.

11.4 Phycobilisome Function

11.4.1 Phycobilisome Binding to PSII and PSI

The phycobilisome is primarily bound to the thylakoid membrane via the L_{CM} subunit that penetrates into the reaction centers of PSII. A typical phycobilisome thus provides the absorptive power of more than 600 chromophores to a single PSII dimer [21]. Taking into account the number of chlorophyll *a* molecules attached to PSII, this brings the total antenna potential for energy absorption to more than 350 chromophores per reaction center. This can be compared to about 100 chlorophylls per cyanobacterial Photosystem I (PSI) monomer [93]. In plant and green algae systems, the number of chromophores is also large [94], but still smaller than that provided by the phycobilisome. The result of this enlarged antenna allows many cyanobacterial species to increase the ratio of PSI/PSII to between three and six [95], as opposed to the green algae and plant situation, where the ratio is close to one [94]. Lowering the amount of PSII may be physiologically beneficial due to the well known requirement of PSII to replace the D1 protein under normal illumination conditions. Under extreme conditions, high light stress or a combination of light and temperature stress, the cells can be photoinhibited, leading to cell death [82, 96]. One method of prevention of photoinhibition can be achieved by rapid replacement of the PSII D1 subunit, something that can be achieved more efficiently by lowering the total amount of PSII.

An additional protective route would be via disconnection of the phycobilisome from PSII, leading to increased fluorescence. Indeed, the phycobilisome has been shown to have the potential to become disconnected from PSII and to functionally associate with Photosystem I (PSI) (see Chapter 2). A brief period of heat treatment also has the effect of disconnecting the phycobilisome from PSII, without direct damage to either complex [97]. Direct measurements using confocal microscopy have indicated that the phycobilisome is quite mobile in vivo, much more than the photosystems [98, 99], indicating that the association between antenna and photosystems in cyanobacteria my be much looser than in other systems. Recent experiments performed on cyanobacterial cells in vivo show that the mobility of the phycobilisome and its disconnection from PSII is light dependent [100]. Disconnection occurs rapidly and thus this phenomenon has been equated to state 1–state 2 transition in plants that occurs via the phosphorylation-induced disconnection of LHCII from PSII. In this study, the product of the gene *rpaC* was found to be required for the occurrence of the state transition at very low light intensities. *rpaC*$^-$ deletion mutants exhibited no state transitions, but were unaffected in comparison to wild-type cells, at high light. Thus, multiple mechanisms for protection of PSII may exist.

Many reports exist showing the ability of the phycobilisome to transfer energy to PSI in vitro [101] and in vivo [102]; however, whether binding to PSI is mediated by specific interactions or by transient association is unclear [103]. Heat treatment appears not to affect the phycobilisome-PSI functional interaction in the same manner as it does the phycobilisome-PSII interaction [97]. The report by Rakhimberdieva and coworkers [102] on *Spirulina* cells indicates that most of the phycobilisome is functionally connected to PSI, both trimers and monomers, and not to PSII. This could correlate with the fact that, while there is a PSI : PSII ratio of ~6, electron micrographs of thylakoid membranes appear to show almost complete coverage of the stromal side of the thylakoid membrane with phycobilisome complexes [15, 25]. Since L_{CM} has been indicated as the source of phycobilisome-PSII binding, it would be unlikely that phycobilisome-PSI binding would occur via the same mechanism. Indeed, there are indications that the phycobilisome binds to PSI via interactions of rod PBPs, mostly PC, with the ferredoxin-NADP$^+$ oxidoreductase (FNR) subunit of PSI [104]. It is certainly possible that these auxiliary-like phycobilisome units not connected to PSII loosely interact with PSI, both structurally and functionally, with the potential to increase its potential for both linear and cyclic electron transfer.

This auxiliary population of phycobilisomes could have a second physiological role, in addition to light absorption. The phycobilisome has been identified as an emergency source of nutrients to be used in the case of nitrogen, sulfur, or carbon starvation [83, 84, 105]. Use of the phycobilisome as a nutrient source requires its ordered disassembly, as described in Section 11.3.2.

11.4.2
Energy Transfer within the Phycobilisome

The existence of crystal structures has revealed many of the structural requirements for the assembly of an efficient energy transfer antenna complex. Relatively high quality images of entire antenna complexes have been obtained from non-oxygenic bacteria [106–109], PSI [93, 110] and PSII [33, 111]. This has enabled the formulation of more complete models of light harvesting [2] that propose the positions of chromophores are determined by universal physical constraints such as photon flux, time scales of photon absorption, energy transfer, and competing processes. The design of antennas has been optimized through evolution to provide maximum energy transfer.

The mechanism of energy transfer in the phycobilisome has been extensively studied by a variety of spectroscopic techniques on the entire complex, but more often on isolated subunits [112]. However, since the crystal structures that exist contain, at most, one $(\alpha\beta)_6$ hexameric unit, the actual method of directed energy transfer down to the reaction center can only be conjectured. Species-specific differences not-withstanding, energy transfer rates are extremely fast [113] and overall quantum yields are high – about 95% [21, 41, 114]. The phycobilisome differs from the chlorophyll-based antennas by having rather large distances between cofactors. In the chlorophyll-based antennas, distances between adjacent cofactors are typically close enough to envision closely interacting absorption rings or aggregates. In the bacterial LH2 complex, the two rings of bacteriochlorophylls have spacing of ~9 Å and ~20 Å, center to center, respectively; the distance between rings is also ~20 Å. The addition of carotenoid molecules makes this antenna quite chromophore dense. In the antenna beds of both PSI and PSII, there are many chlorophyll molecules, positioned with very complex arrangements, which can roughly be separated into two groups, on either face of the membrane – however typical distances are on the order of 10 Å. The densest of all chlorophyll-based systems is the chlorosome found in green sulfur (*Chlorobiaceae*) and green filamentous bacteria (*Chloroflexaceae*). In these, an entire organelle encloses thousands of pigment molecules stacked in close proximity, without intervening protein subunits [115–117].

Within each minimal PC unit, the distances between the α^{84} and both the β^{84} and β^{155} bilins is ~50 Å, while the β^{84} and β^{155} bilins are separated by ~40 Å (Figure 11.1a). Association of the PC monomers into trimers and hexamers (Figure 11.1b) results in a slightly shorter distance between certain bilins, but none are closer than ~20 Å [29, 118]. In assembled PE hexamers, which contain five bilins per monomer, the density is slightly higher, with the nearest approach between two bilins on the order of 15 Å [52]. Thus, it can be assumed that efficient energy transfer does not particularly require the dense packing of cofactors, rather, that dense packing maximizes the potential for light absorption with the lowest expenditure on protein synthesis. Since this is not the case for the phycobilisome, it could be assumed that the secondary role of the phycobilisome, serving as a reservoir of nutrients for starved cells, must be of great enough importance to have preserved this form of antenna system.

The central mechanism for energy transfer proposed by Förster [119] requires the existence of weak coupling between electronic energy levels of nearby chromophores. Due to the relatively large distances between cofactors in the phycobilisome, this coupling is weak; structure-based theoretical energy transfer rate calculations utilizing the Förster mechanism match experimentally obtained values well. Debreczeny and coworkers measured energy transfer rates in isolated PC and obtained values between 50–500 psec for different pathways within the ($\alpha\beta$) monomer [120]. Formation of $(\alpha\beta)_3$ trimers allowed for much faster energy transfer in the 0.5–1 psec range [36], showing the importance of the level of complex formation when dealing with the functional characteristics of the phycobilisome. Ultra fast, two-color pump-probe spectroscopic measurements on different organizational states of APC have revealed energy-transfer at less than 100 fsec, which has been suggested to occur due the formation of dimer-exciton states [37, 38]. These states can only exist between the PCBs on adjacent monomers; the very fast decay component is probably outside of the time-frame accessible to Förster resonance energy transfer. It is still unclear what the effect of further aggregation to full cylinders, cores, or phycobilisomes will have on the energy transfer rates, especially with the presence of linker proteins. However, use of isolated subunit also allows for extremely high quality descriptions of the cofactor/protein/solvent environments during absorption and energy transfer [39, 121]. Detailed discussions on energy transfer within isolated PBP can be found in a number of excellent reviews [21, 41, 113, 122].

11.5 Final Remarks

The details of phycobilisome structure and function described here cannot be deemed complete until a true, experimentally derived structure of the cores, rods, and phycobilisome is determined. This may require the bridging of high resolution crystallography with lower resolution microscopic techniques. It is clear that visualization of the phycobilisome will not only expand our understanding of biological energy harvesting and transfer, but it will also serve as an extraordinary example of nanometric self-assembly and disassembly. Clarification of the chemical rules that lead to the specific association of molecules will be used in the future development and production of true nano-devices.

Acknowledgments

I would like to thank all of the members, past and present, of the Laboratory for Macromolecular Structure Determination of the Schulich Faculty of Chemistry,

Technion for their diligent research and their assistance in the writing of this review. Research described here was supported by the Israel Science Foundation founded by the Israel Academy of Sciences and Humanities (438/02 and 1045/06).

References

1 Schubert, W.D., Klukas, O., Saenger, W., Witt, H.T., Fromme, P. and Krauss, N. (1998) A common ancestor for oxygenic and anoxygenic photosynthetic systems: a comparison based on the structural model of photosystem I. *Journal of Molecular Biology*, **280**, 297–314.

2 Noy, D., Moser, C.C. and Dutton, P.L. (2006) Design and engineering of photosynthetic light-harvesting and electron transfer using length, time, and energy scales. *Biochimica et Biophysica Acta*, **1757**, 90–105.

3 Cogdell, R.J., Gall, A. and Kohler, J. (2006) The architecture and function of the light-harvesting apparatus of purple bacteria: from single molecules to in vivo membranes. *Quarterly Reviews of Biophysics*, **39**, 227–324.

4 Horton, P. and Ruban, A. (2005) Molecular design of the photosystem II light-harvesting antenna: photosynthesis and photoprotection. *Journal of Experimental Botany*, **56**, 365–73.

5 Cruz, J.A., Avenson, T.J., Kanazawa, A., Takizawa, K., Edwards, G.E. and Kramer, D.M. (2005) Plasticity in light reactions of photosynthesis for energy production and photoprotection. *Journal of Experimental Botany*, **56**, 395–406.

6 Cogdell, R.J., Gardiner, A.T., Roszak, A.W., Law, C.J., Southall, J. and Isaacs, N.W. (2004) Rings, ellipses and horseshoes: how purple bacteria harvest solar energy. *Photosynthesis Research*, **81**, 207–14.

7 Ting, C.S., Rocap, G., King, J. and Chisholm, S.W. (2002) Cyanobacterial photosynthesis in the oceans: the origins and significance of divergent light-harvesting strategies. *Trends in Microbiology*, **10**, 134–42.

8 Dekker, J.P. and Boekema, E.J. (2005) Supramolecular organization of thylakoid membrane proteins in green plants. *Biochimica et Biophysica Acta*, **1706**, 12–39.

9 Adir, N. (2005) Elucidation of the molecular structures of components of the phycobilisome: reconstructing a giant. *Photosynthesis Research*, **85**, 15–32.

10 Adir, N., Dines, M., Klartag, M., McGregor, A. and Melamed-Frank, M. (2006) Assembly and disassembly of phycobilisomes, in *Microbiology Monographs: Inclusions in Prokaryotes* (ed. J.M. Shively), Springer Berlin, Heidelberg, pp. 47–77.

11 Blankenship, R.E., Olson, J.M. and Miller, M. (1995) Antenna complexes from green photosynthetic bacteria, in *Anoxygenic Photosynthetic Bacteria* (eds R.E. Blankenship, M.T. Madigan and C.E. Bauer), Kluwer Academic Publishers, Dordrect, The Netherlands, pp. 399–435.

12 Samsonoff, W.A. and MacColl, R. (2001) Biliproteins and phycobilisomes from cyanobacteria and red algae at the extremes of habitat. *Archives of Microbiology*, **176**, 400–5.

13 Tandeau de Marsac, N. (2003) Phycobiliproteins and phycobilisomes: the early observations. *Photosynthesis Research*, **76**, 197–205.

14 Gantt, E. and Conti, S.F. (1966) Granules associated with the chloroplast lamellae of Porphyridium cruentum. *The Journal of Cell Biology*, **29**, 423–34.

15 Gantt, E. and Conti, S.F. (1966) Phycobiliprotein localization in algae. *Brookhaven Symposia in Biology*, **19**, 393–405.

16 Gantt, E. and Lipschultz, C.A. (1972) Phycobilisomes of Porphyridium cruentum. I. isolation. *The Journal of Cell Biology*, **54**, 313–24.

17 Glazer, A.N., Lundell, D.J., Yamanaka, G. and Williams, R.C. (1983) The structure of a "simple" phycobilisome. *Annales de Microbiologie*, **134B**, 159–80.

18 Bryant, D.A., Glazer, A.N. and Eiserling, F.A. (1976) Characterization and structural properties of the major biliproteins of Anabaena sp. *Archives of Microbiology*, **110**, 61–75.

19 Yamanaka, G., Glazer, A.N. and Williams, R.C. (1978) Cyanobacterial phycobilisomes. Characterization of the phycobilisomes of Synechococcus sp. 6301. *The Journal of Biological Chemistry*, **253**, 8303–10.

20 Dobler, M., Dover, S.D., Laves, K., Binder, A. and Zuber, H. (1972) Crystallization and preliminary crystal data of C-phycocyanin. *Journal of Molecular Biology*, **71**, 785–7.

21 Glazer, A.N. (1989) Light guides. Directional energy transfer in a photosynthetic antenna. *The Journal of Biological Chemistry*, **264**, 1–4.

22 Ducret, A., Muller, S.A., Goldie, K.N., Hefti, A., Sidler, W.A., Zuber, H. and Engel, A. (1998) Reconstitution, characterization and mass analysis of the pentacylindrical allophycocyanin core complex from the cyanobacterium Anabaena sp. PCC 7120. *Journal of Molecular Biology*, **278**, 369–88.

23 Apt, K.E., Collier, J.L. and Grossman, A.R. (1995) Evolution of the phycobiliproteins. *Journal of Molecular Biology*, **248**, 79–96.

24 Pastore, A. and Lesk, A.M. (1990) Comparison of the structures of globins and phycocyanins: evidence for evolutionary relationship. *Proteins*, **8**, 133–55.

25 Edwards, M.R. and Gantt, E. (1971) Phycobilisomes of the thermophilic blue-green alga Synechococcus lividus. *The Journal of Cell Biology*, **50**, 896–900.

26 Yi, Z.W., Huang, H., Kuang, T.Y. and Sui, S.F. (2005) Three-dimensional architecture of phycobilisomes from Nostoc flagelliforme revealed by single particle electron microscopy. *FEBS Letters*, **579**, 3569–73.

27 Reuter, W., Wiegand, G., Huber, R. and Than, M.E. (1999) Structural analysis at 2.2 Å of orthorhombic crystals presents the asymmetry of the allophycocyanin-linker complex, AP.LC7.8, from phycobilisomes of Mastigocladus laminosus. *Proceedings of the National Academy of Sciences of the United States of America*, **96**, 1363–8.

28 Anderson, L.K. and Toole, C.M. (1998) A model for early events in the assembly pathway of cyanobacterial phycobilisomes. *Molecular Microbiology*, **30**, 467–74.

29 Schirmer, T., Huber, R., Schneider, M., Bode, W., Miller, M. and Hackert, M.L. (1986) Crystal structure analysis and refinement at 2.5 Å of hexameric C-phycocyanin from the cyanobacterium Agmenellum quadruplicatum. The molecular model and its implications for light-harvesting. *Journal of Molecular Biology*, **188**, 651–76.

30 Schirmer, T., Bode, W. and Huber, R. (1987) Refined three-dimensional structures of two cyanobacterial C-phycocyanins at 2.1 and 2.5 Å resolution. A common principle of phycobilin-protein interaction. *Journal of Molecular Biology*, **196**, 677–95.

31 Brejc, K., Ficner, R., Huber, R. and Steinbacher, S. (1995) Isolation, crystallization, crystal structure analysis and refinement of allophycocyanin from the cyanobacterium Spirulina platensis at 2.3 Å resolution. *Journal of Molecular Biology*, **249**, 424–40.

32 Liu, J.Y., Jiang, T., Zhang, J.P. and Liang, D.C. (1999) Crystal structure of allophycocyanin from red algae Porphyra yezoensis at 2.2-Å resolution. *The Journal of Biological Chemistry*, **274**, 16945–52.

33 Ferreira, K.N., Iverson, T.M., Maghlaoui, K., Barber, J. and Iwata, S. (2004) Architecture of the photosynthetic oxygen-evolving center. *Science*, **303**, 1831–8.

34 Loll, B., Kern, J., Zouni, A., Saenger, W., Biesiadka, J. and Irrgang, K.D. (2005) The antenna system of photosystem II from Thermosynechococcus elongatus at 3.2 Å resolution. *Photosynthesis Research*, **86**, 175–84.

35 Bryant, D.A., Guiglielmi, G., Tandeau de Marsac, N., Castets, A. and Cohen-Bazire, G. (1979) The structure of cyanobacterial

phycobilisomes: a model. *Archives of Microbiology*, **123**, 113–27.

36 Beck, W.F. and Sauer, K. (1992) Energy-transfer and exciton-state relaxation processes in allophycocyanin. *Journal of Physical Chemistry*, **96**, 4658–66.

37 Edington, M.D., Riter, R.E. and Beck, W.F. (1995) Evidence for coherent energy transfer in allophycocyanin trimers. *Journal of Physical Chemistry*, **99**, 15699–704.

38 Edington, M.D., Riter, R.E. and Beck, W.F. (1996) Interexciton-state relaxation and exciton localization in allophycocyanin trimers. *Journal of Physical Chemistry*, **100**, 14206–17.

39 Homoelle, B.J., Edington, M.D., Diffey, W.M. and Beck, W.F. (1998) Stimulated photon-echo and transient-grating studies of protein-matrix solvation dynamics and interexciton-state radiationless decay in alpha phycocyanin and allophycocyanin. *Journal of Physical Chemistry*, **102**, 3044–52.

40 MacColl, R. (2004) Allophycocyanin and energy transfer. *Biochimica et Biophysica Acta*, **1657**, 73–81.

41 MacColl, R. (1998) Cyanobacterial phycobilisomes. *Journal of Structural Biology*, **124**, 311–34.

42 Yu, M.H., Glazer, A.N. and Williams, R.C. (1981) Cyanobacterial phycobilisomes. Phycocyanin assembly in the rod substructures of anabaena variabiliz phycobilisomes. *The Journal of Biological Chemistry*, **256**, 13130–6.

43 Schirmer, T., Bode, W., Huber, R., Sidler, W. and Zuber, H. (1985) X-ray crystallographic structure of the light-harvesting biliprotein C-phycocyanin from the thermophilic cyanobacterium Mastigocladus laminosus and its resemblance to globin structures. *Journal of Molecular Biology*, **184**, 257–77.

44 Adir, N., Dobrovetsky, Y. and Lerner, N. (2001) Structure of C-Phycocyanin from the thermophilic cyanobacterium synechococcus vulcanus at 2.5 Å: structural implications for thermal stability in phycobilisome assembly. *Journal of Molecular Biology*, **313**, 71–81.

45 Adir, N., Vainer, R. and Lerner, N. (2002) Refined structure of C-phycocyanin from the cyanobacterium Synechococcus vulcanus at 1.6 Å: insights into the role of solvent molecules in thermal stability and co-factor structure. *Biochimica et Biophysica Acta*, **1556**, 168–74.

46 Adir, N. and Lerner, N. (2003) The crystal structure of a novel unmethylated form of C-phycocyanin, a possible connector between cores and rods in pycobilisomes. *The Journal of Biological Chemistry*, **278**, 25926–32.

47 Klotz, A.V., Leary, J.A. and Glazer, A.N. (1986) Post-translational methylation of asparaginyl residues. Identification of beta-71 gamma-N-methylasparagine in allophycocyanin. *The Journal of Biological Chemistry*, **261**, 15891–4.

48 Swanson, R.V. and Glazer, A.N. (1990) Phycobiliprotein methylation. Effect of the gamma-N-methylasparagine residue on energy transfer in phycocyanin and the phycobilisome. *Journal of Molecular Biology*, **214**, 787–96.

49 Schmidt, M., Krasselt, A. and Reuter, W. (2006) Local protein flexibility as a prerequisite for reversible chromophore isomerization in alpha-phycoerythrocyanin. *Biochimica et Biophysica Acta*, **1764**, 55–62.

50 Bryant, D.A. (1982) Phycoerythrocyanin and phycoerythrin. Properties and occurrence in cyanobacteria. *Journal of General Microbiology*, **128**, 835–44.

51 Duerring, M., Huber, R., Bode, W., Ruembeli, R. and Zuber, H. (1990) Refined three-dimensional structure of phycoerythrocyanin from the cyanobacterium Mastigocladus laminosus at 2.7 Å. *Journal of Molecular Biology*, **211**, 633–44.

52 Ficner, R., Lobeck, K., Schmidt, G. and Huber, R. (1992) Isolation, crystallization, crystal structure analysis and refinement of B-phycoerythrin from the red alga Porphyridium sordidum at 2.2 Å resolution. *Journal of Molecular Biology*, **228**, 935–50.

53 Ritter, S., Hiller, R.G., Wrench, P.M., Welte, W. and Diederichs, K. (1999) Crystal structure of a phycourobilin-containing phycoerythrin at 1.90-Å resolution. *Journal of Structural Biology*, **126**, 86–97.

54 Grossman, A.R., Bhaya, D. and He, Q. (2001) Tracking the light environment by cyanobacteria and the dynamic nature of light harvesting. *The Journal of Biological Chemistry*, **276**, 11449–52.

55 Grossman, A.R., Schaefer, M.R., Chiang, G.G. and Collier, J.L. (1993) The phycobilisome, a light-harvesting complex responsive to environmental conditions. *Microbiological Reviews*, **57**, 725–49.

56 Wilk, K.E., Harrop, S.J., Jankova, L., Edler, D., Keenan, G., Sharples, F., Hiller, R.G. and Curmi, P.M. (1999) Evolution of a light-harvesting protein by addition of new subunits and rearrangement of conserved elements: crystal structure of a cryptophyte phycoerythrin at 1.63-Å resolution. *Proceedings of the National Academy of Sciences of the United States of America*, **96**, 8901–6.

57 Doust, A.B., Marai, C.N., Harrop, S.J., Wilk, K.E., Curmi, P.M. and Scholes, G.D. (2004) Developing a structure-function model for the cryptophyte phycoerythrin 545 using ultrahigh resolution crystallography and ultrafast laser spectroscopy. *Journal of Molecular Biology*, **344**, 135–53.

58 Tandeau de Marsac, N. and Cohen-Bazire, G. (1977) Molecular composition of cyanobacterial phycobilisomes. *Proceedings of the National Academy of Sciences of the United States of America*, **74**, 1635–9.

59 Liu, L.N., Chen, X.L., Zhang, Y.Z. and Zhou, B.C. (2005) Characterization, structure and function of linker polypeptides in phycobilisomes of cyanobacteria and red algae: an overview. *Biochimica et Biophysica Acta*, **1708**, 133–42.

60 Fairchild, C.D., Zhao, J., Zhou, J., Colson, S.E., Bryant, D.A. and Glazer, A.N. (1992) Phycocyanin alpha-subunit phycocyanobilin lyase. *Proceedings of the National Academy of Sciences of the United States of America*, **89**, 7017–21.

61 Zhao, K.H., Wu, D., Zhang, L., Zhou, M., Bohm, S., Bubenzer, C. and Scheer, H. (2006) Chromophore attachment in phycocyanin. Functional amino acids of phycocyanobilin–alpha-phycocyanin lyase and evidence for chromophore binding. *The FEBS Journal*, **273**, 1262–74.

62 Zhao, K.H., Su, P., Li, J., Tu, J.M., Zhou, M., Bubenzer, C. and Scheer, H. (2006) Chromophore attachment to phycobiliprotein beta-subunits: phycocyanobilin:cysteine-beta84 phycobiliprotein lyase activity of CpeS-like protein from Anabaena Sp. PCC7120. *The Journal of Biological Chemistry*, **281**, 8573–81.

63 Zhao, K.H., Su, P., Tu, J.M., Wang, X., Liu, H., Ploscher, M., Eichacker, L., Yang, B., Zhou, M. and Scheer, H. (2007) Phycobilin:cystein-84 biliprotein lyase, a near-universal lyase for cysteine-84-binding sites in cyanobacterial phycobiliproteins. *Proceedings of the National Academy of Sciences of the United States of America*, **104**, 14300–5.

64 Shen, G., Saunee, N.A., Williams, S.R., Gallo, E.F., Schluchter, W.M. and Bryant, D.A. (2006) Identification and characterization of a new class of bilin lyase: the cpcT gene encodes a bilin lyase responsible for attachment of phycocyanobilin to Cys-153 on the beta-subunit of phycocyanin in Synechococcus sp. PCC 7002. *The Journal of Biological Chemistry*, **281**, 17768–78.

65 Zhao, K.H., Su, P., Bohm, S., Song, B., Zhou, M., Bubenzer, C. and Scheer, H. (2005) Reconstitution of phycobilisome core-membrane linker, LCM, by autocatalytic chromophore binding to ApcE. *Biochimica et Biophysica Acta*, **1706**, 81–7.

66 Tooley, A.J., Cai, Y.A. and Glazer, A.N. (2001) Biosynthesis of a fluorescent cyanobacterial C-phycocyanin holo-alpha subunit in a heterologous host. *Proceedings of the National Academy of Sciences of the United States of America*, **98**, 10560–5.

67 Tooley, A.J. and Glazer, A.N. (2002) Biosynthesis of the cyanobacterial light-harvesting polypeptide phycoerythrocyanin holo-alpha subunit in a heterologous host. *Journal of Bacteriology*, **184**, 4666–71.

68 Duerring, M., Huber, R. and Bode, W. (1988) The structure of gamma-N-

methylasparagine in C-phycocyanin from *Mastigocladus laminosus* and Agmenellum quadriplicatum. *FEBS Letters*, **236**, 167–70.
69 Edwards, M.R., MacColl, R. and Eisele, L.E. (1996) Some physical properties of an unusual C-phycocyanin isolated from a photosynthetic thermophile. *Biochimica et Biophysica Acta*, **1276**, 64–70.
70 Harrison, M.A., Tsinoremas, N.F. and Allen, J.F. (1991) Cyanobacterial thylakoid membrane proteins are reversibly phosphorylated under plastoquinone-reducing conditions in vitro. *FEBS Letters*, **282**, 295–9.
71 Piven, I., Ajlani, G. and Sokolenko, A. (2005) Phycobilisome linker proteins are phosphorylated in Synechocystis sp. PCC 6803. *The Journal of Biological Chemistry*, **280**, 21667–72.
72 Riethman, H.C., Mawhinney, T.P. and Sherman, L.A. (1988) Characterization of phycobilisome glycoproteins in the cyanobacterium Anacystis nidulans R2. *Journal of Bacteriology*, **170**, 2433–40.
73 Fairchild, C.D., Jones, I.K. and Glazer, A.N. (1991) Absence of glycosylation on cyanobacterial phycobilisome linker polypeptides and rhodophytan phycoerythrins. *Journal of Bacteriology*, **173**, 2985–92.
74 Anderson, L.K. and Grossman, A.R. (1990) Structure and light-regulated expression of phycoerythrin genes in wild-type and phycobilisome assembly mutants of Synechocystis sp. strain PCC 6701. *Journal of Bacteriology*, **172**, 1297–305.
75 Plank, T., Toole, C. and Anderson, L.K. (1995) Subunit interactions and protein stability in the cyanobacterial light-harvesting proteins. *Journal of Bacteriology*, **177**, 6798–803.
76 Toole, C.M., Plank, T.L., Grossman, A.R. and Anderson, L.K. (1998) Bilin deletions and subunit stability in cyanobacterial light-harvesting proteins. *Molecular Microbiology*, **30**, 475–86.
77 Glazer, A.N. and Fang, S. (1973) Formation of hybrid proteins form the and subunits of phycocyanins of unicellular and filamentous blue-green algae. *The Journal of Biological Chemistry*, **248**, 663–71.
78 Homoelle, B.J. and Beck, W.F. (1997) Solvent accessibility of the phycocyanobilin chromophore in the alpha subunit of C-phycocyanin: implications for a molecular mechanism for inertial protein-matrix solvation dynamics. *Biochemistry*, **36**, 12970–5.
79 Bolen, D.W. (2004) Effects of naturally occurring osmolytes on protein stability and solubility: issues important in protein crystallization. *Methods*, **34**, 312–22.
80 Wiegand, G., Parbel, A., Seifert, M.H., Holak, T.A. and Reuter, W. (2002) Purification, crystallization, NMR spectroscopy and biochemical analyses of alpha-phycoerythrocyanin peptides. *European Journal of Biochemistry*, **269**, 5046–55.
81 Allen, M.M. and Smith, A.J. (1969) Nitrogen chlorosis in blue-green algae. *Archiv fur Mikrobiologie*, **69**, 114–20.
82 Adir, N., Zer, H., Shochat, S. and Ohad, I. (2003) Photoinhibition – a historical perspective. *Photosynthesis Research*, **76**, 343–70.
83 Collier, J.L. and Grossman, A.R. (1994) A small polypeptide triggers complete degradation of light-harvesting phycobiliproteins in nutrient-deprived cyanobacteria. *The EMBO Journal*, **13**, 1039–47.
84 Richaud, C., Zabulon, G., Joder, A. and Thomas, J.C. (2001) Nitrogen or sulfur starvation differentially affects phycobilisome degradation and expression of the *nblA* gene in Synechocystis strain PCC 6803. *Journal of Bacteriology*, **183**, 2989–94.
85 Sidler, W.A. (1994) Phycobilisome and phycobiliprotein structures, in *The Molecular Biology of Cyanobacteria* (ed. D.A. Bryant), Kluwer Academic Publishers, Dordrect, pp. 139–216.
86 Luque, I., Ochoa De Alda, J.A., Richaud, C., Zabulon, G., Thomas, J.C., and Houmard, J. (2003) The NblAI protein from the filamentous cyanobacterium Tolypothrix PCC 7601: regulation of its expression and interactions with phycobilisome components. *Molecular Microbiology*, **50**, 1043–54.

87 Schwarz, R. and Grossman, A.R. (1998) A response regulator of cyanobacteria integrates diverse environmental signals and is critical for survival under extreme conditions. *Proceedings of the National Academy of Sciences of the United States of America*, **95**, 11008–13.

88 van Waasbergen, L.G., Dolganov, N. and Grossman, A.R. (2002) nblS, a gene involved in controlling photosynthesis-related gene expression during high light and nutrient stress in Synechococcus elongatus PCC 7942. *Journal of Bacteriology*, **184**, 2481–90.

89 Dolganov, N. and Grossman, A.R. (1999) A polypeptide with similarity to phycocyanin alpha-subunit phycocyanobilin lyase involved in degradation of phycobilisomes. *Journal of Bacteriology*, **181**, 610–17.

90 Sendersky, E., Lahmi, R., Shaltiel, J., Perelman, A. and Schwarz, R. (2005) NblC, a novel component required for pigment degradation during starvation in Synechococcus PCC 7942. *Molecular Microbiology*, **58**, 659–68.

91 Bienert, R., Baier, K., Volkmer, R., Lockau, W. and Heinemann, U. (2006) Crystal structure of NblA from Anabaena sp. PCC 7120, a small protein playing a key role in phycobilisome degradation. *The Journal of Biological Chemistry*, **281**, 5216–23.

92 Dines, M., Sendersky, E., Schwarz, R. and Adir, N. (2007) Crystallization of sparingly soluble stress-related proteins from cyanobacteria by controlled urea solublization. *Journal of Structural Biology*, **158**, 116–21.

93 Jordan, P., Fromme, P., Witt, H.T., Klukas, O., Saenger, W. and Krauss, N. (2001) Three-dimensional structure of cyanobacterial photosystem I at 2.5 Å resolution. *Nature*, **411**, 909–17.

94 Danielsson, R., Albertsson, P.A., Mamedov, F. and Styring, S. (2004) Quantification of photosystem I and II in different parts of the thylakoid membrane from spinach. *Biochimica et Biophysica Acta*, **1608**, 53–61.

95 Shen, G., Boussiba, S. and Vermaas, W.F. (1993) Synechocystis sp PCC 6803 strains lacking photosystem I and phycobilisome function. *Plant Cell*, **5**, 1853–63.

96 Prasil, O., Adir, N. and Ohad, I. (1992) Dynamics of Photosystem II: mechanism of photoinhibition and recovery processes, in *The Photosystems: Structure, Function and Molecular Biology* (ed. J. Barber), Elsievier Science Publishers, B.V., Amsterdam, pp. 295–348.

97 Wen, X., Gong, H. and Lu, C. (2005) Heat stress induces an inhibition of excitation energy transfer from phycobilisomes to photosystem II but not to photosystem I in a cyanobacterium Spirulina platensis. *Plant Physiology and Biochemistry*, **43**, 389–95.

98 Mullineaux, C.W., Tobin, M.J. and Jones, G.R. (1997) Mobility of photosynthetic complexes in thylakoid membranes. *Nature*, **390**, 421–4.

99 Sarcina, M., Tobin, M.J. and Mullineaux, C.W. (2001) Diffusion of phycobilisomes on the thylakoid membranes of the cyanobacterium Synechococcus 7942. Effects of phycobilisome size, temperature, and membrane lipid composition. *The Journal of Biological Chemistry*, **276**, 46830–4, Epub 42001 Oct 46834. Order.

100 Mullineaux, C.W. and Emlyn-Jones, D. (2005) State transitions: an example of acclimation to low-light stress. *Journal of Experimental Botany*, **56**, 389–93.

101 Kirilovsky, D. and Ohad, I. (1986) Functional assembly in vitro of phycobilisomes with isolated photosystem II particles of eukaryotic chloroplasts. *The Journal of Biological Chemistry*, **261**, 12317–23.

102 Rakhimberdieva, M.G., Boichenko, V.A., Karapetyan, N.V. and Stadnichuk, I.N. (2001) Interaction of phycobilisomes with photosystem II dimers and photosystem I monomers and trimers in the cyanobacterium Spirulina platensis. *Biochemistry*, **40**, 15780–8.

103 Aspinwall, C.L., Sarcina, M. and Mullineaux, C.W. (2004) Phycobilisome mobility in the cyanobacterium Synechococcus sp. PCC7942 is influenced by the trimerisation of Photosystem I. *Photosynthesis Research*, **79**, 179–87.

104 Gomez-Lojero, C., Perez-Gomez, B., Shen, G., Schluchter, W.M. and Bryant, D.A. (2003) Interaction of ferredoxin: NADP+ oxidoreductase with phycobilisomes and phycobilisome substructures of the cyanobacterium Synechococcus sp. strain PCC 7002. *Biochemistry*, **42**, 13800–11.

105 Li, H. and Sherman, L.A. (2002) Characterization of Synechocystis sp. strain PCC 6803 and deltanbl mutants under nitrogen-deficient conditions. *Archives of Microbiology*, **178**, 256–66.

106 Koepke, J., Hu, X., Muenke, C., Schulten, K. and Michel, H. (1996) The crystal structure of the light-harvesting complex II (B800-850) from Rhodospirillum molischianum. *Structure*, **4**, 581–97.

107 Prince, S.M., Papiz, M.Z., Freer, A.A., McDermott, G., Hawthornthwaite-Lawless, A.M., Cogdell, R.J. and Isaacs, N.W. (1997) Apoprotein structure in the LH2 complex from Rhodopseudomonas acidophila strain 10050: modular assembly and protein pigment interactions. *Journal of Molecular Biology*, **268**, 412–23.

108 Roszak, A.W., Howard, T.D., Southall, J., Gardiner, A.T., Law, C.J., Isaacs, N.W. and Cogdell, R.J. (2003) Crystal structure of the RC-LH1 core complex from Rhodopseudomonas palustris. *Science*, **302**, 1969–72.

109 McLuskey, K., Prince, S.M., Cogdell, R.J. and Isaacs, N.W. (2001) The crystallographic structure of the B800-820 LH3 light-harvesting complex from the purple bacteria Rhodopseudomonas acidophila strain 7050. *Biochemistry*, **40**, 8783–9.

110 Ben-Shem, A., Frolow, F. and Nelson, N. (2003) Crystal structure of plant photosystem I. *Nature*, **426**, 630–5.

111 Liu, Z., Yan, H., Wang, K., Kuang, T., Zhang, J., Gui, L., An, X. and Chang, W. (2004) Crystal structure of spinach major light-harvesting complex at 2.72 Å resolution. *Nature*, **428**, 287–92.

112 Sauer, K. and Scheer, H. (1988) Exitation transfer in C-phycocyanin. Forster transfer rate and exciton calculations based on new crystal structure data for C-phycocyanins from *Agmenellum quadruplaticum* and Mastigocladus laminosus. *Biochimica et Biophysica Acta*, **936**, 157–70.

113 Knox, R.S. (1999) Ultrashort processes and biology. *Journal of Photochemistry and Photobiology. B, Biology*, **49**, 81–8.

114 Searle, G.F., Barber, J., Porter, G. and Tredwell, C.J. (1978) Picosecond time-resolved energy transfer in Porphyridium cruentum. Part II. In the isolated light harvesting complex (phycobilisomes). *Biochimica et Biophysica Acta*, **501**, 246–56.

115 Frigaard, N.-U., Vassilieva, E.V., Li, H., Milks, K.J., Zhao, J. and Bryant, D.A. (2001) *The Remarkable Chlorosome*. In Proc Int Cong Photosyn 12:S1-003 CSIRO Publishing, Collingwood, Australia, Brisbane **Vol. S1**.

116 Vassilieva, E.V., Stirewalt, V.L., Jakobs, C.U., Frigaard, N.U., Inoue-Sakamoto, K., Baker, M.A., Sotak, A. and Bryant, D.A. (2002) Subcellular localization of chlorosome proteins in Chlorobium tepidum and characterization of three new chlorosome proteins: CsmF, CsmH, and CsmX. *Biochemistry*, **41**, 4358–70.

117 Montano, G.A., Bowen, B.P., LaBelle, J.T., Woodbury, N.W., Pizziconi, V.B. and Blankenship, R.E. (2003) Characterization of Chlorobium tepidum chlorosomes: a calculation of bacteriochlorophyll c per chlorosome and oligomer modeling. *Biophysical Journal*, **85**, 2560–5.

118 Nield, J., Rizkallah, P.J., Barber, J. and Chayen, N.E. (2003) The 1.45 Å three-dimensional structure of C-phycocyanin from the thermophilic cyanobacterium Synechococcus elongatus. *Journal of Structural Biology*, **141**, 149–55.

119 Forster, T. (1948) Zwischenmolekulare Energiewanderung und Fluoreszenz. *Annals of Physics (Leipzig)*, **2**, 55–75.

120 Debreczeny, M.P., Sauer, K., Zhou, J. and Bryant, D.A. (1993) Monomeric C-phycocyanin at room temperature and 77K: resolution of the absorption and fluorescence spectra of the individual chromophores and the energy-transfer rate constants. *Journal of Physical Chemistry*, **97**, 9852–62.

121 Homoelle, B.J. and Beck, W.F. (1997) Solvent accessibility of the phycocyanobilin chromophore in the R subunit of C-Phycocyanin: implications for a molecular mechanism for inertial protein-matrix solvation dynamics. *Biochemistry*, **36**, 12970–5.

122 Huber, R. (1989) Nobel lecture. A structural basis of light energy and electron transfer in biology. *The EMBO Journal*, **8**, 2125–47.

12
Reaction Centers from Purple Bacteria

James P. Allen and JoAnn C. Williams

12.1
Introduction

In photosynthetic bacteria, light is captured by light-harvesting complexes and its energy is transferred to the protein–pigment complex termed the reaction center. In the reaction center, the formation of an excited state of the primary electron donor is followed by the transfer of an electron to a series of electron acceptors. These electron transfer reactions occur on timescales ranging from femtoseconds to microseconds with essentially every photon of light resulting in electron transfer, that is, with a quantum yield of nearly unity. By coupling the reaction center to mobile electron donors and acceptors, the electrons and accompanying protons are transferred to other components of the photosynthetic apparatus, ultimately to be converted into chemically rich compounds such as ATP and NADPH.

In this chapter, the properties of the reaction center are presented with an emphasis on those structural features that are conserved among purple bacteria. The overall architecture is first discussed, namely the arrangements of the protein subunits and the embedded pigments that serve as the electron transfer chain participants, including the primary electron donor. Many of the presented aspects are based upon extensive mutagenesis studies that have been performed on the reaction center from purple bacteria, for which the influence of specific protein-cofactor interactions can be probed due to our ability to perform a wide variety of spectroscopic investigations as well as the availability of high-resolution X-ray structures. Once the protein and cofactor framework is established, the bacterial reaction center is compared to the evolutionarily related complex, Photosystem II, the site for the oxidation of water in oxygenic photosynthesis (Chapter 4). Finally, the use of bacterial reaction centers as key components in biologically inspired technological devices is briefly summarized.

Photosynthetic Protein Complexes: A Structural Approach. Edited by P. Fromme

ISBN: 978-3-527-31730-1

12.2 The Overall Structure

All bacterial reaction centers contain at least two protein subunits, termed the L and M subunits, that surround the cofactors [1, 2]. These two subunits each contain five transmembrane helices that are organized around a twofold symmetry axis of the protein [3–9] (Figure 12.1a). The protein is largely composed of α helices, most notably the transmembrane helices that are composed of 24 to 31 amino acid residues with a pronounced hydrophobic character. At the center of the protein are four intertwining helices, two from each subunit, forming a four helical bundle that spans the cell membrane (Figure 12.1b). These helices can be thought of as being the core of the protein with the remaining helices as more peripheral. The cofactors that participate in electron transfer all lie within this core and are arranged in symmetry-related A and B branches. In addition to the transmembrane helices, there are several other smaller α helices. Three of these helices in each subunit are aligned on the periplasmic surface of the protein along the edge of the cell membrane. In each subunit, another helix is on the cytoplasmic side of the protein and largely forms the binding site for one of the quinone cofactors.

Reaction centers from purple bacteria have an additional subunit termed the H subunit. The H subunit has one transmembrane helix and a large extramembranous domain on the cytoplasmic side of the reaction center. While the H subunit

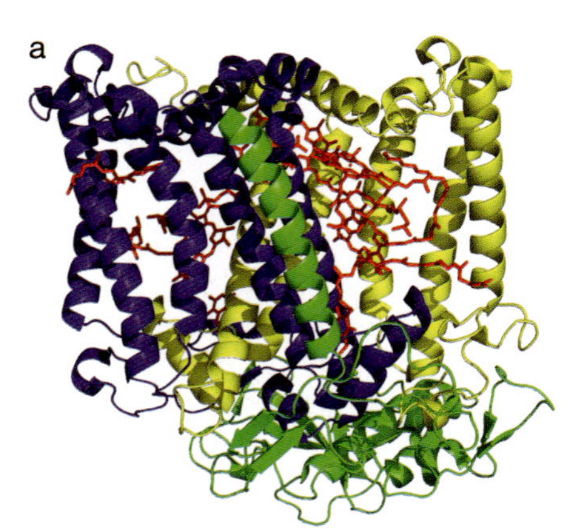

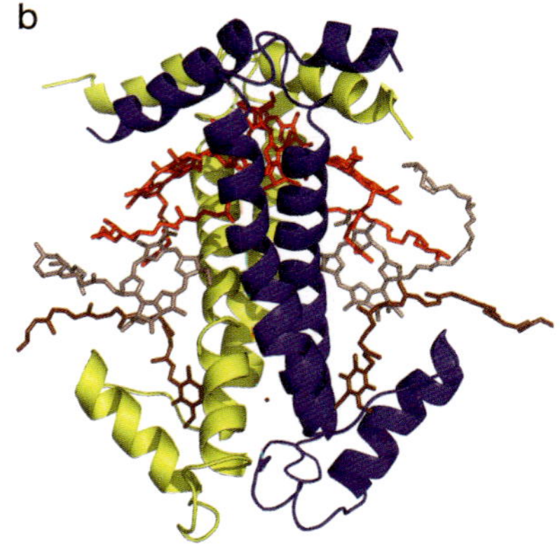

Figure 12.1 **(a)** Structure of the reaction center from *R. sphaeroides* showing the three protein subunits: L (yellow), M (blue), and H (green) and the cofactors (red). **(b)** Structure of the core motif found in photosynthetic complexes formed by the L (yellow) and M (blue) subunits: the four central transmembrane helices and six smaller helices surrounding the A (right hand side) and B (left hand side) branches of cofactors consisting of the bacteriochlorophyll dimer (red), two bacteriochlorophyll monomers (purple), two bacteriopheophytins (orange), quinones and non-heme iron (brown). Coordinates from 4RCR.pdb [5].

does not have any direct interactions with the cofactors, its presence stabilizes the protein [10] and is required for assembly of the functional complex [11–13].

While all reaction centers from purple bacteria have three subunits, L, M, and H, the total number of subunits is either three or four. Some reaction centers, including the well-characterized species *Rhodobacter sphaeroides* and *Rhodobacter capsulatus*, have only the three subunits. However, reaction centers from many other purple bacteria, including those from *Blastochloris viridis*, have a bound cytochrome subunit that contains four hemes (Figure 12.2a). This subunit is located on the periplasmic side of the protein complex and forms a large globular domain. After light excites the bacteriochlorophyll dimer, an electron is transferred to the A branch bacteriopheophytin and then to the primary and secondary quinone acceptors. The presence of the tetraheme subunit near the bacteriochlorophyll dimer allows the closest heme to serve as a rapid secondary electron donor for the oxidized bacteriochlorophyll dimer.

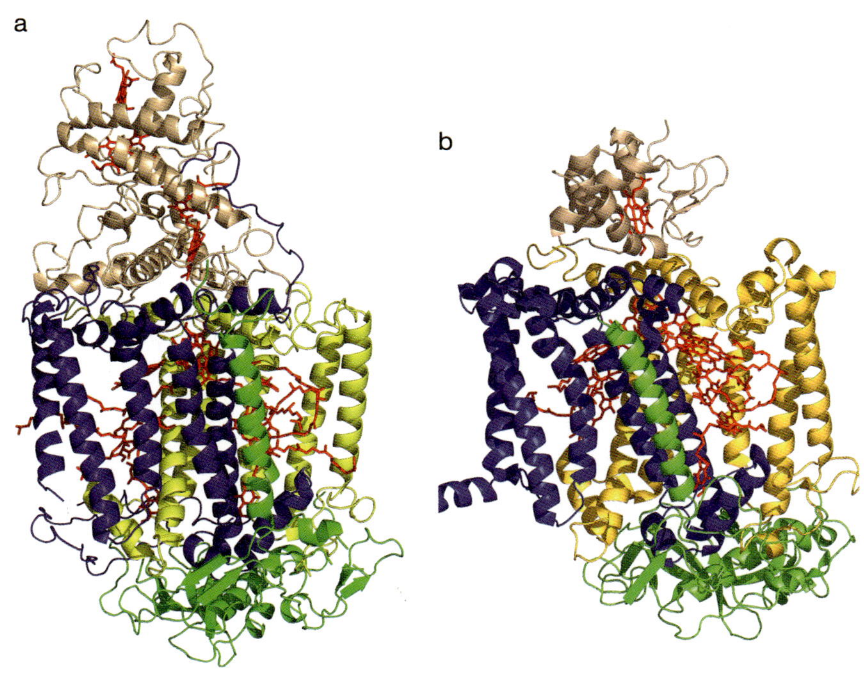

Figure 12.2 **(a)** Structure of the reaction center from *Blastochloris viridis* showing the cofactors (red), including the four hemes (pink), and the four protein subunits: L (yellow), M (blue), H (green), and tetraheme (wheat). Coordinates from 4PRC.pdb [4]. **(b)** Structure of the reaction center-cytochrome c_2 complex from *R. sphaeroides* showing the cofactors (red), including the heme (pink) and the protein subunits: L (yellow), M (blue), H (green), and the cytochrome (wheat). The binding of the cytochrome essentially leaves the structure of the reaction center unchanged, except for the amino acid side chains participating at the binding site and the amino terminus region of the M subunit. Coordinates from 1L9B.pdb [14].

Cyclic electron transfer is achieved through a series of subsequent electron and proton transfer processes involving the reaction center and the cytochrome bc_1 complex (see Chapter 7). Critical to the cyclic process is the involvement of the secondary quinone, as discussed below, and a water-soluble cytochrome c_2. These two cellular components serve as electron and proton carriers between the reaction center and cytochrome bc_1 complex. Unlike the tetraheme subunit, the cytochrome c_2 binds only transiently to the reaction center in order to perform electron transfer [15]. Once the water-soluble cytochrome c_2 is bound to the reaction center, electron transfer proceeds in 1 μs. The binding of the cytochrome to the periplasmic surface of the reaction center of *R. sphaeroides* has been shown by protein crystallography to result in the heme residing directly over the bacteriochlorophyll dimer [14] (Figure 12.2b). While electrostatic interactions are the primary factors that determine the binding, other interactions such as hydrophobic interactions establish the final configuration of the bound complex [16]. For bacterial reaction centers with a bound tetraheme subunit, after an electron is transferred to the oxidized bacteriochlorophyll dimer, the tetraheme is subsequently reduced by cytochrome c_2.

In contrast to the reaction centers from purple bacteria, the reaction center from the green bacterium *Chloroflexus auranticus* has only two subunits, corresponding to the L and M subunits [17]. The cofactors and their properties are similar to those in purple bacteria except that in place of the B-side bacteriochlorophyll monomer there is a bacteriopheophytin. The difference in protein composition reflects the more extensive light-harvesting complexes found in green bacteria. Purple bacteria have their light-harvesting complexes located exclusively in the cell membrane. Energy transfer then occurs from the bacteriochlorophylls of the light-harvesting complex to the reaction center, all within the membrane (see Chapter 14). Green bacteria have a large light-harvesting complex termed a chlorosome that is attached to the cell membrane. The smaller size of the reaction center of green bacteria allows it to pack in the membrane close to the chlorosome, perhaps being attached to it, and to accept the captured light energy.

The L and M subunits in wild-type reaction centers encase a total of 10 cofactors. At the periplasmic side are two closely associated bacteriochlorophylls that serve as the primary electron donor. Each branch also has a bacteriochlorophyll monomer, a bacteriopheophytin, and a quinone. The single non-heme iron atom lies on the symmetry axis between the two quinones. The only cofactor that does not follow the symmetry pattern is the carotenoid that is asymmetrically positioned near one of the bacteriochlorophyll monomers. The properties of these cofactors are discussed below.

12.3 Bacteriochlorophyll Dimer

All photosynthetic complexes have bacteriochlorophyll or chlorophyll cofactors that are the focal point for light excitation and serve as the primary electron donor.

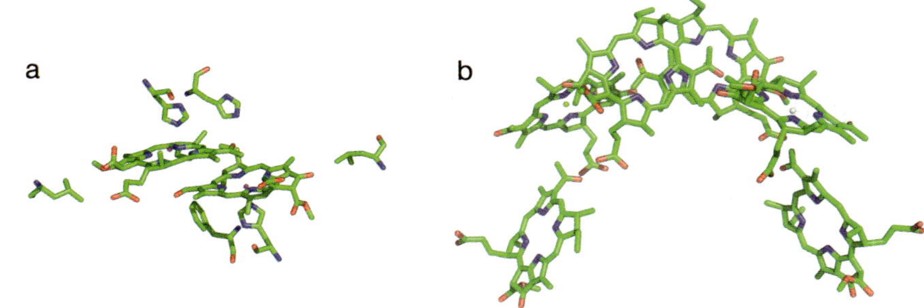

Figure 12.3 **(a)** Structure of the bacteriochlorophyll dimer from *R. sphaeroides* and some surrounding amino acid residues, including His L173 and M202 that each coordinate one of the Mg, His L168 that is hydrogen bonded to the acetyl group of ring A on one side, and Leu L131, Leu M160, and Phe M197, that, when changed to histidines, form hydrogen bonds with the dimer. Coordinates from 1M3X.pdb [9]. **(b)** Structure of the bacteriochlorophyll dimer (P_A, P_B), bacteriochlorophyll monomers (B_A, B_B), and bacteriopheophytins (H_A, H_B) of the reaction center from *R. sphaeroides* (colored by atom type). Coordinates from 4RCR.pdb [5]. View in (a) is rotated by 90° compared to Figure 12.1; view in (b) is same as Figure 12.1.

In reaction centers from purple bacteria, a wide variety of measurements have established that the two closely overlapping bacteriochlorophylls evident in the three-dimensional structure serve as the primary electron donor (Figure 12.3a). The separation between the conjugated ring systems of these two bacteriochlorophylls is approximately 3 Å with the overlap being at the ring A position.

Associated with the close distance are interactions between the two bacteriochlorophylls that result in sharing of electrons between both tetrapyrroles. The distribution of the electrons over two bacteriochlorophylls rather than one consequently alters the properties of the dimer, compared to a monomer. For example, the redistribution of electrons causes a shift in the energies of the electronic orbitals, resulting in the dimer being easier to oxidize. The interactions between the two bacteriochlorophylls also result in the presence of a distinct optical absorption band in the near-infrared region for the donor of purple bacteria, which is found at 865 nm in reaction centers from *R. sphaeroides* compared to 800 nm for the bacteriochlorophyll monomers. Thus, the transition energy of the dimer is lower than that of the bacteriochlorophyll monomers. For optical spectroscopic studies, the presence of the distinct band for the bacteriochlorophyll dimer provides a means of distinguishing the involvement of the bacteriochlorophyll dimer in electron transfer reactions from the bacteriochlorophyll monomers.

Each of the central Mg atoms of the bacteriochlorophylls forming the dimer are five-coordinated with four in-plane ligands provided by the nitrogens of the tetrapyrrole and the fifth being an out-of-plane nitrogen ligand from a histidine side chain, His L173 and His M202 in *R. sphaeroides*. Replacement of either of these histidine residues with leucine by site-directed mutagenesis results in reaction centers with an altered cofactor composition, with a bacteriopheophytin replacing one of the bacteriochlorophylls. The resulting strain containing a bacteriochlorophyll-bacteriopheophytin dimer is referred to as a heterodimer mutant [18–24]. While reaction centers from the heterodimer mutants are structurally little changed compared to wild-type reaction centers (except for the cofactor substitution and the changes at the mutation site), the electronic structure of the dimer is significantly altered. Due to the difference in the chemical nature of bacteriopheophytin compared to bacteriochlorophyll, the electrons are no longer shared, the oxidation/reduction midpoint potential is increased by approximately 130 mV, and the optical absorption band associated with the dimer is greatly broadened. When His M202 is substituted by Gly, the Mg atom of the bacteriochlorophyll is inferred to have a water molecule that serves as an out-of-plane ligand and very little is changed in the properties of the dimer [23]. Thus, the changes observed for the heterodimer mutant arise from the cofactor substitution rather than the change in ligand. In contrast, substitution of the ligand in the central iron atom of hemes results in significant alteration of the protein properties, such as a change in the oxidation/reduction midpoint potential by over 500 mV [25].

Hydrogen bond interactions between the conjugated carbonyl groups of the dimer and amino acid residues also contribute substantially to the properties of the bacteriochlorophyll dimer. For example, the oxidation/reduction midpoint potential of the bacteriochlorophyll dimer is sensitive to the number of hydrogen bonds between the dimer and the surrounding protein environment. Loss of the hydrogen bond between His L168 and the bacteriochlorophyll dimer in wild type (Figure 12.3) results in a 95 mV decrease in the oxidation/reduction midpoint potential. Substitution of histidines at Leu L131, Leu M160, and Phe M197 creates up to three new hydrogen bonds. The oxidation/reduction midpoint potential increases with each new hydrogen bond in an additive fashion leading to increases up to 260 mV [26]. While alteration of the hydrogen-bonding pattern leads to dramatic changes in the oxidation/midpoint potential, the optical absorption spectra of most mutants are very similar to wild type, showing that the transition energy is essentially the same. Small shifts of the absorption band, up to 15 nm to lower wavelengths, are associated with mutations that remove one of the hydrogen bonds to the acetyl groups, with the shifts presumably arising from rotation after loss of the bond.

The electron density distributions of the bacteriochlorophyll dimer in mutants with altered hydrogen-bonding patterns were measured using electron nuclear double resonance and found to be systematically shifted [27]. For example, the addition of each hydrogen bond to the dimer of the reaction center from purple bacteria was found to stabilize the bacteriochlorophyll containing the proton-accepting carbonyl; this preferential stabilization of one side of the dimer resulted in a change in the symmetry of the electron density distribution. By correlating the changes in electron distributions with the shifts in the oxidation/reduction midpoint potential, a detailed molecular model of the electronic structure of the dimer was developed that could quantitatively explain the effect of protein interactions on the properties of the bacteriochlorophyll dimer [28, 29].

The properties of the bacteriochlorophyll dimer, in particular the energy of the oxidized state, are also sensitive to electrostatic interactions with charged amino acid residues. For example, the oxidation/reduction midpoint potential of the dimer was found to decrease up to 60 mV due to the introduction of a negative charge located approximately 10 Å from the bacteriochlorophyll dimer, or increase up to 50 mV due to a positive charge [30–33]. The changes of the oxidation/reduction midpoint potentials are in agreement with electrostatic models, provided that the introduced charges are largely screened by the protein [33]. Other interactions that also contribute to the properties of the dimer, although to a lesser extent, are electrostatic interactions from polar groups of amino acids and van der Waals contacts involving aromatic side chains.

12.4 Bacteriochlorophyll and Bacteriopheophytin Monomers

The lack of the central Mg in bacteriopheophytin, compared to bacteriochlorophyll, results in differences in the electronic properties of these two molecules [34]. For example, bacteriopheophytin has a different oxidation/reduction potential, which makes it a better electron acceptor. Also, the optical absorption band of the bacteriopheophytins in reaction centers is present at a wavelength of 760 nm, compared to 800 nm for the bacteriochlorophyll monomers [35]. This distinct band provides a unique spectral signature that firmly establishes a bacteriopheophytin as receiving an electron from the primary electron donor approximately 3 ps after light excitation. The identity of the bacteriopheophytin serving as the electron acceptor was established by examination of the visible region of the absorption spectrum of the reaction center. The absorption bands of the two bacteriopheophytins in the visible region overlap near 540 nm at room temperature but separate into two distinctive bands at low temperature. The differences in the optical changes in the 540 nm region after excitation are a clear marker that the bacteriopheophytin on the A branch serves preferentially as the electron acceptor.

The three-dimensional structure shows that on each branch of cofactors a bacteriochlorophyll monomer bridges the bacteriochlorophyll dimer and the bacteriopheophytin (Figure 12.3b). The close interaction of the bacteriochlorophyll monomer with the bacteriochlorophyll dimer and bacteriopheophytin on the A-branch allows this bacteriochlorophyll monomer to serve as a transient acceptor when an electron is transferred from the excited state of the bacteriochlorophyll dimer to the bacteriopheophytin. Transient optical measurements show only small absorption changes in the absorption band for the bacteriochlorophyll monomer, indicating that electron transfer from the bacteriochlorophyll monomer to the bacteriopheophytin is very rapid. Assignment of the optical changes to individual cofactors is difficult during the initial electron transfer as the excited states are highly coupled and so act together as a supermolecule until the transfer is complete [36, 37].

12.5
Primary Quinone, Secondary Quinone, and Non-heme Iron

The primary quinone lies near the A-branch bacteriopheophytin and accepts an electron in approximately 200 ps from the reduced bacteriopheophytin. After the primary quinone is reduced, electron transfer continues to the secondary quinone in approximately 200 μs [38, 39]. Once the oxidized bacteriochlorophyll dimer is reduced by cytochrome, the bacteriochlorophyll dimer is able to be excited again and to transfer a second electron. Unlike the case for the first electron, the second reduction of the secondary quinone is coupled to the transfer of two protons. The quinone then carries the electrons and protons to the membrane and is replaced with another quinone.

The primary and secondary quinones have distinct functional roles despite being identical molecules in reaction centers from *R. sphaeroides* (Figure 12.4a). The functional differences can be understood by a comparison of the protein environments for the two quinones. The primary quinone has a much more hydrophobic environment than the secondary quinone. For example, the primary quinone interacts with a close aromatic amino acid residue, Trp M252, while the secondary quinone has two nearby ionizable residues, Glu L212 and Asp L213. The accessibility of protonatable amino acid residues near the secondary quinone allows protons to be transferred to the quinone in conjunction with the transfer of the second electron. Proton transfer pathways, which involve a series of closely interacting water molecules and protonatable amino acid residues, provide the means to transfer protons from the solvent to those residues and reset the protonation states. Another structural difference between the quinones is the presence of two distinct binding sites for the secondary quinone. The possible movement of the secondary quinone during the transfer of electrons to this quinone provides an attractive, but unproven, gating mechanism for regulating the process [40–42].

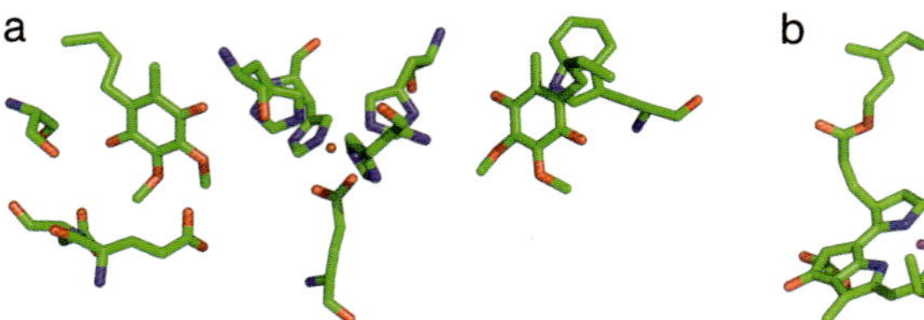

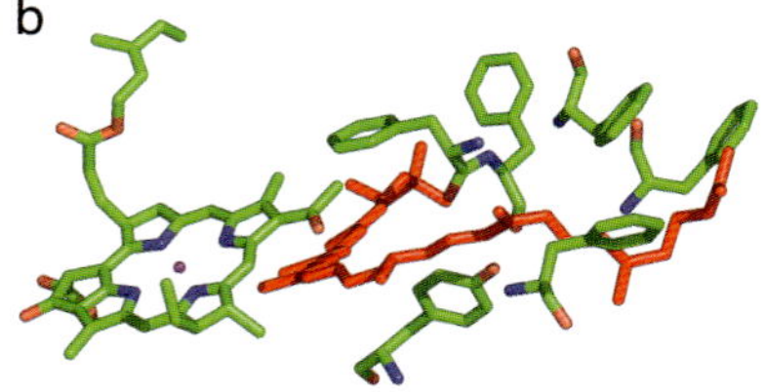

Figure 12.4 **(a)** Structure of the reaction center from *R. sphaeroides* showing the primary quinone (Q_A), secondary quinone (Q_B), and non-heme iron with some surrounding amino acid residues: Trp M252 near the primary quinone, the five residues coordinating the iron (His L190, His L230, His M219, His M266, and Glu M234), and Glu L212, Asp L213, and Ser L223 near the secondary quinone. Atoms are colored by atom type. Coordinates from 4RCR.pdb [5]; **(b)** Structure of the reaction center from *R. sphaeroides* showing the carotenoid (red) and several nearby residues (colored by atom type) (Phe M67, Phe M68, Phe M120, Phe M123, Phe 162, and Tyr M177) and the bacteriochlorophyll monomer on the B branch. Coordinates from 1M3X.pdb [9]. Views are approximately the same as Figure 12.1.

Between the two quinones is a non-heme iron coordinated by four histidine residues and a glutamate residue. Two of the histidine ligands and the iron structurally bridge the 10 Å distance between the two quinones (Figure 12.4a). Although it does interact magnetically with the reduced quinones, the iron does not change oxidation state during electron transfer. Thus the iron does not serve as an electron acceptor, although the non-heme iron and the histidines may play a role in the process [1, 43–49]. The iron can be replaced with other metals, such as zinc, either biochemically or through alteration of one of the histidine residues; the metal-replaced reaction centers have very similar properties to wild type, showing that the electron transfer is not dependent upon the properties of the iron or the ligands [50, 51]. If the iron is removed from the reaction center, electron transfer still occurs, although the complex is less stable, and the rates involving the quinones are significantly altered. Thus, the primary role of the non-heme iron is to stabilize the structure of the complex through ligation to histidine residues from each of the four core transmembrane helices.

12.6 Carotenoid

The wild-type reaction center has a carotenoid molecule located near the bacteriochlorophyll monomer on the B branch (Figure 12.4b). The carotenoid, which is a spheroidene in *R. sphaeroides*, is located in a very hydrophobic region of the reaction center formed largely of aromatic amino acid residues. These residues form a well-defined binding pocket for the carotenoid when it adopts a 15,15′-cis configuration. The carotenoid is not required for either reaction center assembly or function, as strains that lack the carotenoid, such as the well-characterized *R. sphaeroides* R-26 strain, have been isolated and are capable of photosynthetic growth. Reaction centers from the carotenoid-less *R. sphaeroides* R-26 strain have electron transfer rates that are very similar to those of the wild type, showing that the carotenoid does not participate in the electron transfer process. Although not required for electron transfer, the carotenoid does serve in a photoprotection role by quenching potentially harmful excited states of the bacteriochlorophyll dimer [52]. The bacteriochlorophyll monomer on the B-branch, which lies between the bacteriochlorophyll dimer and the carotenoid, can mediate this energy transfer from the bacteriochlorophyll dimer to the carotenoid. However, the efficiency of the transfer process is sensitive to the energetics of the electronic structure of the carotenoid, and different carotenoids are found in different purple bacteria.

12.7 Electron Transfer

In order to establish a high quantum yield of electron transfer, that is a conversion of nearly every photon of light into a charge-separated state, bacterial reaction centers have a series of closely spaced acceptors [35]. The presence of several acceptors close together allows for a rapid forward electron transfer with a minimum of unproductive charge recombination reactions. The rates of electron transfer between these cofactors, as delineated by Marcus and coworkers [53], can be expressed in terms of several factors, including the free energy difference, the reorganization energy, the coupling between the electron donor and acceptor, and protein dynamics.

For most processes, increasing the free energy difference between the final and initial states increases the rate. Experimentally, it can be difficult to measure such energy differences. However, reaction center mutants with different oxidation/reduction midpoint potentials for the bacteriochlorophyll dimer provide the opportunity to determine the effect of shifting the energies on the electron transfer rates [54]. For example, using transient optical spectroscopy, the charge recombination rates for the primary and secondary quinones were measured and found to increase with increasing free energy differences [55].

In the Marcus theory, electron transfer rates are exponentially dependent on the difference between the free energy difference and the reorganization energy, which represents the energy needed to change the initial state into the final state. The difference between these two factors is a measure of the activation energy of a process. When the reorganization energy is much larger than the free energy difference, the activation energy is large and the rate is slow. As the free energy difference approaches the reorganization energy, the activation energy decreases and the rate becomes faster. In wild-type reaction centers, the charge recombination rate from the secondary quinone is much slower than that from the primary quinone because the reorganization energy for charge recombination from the secondary quinone is much larger, due to its more hydrophilic environment.

Electron transfer rates are also proportional to the coupling, which is a measure of the interactions between the electron donor and acceptor. While the coupling depends upon a number of factors, the predominant determinant is the distance between donor and acceptor. To avoid the impact of the free energy difference, the maximal rate (which occurs when the difference between the free energy difference and the reorganization energy is zero), is used when comparing rates for different reactions. In proteins, the maximal rates of electron transfer show an approximate exponential dependence upon distance for proteins [56, 57]. For the primary and secondary quinones in the reaction center, the maximal charge recombination rates were estimated to be equal, which is consistent with their nearly identical distances to the bacteriochlorophyll dimer.

The initial rate of electron transfer from the excited bacteriochlorophyll dimer to the bacteriopheophytin has a weak dependence on the free energy difference, with wild-type reaction centers having a small activation energy [31]. While the presence of two nearly symmetrical branches would predict comparable couplings, and hence rates along the two branches, in wild-type reaction centers the initial electron transfer is dominantly along the A-branch of cofactors. Key to the directionality is the initial electron transfer of an electron from the excited state of the bacteriochlorophyll dimer preferentially to the A-side bacteriopheophytin in 3 ps. The reason for the functional asymmetry of this initial electron transfer has been investigated extensively using transient optical spectroscopic measurements of wild type and mutant reaction centers [35]. These studies indicate that the energy difference for transfer along the A-branch is more favorable than for the B-branch and that this energy difference can be manipulated by mutagenesis. Although the yield of electron transfer directly along the B-branch is very low in wild type, mutations of the reaction center can increase the direct reduction of Q_B up to 30% [58, 59]. The influence of the protein environment on the initial electron transfer is also illustrated by the observation that charge separation along the A branch or the B branch can be switched by using the pH to tune the ionization state of protonatable residues [60].

Finally, protein dynamics also play a role in the electron transfer process, although direct measurement of the dynamics has been experimentally difficult. By monitoring the tryptophan absorption change, which is a marker of protein relaxation in the reaction center, the dynamics associated with the initial transfer of the electron from the bacteriochlorophyll dimer to the bacteriopheophytin have been shown to modulate the initial rate of electron transfer [61].

12.8
Evolution from Anoxygenic to Oxygenic Photosynthesis

Compared to the present day, conditions on the early earth were significantly different, with the primitive photosynthetic organisms growing in an anaerobic environment [62–65]. The earliest phototrophs presumably contained a simple photosystem that was capable of absorbing light and creating a charge-separated state. These primitive anaerobic phototrophs evolved into the current photosynthetic organisms, namely purple and green bacteria, cyanobacteria, algae, and plants. Although the detailed relationships among the photosynthetic organisms are still debated, the consensus is that the reaction center from purple bacteria is evolutionarily related to Photosystem II.

The genes encoding the L and M subunits of the reaction center are homologous to the genes encoding the D1 and D2 subunits of Photosystem II [66–70]. Although the overall similarity of the sequences of the L and M subunits to those of the D1 and D2 subunits is fairly low, many of the key amino acid residues observed near the cofactors in the structure of the reaction center are conserved in Photosystem II [71–74].

The homology between the sequences of the L and M subunits of the reaction center and the D1 and D2 subunits of Photosystem II is reflected in their structural homology (see also Chapter 4). A common feature in the structures of bacterial reaction centers and Photosystem II is the presence of a twofold symmetry axis relating both the cofactors essential for photochemistry and the core protein subunits, L and M in purple bacteria, and D1 and D2 in Photosystem II. The D1 and D2 subunits of Photosystem II are significantly larger, but the pattern of five symmetrically related transmembrane helices is still present and the transmembrane helices of the two complexes can be approximately overlaid. Not only are the protein subunits homologous, but the arrangement of the core cofactors is also similar, as evident in the conserved division of the cofactors into two branches related by a twofold symmetry axis. Although differences are found at a detailed level, in general both complexes contain two tetrapyrrole molecules flanked by four monomeric tetrapyrrole molecules and an iron-binding site on the opposite side. From these comparisons emerges a common reaction center motif of two symmetric branches of cofactors that are bounded by a dimeric tetrapyrrole and a metal site and surrounded by five helices, each from two symmetric core subunits. The conservation of the central structural motif supports the idea that the earliest photosynthetic complex was originally a homodimeric complex [75].

In addition to describing relationships among the photosynthetic complexes, evolutionary schemes must explain how complex electron and proton transfer processes were developed. While the purple bacterial reaction center and Photosystem II share a common structural motif, only Photosystem II can oxidize water, as evidenced by three functional reasons. First, only Photosystem II contains a primary electron donor that is sufficiently oxidizing for water oxidation. Second, water oxidation is a complex process that requires the electron transfer process to be coupled to proton transfer, which in Photosystem II involves the tyrosyl radical $Tyr_Z^{\bullet}$. Finally, the overall reaction consists of a series of steps with four electron equivalents being collected on a special cofactor assembly, the manganese cluster.

Currently, reaction centers are being developed with some of the functional properties of Photosystem II. One of the reasons that the primary electron donor of Photosystem II is a very strong oxidant is that the donor is composed of chlorophylls which are more difficult to oxidize than bacteriochlorophylls by about 200 mV [34]. However, the oxidation potential is also sensitive to the protein environment; incorporation of hydrogen bonds to the bacteriochlorophyll dimer yields a highly oxidizing electron donor in the bacterial reaction center [26]. The introduction of a tyrosine in the highly oxidizing reaction center at a site homologous to that of Tyr_Z results in a light-induced tyrosyl radical [76]. The presence of bound manganese capable of reducing the oxidized bacteriochlorophyll dimer was achieved by incorporating a metal binding site at a location homologous to that of Photosystem II [77]. The transition from a single bound ion to a bound manganese cluster capable of water oxidation presumably involved complexes that could perform intermediate reactions, such as using bicarbonate as a substrate [78]. The manganese cluster is a cofactor assembly unique to Photosystem II; its development may have involved the direct incorporation of manganese clusters that are naturally occurring in the environment, rather than modification of biosynthetic pathways used in the assembly of manganese enzymes [79].

12.9 Biotechnological Applications

In principle, the photosynthetic system serves as a light-driven electronic device with a quantum conversion efficiency of nearly 100%. The circuitry is organized in a protein scaffold embedded in a bilayer interface allowing the accumulation of charges until needed. Such nanoscale circuits would be attractive in real applications if they could be integrated into solid-state electronics. A number of obstacles prevent their usage, notably the difficulties in manipulating the complexes and coupling the transfer of electrons into an electronic current. However, the potential promise has led a number of scientists to begin the development of such biodevices.

In a sense, the development of photoactivated biodevices began in the 1970s with the incorporation of the bacterial reaction center into artificial Langmuir-Blodgett monolayers and multiple bilayer-type structures [80, 81]. Once incorporated, it was possible to characterize the electron transfer in the bacterial reaction center in a membrane. Bacterial reaction centers can also be incorporated into liposomes and are more stable when light-harvesting complexes are also included [82]. The limited stability of these membrane systems has restricted practical usage, but more robust bilayers became available with the development of supported lipid membranes [83]. Supported membranes are formed by the fusion of proteoliposomes onto a solid support, typically a glass surface. In these membranes, the lipids retain their mobility with a thin water layer of ~10 Å separating the membrane from the surface. When incorporated into the membrane, bacterial reaction centers are highly oriented and retain their functionality [84]. While useful in probing the properties of the reaction center in the membrane, proteins in supported membranes are difficult to couple to other circuit components. More recently, reaction centers have been incorporated into alternative synthetic biomembrane mimics while retaining protein function. In one approach, reaction centers are placed into assemblies formed from phospholipids, a lipopolymer, and detergent surfactant [85]. By controlling the temperature, such mixtures can support functional reaction centers while reversibly forming either lamellar or cubic phases.

Different strategies have been used to couple the electron transfer process of the bacterial reaction center to external devices [86]. In one approach, the reaction center is bound to a nickel layer on a gold surface through a polyhistidine tag at the end of the M subunit [87]. When incorporated into a photovoltaic cell, the photocurrent response of the cell was found to vary with the wavelength of the incident laser pulse, with a spectral dependence that mapped to the optical spectrum of the reaction center in the near-infrared region. These studies demonstrate that the reaction center could be integrated into a photodevice, although the overall quantum efficiency of the device was limited at 12%. By incorporating cytochrome *c* into a similar system, an enhancement in the current was observed, suggesting that the water-soluble cytochrome could facilitate electron transfer from the reaction center to the conductor of the photocell [88]. To improve the efficiency of transfer to an electrode, reaction centers can be embedded in mesoporous titanium dioxide films attached to the electrode surface, or attached to nanotubes that, in principle, could serve as one-dimensional conductors [89, 90].

Acknowledgments

The work described is supported by a grant from the National Science Foundation, MCB 0640002.

References

1 Feher, G., Allen, J.P., Okamura, M.Y. and Rees, D.C. (1989) Structure and function of bacterial photosynthetic reaction centres. *Nature*, **339**, 111–16.

2 Blankenship, R.E., Madigan, M.T. and Bauer, C.E. (eds) (1995) *Anoxygenic Photosynthetic Bacteria*, Kluwer Academic Publishers, Dordrecht, The Netherlands.

3 Deisenhofer, J., Epp, O., Miki, K., Huber, R. and Michel, H. (1985) Structure of the protein subunits in the photosynthetic reaction centre of *Rhodopseudomonas viridis* at 3 Å resolution. *Nature*, **318**, 618–24.

4 Deisenhofer, J., Epp, O., Sinning, I. and Michel, H. (1995) Crystallographic refinement at 2.3 Å resolution and refined model of the photosynthetic reaction centre from *Rhodopseudomonas viridis*. *Journal of Molecular Biology*, **246**, 429–57.

5 Allen, J.P., Feher, G., Yeates, T.O., Komiya, H. and Rees, D.C. (1987) Structure of the reaction center from *Rhodobacter sphaeroides* R-26: the cofactors. *Proceedings of the National Academy of Sciences of the United States of America*, **84**, 5730–4.

6 Chang, C.H., El-Kabbani, O., Tiede, D., Norris, J. and Schiffer, M. (1991) Structure of the membrane-bound protein photosynthetic reaction center from *Rhodobacter sphaeroides*. *Biochemistry*, **30**, 5352–60.

7 Ermler, U., Fritzsch, G., Buchanan, S.K. and Michel, H. (1994) Structure of the photosynthetic reaction centre from *Rhodobacter sphaeroides* at 2.65 Å resolution: cofactors and protein-cofactor interactions. *Structure*, **2**, 925–36.

8 McAuley, K.E., Fyfe, P.K., Ridge, J.P., Isaacs, N.W., Cogdell, R.J. and Jones, M.R. (1999) Structural details of an interaction between cardiolipin and an integral membrane protein. *Proceedings of the National Academy of Sciences of the United States of America*, **96**, 14706–11.

9 Cámara-Artigas, A., Brune, D. and Allen, J.P. (2002) Interactions between lipids and bacterial reaction centers determined by protein crystallography. *Proceedings of the National Academy of Sciences of the United States of America*, **99**, 11055–60.

10 Debus, R.J., Feher, G. and Okamura, M.Y. (1985) LM complex of reaction centers from *Rhodopseudomas sphaeroides* R-26: Characterization and reconstitution with the H subunit. *Biochemistry*, **24**, 2488–500.

11 Cheng, Y.S., Brantner, C.A., Tsapin, A. and Collins, M.L.P. (2000) Role of the H protein in assembly of the photochemical reaction center and intracytoplasmic membrane in *Rhodosprillum rubrum*. *Journal of Bacteriology*, **182**, 1200–7.

12 Tehrani, A., Prince, R.C. and Beatty, J.T. (2003) Effects of photosynthetic reaction center H protein domain mutations on photosynthetic properties and reaction center assembly in *Rhodobacter sphaeroides*. *Biochemistry*, **42**, 8919–28.

13 Lupo, D. and Ghosh, R. (2004) The reaction center H subunit is not required for high levels of light-harvesting complex I in *Rhodosprillum rubrum* mutants. *Journal of Bacteriology*, **186**, 5585–95.

14 Axelrod, H.L., Abresch, E.C., Okamura, M.Y., Yeh, A.P., Rees, D.C. and Feher, G. (2002) X-ray structure determination of the cytochrome c_2 reaction center electron transfer complex from *Rhodobacter sphaeroides*. *Journal of Molecular Biology*, **319**, 501–15.

15 Axelrod, H.L. and Okamura, M.Y. (2005) The structure and function of the cytochrome c_2: reaction center electron transfer complex from *Rhodobacter sphaeroides*. *Photosynthesis Research*, **85**, 101–14.

16 Miyashita, O., Okamura, M.Y. and Onuchic, J.N. (2005) Interprotein electron transfer from cytochrome c_2 to photosynthetic reaction center: tunneling across an aqueous interface. *Proceedings of the National Academy of Sciences of the United States of America*, **102**, 3558–63.

17 Feick, R., Shiozawa, J.A. and Ertimaier, A. (1995) Biochemical and spectroscopic properties of the reaction center of the green filamentous bacterium *Chloroflexus aurantiacus*, in *Anoxygenic Photosynthetic Bacteria* (eds R.E. Blankenship, M.T. Madigan and C.E. Bauer), Kluwer,

Dordrecht, The Netherlands, pp. 699–708.

18 Bylina, E.J. and Youvan, D.C. (1988) Directed mutations affecting spectroscopic and electron transfer properties of the primary donor in the photosynthetic reaction center. *Proceedings of the National Academy of Sciences of the United States of America*, **85**, 7226–30.

19 McDowell, L.M., Gaul, D., Kirmaier, C., Holten, D. and Schenck, C.C. (1991) Investigation into the source of electron transfer asymmetry in bacterial reaction centers. *Biochemistry*, **30**, 8315–22.

20 Allen, J.P., Artz, K., Lin, X., Williams, J.C., Ivancich, A., Albouy, D., Mattioli, T.A., Fetsch, A., Kuhn, M. and Lubitz, W. (1996) Effects of hydrogen bonding to a bacteriochlorophyll-bacteriopheophytin dimer in reaction centers from *Rhodobacter sphaeroides*. *Biochemistry*, **35**, 6612–19.

21 Laporte, L.L., Palaniappan, V., Davis, D.G., Kirmaier, C., Schenck, C.C., Holten, D. and Bocian, D.F. (1996) Influence of electronic asymmetry on the spectroscopic and photodynamic properties of the primary electron donor in the photosynthetic reaction center. *The Journal of Physical Chemistry*, **100**, 17696–707.

22 van Brederode, M.E., van Stokkum, I.H.M., Katilius, E., van Mourik, F., Jones, M.R. and van Grondelle, R. (1999) Primary charge separation routes in the BChl:BPhe heterodimer reaction centers of *Rhodobacter sphaeroides*. *Biochemistry*, **38**, 7545–55.

23 Goldsmith, J.O., King, B. and Boxer, S.G. (1996) Mg coordination by amino acid side chains is not required for assembly and function of the special pair in bacterial photosynthetic reaction centers. *Biochemistry*, **35**, 2421–8.

24 Cámara-Artigas, A., Magee, C., Goetsch, A. and Allen, J.P. (2002) The structure of the heterodimer reaction center from *Rhodobacter sphaeroides* at 2.55 Å resolution. *Photosynthesis Research*, **74**, 87–93.

25 Wuttke, D.S. and Gray, H.B. (1993) Protein engineering as a tool for understanding electron transfer. *Current Opinion in Structural Biology*, **3**, 555–63.

26 Lin, X., Murchison, H.A., Nagarajan, V., Parson, W.W., Allen, J.P. and Williams, J.C. (1994) Specific alteration of the oxidation potential of the electron donor in reaction centers from *Rhodobacter sphaeroides*. *Proceedings of the National Academy of Sciences of the United States of America*, **91**, 10265–9.

27 Artz, K., Williams, J.C., Allen, J.P., Lendzian, F., Rautter, J. and Lubitz, W. (1997) Relationship between the oxidation potential and electron spin density of the primary electron donor in reaction centers from *Rhodobacter sphaeroides*. *Proceedings of the National Academy of Sciences of the United States of America*, **94**, 13582–7.

28 Müh, F., Lendzian, F., Roy, M., Williams, J.C., Allen, J.P. and Lubitz, W. (2002) Pigment-protein interactions in bacterial reaction centers and their influence on oxidation potential and spin density distribution of the primary donor. *The Journal of Physical Chemistry B*, **106**, 3226–36.

29 Reimers, J.R. and Hush, N.S. (2004) A unified description of the electrochemical, charge distribution, and spectroscopic properties of the special-pair radical cation in bacterial photosynthesis. *Journal of the American Chemical Society*, **126**, 4132–44.

30 Williams, J.C., Haffa, A.L.M., McCulley, J.L., Woodbury, N.W. and Allen, J.P. (2001) Electrostatic interactions between charged amino acid residues and the bacteriochlorophyll dimer in reaction centers from *Rhodobacter sphaeroides*. *Biochemistry*, **40**, 15403–7.

31 Haffa, A.L.M., Lin, S., Katilius, E., Williams, J.C., Taguchi, A.K.W., Allen, J.P. and Woodbury, N.W. (2002) The dependence of the initial electron transfer rate on driving force in *Rhodobacter sphaeroides* reaction centers. *The Journal of Physical Chemistry B*, **106**, 7376–84.

32 Johnson, E.T. and Parson, W.W. (2002) Electrostatic interactions in an integral membrane protein. *Biochemistry*, **41**, 6483–94.

33 Johnson, E.T., Müh, F., Nabedryk, E., Williams, J.C., Allen, J.P., Lubitz, W., Breton, J. and Parson, W.W. (2002) Electronic and vibronic coupling of the special pair of bacteriochlorophylls in photosynthetic reaction centers from wild-type and mutant strains of *Rhodobacter sphaeroides*. *The Journal of Physical Chemistry B*, **106**, 11859–69.

34 Grimm, B., Porra, R.J., Rudiger, W. and Scheer, H. (eds) (2006) *Chlorophylls and Bacteriochlorophylls: Biochemistry, Biophysics, Functions and Applications*, Springer, Dordrecht, The Netherlands.

35 Woodbury, N.W. and Allen, J.P. (1995) The pathways, kinetics, and thermodynamics of electron transfer in the reaction centers of purple nonsulfur bacteria, in *Anoxygenic Photosynthetic Bacteria* (eds R.E. Blankenship, M.T. Madigan and C.E. Bauer), EdsKluwer, Dordrecht, The Netherlands, pp. 527–57.

36 Parson, W.W. (2007) Long live electronic coherence! *Science*, **316**, 1438–9.

37 Lee, H., Cheng, Y.C. and Fleming, G.R. (2007) Coherence dynamics in photosynthesis: Protein protection of excitation coherence. *Science*, **316**, 1462–5.

38 Okamura, M.Y., Paddock, M.L., Graige, M.S. and Feher, G. (2000) Proton and electron transfer in bacterial reaction centers. *Biochimica et Biophysica Acta*, **1458**, 148–63.

39 Zhu, Z. and Gunner, M.R. (2005) Energetics of quinone-dependent electron and proton transfers in *Rhodobacter sphaeroides* photosynthetic reaction centers. *Biochemistry*, **44**, 82–6.

40 Stowell, M.H.B., McPhillips, T.M., Rees, D.C., Soltis, S.M., Abresch, E. and Feher, G. (1997) Light-induced structural changes in photosynthetic reaction center: Implications for mechanism of electron-proton transfer. *Science*, **276**, 812–16.

41 Li, J., Gilroy, D., Tiede, D.M. and Gunner, M.R. (1998) Kinetic phases in the electron transfer from $P^+Q_A^-Q_B$ to $P^+Q_AQ_B^-$ and the associated processes in *Rhodobacter sphaeroides* R-26 reaction centers. *Biochemistry*, **37**, 2818–29.

42 Breton, J. (2004) Absence of large-scale displacement of quinone Q_B in bacterial photosynthetic reaction centers. *Biochemistry*, **43**, 3318–26.

43 Ito, H. and Nakatsuji, H. (2001) Roles of proteins in the electron transfer in the photosynthetic reaction center of *Rhodopseudomonas viridis*: bacteriopheophytin to ubiquinone. *Journal of Computational Chemistry*, **22**, 265–72.

44 Remy, A. and Gerwert, K. (2003) Coupling of light-induced electron transfer to proton uptake in photosynthesis. *Nature Structural Biology*, **10**, 637–44.

45 Di Donato, M., Correa, A. and Peluso, A. (2003) The role of the iron-histidine bridge in the early steps of photosynthesis. *Chemical Physics Letters*, **369**, 549–55.

46 Ishikita, H. and Knapp, E.W. (2006) Electrostatic role of the non-heme iron complex in bacterial photosynthetic reaction center. *FEBS Letters*, **580**, 4567–70.

47 Hermes, S., Bremm, O., Garczarek, F., Derrien, V., Liebisch, P., Loja, P., Sebban, P., Gerwert, K. and Haumann, M. (2006) A time-resolved iron-specific X-ray absorption experiment yields no evidence for an Fe^{2+} to Fe^{3+} transition during Q_A^- to Q_B electron transfer in the photosynthetic reaction center. *Biochemistry*, **45**, 353–9.

48 Breton, J. (2007) Steady-state FTIR spectra of the photoreduction of Q_A and Q_B in *Rhodobacter sphaeroides* reaction centers provide evidence against the presence of a proposed transient electron acceptor X between the two quinones. *Biochemistry*, **46**, 4459–65.

49 Breton, J., Lavergne, J., Wakeham, M.C., Nabedryk, E. and Jones, M.R. (2007) The unusually strong hydrogen bond between the carbonyl of Q_A and His M219 in the *Rhodobacter sphaeroides* reaction center is not essential for efficient electron transfer from Q_A^- to Q_B. *Biochemistry*, **46**, 6468–76.

50 Debus, R.J., Feher, G. and Okamura, M.Y. (1986) Iron-depleted reaction centers from *Rhodopseudomonas sphaeroides* R-26.1: characterization and reconstitution with Fe^{2+}, Mn^{2+}, Co^{2+}, Ni^{2+}, Cu^{2+}, and Zn^{2+}. *Biochemistry*, **25**, 2276–87.

51 Williams, J.C., Way, Y.P., Paddock, M.L. and Allen, J.P. (2007) Changes in metal specificity due to iron ligand substitutions in reaction centers from Rhodobacter sphaeroides. *Applied Magnetic Resonance*, **31**, 45–58.

52 Hashimoto, H., Fujii, R., Yanagi, K., Kusumoto, T., Gardiner, A.T., Cogdell, R.J., Roszak, A.W., Isaacs, N.W.,

Pendon, Z., Niedzwiedski, D. and Frank, H.A. (2006) Structures and function of carotenoids bound to reaction centers from purple photosynthetic bacteria. *Pure and Applied Chemistry*, **78**, 1505–18.

53 Marcus, R.A. and Sutin, N. (1985) Electron transfers in chemistry and biology. *Biochimica et Biophysica Acta*, **811**, 265–322.

54 Allen, J.P. and Williams, J.C. (1995) Relationship between the oxidation potential of the bacteriochlorophyll dimer and electron transfer in photosynthetic reaction centers. *Journal of Bioenergetics and Biomembranes*, **27**, 275–83.

55 Allen, J.P., Williams, J.C., Graige, M.S., Paddock, M.L., Labahn, A., Feher, G. and Okamura, M.Y. (1998) Free energy dependence of the direct charge recombination from the primary and secondary quinones in reaction centers from *Rhodobacter sphaeroides*. *Photosynthesis Research*, **55**, 227–33.

56 Gray, H.B. and Winkler, J.R. (2005) Long-range electron transfer. *Proceedings of the National Academy of Sciences of the United States of America*, **102**, 3534–9.

57 Noy, D., Moser, C.C. and Dutton, P.L. (2006) Design and engineering of photosynthetic light-harvesting and electron transfer using length, time, and energy scales. *Biochimica et Biophysica Acta*, **1757**, 90–105.

58 Kee, H.L., Laible, P.D., Bautista, J.A., Hanson, D.K., Holten, D. and Kirmaier, C. (2006) Determination of the rate and yield of B-side quinone reduction in *Rhodobacter capsulatus* reaction centers. *Biochemistry*, **45**, 7314–22.

59 Paddock, M.L., Flores, M., Isaacson, R., Chang, C., Abresch, E.C., Selvaduray, P. and Okamura, M.Y. (2006) Trapped conformational states of semiquinone $(D^{+\cdot}Q_A^{-\cdot}$ formed by B-branch electron transfer at low temperature in *Rhodobacter sphaeroides* reaction centers. *Biochemistry*, **45**, 14032–42.

60 Haffa, A.L.M., Lin, S., Williams, J.C., Bowen, B.P., Taguchi, A.K.W., Allen, J.P. and Woodbury, N.W. (2004) Controlling the pathway of photosynthetic charge separation in bacterial reaction centers. *The Journal of Physical Chemistry B*, **108**, 4–7.

61 Wang, H., Lin, S., Allen, J.P., Williams, J.C., Blankert, S., Laser, C. and Woodbury, N.W. (2007) Protein dynamics control the kinetics of initial electron transfer in photosynthesis. *Science*, **316**, 747–50.

62 Schopf, J.W. (ed.) (1983) *Earth's Earliest Biosphere: Its Origin and Evolution*, Princeton University Press, Princeton, NJ.

63 Woese, C.R. (1987) Bacterial evolution. *Microbiology Reviews*, **51**, 221–71.

64 Olson, J.M. and Pierson, B.K. (1987) Origin and evolution of photosynthetic reaction centers. *Origins of Life and Evolution of the Biosphere*, **17**, 419–30.

65 Blankenship, R.E. and Hartman, H. (1998) The origin and evolution of oxygenic photosynthesis. *Trends in Biochemical Sciences*, **23**, 94–7.

66 Williams, J.C., Steiner, L.A., Ogden, R.C., Simon, M.I. and Feher, G. (1983) Primary structure of the M subunit of the reaction center from *Rhodopseudomonas sphaeroides*, *Proceedings of the National Academy of Sciences of the United States of America*, **80**, 6505–9.

67 Williams, J.C., Steiner, L.A. and Feher, G. (1986) Primary structure of the reaction center from *Rhodopseudomonas sphaeroides*. *Proteins*, **1**, 312–25.

68 Youvan, D.C., Bylina, E.J., Alberti, M., Begusch, H. and Hearst, J.E. (1984) Nucleotide and deduced polypeptide sequences of the photosynthetic reaction-center, B870 antenna, and flanking polypeptides from *R. capsulata*. *Cell*, **37**, 949–57.

69 Michel, H., Weyer, K.A., Gruenberg, H., Dunger, I., Oesterhelt, D. and Lottspeich, F. (1986) The "light" and "medium" subunits of the photosynthetic reaction centers from *Rhodopseudomonas viridis*: Isolation of the genes, nucleotide and amino acid sequence. *The EMBO Journal*, **5**, 1149–58.

70 Belanger, G., Berand, J., Corriveau, P. and Gingras, G. (1988) The structural genes coding for the L subunit and M subunit of *Rhodospirillum rubrum* photoreaction center. *The Journal of Biological Chemistry*, **263**, 7632–8.

71 Rochaix, J.D., Dron, M., Rahire, M. and Malnoe, P. (1984) Sequence homology between the 32K dalton and the D2 chloroplast membrane polypeptides of

Chlamydomonas reinhardtii. *Plant Molecular Biology*, **3**, 363–70.

72 Trebst, A. (1986) The topology of the plastoquinone and herbicide binding peptides of Photosystem II in the thylakoid membrane. *Zeitschrift Fur Naturforschung. C, Journal of Biosciences*, **41**, 240–5.

73 Deisenhofer, J. and Michel, H. (1989) The photosynthetic reaction center from the purple bacterium *Rhodopseudomonas viridis*. *Science*, **245**, 1463–73.

74 Komiya, H., Yeates, T.O., Rees, D.C., Allen, J.P. and Feher, G. (1988) Structure of the reaction center from *Rhodobacter sphaeroides* R26 and 2.4.1: symmetry relations and sequence comparisons between different species. *Proceedings of the National Academy of Sciences of the United States of America*, **85**, 9012–16.

75 Sadekar, S., Raymond, J. and Blankenship, R.E. (2006) Conservation of distantly related membrane proteins: Photosynthetic reaction centers share a common structural core. *Molecular Biology and Evolution*, **23**, 2001–7.

76 Kálmán, L., LoBrutto, R., Allen, J.P. and Williams, J.C. (1999) Modified reaction centres oxidize tyrosine in reactions that mirror photosystem II. *Nature*, **402**, 696–9.

77 Thielges, M., Uyeda, G., Cámara-Artigas, A., Kálmán, L., Williams, J.C. and Allen, J.P. (2005) Design of a redox-linked active metal site: Manganese bound to bacterial reaction centers at a site resembling that of photosystem II. *Biochemistry*, **44**, 7389–94.

78 Dismukes, G.C., Klimov, V.V., Baranov, S.V., Kozlov, Y.N., DasGupta, J. and Tyryshkin, A. (2001) The origin of atmospheric oxygen on earth: The innovation of oxygenic photosynthesis. *Proceedings of the National Academy of Sciences of the United States of America*, **98**, 2170–5.

79 Sauer, K. and Yachandra, V.K. (2002) A possible evolutionary origin for the Mn_4 cluster of the photosynthetic water oxidation complex from natural MnO_2 precipitates in the early ocean. *Proceedings of the National Academy of Sciences of the United States of America*, **99**, 8631–6.

80 Gopher, A., Blatt, Y., Schonfeld, M., Okamura, M.Y., Feher, G. and Montal, M. (1985) The effect of an applied electric field on the charge recombination kinetics in reaction centers reconstituted in planar lipid bilayers. *Biophysical Journal*, **48**, 311–20.

81 Alegria, G. and Dutton, P.L. (1991) Langmuir-Blodgett monolayer films of bacterial photosynthetic membranes and isolated reaction centers: preparation, spectrophotometric and electrochemical characterization. *Biochimica et Biophysica Acta*, **1057**, 239–57.

82 Kobayashi, M., Fujioka, Y., Mori, T., Terashima, H., Suzuki, H., Shimada, Y., Saito, T., Wang, Z.Y. and Nozawa, T. (2005) Reconstitution of photosynthetic reaction centers and core antenna-reaction center complexes in liposomes and their thermal stability. *Bioscience, Biotechnology and Biochemistry*, **69**, 1130–6.

83 Sackmann, E. (1996) Supported membranes: scientific and practical applications. *Science*, **271**, 43–8.

84 Salafsky, J., Groves, J.T. and Boxer, S.G. (1996) Architecture and function of membrane proteins in planar supported bilayers: a study with photosynthetic reaction centers. *Biochemistry*, **35**, 14773–81.

85 Laible, P.D., Kelly, R.F., Wasielewski, M.R. and Firestone, M.A. (2005) Electron-transfer dynamics of photosynthetic reaction centers in thermoresponsive soft materials. *The Journal of Physical Chemistry B*, **109**, 23679–86.

86 Lu, Y., Xu, J., Liu, B. and Kong, J. (2006) Photosynthetic reaction center functionalized nano-composite films: Effective strategies for probing and exploiting the photo-induced electron transfer of photosensitive membrane protein. *Biosensors and Bioelectronics*, **22**, 1173–85.

87 Das, R., Kiley, P.J., Segal, M., Norville, J., Yu, A.A., Wang, L., Trammel, S.A., Reddick, L.E., Kumar, R., Stellacci, F., Lebedev, N., Schnur, J., Bruce, B.D., Zhang, S. and Baldo, M. (2004) Integration of photosynthetic protein molecular complexes in solid-state electronic devices. *Nano Letters*, **4**, 1079–83.

88 Lebedev, N., Trammell, S.A., Spano, A., Lukashev, E., Griva, I. and Schnut, J. (2006) Conductive wiring of immobilized photosynthetic reaction center to electrode by cytochrome c. *Journal of the American Chemical Society*, **128**, 12044–5.

89 Lu, Y., Yuan, M., Liu, Y., Tu, B., Xu, C., Liu, B., Zhao, D. and Kong, J. (2005) Photoelectric performance of bacteria photosynthetic proteins entrapped on tailored mesoporous WO_3-TiO_2 films. *Langmuir: The ACS Journal of Surfaces and Colloids*, **21**, 4071–6.

90 Dorogi, M., Balint, Z., Miko, C., Vileno, B., Milas, M., Hernadi, K., Forro, L., Varo, G. and Nagy, L. (2006) Stabilization effect of single-walled carbon nanotubes on the functioning of photosynthetic reaction centers. *The Journal of Physical Chemistry B*, **110**, 21473–9.

13
Anoxygenic Type-I Photosystems and Evolution of Photosynthetic Reaction Centers

Martin F. Hohmann-Marriott and Robert E. Blankenship

13.1
Introduction

Less than one billionth of the total light emitted by the sun reaches the earth. However, this scrap of light was the incentive to develop one of nature's most extraordinary machines: the photosystems with a FeS-type RC. These complex machines are at the heart of a coordinated series of reactions that convert sun energy into chemical energy that fuels and shapes life on earth.

Green sulfur bacteria (GSB) live in anaerobic, sulfur-rich conditions. Their photosynthetic machinery has adapted to living in an extremely low light environment. The infrared part of the black body radiation emitted by black smokers found on the ocean floor [1] and the little sunlight they receive in the Black Sea a hundred meters below the surface [2, 3] are enough light to sustain their life functions.

Heliobacteria also live in anaerobic conditions. They can be found in such diverse environments as soil and hot springs [4]. These bacteria use light energy to convert atmospheric nitrogen into ammonia [5], a feat that humans only accomplished in the early part of the twentieth century [6] and which lead to a revolution in agriculture [7].

The other major photosynthetic nitrogen-fixers are the cyanobacteria that humans relied on for centuries to fertilize rice fields [8]. Cyanobacteria and their endosymbiotic relatives, the plastids of eukaryotic photosynthetic organisms, are capable of oxygenic photosynthesis. In this process a FeS-type RC containing Photosystem I (PSI) (see Chapter 2), works in tandem with another type of photosystem, Photosystem II (PSII) (see Chapter 4), that contains a type-II RC (quinone type RC), to extract electrons from water, thereby liberating molecular oxygen, O_2. PSI photoactivates these electrons so the captured energy can be used to convert oxidized inorganic molecules into reduced organic molecules. The energy that can be obtained when these organic molecules are reunited with the produced oxygen provides the fuel – directly or indirectly – for most ecosystems and has done so since ancient times. A first fingerprint of the large scale production of oxygen by

Photosynthetic Protein Complexes: A Structural Approach. Edited by P. Fromme

ISBN: 978-3-527-31730-1

the ancestors of cyanobacteria is probably the massive accumulations of iron oxide known as the Banded Iron Formations (BIFs). Together, with fossil and isotopic evidence, these iron oxide bands indicate that oxygenic photosynthesis existed at least 2.4 billion years ago [9, 10].

Interestingly, PSI and PSII have functional and structural similarities with the RCs of organisms that carry out anoxygenic photosynthesis, pointing to a common RC ancestry. What evolutionary path led from this Ur-RC to RCs found today in cyanobactria, GSB, and heliobacteria? The answer is encrypted within the structural and functional characteristics of the RCs and photosystems and the genomic information of the organisms that house them. In this chapter, the current thinking and speculation that tries to decipher the path that led from the first successful capture of light energy to the modern photosystems with FeS-type RC is compiled.

13.2
The Photosynthetic Way of Life

Making a living off light and air is a tempting concept. It is therefore not surprising that the photosynthetic lifestyle, once established, has been extremely successful. Photosynthetic bacteria live in anaerobic conditions (green sulfur bacteria, heliobacteria), semiaerobic conditions (purple bacteria and filamentous anoxygenic phototrophs (FAPs)), and aerobic conditions (cyanobacteria). Cyanobacteria establish a permanent symbiotic relationship with eukaryotic cells and live as plastids within their symbionts. As diverse as these environments are the phylogenetic groups to which these organisms belong. Before extensive genomic information and the inferred evolutionary relationships between these organisms became available, researchers devised a system to categorize the reaction centers found in these organisms, based on the chemical nature of the early electron acceptors.

Photosynthetic RCs are categorized into two broad groups [11–13]. RCs that give electrons to a mobile quinone are called Q-type reaction centers, while RCs that give electrons to a series of iron sulfur clusters are called FeS-type RCs (Figure 13.1). An alternate name for Q-type RCs is Type II, as this group includes PSII in oxygen-evolving systems. Similarly, an alternate name for FeS-type RCs is Type I, as this group includes PSI in oxygen-evolving systems. Green sulfur bacteria and heliobacteria possess FeS-type reaction centers. A newly discovered group of photosynthetic bacteria, the Chloracidobacteria [14], also appears to have a Type I RC, although it has not yet been characterized biochemically. Purple bacteria and FAPs harbor a Q-type RC. The oxygenic cyanobacteria and plastid-containing eukaryotes possess a FeS-type reaction center (PSI) and a Q-type reaction center (PSII). While the simple categorization into FeS-type and Q-type RCs generally makes sense in a functional context, this categorization may not reflect the complex evolutionary relationship between these different types of RCs. When RC features other than the electron acceptors are investigated, similarities between RCs in both classes emerge, indicating a common ancestor and billions of years of complex evolution [15–20](see Section 13.6 on the Ur-Reaction center).

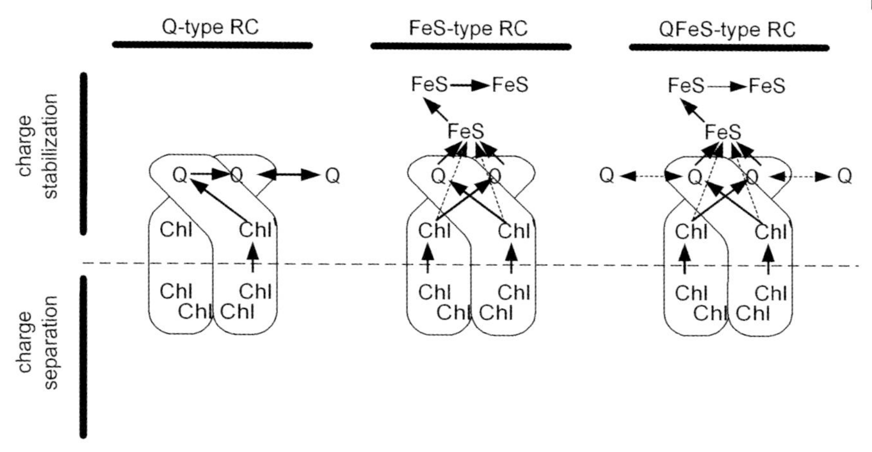

Figure 13.1 Common features of different types of RCs. Q: quinone; Chl: chlorophyll(-like) molecule; QFeS-type RC: a theoretical RC that has the functionality of both a Q-type RC and FeS-type RC.

All known RCs consist of a dimeric integral membrane protein core with five transmembrane helices (TMHs) (see Figure 13.2). This RC core is fused with an antenna domain, consisting of 6 TMH's leading to a type-I photosystem, that is, the core proteins of type-I RC's contain 11 TMH's. In green sulfur bacteria and heliobacteria, only one gene codes for the core reaction center protein with the result that the core RC is a homodimer of two identical copies of a single large integral membrane protein. In all other organism groups, purple bacteria, FAP, cyanobacteria, and photosynthetic eukaryotes, the RC is encoded by two different genes leading to a heterodimeric reaction center consisting of two distinct, but related, integral membrane proteins. In addition to the dimeric protein core, all RCs also contain additional protein subunits to form the final active complex.

Light conversion requires two steps, the capture of light and the conversion of this captured energy into redox energy. These two steps are compartmentalized into different protein domains. In all reaction centers (including the photosystems), the conversion of redox energy takes place within the central RC dimer interface made up of 10 TMH, (5 from each core dimer). Light capture is accomplished by a pigment collective that constitutes the light-harvesting system. Interfacing of the RC core with the light-harvesting systems is accomplished differently in different organism groups.

All photosystems featuring a FeS-type RC and consist of a 22 TMH dimer. Each of the photosystem monomers in heliobacteria (PshA protein), green sulfur bacteria (PscA protein), and PSI (PsaA and PsaB proteins) has 11 TMH, 5 of which house the core electron transfer domain (RC); the remaining 6 TMH act as a light-harvesting system called the core antenna domain. Like these photosystems, PSII also consists of a RC domain and antenna domain. However, in PSII, the part that houses the electron transport chain (RC) and core antenna consists of two different proteins; the PSII is encoded by at least four genes. The five TMH

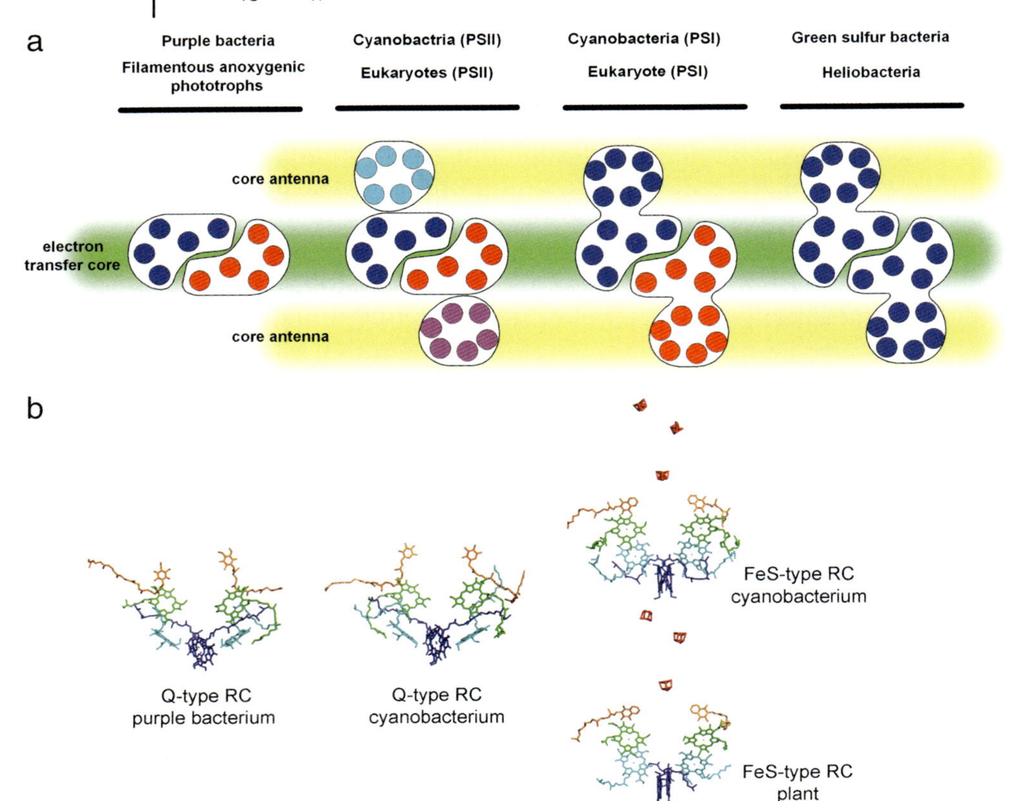

Figure 13.2 **(a)** Schematic diagram indicating the transmembrane helical composition of photosynthetic reaction centers. Purple bacteria and filamentous anoxygenic phototrophs possess a heterodimeric (L, M) quinone-type reaction center (see Chapter 12). The quinone-type reaction center (PSII) of cyanobacteria and photosynthetic eukaryotes consists of a heterodimeric electron transfer core (D1, D2) and two homologous subunits (CP47, CP43) that act as a core antenna (see Chapter 4). The iron sulfur-type reaction center (PSI) of cyanobacteria and higher plants is heterodimeric (see Chapters 2 and 3). Heliobacteria and green sulfur bacteria possess a homodimeric iron sulfur-type reaction center. The core antennae of heliobacteria, green sulfur bacteria, and PSI is homologous to the separately encoded core antenna of PSII of oxygenic eukaryotes and cyanobacteria. Transmembrane helixes indicated by circles. Transmembrane helixes encoded by separate genes are shown in different colors and are indicated by circles; **(b)** Electron transport cofactors involved in the charge separation and stabilization of Q-type and FeS-type RCs. The structures were obtained from solved crystal structures: Q-type purple bacterium RC (*Rhodobacter sphaeroides,* PDB 1aij), Q-type cyanobacterium RC (*Thermosynechococcus elongatus* PDB 1s5l), FeS-type cyanobacterial RC (*Thermosynechococcus elongatus*, PDB1jb0), FeS-type plant RC (*Pisum sativum*, PDB 2o01). The special set of chlorophylls that is involved in charge separation is colored blue (special pair) and cyan. The green colored tetrapyrroles are phaeophytins in Q-type RCs and chlorophylls (A_1) in FeS-type RCs indicated in green. Ubiquinone (purple bacterium), plastoquinones (Q-type RC of cyanobacteria) and phylloquinone (FeS-type RC of cyanobacteria and plant) are indicated in orange. The iron sulfur clusters of the FeS-type RCs are colored in red.

PSII RC containing the elements of the electron transfer are housed in by the PsbA and PsbD proteins; the six TMH core antenna are housed in the PsbB and PsbC proteins. There is little doubt that the electron transfer domains of heliobacteria, green sulfur bacteria, PSI, and PSII are homologous, that is, evolutionarily related structures. There is also little doubt that the core light-harvesting systems of PSII and the PSI of oxygenic bacteria and plastids, along with the core light-harvesting systems of the photosystems of GSB and heliobacteria, are homologous structures. However, the current functional distribution within the phylogenic tree cannot easily be explained by a single fusion or splitting event between the core antenna and the electron transfer core.

Charge separation and stabilization are the key functions of all RCs. FeS-type and Q-type RCs share a substantial similarity in the organization and sequence of cofactors that enable these functions (Figures 13.1 and 13.2). At the interface of each dimer is a special pair of two Chls or BChls that is energetically coupled to two or more Chls or BChls within each monomer. These pigments are able to accomplish charge separation when excited. The expelled electron is transferred to a series of electron carriers that accomplish charge stabilization. In FeS-type RCs, the special set of chlorophylls donates an electron to chlorophyll (A_0). From here, the electron reaches a quinone (A_1), followed by iron sulfur clusters (F_X, F_A, F_B). The chemical natures of the chlorophylls and quinones are summarized in Table 13.1 and further detail on their characterization is provided in the section on individual photosystems (see Sections 13.3, 13.4 and 13.5). In Q-type RCs a special pair of chlorophyll *a* molecules donates an electron to a pheophytin molecule. From pheophytin, the electron is transferred to the quinone Q_A, followed by a second quinone, Q_B. Q_B can leave the RC and exchange position with a member of a pool of quinones that are chemically identical to Q_B. In purple bacteria and FAPs, the quinone is either exclusively ubiquinone or menaquinone, or one ubiquinone and one menaquinone; in PSII, it is exclusively plastoquinone (see Chapters 4 and 12).

Charge separation and stabilization follow a common theme in both FeS-type and Q-type RCs. A set of four BChl/Chl molecules carries out initial charge separation, which is followed by electron transfer to another chlorophyll-like molecule, followed by electron transfer to a quinone (Figure 13.1). The greatest difference between the two types of reaction centers occurs after the first quinone acceptor. Q-type RCs appear to be missing the three FeS cluster present in Fe-type RCs; or, interpreted alternatively, the electron transport chain in FeS-type reaction centers is expanded by three FeS clusters. Arguments concerning which reaction center type is akin to the common ancestor continue, and will be evaluated in the Section 13.6 on the Ur-reaction center.

13.3 The Photosystem of Heliobacteria

In 1983, photosynthetic heliobacteria were discovered by Gest and Favinger [21–23]. These bacteria showed new features in pigmentation and life cycle that

Table 13.1 Subunit composition of FeS-type reaction centers in cyanobacteria, plants, green sulfur bacteria and heliobacteria. Homologous subunits are shown in the upper half of the table in the same row, subunits that have no homologous counterpart are shown in the lower part of the table, where position in the same row does not indicate homology.

	Cyanobacteria		Plants		Green sulfur bacteria		Heliobacteria	
	Subunit	**Cofactors**	**Subunit**	**Cofactors**	**Subunit**	**Cofactors**	**Subunit**	**Cofactors**
Homologous	PsaA	40 Chl*a*, PhylQ, *	PsaA	40 Chl*a*, PhylQ, *	PscA	8 Bchl*a*, 2 Chl *a*, MQ, *	PshA	~18 Bchl*a*, 2 Chl *a*, MenQ, *
	PsaB	39 Chl*a*, PhylQ, *	PsaB	39 Chl*a*, PhylQ, *				
	PsaC	2 FeS	PsaC	2 FeS	PscB[g)]	2 FeS	PshB[g)]	2 FeS
	PsaD		PsaD					
	PsaE		PsaE					
	PsaF		PsaF					
	PsaI[a)]		PsaI					
	PsaJ[a)]	3 Chl*a*	PsaJ	2 Chl*a*				
	PsaK[a)]	2 Chl*a*	PsaK	2 Chl*a*				
	PsaL	3 Chl*a*	PsaL	3 Chl*a*				
Non homologous	PsaM[b)]	1 Chl*a*	PsaG	1 Chl*a*	PscC	1 cyt *c*		
	PsaX	1 Chl*a*	PsaH	1 Chl*a*	PscD			
	PsaZ[c)]		PsaO[d)]		FMO	7 Bchl*a*		
			PsaP[e)]					
			PsaY[f)]					

PhylQ; Phylloquinone, MenQL: Menaquinone.
Chl: Chlorophyll a, Bchl Bacteriochlorophyll.
* FeS bound by A and subunit or A and A subunits.
a) absent in *Gloeobacter violoaceus*.
b) only in *T. elongatus* and *A. variabilis*.
c) only in Gloeobacter violoaceus.
d) only biochemical evidence.
e) possibly present in crystal structure.
f) only biochemical evidence.
g) PscB and PshB may not be homologous to PsaC.

separated them from all other known photosynthetic bacteria. Heliobacteria contain BChl g, form endospores [24], and can be classified as gram-positive. Heliobacteria are obligate anaerobic bacteria with a mesophilic photoheterotrophic growth pattern [21]. Features of the heliobacterial photosystem have been recently reviewed [25].

13.3.1
Reaction Center Core

The photosystem of heliobacteria consists of a core of homodimeric proteins that, like PSI and the photosystems of GSB, contains 11 TMH per monomer [26]. From the homodimeric organization it is tempting to conclude that heliobacterial photosystems are closely related to the homodimeric photosystems of green sulfur bacteria (see Section 13.4). When respective sequences are compared, however, heliobacterial photosystems appear slightly more related to the photosystems of GSB (17.4% sequence identity) than to PsaA/PsaB of PSI (16.3% and 16.8% sequence similarity) found in oxygenic phototrophs [27]. However, the heliobactrial RC shows greater similarity to the PSI (PsaA/PsaB) of oxygenic phototrophs, and parts of the core antennae show remarkable similarities to the core antennae protein CP47 of PSII, which contains a Q-type RC. [27]. Interestingly other phylogenetic indicators also assign cyanobacteria as the closest photosynthetic relatives of heliobacteria [27]. This position – having the simplest of all photosynthetic RCs in composition, but being closely related to organisms with the most diversified photosynthetic apparatus – makes the evolution of the heliobactrial RC a much discussed topic. The homodimeric photosystem proteins, and their overall weak identity to either PsaA or PsaB, implies that the green sulfur bacterial photosystem split before the gene duplication that lead to PsaA/ PsaB of PSI. Is the heliobactrial photosystem a living fossil that closely resembles the Ur-RC? One sign that this may be the case is that heliobacteria did not acquire an external light-harvesting system, making heliobacteria the only photosynthetic group of organisms that has not developed pigmented proteins to extend the absorption cross section of the RC. Heliobacteria seem to prefer heterotrophic conditions and are not known to be able to perform inorganic carbon fixation.

13.3.2 External Subunit

The only identified subunit associated with the RC (PshA) is the F_A and F_X iron sulfur cluster-containing PshB subunit, which shows significant sequence identity with the PscB subunit of GSB (further discussed in Section 13.7). The heliobacterial photosystem is, thus, the simplest photosystem/RC known in terms of overall protein composition.

13.3.3 Pigments

The photosystem of *Heliobacterium chlorum* contains ~35 Bchl g (or derived BChl g') per reaction center [28]. With 35 porphyrins, the heliobacterial photosystem has more porphyrins than the GSB core proteins (~8 BChl *a* and 2 Chl *a* per core), but also fewer than PSI (80 Chl *a* per core). An interesting feature of BCh g is that it can be easily converted into Chl *a* [29]. It has therefore been suggested that BChl g may have been a precursor that led to the formation of Chl *a*. This makes BChl g a potential candidate for the pigment that may have been used by the Ur-RC. To further this idea, the still obscure steps that lead to the biogenesis of BChl need to be established in this organism. The genome sequence of *Heliobacterium modesticaldum* has recently been completed (http://genomes.tgen.org).

13.3.4
Electron Donor

The oxidized heliobacterial photosystem is replenished with electrons from a membrane-bound cytochrome *c* [30]. The protein sequence of this cytochrome (PetJ) indicates that the membrane-association is not accomplished by a hydrophophic protein membrane anchor. Membrane association is accomplished by attachment of fatty acids to the N-terminal domain of the cytochrome [31].

13.3.5
Charge Separation

Heliobacteria are the only organisms in which the special set of porphyrins that accomplishes charge separation is BChl *g* [32]. It has been proposed that the special pair P798 pigments are actually its 13^2 (BChl g') epimer [28].

13.3.6
Charge Stabilization

The first electron acceptor after charge separation (A_0) has been identified as a BChl *g* derivative. A_0 is a 81-OH Chl *a* [33, 34], similar to the A_0 in cyanobacterial PSI and green sulfur bacterial photosystems that are both Chl *a*. The nature of the next electron acceptor remains somewhat controversial. Heliobacteria contain, on average, 1.5 menaquinone molecules per reaction center [35], similar to PSI containing two phylloquinones with a clear role as electron acceptors of A_0. However, when the quinones are extracted from heliobacteria, no change in electron transfer kinetics from A_0 is observed [36]. Furthermore, an investigation that tried to detect spectroscopic evidence for the formation of semiquinones showed little, if any, photoinduced semiquinone production [37]. While the electron acceptor role of the quinone remains unclear, a functional role of the first iron sulfur cluster (F_X) has recently been demonstrated [38]. Heliobacteria also possess two additional iron sulfur cluster in a protein separate from the homodimeric core. F_A and F_B housed in PshB have clearly been identified as accepting electrons after RC photoxidation [39–41].

13.4
The Photosystem of Green Sulfur Bacteria

The second group of photosynthetic organisms that possesses a FeS-type photosystem is green sulfur bacteria (GSB). These organisms comprise a phylogentic group of photosynthetic gram-negative bacteria [42]. In 1887, Winogradsky [43] suggested naming the group of organisms that can oxidize hydrogen sulfide into sulfur “Schwefelbacterien” (Engl. sulfur bacteria). The distinct pigments found in green sulfur bacteria pointed for a long time at a distinct photosynthetic apparatus; but it was only when biochemical and electron microscopy studies linked the pigments with the membranous vesicles that the true uniqueness of green sulfur bacterial photosynthesis became apparent [44]. Chlorosomes are the dominating light-harvesting systems in GSB and contain hundreds of thousands of BChl *c*,d and/or e molecules. Chlorosomes are enclosed by a unilayer membrane facing the cytoplasm and an array of BChl *a* containing proteins, which constitute the baseplate. The baseplate energetically couples the chlorosomes to the reaction center via the Fenna-Matthews-Olson (FMO)-protein. The FMO-proteins form trimers that are very tightly linked to the FeS-type RC. The RC of green sulfur bacteria has been reviewed in detail [45].

13.4.1 Photosystem Center Core

The photosystem of GSB is composed of 2 homodimers (PscA) with 11 TMH each [46]. Phylogenetic analysis indicates that green sulfur bacteria diverged before the heliobacteria and cyanobacteria lineages separated. The occurrence of homodimers in heliobacteria and GSB can, therefore, be seen as a confirmation that a photosynthetic apparatus of heliobacteria, or green sulfur bacteria, is representative of the most underived RC that is known.

13.4.2 External Subunits

As mentioned before, the FMO-protein is intimately associated with the green sulfur bacterial photosystem and often copurifies with the photosystems in biochemical preparations. Single particle analysis of isolated photosystems shows one or two FMO trimers associated with each photosystem [47], confirming previous biochemical observations. Given the homodimeric character of the photosystem the FMO trimers should be as likely to interface with each PsaA, so a FMO-photosystem complex is likely to be composed of two PscA and two FMO trimers [45]. Each FMO contains seven BChl *a* [48, 49]. FMO proteins are unique and do not have an easily identifiable evolutionary precursor. Only one evolutionary scenario has been proposed; in it the FMO-protein is derived from the RC itself [50]. If this idea proves true, it suggests something very interesting–that great evolutionary pressure was exerted to make the RCs connect to the chlorosomes. An initially mutated and misfolded photosystem monomer was probably a very poor energetic couple to the chlorosome, and it should have taken a long time of refinement to fit, as an analogy, this square peg into this round hole. This evolutionary pressure may have been provided by lateral gene transfer of chlorosome functionality [51, 52] and the great benefit of coupling the RC to this enormous light-harvesting complex.

Two iron sulfur clusters, F_A and F_B, are housed within PscB [53]. This configuration is similar to the arrangement in the photosystem with FeS-type RC (PSI) of cyanobacteria and chloroplasts. However, both the green sulfur bacterial PscB and the cyanobacterial PsaC appear not to be closely related, and may have been acquired independently from each other.

Two cytochrome *c* subunits encoded by pscC are associated with each RC of *Chlorobium tepidum*, the most studied green sulfur bacterium. A striking feature of these cytochromes is the presence of three TMHs. These TMH not only anchor PscC to the membrane, but also anchor it to the RC, a link that is not easily disrupted during RC isolation [54]. In *Chlorobium limicola*, however, PscC appears less tightly associated with the RC, and photosynthetically competent RCs can be obtained that do not contain PscC [55].

There are also two subunits of PscD per RC. There is some evidence that PscD has a function on the electron acceptor side and is implicated in stabilizing PscB [55], mediating ferredoxin interaction with the PscB [56], or enhancing energy transfer from the FMO protein to the RC [57]. The function of improving electron transport at the donor side combined with a low but significant sequence similarity to PsaD may indicate that PscD and PsaD have a common origin. Additionally, this could show that PscD also had a role in stabilizing RC-FMO interaction, thereby enhancing energy transfer from the chlorosome to the photosystem [57].

13.4.3 Pigments

Besides the chlorosome BChls (*c,d,e*), BChl *a* is the primary pigment found in the baseplate, FMO protein, RC, and core antenna domain. Each PsaA is associated with eight BChl, the lowest number of any FeS-type RC core [45]. One could speculate that having the chlorosome (a light-harvesting system with an extremely large cross-section), available, drove photosystem evolution in GSB to refine coupling to the chlorosome. Heliobacteria and cyanobacteria, in contrast, were driven to accumulate pigments within the RC core antennae. The biosynthesis of Bchl *a*, *c*, *d*, and *e* of green sulfur bacteria has been clarified by functional genomics [51].

13.4.4 Charge Separation

Charge separation in the GSB RC is accomplished by a set of BChl *a*. Of interest is the presence of a 13^2 epimer, analogous to Chl *a* 13^2 epimer found in cyanobacteria and the heliobacterial BChl g 13^2 epimer, located at the interface between the PsaA and constituting the special pair with an absorbtion maximum at 798 nm.

13.4.5 Charge Stabilization

The first electron acceptors have been identified as two Chl *a* molecules, in analogy to the FeS-type RCs contained in the photosystems of oxygenic phototrophs and heliobacetria. There are indications that the electron acceptor after A_0 may be a menaquinone [58]. However, even in reaction centers that lack menaquinone, electrons can reach the FeS cluster in the PscC subunit [59]. Similar to heliobacteria, there is presently no certainty of the role of menaquinone in the electron transport within the green sulfur bacterial RC. On this note, another interesting finding is that the menaquinones appear only loosely bound to the RC and are easily lost during purification [45]. This may indicate that the GSB RC has the flexibility of donating electrons to either quinone or F_X, retaining this feature from a flexible Ur-RC (see Section 13.6 on the Ur-RC section).

13.5 Photosystem I of Cyanobacteria and Plastids

The development of oxygenic photosynthesis within cyanobacteria is one of the main events in earth's history. Another event that changed the biosphere was the permanent incorporation of a cyanobacterial-like organism within a eukaryotic cell, a process called endosymbiosis [60–62]. This endosymbiotic event gave rise to all photosynthetic eukaryotes that contain a plastid – a cyanobacterial cell within a cell. The photosynthetic machineries of current cyanobacteria and photosynthetic eukaryotes have developed independently since this event; giving us the opportunity to compare these systems and deduce which system features may be more derived or consistent with the PSI present in the last common cyanobacterial ancestor [63].

The great similarity of cyanobacterial and plastid PSI core proteins (PsaA and PsaB) indicates that the common ancestor of all known modern cyanobacteria and plastids had two genes coding for a heterodimeric proto-PSI. A gene duplication may have been a necessary step to overcome the harmful effect of oxygen on a homodimeric proto-PSI. Having extra copies of a gene allows for more creativity, that is, more random tries, to solve an environmental challenge, such as oxygen probably posed. A single gene, coding for a homodimeric photosystem, may be one of the factors restricting GSB and heliobacteria to anaerobic conditions (see Sections 13.3.1 and 13.4.1). The heterodimeric PSI retained a stronger symmetry between the two RC core proteins, compared to Q-type RCs [64]; current functional evidence indicates that both branches of the electron transfer pathways are almost equally functional in plant PSI [65], and at least to some extent in cyanobacteria [66].

While significant diversification between cyanobacteria and eukaryotic phototrophs occurred at the level of light-harvesting systems and external subunits, the photosystem (RC and core antennae domain) remained virtually unchanged (Figure 13.3), testifying to the sophistication already achieved by ancient cyanobacteria. In the following section, some similarities and differences are pointed out between cyanobacterial and eukayotic PSI that may provide clues about the ancestral PSI.

13.5.1 Reaction Center Core

Genomic information indicates that the core structure of PSI of cyanobacteria and photosynthetic eukaryotes are remarkably similar [64]. Recent advances in crystallography allow a direct comparison of PSI of *(Thermo)Synechococcus elongatus* (cyanobacterium) [67] and *Pisum sativum* (eukaryote) [68, 69] structure at an atomic

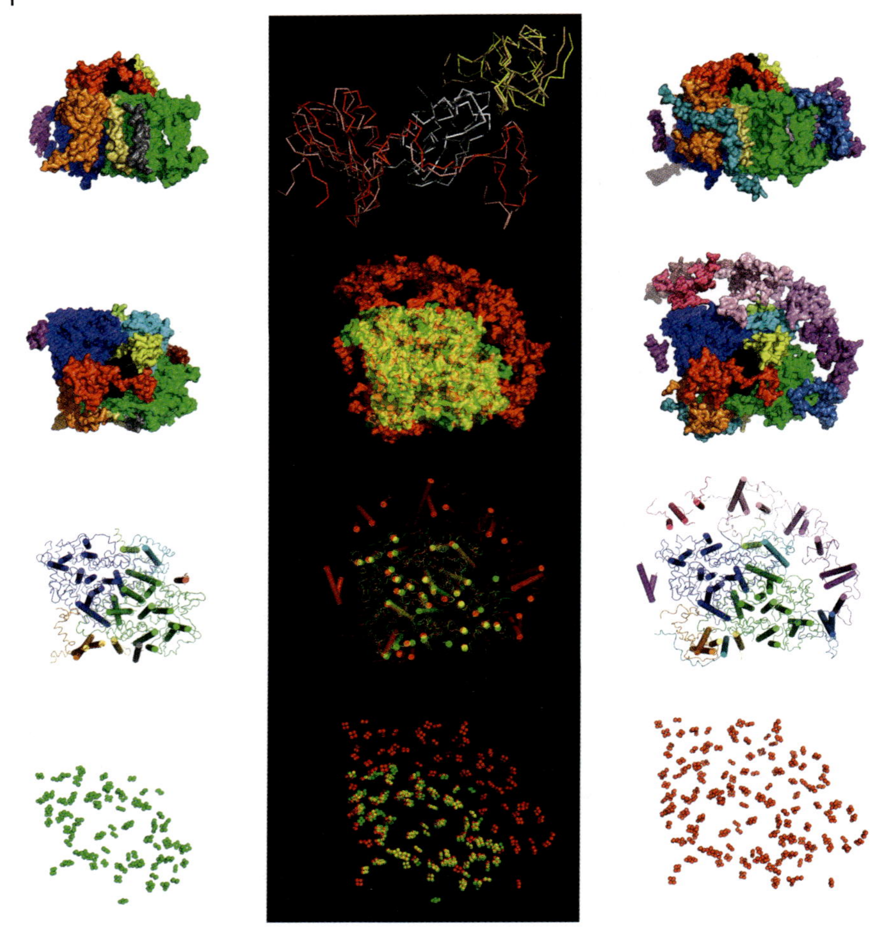

Figure 13.3 Structural comparison of PSI from cyanobacteria (*Thermosynechococcus elongatus*, PDB1jb0, left column) and plants (*Pisum sativum*, PDB 2o01, right column). First row, side view. Note that the superimposed side view (middle) shows the subunit C (white, gray), D (beige, yellow) and E (rose, red) of cyanobacteria and plant subunits respectively. Second row, top view (space filled model). Third row, top view (cartoon). Forth row, top view chlorophylls. Cyanobacterial structures are shown in green, plant structures are shown in red. Superimposition (middle column) of green (cyanobacteria) and red (plant) results in yellow.

level. The structural data reveals that the photosystem core proteins (PsaA, PsaB) and cofactors of cyanobacteria and plastids superimpose nearly perfectly (Figure 13.3).

13.5.2 External Subunits

In addition to PsaA and PsaB that encode the PSI heterodimer, PSI of cyanobacteria and photosynthetic eukaryotes interface with several additional subunits.

Subunits in common between cyanobacteria and photosynthetic eukaryotes PSI are either membrane-bound subunits (PsaF, PsaI, PsaJ, PsaK, PsaL) or stromal/ cytoplasmic subunits (PsaC, PsaD, and PsaE). The stromal/cytoplasmic subunits provide the electron acceptor side of PSI. The function of the membrane-bound subunits is to interact with additional light-harvesting systems and to participate in the formation of PSI trimers in cyanobacteria.

The analysis of 16S rRNA indicates that within the oxygenic clade, the cyanobacterium *Gloeobacter violaceus* diverges prior to other known cyanobacteria and plastids [70]. This special position is supported by its ultrastructural features (lack of thylakoid stacks) [71] and its genome analysis that shows many distinct features [72]. Interesting for PSI evolution is PsaZ, a unique subunit [73] and the lack of PsaI, PsaJ, PsaK, and PsaX subunits [72], compared to other cyanobacteria and plastids. Whether these subunits were lost by *Gloeobacter violaceus*, or newly developed by the common ancestor of cyanobacteria and plastids, remains an open question.

13.5.3 Light-Harvesting Systems

In contrast to cyanobacteria, plastids possess peripheral light-harvesting systems, which are related to PSII-associated light-harvesting systems (LHCb), and LHCa which is exclusively attached to PSI. The number of LHCa complexes per PSI differs in different organisms; expression is probably related to specific growth conditions. There is no specific PSI-associated peripheral light-harvesting system in most cyanobacteria, under optimal growth conditions. However, under some environmental conditions, such as limited availability of iron, a ring of pigmented proteins that shows homology to the CP 43, that is part of PSII, is formed [74, 75]. The proteins forming this ring are called IsiA (for iron-stress-induced protein A); 18 subunits of IsiA completely surround a trimeric PSI supercomplex (see Figure 13.4). The use of the IsiA antenna ring is not always limited to iron-starved conditions. It may be common amongst cyanobacteria, as demonstrated by *Prochlorococcus*, which uses a IsiA homologue, the so-called Pcb protein, in low light conditions [76].

What may be the reasons for the different light-harvesting systems in cyanobacteria and eukaryotes? The light-harvesting systems in photosynthetic eukaryotes are membrane-embedded and enable redistribution of light energy to achieve a balance between the generation of electrons by PSII and the donation of electrons to the carbon fixation by PSI and cyclic electron transport to generate an additional proton gradient for ATP synthesis (see Chapter 9. In this process, called "state transition," LHC-II light-harvesting complexes can exchange between PSI and PSII [77–79] (see Chapter 6). The reaction centers were modified to interface with these LHCs, leading to the monomeric PSI in eukaryotes (see Chapter 3). Redistribution of light energy also occurs in cyanobacteria by state transition. However, in cyanobacteria the membrane-external phycobilisomes can change association from PSII to PSI [80–82] (see Chapter 11). Excess excitation energy can be diverted in this way from PSII to PSI, which is able to quench excitation more efficiently than PSII, without becoming damaged.

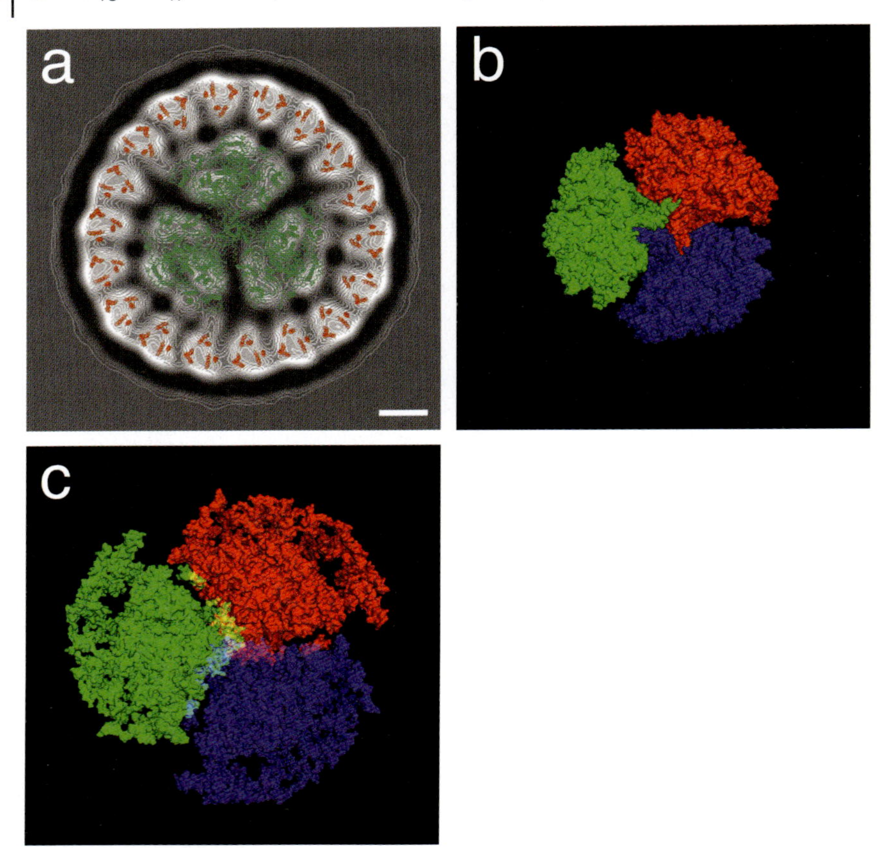

Figure 13.4 **(a)** electron microscopy data confirms the presence of a PSI trimer within the centre of an 18-member ring of CP43′. (Bibby et al., Science, 2001, with permission); **(b)** Space fill model of a cyanobacterial PSI trimer. Each PSI is shown in a different color); **(c)** Space fill model of three plant PSI subunits, which coordinates had been fitted to the the cyanobacterial PSI trimer coordinates. Each monomer is shown in a different color. Cyan, purple, yellow and white areas where the superimposed subunits interfere, making the formation of a PSI trimer impossible.

13.5.4
Interfacing with the Light-Harvesting System

The need of PSI for an extended antenna system is a main driving force for structural changes through the rearrangement, addition, and loss of subunits. This driving force was strong enough to induce the few substantial modifications in amino acids sequence that are observed between PSI of cyanobacteria and plastids. These modifications occur in the solvent-exposed loops of PsaA and PsaB. Modifications of PsaB provide the ability to bind additional chlorophylls that probably enable energy transfer from a proximal LHC-I complex in eukaryotes, whereas, in cyanobacteria, this loop interfaces with the unique cyanobacterial subunit PsaX. Another modification in eukaryotic PSI is that the PsaB loop allows interfacing with a LHC-II complex [68] (see Chapters 3 and 6).

Connecting to external light-harvesting complexes is the likely function of PsaF and PsaJ, PsaK in eukaryotic PSI. PsaF is associated (not coordinated) with several chlorophylls and carotenoids and may help to funnel excitation from external antenna systems (isiA and phycobilisomes in cyanobacteria and LHC in eukaryotes) to the PSI core. In photosynthetic eukaryotes, PsaF contains a 25 amino acid insertion shown to interface with plastocyanin [83, 84]. Connecting the reaction center core with external antenna systems is likely to be a function of PsaJ (two chlorophylls in eukaryotic, three in cyanobacterial PSI), PsaK (two chlorophylls), and the unique plant subunit PsaG.

Sharing of light-harvesting systems is also accomplished by the PSI trimer formation that is unique to cyanobacteria. PsaL forms the trimerization domain in PSI, stabilized by PsaI. In addition, PsaM possesses a single chlorophyll and may have a role in energy transfer between trimers. The presumed function of PsaL, which binds three chlorophylls, is to form the trimer and facilitate excitation energy transfer between the monomers. Cyanobacteria and plastids have executed different strategies for extended PSI light-harvesting system. In cyanobacteria the PSI trimer provide a flexible way of coupling PSI to isiA and Pcb antennae, while the strategy for increasing light capture in photosynthetic eukaryotes relies on optimized interactions with LHC complexes The subunit PsaH that interfaces with additional LHC in eukaryotic PSI is located exactly at the position where the three trimers intersect in cyanobacteria, and so PSI timer formation in photosynthetic eukaryotes is not possible. Another subunit that is unique to eukaryotic PSI is PsaG, which may also have a function in interfacing PSI with LHC. PsaG probably arose through duplication of PsaK [85].

PsaX is unique to cyanobacteria. The existence of PsaX was confirmed from the crystal structure of cyanobacterial PSI [67]. PsaX binds a single molecule of chlorophyll and may mediate energy transfer from isiA. The evolutionary significance of PsaX is unclear.

13.5.5
Pigments

The PS-core formed by PsaA and PsaB of cyanobacteria and plant PSI contain 87 (cyanobacterial) and 86 (eukaryotic) chlorophylls. PsaA coordinates 40 and PsaB coordinates 39 chlorophylls. Six of the chlorophylls are involved in charge separation and stabilization. All chlorophylls are exclusively chlorophyll *a*, with the sole exception of the chlorophyll *d* containing cyanobacterium *Acaryochloris marina*, in which chlorophyll *d* substitutes for chlorophyll *a*. In all cases, one molecule of the epimer Chl *a*' (or Chl *d*'), in which the stereochemistry of the C-13^2 position is reversed, are found as part of the special pair of PSI. Subunits PsaF and PsaI do not have any coordinated chlorophylls. PsaJ coordinates two chlorophylls and PsaL three. The difference in chlorophylls bound by cyanobacterial and plastid PSI is due to PsaJ, which has two chlorophylls in cyanobacteria and three in eukaryotes. The PsaM and PsaX subunits are specific to cyanobacteria and bind one chlorophyll each. Subunits PsaG and PsaH are unique to eukaryotes and bind one chlorophyll.

The conservation of the chlorophyll binding sites of PSI of cyanobacteria and plastids is truly remarkable, indicating that the common ancestor to all PSI complexes already possessed mature machinery for funneling light energy to the RC core to accomplish charge separation. In addition to chlorophylls, there are 22 carotenoids [67–69] in both systems.

13.5.6 Electron Donors

The electron donors to PSI are soluble luminal proteins. In plants, plastocyanin is the only electron donor to PSI. In addition to plastocyanin, cyanobacteria and green algae also possess a cytochrome (c_{553}, a.k.a cyt c_6) [86] that can donate electrons to PSI [87]. Plastocyanin is the dominant electron carrier under normal nutritional conditions. However, the copper-containing plastocyanin is replaced by cytochrome under copper-limiting conditions [88, 89] (see Chapter 8).

Plastocyanins show significant, yet distant, homology to other copper-containing mobile redox carriers found in FAP (auracyanin) [90] and *Pseudomonas* (azurin) with beta barrel structure [91]. The presence of plastocyanin-related proteins in distantly related organisms may be indicative of a lateral gene transfer between photosynthetic organisms [92]. The use of plastocyanin may have allowed a separation and diversification of photosynthetic and respiratory electron transport, which share protein complexes in cyanobacteria. A diversification into different electron transport pathways is also postulated within the "redox switch hypothesis" [93] and could, in part, have been accomplished by different electron carriers to the RCs (see Section 13.7). There is a significant difference in the rate of electron donation from plastocyanin to PSI in cyanobacteria and plastids. The electron donation from plastid plastocyanin to PSI is twofold faster than in cyanobacteria [84]. This increased efficiency may be due an 18 amino acid extension of the eukaryotic PsaF protein, which is part of the plastocyanin binding site. The protein extension may constitute one of the few improvements in function that have been made in plastids compared to the proto-cyanobacterial PSI.

13.5.7 Electron Acceptors

The physiological function of Photosystem I (see Chapter 2) is to generate reduced electron carriers by oxidizing reduced electron carriers of the photosynthetic electron transport chain. The soluble electron carriers reduced by PSI (ferredoxin or flavodoxin) (see Chapter 15) participate directly as electron donors in many reactions, such as the reduction of nitrogen and sulfur, or the evolution of hydrogen. A substantial amount of photogenerated reduced electron carriers is oxidized to generate NADPH, which acts as the reducing agent in the conversion of inorganic carbon into sugars.

The three 4Fe4S iron sulfur clusters of PSI are coordinated by three proteins. F_X is coordinated by four cysteine residues (two from PsaA and two from PsaB). F_A and F_B are both coordinated within PsaC, a PSI subunit that protrudes into the cytoplasm. There are spectroscopic indications that GSB and heliobacteria also contain an F_X cluster and two FeS clusters within a single protein on the cytoplasmic interface, in analogy to PSI. The high homology of the PsaC with bacterial-type ferredoxins was suggested as an indication that the PsaC constitutes a permanently bound ferredoxin that originally was a mobile electron acceptor of a proto-PSI. Interestingly, the protein that binds the FeS clusters of heliobacteria is only loosely bound and is easily lost during photosystem preparations [41, 94]. The protein that binds the two FeS clusters in GSB (PscB) only shows a distant homology to PSI PsaC [95]. It appears, therefore, that the association with secondary FeS clusters may have been accomplished independently in heliobacteria, GSB, and ancient cyanobacteria.

PSI shows flexibility at the electron acceptor side, which can be triggered by environmental stress. PSI of iron-stressed cells form trimers that are surrounded by a ring of light-harvesting systems composed of isiA proteins [74, 75]. Interestingly, the structure of this supercomplex obtained by electron microscopy does not show the electron densities expected for PsaC. Iron-stress-induced (isi) proteins are also induced in other stress conditions, including high light and oxidative stress [96], and form, in most species, an operon that also includes a flavodoxin (isiB). The depletion of PsaC and coexpression of flavodoxin could indicate that PSI may donate electrons from F_X directly to flavodoxin in iron-limited conditions. The iron usually found in PsaC FeS-clusters may then be used by crucial enzymes where iron cannot be functionally replaced as easily. The PsaE subunit of PSI has been demonstrated to interact with flavodoxin [97, 98]. Depending on whether the PsaC iron sulfur cluster still participates in delivering electrons to flavodoxin, a shortcut, or an alternate, electron transport pathway in PSI exists. A different PSI-associated extension of the electron transport exists in eukaryotic photosynthetic organisms. In these organisms, the ferredoxin NADPH oxidoreductase forms a complex with PsaE [99]. While not demonstrated biochemically, there are also kinetic indications that the ferredoxin NADPH oxidoreductase may be coupled to PSI in cyanobacteria [100].

13.5.8
Charge Stabilization

Although PSI is a heterodimer, there is remarkable symmetry between PsaA and PsaB that also includes the location of redox active cofactors engaged in charge stabilization. Given the homodimeric photosystems with FeS-type RCs in GSB and heliobacteria, a question that immediately arises is: do electrons use both branches of cofactors located in the PsaA and PsaB to reduce F_X, or do they use only one branch? There are strong indications that the answer is more nuanced. In plastid PSI, both branches are active, albeit to a different extent [101–103], with the A branch slightly favored (55–65%). A significant decrease in the B branch, however, is reported from cyanobacterial PSI [66, 104]. Did the ancestral PSI of current cyanobacteria and plastids use both branches as well, or was there already a preferred branch as in current cyanobacteria? An interesting clue to answer this question may be found in *Gloeobacter violaceus*, an early branching cyanobacteria [70]. Unfortunately, no detailed analysis of the use of the A and B branches in *Gloeobacter violaeceus* is available at this time, but biochemical analysis indicates that the quinones in PSI of *Gloeobacter violaceus* are not phylloquinones found in other oxygenic phototrophs, but menaquinone-4 [105]. Because many proteins required for the biosynthesis pathway of phylloquinone are homologous to proteins required for menaquinone synthesis in prokaryotes, the presence of menaquinone in *Gloeobacter violaeceus* may indicate that menaquinone is the quinone used as the A_1 electron acceptors in proto-PSI, the same quinone possibly functioning as A_1 in homodimeric RCs of heliobacteria and green sulfur bacteria, in which both branches deliver electrons to F_X.

13.6
The Ur-Reaction Center

What was the nature of the first photosynthetic reaction centers? How similar were they to modern RCs? Were they integral membrane proteins similar to the ones we study in organisms that are alive today? Unfortunately, there is almost nothing that can be concluded for certain about these earliest RCs, which almost certainly were found in organisms that are long extinct. However, some informed speculations about these first RCs [18, 106] can be made, but the uncertainties far outweigh any firm conclusions that can be drawn.

While information on the very earliest reaction centers may be lost forever, some of the story can be picked up and some reasonably reliable conclusions can be drawn by doing a comparative analysis of the structures and gene sequences of all known reaction centers, and by looking for conserved elements. Assuming that these complexes are products of divergent, and not convergent, evolution it has been concluded that these conserved elements are derived from an ancestor of all known reaction centers. This ancestral complex may have been very different from the earliest RCs, presumably much more sophisticated and more structurally complex. This ancestral complex, having a clear relationship to the modern RCs, will be called the Ur-reaction center. The properties of this Ur-RC are discussed in this section.

Figure 13.1 summarizes the general structural features of known modern RCs, especially with respect to the protein subunits. The Ur-RC was a dimeric integral membrane protein. It almost certainly was a homodimer, such as is found in the modern heliobacteria and GSB. Three independent gene duplication and divergence events have given rise to the heterdimeric RCs found in most modern RCs; the simplest conclusion is that the form of the Ur-RC was a homodimer. Whether an even earlier monomeric RC ever existed is uncertain, as the dimeric nature of all known modern RCs is essential to their function. Perhaps the RC only became a functional electron transfer complex when it dimerized, having some other function when it was a monomer. This would represent a case of gene recruitment, in that the more ancient protein, whose gene was recruited to become the reaction center, had a different function. Currently, there are no good candidates for what might have been this precursor. Meyer [107] and Xiong and Bauer [108] have suggested that it might have been the cytochrome *b* subunit of the cytochrome bc_1 complex (see Chapter 7), but detailed structural comparisons do not reveal any significant similarity in structure, so this now seems unlikely [20]. The core structure of the RC has been remarkably well conserved during the course of evolution from the Ur-RC to the present day. This is shown in Figure 13.5, in which the known structures of ten protein chains from five RC complexes are superimposed. However, it is important to keep in mind that there are no structures for the RCs of three of the groups of photosynthetic organisms, the GSB, the FAP, and the heliobacteria, which are all among the simplest RCs in terms of protein composition and, therefore, possibly closer to the Ur-RC. There is still a long way to go to accumulate a robust dataset from which to draw firm conclusions.

The Ur-RC probably had a bidirectional electron transfer chain, using chlorophyll-type pigments, although likely not the exact pigments that we know today. These pigments are clearly the product of a long evolutionary process that has optimized them for light absorption and efficient electron transfer properties. There is every reason to expect that the impressive quantum efficiencies of close to one for modern RCs developed over a long time, and that earlier complexes were significantly less efficient. As Erasmus said, "In the land of the blind, the one-eyed man is king," so even an inefficient light-storage system is better than none at all. Evolution subsequently relentlessly improved the quantum efficiency so that there remains little more room for improvement. The same cannot be said for the energy efficiency of photochemistry, which is significantly less – in most cases, it is only a few tenths of percent. While the dramatic difference in the evolutionary optimization of quantum efficiency, but not energy efficiency, is not well understood, it probably at least in part represents some fundamental limits imposed by the second law of thermodynamics.

Could the similar structures of the reaction centers shown in Figure 13.5 have arisen by convergent evolution? Convergent evolution occurs when organisms independently evolve similar traits or structures, rather than descending from a common ancestor, called divergent evolution. Many clear cases of convergent evolution are known, for example, wing development in insects and birds. However, when they can be seen at an increasing resolution, especially down to the molecular level, their independent origin usually becomes readily apparent. Relatively few cases of convergent evolution at the molecular level are known and these are readily recognizable. The classic example is the apparently independent evolution of the active site catalytic triad of the two broad classes of serine protease enzymes found in bacteria and eukaryotes [109, 110]. Here, the active site geometries of the critical triad of amino acids that form the active site are similar, but the protein folds that position the active site residues are entirely distinct. The proteins have no apparent structural or sequence similarity, with the active site residues coming from different parts of the primary sequence in the two different classes of proteases. In this case, it appears that the chemistry of the reaction that the enzymes

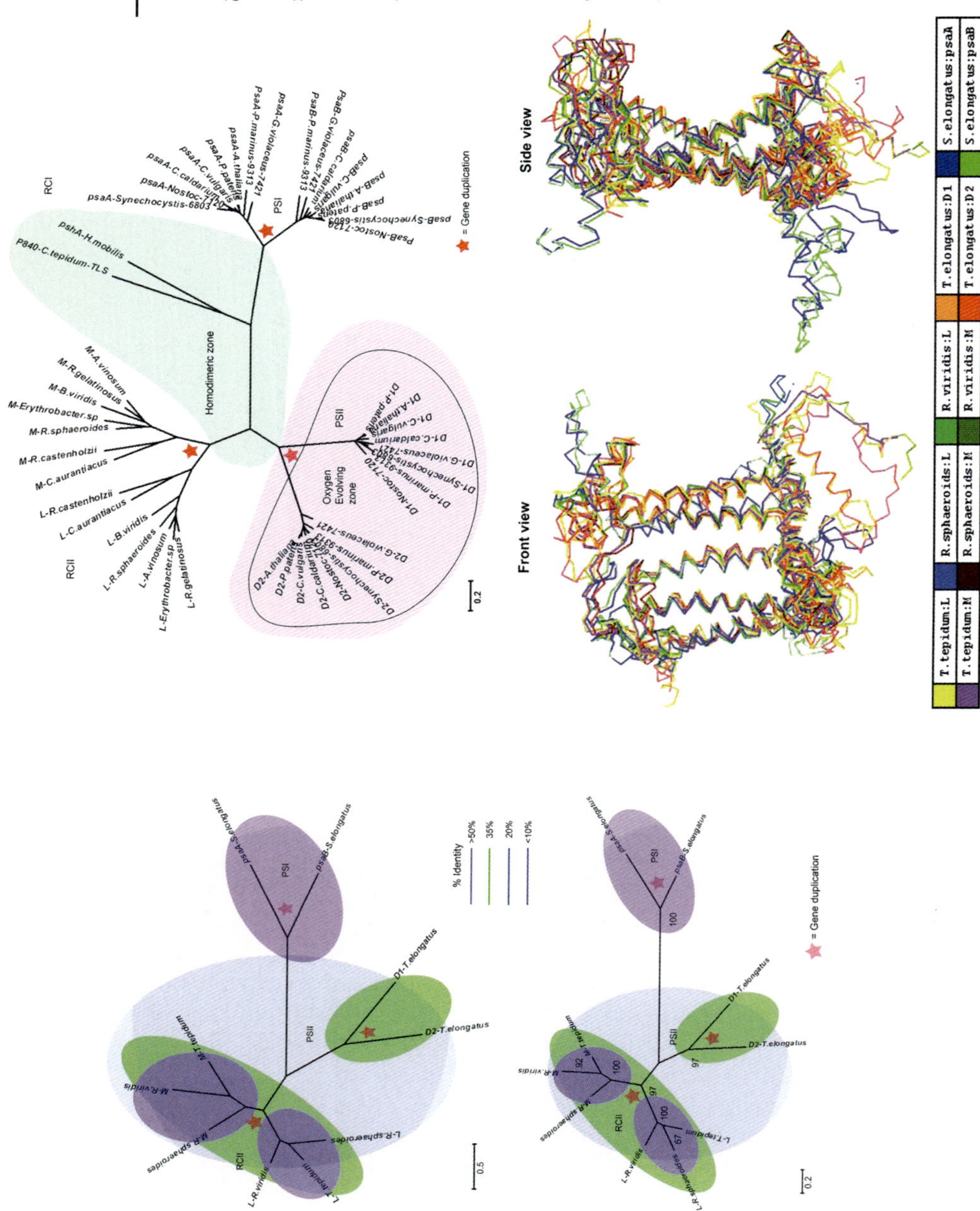

= Gene duplication
PSI
RCI
Homodimeric zone
pshA-H.mobilis
P840-C.tepidum-TLS
PSII
Oxygen Evolving zone
RCII
Side view
Front view
T.tepidum:L
T.tepidum:M
R.sphaeroids:L
R.sphaeroids:M
R.viridis:L
R.viridis:M
T.elongatus:D1
T.elongatus:D2
S.elongatus:psaA
S.elongatus:psaB
% Identity
>50%
35%
20%
<10%
D1-T.elongatus
D2-T.elongatus

Figure 13.5 **(Left)** Unrooted phylogenetic tree constructed using RMSDs derived from structural alignments directly as proxies for evolutionary distances. (*B*) Unrooted Neighbor-Joining phylogenetic tree of photosynthetic reaction centers based on a sequence alignment derived from the structural alignments shown at the bottom right. The red stars represent inferred gene duplication events. The colored boundaries enclose proteins sharing a particular percentage of similarity. **(Bottom right)** Structural alignments of all photosynthetic reaction center proteins: α-proteobacteria: *Rhodobacter sphaeroides* (1AIJ), L, M chains; *Rhodopseudomonas viridis* (1DXR), L, M chains; *Thermochromatium tepidum* (1EYS), L, M chains; Cyanobacteria: *Thermosynechococcus elongatus* (1S5L), D1, D2 chains of photosystem II; and *Synechococcus elongatus* (1JB0), A1, A2 chains of photosystem I. The 6 N-terminal helices that constitute the antenna domain of the photosystem I complex are not shown but were included in the data set used for alignment. The unaligned thread-like portions on the top and the bottom are the loops outside the membranes, joining the transmembrane helices. The left figure shows a front view of the 10 overlaid structures, whereas the right figure shows the side view of the same complexes rotated by 90°. **(Top right)** Unrooted Neighbor-Joining phylogenetic tree based on an extensive set of sequences of photosynthetic reaction centers. Only the C-terminal electron transfer domains of the Type I reaction centers were used in the analysis. The red stars represent inferred gene duplication events. The blue colored region represents sequence space of reaction centers either known or inferred to have a homodimeric core protein structure, whereas all others have a heterodimeric structure. The red colored region represents sequence space of those reaction centers that evolve oxygen. The dashed line indicates a development of oxygen evolution capability that is well after the gene duplication event that led to a heterodimeric reaction center, whereas the solid line indicates an earlier development of oxygen evolution capability. Figures from Sadekar *et al.*, Mol. Biol. Evol, 2006 [20], with permission.

carry out has dictated a narrow range of potential structures that will do the job; two different evolutionary paths have led to the same cofactor arrangement, but most other aspects of the proteins are entirely different.

In sharp contrast, the RC structures have all the hallmarks of purely divergent evolution – they have not only the same general structure at the level of the protein fold and cofactor placement, but they also have conservation of the order of cofactor binding residues in the primary sequence. One feature that does, however, seem to be describable as convergent is the gene duplication events that produced heterodimeric RCs from homodimeric ones. All evidence suggests that this took place on at least two, and, more likely, three, separate occasions. The gene duplication that gave rise to the heterodimer in PSI is clearly the most recent of these events, as both the structure of the heterodimer PSI subunits and the primary sequences of the core PSI proteins are very similar. Also, as described above, current evidence suggests that electron transfer does go down both sides of PSI, so even its function is closer to the homodimeric case than the type-II RCs, which appear to have a very strong preference for one electron transfer pathway. Presently, no sound understanding exists of the evolutionary pressures that gave rise to heterodimeric reaction centers, especially in PSI where electrons from the two pathways converge at the F_X center.

An essential aspect of the Ur-RC that has been much discussed is whether it had 5 or 11 TMH in each half of the protein core. Modern RCs and photosystems with both arrangements are known, and it is not clear which model is the ancestral and which is the derived trait. The last 5 TMH of the 11 helix type-I photosystems are clearly functionally and structurally homologous with the 5 helix type-II RCs. The six N-terminal TMH of the type-I photosystems are also similar to the core antenna proteins in PSII, but nothing similar is found in any of the anaerobic type-II RCs. There are basically two different schools of thought on this question, which can be summarized into the representations of 11 − 6 = 5 or 5 + 6 = 11. The 11 − 6 = 5 scenario imagines that the Ur-RC had 11 TMH with the first 6 functioning as an antenna and the last 5 as the electron transfer domain, much as in modern PSI. A putative gene fission event separated the two domains into separate proteins. The antenna domain was retained as the CP43 and CP47 core antenna proteins in PSII, but was lost in the anoxygenic RCs, leaving a different type of core antenna to take its place. The 5 + 6 = 11 scenario imagines that the Ur-RC had 5 TMH and a gene fusion of the RC domain with the separately evolved 6 helix antenna domain gave rise to the 11 TMH complexes found in type-I photosystems, while the type-II RCs are closer to the ancestral state. Both scenarios have been extensively discussed by numerous authors [111–119].

There is one significant argument in favor of the 11–6 = 5 scenario. In the PSI and PSII, the antenna domains are remarkably similar to each other, both in terms of overall structure and positions of some of the pigments. Evolutionary analysis of the antenna domains of PSI, PSII, heliobacteria, and GSB show trees with similar topology to those derived from the core electron transfer domains of the RCs [20, 118]. This suggests that the antenna domains have followed a similar evolutionary trajectory as the electron transfer domain. This, in turn, seems most consistent with the scenario that has these 2 domains developing together in an 11 TMH complex. This further suggests that the 11 TMH complex is the more ancestral version. However, it is important not to take this argument too seriously, as the sequence identity in these comparisons is extremely low and could be confounded by tree construction artifacts such as long-branch attraction.

If the Ur-RC was already a photosystem and had 11 TMH, as many authors have suggested and the above argument supports, then a difficult question arises regarding whether or not that complex was, itself, derived from a fusion of an even more ancient antenna complex with an electron transfer complex. This is well beyond the scope of current knowledge, and may well be lost forever in the long evolutionary development of photosynthesis from the earliest systems to the Ur-RC.

Do all the antenna domains of the type-I photosystems have similar antenna pigment arrangements? How do these relate to the antenna pigments in PSII? PSI has, by far, the most pigments of any of the RCs, ~100 vs 14 for the GSB RC, ~35 in heliobacteria, and ~36 in PSII. In addition, PSI and PSII have 26 Chl molecules in very similar positions in the antenna domain [120], although PSI has a significantly larger number of pigments, especially in the region near to the electron transfer domain. Fyfe et al. [121] have carried out modeling of all the type-I photosystems based on the structure of PSI. Sequence comparisons and measurements of the pigment contents of these photosystems led these authors to propose that 14 BChl in the GSB RCs, and somewhat more in the heliobacterial complex, bind to conserved ligands, principally histidines, in the N-terminal domains. So, there seems to be a core structure of the antenna domains with a modest number of pigments, possibly in similar positions that are common to all type-I photosystems and PSII. However, until such time as detailed structures of the 11 helix photosystems from GSB and heliobacteria are available to compare to those from PSI and PSII, all of these comparisons will remain uncertain and should not be taken too seriously.

13.7 Conclusions

Life found a way to capture and use sunlight, the most ubiquitous, inexhaustible, and reliable source of energy available on a young Earth, and has held on to it ever since. The RCs found in contemporary organisms are the crowning achievement of this evolutionary development that commenced billions of years ago.

Photosystems with FeS-type RC have been identified in three distinct groups of organisms: cyanobacteria and their progeny, the plastids, as well as in heliobacteria, and GSB. All these photosystems centers can trace their ancestry to a common photosystem. This proto-photosystem featuring a FeS-type RC probably shared many characteristics with the photosystem of current heliobacteria and green sulfur bacteria, while the photosystem with FeS-type RC of cyanobacteria and plastids appears most derived from the original design. The proto-photosystem with Fe-type RC probably featured an 11 TMH homodimeric core. The ancestral photosystem probably received electrons from a cytochrome. Cytochromes are the only identified electron donors to the photosystems of heliobacteria and GSB. Cyanobacteria; the plastids of some algae have retained the ability to use cytochromes, while the standard electron donor to the more derived photosynthetic eukaryotes is plastocyanin.

Charge separation in all FeS-type RCs is accomplished by a special set of chlorophyll-type pigments, one of which posses a 13^2 epimeric structure. The first electron acceptor (A_0) is a Chl *a*-like molecule. The proto-FeS RC may have had the flexibility of donating electrons directly to an iron sulfur cluster (F_X) or to a quinone (A_1), or to the iron sulfur cluster via a quinone. This flexibility may still be present in GSB and heliobactria but was probably lost in PSI of oxygenic phototrophs. In all three groups of organisms, an originally intermittently bound ferredoxin became permanently associated with the photosystem. This association, however, may have taken place on more than one occasion, as indicated by the small sequence identity between PscB of GSB, PshB of heliobactria, and PscC of cyanobacteria.

The need for a large absorption cross-section drove pigmentation and the development/acquisition of a light-harvesting system in green sulfur bacteria and cyanobactria. In cyanobacteria, an efficient light-harvesting system was achieved by packing the core antennae with pigments and by the formation of PSI trimers. This core harvesting system was extended by a membrane-intrinsic IsiA ring. Plastids developed a unique membrane-bound antenna system (LHCs) that interfaces with a monomeric PSI. GSB developed the FMO-protein to interface with enormous amounts of pigments within chlorosomes, while no great pressure existed to increase the core antennae. The apparent lack of evolutionary pressure in heliobacteria to develop a substantial light-harvesting system within the core antennae, or external light-harvesting structures, remains as enigmatic as the identity of the original pigment used by a proto-photosystem with a FeS-type RC.

There is near certainty that all RCs found today can trace their roots to a single Ur-RC. Significant progress has been made in tracing the origin of RCs and in understanding their development. The reaction centers encountered today are the few that stood the test of time, adapted to specialized conditions shaped by, and shaping, the environment and the genetic capabilities of the organisms that house them. Understanding RC and photosystem evolution in a genetic and environmental context is, therefore, crucial. The availability of a recent plethora of genetic information gave the current understanding of RC and photosystem evolution a tremendous leap. The crystal structures of RCs and photosystems of more representatives of photosynthetic groups, and continued biochemical and biophysical characterizations, put this genomic data into a functional context. An increased understanding of the geological and geochemical environment and the physiological challenges and capabilities of past and current photosynthetic organisms may lead to a parsimonious model of RC and photosystem evolution.

Acknowledgments

Martin F. Hohmann-Marriott is supported by a postdoctoral fellowship from the National Research Council (USA). This work has been supported by a grant from the Exobiology program of the National Aeronautics and Space Administration (NASA) to REB.

References

1 Beatty, J.T., Overmann, J., Lince, M.T., Manske, A.K., Lang, A.S., Blankenship, R.E., van Dover, C.L., Martinson, T.A. and Plumley, F.G. (2005) An obligately photosynthetic bacterial anaerobe from a deep-sea hydrothermal vent. *Proceedings of the National Academy of Sciences of the United States of America*, **102**, 9306–10.

2 Repeta, D.J., Simpson, D.J., Jorgensen, B.B. and Jannasch, H.W. (1989) Evidence for anoxygenic photosynthesis from the distribution of bacteriochlorophylls in the Black Sea. *Nature*, **342**, 69–72.

3 Overmann, J., Cypionka, H. and Pfennig, N. (1992) An extremely low-light-adapted phototrophic sulfur bacterium from the Black Sea. *Limnology and Oceanography*, **37**, 150–5.

4 Stevenson, A.K., Kimble, L.K., Woese, C.R. and Madigan, M.T. (1997) Characterization of new phototrophic heliobacteria and their habitats. *Photosynthesis Research*, **53**, 1–12.

5 Kimble, L.K. and Madigan, M.T. (1992) Nitrogen fixation and nitrogen metabolism in heliobacteria. *Archives of Microbiology*, **158**, 155–61.

6 Haber, F. (1922) Über die Darstellung des Ammoniaks aus Stickstoff und Wasserstoff. *Die Naturwissenschaften*, **10**, 1041–49.

7 Smil, V. (2001) *Enriching the Earth: Fritz Haber, Carl Bosch, and the Transformation of World Food Production*, Vol. **338**, MIT Press.

8 Lumpkin, T.A. and Plucknett, D.L. (1980) Azolla: botany, physiology and use as a green manure. *Economic Botany*, **34**, 111–53.
9 Des Marais, D.J. (2000) Evolution. When did photosynthesis emerge on Earth? *Science*, **289**, 1703–5.
10 Farquhar, J., Bao, H. and Thiemens, M. (2000) Atmospheric influence of Earth's earliest sulfur cycle. *Science*, **289**, 756–8.
11 Nitschke, W. and Rutherford, A.W. (1991) Photosynthetic reaction centers: variations on a common structural theme? *Trends in Biochemical Sciences*, **16**, 241–5.
12 Blankenship, R.E. (1992) Origin and early evolution of photosynthesis. *Photosynthesis Research*, **33**, 91–111.
13 Pierson, B.K. and Olson, J.M. (1987) Evolution of photosynthesis in anoxygenic photosynthetic prokaryotes, in *Microbial Mats* (eds Y. Cohen and E. Rosenberg), American Society of Microbiologists, Washington, DC, pp. 402–27.
14 Bryant, D.A., Garcia Costas, A.M., Maresca, J.A., Gomez Maqueo Chew, A., Klatt, C.G., Bateson, M.M., Tallon, L.J., Hostetler, J., Nelson, W.C., Heidelberg, J.F. and Ward, D.M. (2007) "*Candidatus* Chloroacidobacterium thermophilum": an aerobic phototrophic acidobacterium. *Science*, **317**, 523–6.
15 Xiong, J., Fischer, W.M., Inoue, K., Nakahara, M. and Bauer, C.E. (2000) Molecular evidence for the early evolution of photosynthesis. *Science*, **289**, 1724–30.
16 Raymond, J., Zhaxybayeva, O., Gogarten, J.P., Gerdes, S.Y. and Blankenship, R.E. (2002) Whole-genome analysis of photosynthetic prokaryotes. *Science*, **298**, 1616–20.
17 Xiong, J. and Bauer, C.E. (2002) Complex evolution of photosynthesis. *Annual Review of Plant Biology*, **53**, 503–21.
18 Olson, J.M. and Blankenship, R.E. (2004) Thinking about the evolution of photosynthesis. *Photosynthesis Research*, **80**, 373–86.
19 Mulkidjanian, A.Y., Koonin, E.V., Makarova, K.S., Mekhedov, S.L., Sorokin, A., Wolf, Y.I., Dufresne, A., Partensky, F., Burd, H., Kaznadzey, D., Haselkorn, R. and Galperin, M.Y. (2006) The cyanobacterial genome core and the origin of photosynthesis. *Proceedings of the National Academy of Sciences of the United States of America*, **103**, 13126–31.
20 Sadekar, S., Raymond, J. and Blankenship, R.E. (2006) Conservation of distantly related membrane proteins: photosynthetic reaction centers share a common structural core. *Molecular Biology and Evolution*, **23**, 2001–7.
21 Gest, H. and Favinger, J.L. (1983) *Heliobacterium chlorum*, an anoxygenic brownish-green photosynthetic bacterium containing a "new" form of bacteriochlorophyll. *Archives of Microbiology*, **136**, 11–16.
22 Gest, H. (1994) Discovery of the heliobacteria. *Photosynthesis Research*, **41**, 17–21.
23 Gest, H. and Blankenship, R.E. (2004) Time line of discoveries: anoxygenic bacterial photosynthesis. *Photosynthesis Research*, **80**, 59–70.
24 Ormerod, J.G., Kimble, L.K., Nesbakken, T., Torgersen, Y.A., Woese, C.R. and Madigan, M.T. (1996) *Heliophilum fasciatum* gen. nov. sp. nov. and *Heliobacterium gestii* sp. nov.: endospore-forming heliobacteria from rice field soils. *Archives of Microbiology*, **165**, 226–34.
25 Oh-Oka, H. (2007) Type 1 reaction center of photosynthetic heliobacteria. *Photochemistry and Photobiology*, **83**, 177–86.
26 Liebl, U., Mockensturm-Wilson, M., Trost, J.T., Brune, D.C., Blankenship, R. E. and Vermaas, W. (1993) Single core polypeptide in the reaction center of the photosynthetic bacterium *Heliobacillus mobilis*: Structural implications and relations to other photosystems. *Proceedings of the National Academy of Sciences of the United States of America*, **90**, 7124–8.
27 Vermaas, W.F.J. (1994) Evolution of heliobacteria: implications for photosynthetic reaction center complexes. *Archives of Microbiology*, **41**, 285–94.
28 Kobayashi, M., van de Meent, E.J., Erkelens, C., Amesz, J., Ikegami, I. and Watanabe, T. (1991) Bacteriochlorophyll g epimer as a possible reaction center component of heliobacteria. *Biochimica et Biophysica Acta*, **1057**, 89–96.

29 Kobayashi, M., Hamano, T., Akiyama, M., Watanabe, T., Inoue, K., Oh-oka, H., Amesz, J., Yamamura, M. and Kise, H. (1998) Light-independent isomerization of bacteriochlorophyll g to chlorophyll a catalyzed by weak acid in vitro. *Analytica Chimica Acta*, **365**, 199–203.

30 Oh-oka, H., Iwaki, M. and Itoh, S. (2002) Electron donation from membrane-bound cytochrome c to the photosynthetic reaction center in whole cells and isolated membranes of *Heliobacterium gestii*. *Photosynthesis Research*, **71**, 137–47.

31 Albert, I., Rutherford, A.W., Grav, H., Kellermann, J. and Michel, H. (1998) The 18 kDa cytochrome *c*553 from *Heliobacterium gestii*: gene sequence and characterization of the mature protein. *Biochemistry*, **37**, 9001–8.

32 Prince, R.C., Gest, H. and Blankenship, R.E. (1985) Thermodynamic properties of the photochemical reaction center of *Heliobacterium chlorum*. *Biochimica et Biophysica Acta*, **810**, 377–84.

33 van de Meent, E.J., Kobayashi, M., Erkelens, C., van Veelen, P.A., Amesz, J. and Watanabe, T. (1991) Identification of 8^1-hydroxychlorophyll *a* as a functional reaction center pigment in heliobacteria. *Biochimica et Biophysica Acta*, **1058**, 356–62.

34 Mizoguchi, T., Oh-oka, H. and Tamiaki, H. (2005) Determination of stereochemistry of bacteriochlorophyll g_F and 8^1-hydroxychlorophyll a_F from *Heliobacterium modesticaldum*. *Photochemistry and Photobiology*, **81**, 666–73.

35 Trost, J.T. and Blankenship, R.E. (1989) Isolation of a photoactive photosynthetic reaction center-core antenna complex from *Heliobacillus mobilis*. *Biochemistry*, **28**, 9898–904.

36 Kleinherenbrink, F.A.M., Ikegami, I., Hiraishi, A., Otte, S.C.M. and Amesz, J. (1993) Extraction of menaquinone does not change electron transport considerably. *Biochimica et Biophysica Acta*, **1142**, 69–73.

37 Lin, S., Chiou, H.-C. and Blankenship, R.E. (1995) Secondary electron transfer processes in membranes of *Heliobacillus mobilis*. *Biochemistry*, **34**, 12761–7.

38 Miyamoto, R., Iwaki, M., Mino, H., Harada, J., Itoh, S. and Oh-oka, H. (2006) ESR signal of the iron–sulfur center F_X and its function in the homodimeric reaction center of *Heliobacterium modesticaldum*. *Biochemistry*, **45**, 6306–16.

39 Nitschke, W., Setif, P., Liebl, U., Feiler, U. and Rutherford, A.W. (1990) Reaction center photochemistry of *Heliobacterium chlorum*. *Biochemistry*, **29**, 11079–88.

40 Heinnickel, M., Agalarov, R., Svensen, N., Krebs, C. and Golbeck, J.H. (2006) Identification of F_X in the heliobacterial reaction center as a [4Fe–4S] cluster with an S = 3/2 ground spin state. *Biochemistry*, **45**, 6756–64.

41 Heinnickel, M., Shen, G. and Golbeck, J.H. (2007) Identification and characterization of PshB, the dicluster ferredoxin that harbors the terminal electron acceptors F_A and F_B in *Heliobacterium modesticaldum*. *Biochemistry*, **46**, 2530–6.

42 Overmann, J. and Tuschak, C. (1997) Phylogeny and molecular fingerprinting of green sulfur bacteria. *Archives of Microbiology*, **167**, 302–9.

43 Winogradsky, S. (1887) Über Schwefelbakterien. *Botanische Zeitung*, **45**, 489–507.

44 Cohen Bazire, G., Pfennig, N. and Kunisawa, R. (1964) The fine structure of green bacteria. *The Journal of Cell Biology*, **22**, 207–25.

45 Hauska, G., Schoedl, T., Remigy, H. and Tsiotis, G. (2001) The reaction center of green sulfur bacteria. *Biochimica et Biophysica Acta*, **1507**, 260–77.

46 Buttner, M., Xie, D.L., Nelson, H., Pinther, W., Hauska, G. and Nelson, N. (1992) Photosynthetic reaction center genes in green sulfur bacteria and in photosystem. *Proceedings of the National Academy of Sciences of the United States of America*, **89**, 8135–39.

47 Remigy, H.W., Stahlberg, H., Fotiadis, D., Muller, S.A., Wolpensinger, B., Engel, A., Hauska, G. and Tsiotis, G. (1999) The reaction center complex from the green sulfur bacterium *Chlorobium tepidum*: A structural analysis by scanning transmission electron microscopy. *Journal of Molecular Biology*, **290**, 851–8.

48 Fenna, R.E. and Matthews, B.W. (1975) Chlorophyll arrangement in a bacteriochlorophyll protein from *Chlorobium limicola*. *Nature*, **258**, 573–7.

49 Li, Y.F., Zhou, W., Blankenship, R.E. and Allen, J.P. (1997) Crystal structure of the bacteriochlorophyll *a* protein from Chlorobium tepidum. *Journal of Molecular Biology*, **271**, 456–71.

50 Olson, J.M. and Raymond, J. (2003) The FMO protein is related to PscA in the reaction center of green sulfur bacteria. *Photosynthesis Research*, **75**, 277–85.

51 Frigaard, N.U. and Bryant, D.A. (2004) Seeing green bacteria in a new light: genomics-enabled studies of the photosynthetic apparatus in green sulfur bacteria and filamentous anoxygenic phototrophic bacteria. *Archives of Microbiology*, **182**, 265–76.

52 Hohmann-Marriott, M.F. and Blankenship, R.E. (2007) Hypothesis on chlorosome biogenesis in green photosynthetic bacteria. *FEBS Letters*, **581**, 800–3.

53 Illinger, N., Xie, D.L., Hauska, G. and Nelson, N. (1993) Identification of the subunit carrying FeS-centers A and B in the P840-reaction center preparation of *Chlorobium limicola*. *Photosynthesis Research*, **38**, 111–14.

54 Kusumoto, N., Inoue, K., Nasu, H. and Sakurai, H. (1994) Preparation of a photoactive reaction center complex containing photo-reducible Fe-S centers and photooxidizable cytochrome *c* from the green sulfur bacterium *Chlorobium tepidum*. *Plant and Cell Physiology*, **35**, 17–25.

55 Hager-Braun, C. (1995) Stable photobleaching of P840 in *Chlorobium* reaction center preparations: Presence of the 42-kDa bacteriochlorophyll *a* protein and a 17-kda polypeptide. *Biochemistry*, **34**, 9617–24.

56 Seo, D., Tomioka, A., Kusumoto, N., Kamo, M., Enami, I. and Sakurai, H. (2001) Purification of ferredoxins and their reaction with purified reaction center complex from the green sulfur bacterium *Chlorobium tepidum*. *Biochimica et Biophysica Acta*, **1503**, 377–84.

57 Tsukatani, Y., Miyamoto, R., Itoh, S. and Oh-oka, H. (2004) Function of a PscD subunit in a homodimeric reaction center complex of the photosynthetic green sulfur bacterium *Chlorobium tepidum* studied by insertional gene inactivation: Regulation of energy transfer and ferredoxin-mediated NADP+ reduction on the cytoplasmic side. *The Journal of Biological Chemistry*, **279**, 51122–30.

58 Kjaer, B., Frigaard, N.-U., Yang, F., Zybailov, B., Miller, M., Golbeck, J.H. and Scheller, H.V. (1998) Menaquinone-7 in the reaction center complex of the green sulfur bacterium *Chlorobium vibrioforme* functions as the electron acceptor A1. *Biochemistry*, **37**, 3237–42.

59 Hager-Braun, C., Jarosch, U., Hauska, G., Nitschke, W. and Riedel, A. (1997) EPR studies of the terminal electron acceptors of the green sulfur bacterial reaction centre. Revisited. *Photosynthesis Research*, **51**, 127–36.

60 McFadden, G.I. (1999) Endosymbiosis and evolution of the plant cell. *Current Opinion in Plant Biology*, **2**, 513–19.

61 Mereschkowsky, C. (1905) Über Natur und Ursprung der Chromatophoren im Pflanzenreiche. *Biologisches Centralblatt*, **25**, 593–604.

62 Schimper, A.F.W. (1883) Über, Die Entwicklung der Chlorophyllkörner und Farbkörper. *Botanische Zeitung*, **41**, 105–14.

63 Larkum, A.W.D., Lockhart, P.J. and Howe, C.J. (2007) Shopping for plastids. *Trends in Plant Science*, **12**, 189–95.

64 Baymann, F., Brugna, M., Muhlenhoff, U. and Nitschke, W. (2001) Daddy, where did (PS)I come from? *Biochimica et Biophysica Acta*, **1507**, 291–310.

65 Ramesh, V.M., Gibasiewicz, K., Lin, S., Bingham, S.E. and Webber, A.N. (2004) Bidirectional electron transfer in Photosystem I: accumulation of A_0- in A-side or B-side mutants of the axial ligand to chlorophyll A_0. *Biochemistry*, **43**, 1369–75.

66 Xu, W., Chitnis, P.R., Valieva, A., van der Est, A., Brettel, K., Guergova-Kuras, M., Pushkar, Y.N., Zech, S.G., Stehlik, D., Shen, G., Zybailov, B. and Golbeck, J.H. (2003) Electron transfer in cyanobacterial Photosystem I. II. Determination of forward electron transfer rates of site-directed mutants in a putative electron transfer pathway from A_0 through A_1 to F_X. *The Journal of Biological Chemistry*, **27**, 27876–87.

67 Jordan, P., Fromme, P., Witt, H.T., Klukas, O., Saenger, W. and Krauss, N. (2001) Three-dimensional structure of cyanobaoterial photosystem I at 2.5 Å resolution. *Nature*, **411**, 909–17.

68 Ben-Shem, A., Frolow, F. and Nelson, N. (2003) Crystal structure of plant photosystem I. *Nature*, **426**, 630–5.

69 Amunts, A., Drory, O. and Nelson, N. (2007) The structure of a plant photosystem I supercomplex at 3.4 Å resolution. *Nature*, **447**, 58–63.

70 Nelissen, B., van de Peer, Y., Wilmotte, A. and de Wachter, R. (1995) An early origin of plastids within the cyanobacterial divergence is suggested by evolutionary trees based on complete 16S rRNA sequences. *Molecular Biology and Evolution*, **12**, 1166–73.

71 Rippka, R., Waterbury, J. and Cohen-Bazire, G. (1974) A cyanobacterium which lacks thylakoids. *Archives of Microbiology*, **100**, 419–36.

72 Nakamura, Y., Kaneko, T., Sato, S., Mimuro, M., Miyashita, H., Tsuchiya, T., Sasamoto, S., Watanabe, A., Kawashima, K., Kishida, Y., Kiyokawa, C., Kohara, M., Matsumoto, M., Matsuno, A., Nakazaki, N., Shimpo, S., Takeuchi, C., Yamada, M. and Tabata, S. (2003) Complete genome structure of *Gloeobacter violaceus* PCC 7421, a cyanobacterium that lacks thylakoids. *DNA Research*, **10**, 137–45.

73 Inoue, H., Tsuchiya, T., Satoh, S., Miyashita, H., Kaneko, T., Tabata, S., Tanaka, A. and Mimuro, M. (2004) Unique constitution of photosystem I with a novel subunit in the cyanobacterium *Gloeobacter violaceus* PCC 7421. *FEBS Letters*, **578**, 275–9.

74 Boekema, E.J., Hifney, A., Yakushevska, A.E., Piotrowski, M., Keegstra, W., Berry, S., Michel, K.-P., Pistorius, E.K. and Kruip, J. (2001) A giant chlorophyll-protein complex induced by iron deficiency in cyanobacteria. *Nature*, **412**, 745–8.

75 Bibby, T.S., Nield, J. and Barber, J. (2001) Iron deficiency induces the formation of an antenna ring around trimeric photosystem I in cyanobacteria. *Nature*, **412**, 743–5.

76 Bibby, T.S., Mary, I., Nield, J., Partensky, F. and Barber, J. (2003) Low-light-adapted *Prochlorococcus* species possess specific antennae for each photosystem. *Nature*, **424**, 1051–4.

77 Allen, J.F. (2003) State transitions–a question of balance. *Science*, **299**, 1530–2.

78 Haldrup, A., Jensen, P.E., Lunde, C. and Scheller, H.V. (2001) Balance of power: a view of the mechanism of photosynthetic state transitions. *Trends in Plant Science*, **6**, 301–5.

79 Bonaventura, C. and Myers, J. (1969) Fluorescence and oxygen evolution from *Chlorella pyrenoidosa*. *Biochimica et Biophysica Acta*, **189**, 366–83.

80 Mullineaux, C.W. (1992) Excitation energy transfer from phycobilisomes to Photosystem I in a cyanobacterium. *Biochimica et Biophysica Acta*, **1100**, 285–92.

81 Murata, N. (1970) Control of excitation transfer in photosynthesis. IV. Kinetics of chlorophyll a fluorescence in *Porphyra yezoensis*. *Biochimica et Biophysica Acta*, **205**, 379–89.

82 Murata, N. (1969) Control of exitation transfer in photosynthesis. I. Light induced change of chlorophyll a fluorescence in *Porphyridium cruentum*. *Biochimica et Biophysica Acta*, **172**, 242–51.

83 Hippler, M., Reichert, J., Sutter, M., Zak, E., Altschmied, L., Schroer, U., Herrmann, R.G. and Haehnel, W. (1996) The plastocyanin binding domain of photosystem I. *The EMBO Journal*, **15**, 6374–84.

84 Hippler, M., Drepper, F., Haehnel, W. and Rochaix, J.D. (1998) The N-terminal domain of PsaF: Precise recognition site for binding and fast electron transfer from cytochrome c_6 and plastocyanin to photosystem I of *Chlamydomonas reinhardtii*. *Proceedings of the National Academy of Sciences of the United States of America*, **95**, 7339–44.

85 Kjaerulff, S., andersen, B., Nielsen, V.S., Moller, B.L. and Okkels, J.S. (1993) The PSIK subunit of photosystem I from barley (*Hordeum vulgare L.*). Evidence for a gene duplication of an ancestral PSI-G/K gene. *The Journal of Biological Chemistry*, **268**, 18912–16.

86 Zhang, L., Pakrasi, H.B. and Whitmarsh, J. (1994) Photoautotrophic growth of the cyanobacterium *Synechocystis* sp. PCC 6803 in the absence of cytochrome c_{553} and plastocyanin. *The Journal of Biological Chemistry*, **269**, 5036–42.

87 Wood, P.M. (1978) Interchangeable copper and iron proteins in algal photosynthesis. Studies on plastocyanin and cytochrome c-552 in Chlamydomonas. *European Journal of Biochemistry*, **87**, 9–19.

88 Merchant, S. and Bogorad, L. (1987) Metal ion regulated gene expression: use of a plastocyanin-less mutant of *Chlamydomonas reinhardtii* to study the Cu(II)-dependent expression of cytochrome c-552. *The EMBO Journal*, **6**, 2531–5.

89 Briggs, L.M., Pecoraro, V.L. and McIntosh, L. (1990) Copper-induced expression, cloning, and regulatory studies of the plastocyanin gene from the cyanobacterium *Synechocystis* sp. PCC 6803. *Plant Molecular Biology*, **15**, 633–42.

90 Bond, C.S., Blankenship, R.E., Freeman, H.C., Guss, J.M., Maher, M. J., Selvaraj, F.M., Wilce, M.C. and Willingham, K.M. (2001) Crystal structure of auracyanin, a "blue" copper protein from the green thermophilic photosynthetic bacterium Chloroflexus aurantiacus. *Journal of Molecular Biology*, **306**, 47–67.

91 van Driessche, G., Hu, W., van de Werken, G., Selvaraj, F., McManus, J. D., Blankenship, R.E. and van Beeumen, J.J. (1999) Auracyanin A from the thermophilic green gliding photosynthetic bacterium *Chloroflexus aurantiacus* represents an unusual class of small blue copper proteins. *Protein Science*, **8**, 947–57.

92 Aoki, M. and Katoh, S. (1982) Oxidation and reduction of plastoquinone by photosynthetic and respiratory electron transport in a cyanobacterium *Synechococcus* sp. *Biochimica et Biophysica Acta*, **682**, 307–14.

93 Allen, J.F. (2005) A redox switch hypothesis for the origin of two light reactions in photosynthesis. *FEBS Letters*, **579**, 963–8.

94 Grotjohann, I. and Fromme, P. (2005) Structure of cyanobacterial photosystem I. *Photosynthesis Research*, **85**, 51–71.

95 Raymond, J. and Blankenship, R.E. (2006) *Evolutionary Relationships among Type I Photosynthetic Reaction Centers in: Photosystem I: The Light-Driven Plastocyanin: Ferredoxin Oxidoreductase* (ed. J. Golbeck), Springer, The Netherlands, pp. 669–81.

96 Havaux, M., Guedeney, G., Hagemann, M., Yeremenko, N., Matthijs, H.C. and Jeanjean, R. (2005) The chlorophyll-binding protein IsiA is inducible by high light and protects cyanobacterium *Synechocystis* PCC 6803 from photooxidative stress. *FEBS Letters*, **579**, 2289–93.

97 Muehlendorf, U., Zhao, J. and Bryant, D.A. (1996) Interaction between Photosystem I and flavodoxin from the cyanobacterium *Synechococcus* sp. PCC 7002 as revealed by chemical cross-linking. *European Journal of Biochemistry*, **235**, 324–31.

98 Meimberg, K., Lagoutte, B., Bottin, H. and Muhlenhoff, U. (1998) The PsaE subunit is required for complex formation between Photosystem I and flavodoxin from the cyanobacterium *Synechocystis* sp. PCC 6803. *Biochemistry*, **37**, 9759–67.

99 andersen, B., Scheller, H.V. and Lindberg Moller, B. (1992) The PSI-E subunit of photosystem I binds ferredoxin: NADP+ oxidoreductase. *FEBS Letters*, **311**, 169–73.

100 van Thor, J.J., Geerlings, T.H., Matthijs, H.C.P. and Hellingwerf, K.J. (1999) Kinetic evidence for the PsaE-dependent transient ternary complex Photosystem I/ ferredoxin/ferredoxin:NADP$^+$ reductase in a cyanobacterium. *Biochemistry*, **38**, 12735–46.

101 Guergova-Kuras, M., Boudreaux, B., Joliot, A., Joliot, P. and Redding, K. (2001) Evidence for two active branches for electron transfer in photosystem I. *Proceedings of the National Academy of Sciences of the United States of America*, **98**, 4437–42.

102 Fairclough, W.V., Forsyth, A., Evans, M.C.W., Rigby, S.E.J., Purton, S. and Heathcote, P. (2003) Bidirectional electron transfer in photosystem I: electron transfer on the PsaA side is not essential for phototrophic growth in *Chlamydomonas*. *Biochimica et Biophysica Acta*, **1606**, 43–55.

103 Muhiuddin, I.P., Heathcote, P., Carter, S., Purton, S., Rigby, S.E.J. and Evans, M.C.W. (2001) Evidence from time resolved studies of the P700$^+$/A$_1$ radic,al

pair for photosynthetic electron transfer on both the PsaA and PsaB branches of the Photosystem I reaction centre. *FEBS Letters*, **503**, 56–60.

104 Xu, W., Chitnis, P.R., Valieva, A., van der Est, A., Pushkar, Y.N., Krzystyniak, M., Teutloff, C., Zech, S.G., Bittl, R., Stehlik, D., Zybailov, B., Shen, G. and Golbeck, J.H. (2003) Electron transfer in cyanobacterial photosystem I: I. Physiological and spectroscopic characterization of site-directed mutants in a putative electron transfer pathway from A_0 through A_1 to F_X. *The Journal of Biological Chemistry*, **278**, 27864–75.

105 Mimuro, M., Tsuchiya, T., Inoue, H., Sakuragi, Y., Itoh, Y., Gotoh, T., Miyashita, H. and Bryant, D.A. (2005) The secondary electron acceptor of photosystem I in *Gloeobacter violaceus* PCC 7421 is menaquinone-4 that is synthesized by a unique but unknown pathway. *FEBS Letters*, **579**, 3493–6.

106 Olson, J.M. (2001) "Evolution of Photosynthesis" (1970), re-examined thirty years later. *Photosynthesis Research*, **68**, 95–112.

107 Meyer, T.E. (1994) Evolution of photosynthetic reaction centers and light harvesting chlorophyll proteins. *BioSystems*, **33**, 167–75.

108 Xiong, J. and Bauer, C.E. (2002) A cytochrome b origin of photosynthetic reaction centers: an evolutionary link between respiration and photosynthesis. *Journal of Molecular Biology*, **322**, 503–21.

109 Doolittle, R.F. (1994) Convergent evolution: the need to be explicit. *Trends in Biochemical Sciences*, **19**, 15–18.

110 Jambon, M., Imberty, A., Deleage, G. and Geourjon, C. (2003) A new bioinformatic approach to detect common 3D sites in protein structures. *Proteins: Structure, Function, and Genetics*, **52**, 137–45.

111 Vermaas, W.F. (1994) Evolution of heliobacteria: implications for photosynthetic reaction center complexes. *Photosynthesis Research*, **41**, 285–94.

112 Nitschke, W., Mattioli, T. and Rutherford, A.W. (1996) The FeS-type photosystems and the evolution of photosynthetic reaction centers, in *Origin and Evolution of Biological Energy Conversion* (ed. H. Baltscheffsky), VCH Publishers, pp. 177–204.

113 Mulkidjanian, A.Y. and Junge, W. (1997) On the origin of photosynthesis as inferred from sequence analysis. A primordial UV-protector as common ancestor of reaction centers and antenna proteins. *Photosynthesis Research*, **51**, 27–42.

114 Schubert, W.D., Klukas, O., Saenger, W., Witt, H.T., Fromme, P. and Krauss, N. (1998) A common ancestor for oxygenic and anoxygenic photosynthetic systems: a comparison based on the structural model of photosystem I. *Journal of Molecular Biology*, **280**, 297–314.

115 Xiong, J., Inoue, K. and Bauer, C.E. (1998) Tracking molecular evolution of photosynthesis by characterization of a major photosynthesis gene cluster from *Heliobacillus mobilis. Proceedings of the National Academy of Sciences of the United States of America*, **95**, 14851–6.

116 Nitschke, W., Mühlenhoff, U. and Liebl, U. (1998) Evolution, in *Photosynthesis: A Comprehensive Treatise* (ed. A. Raghavendra), Cambridge University Press, pp. 285–304.

117 Blankenship, R.E. (2001) Molecular evidence for the evolution of photosynthesis. *Trends in Plant Science*, **6**, 4–6.

118 Mix, L.J., Haig, D. and Cavanaugh, C.M. (2005) Phylogenetic analyses of the core antenna domain: Investigating the origin of photosystem I. *Journal of Molecular Evolution*, **60**, 153–63.

119 Nelson, N. and Ben-Shem, A. (2005) The structure of photosystem I and evolution of photosynthesis. *BioEssays: News and Reviews in Molecular, Cellular and Developmental Biology*, **27**, 914–22.

120 Vasil'ev, S. and Bruce, D. (2004) Optimization and evolution of light harvesting in photosynthesis: The role of antenna chlorophyll conserved between photosystem II and photosystem I. *Plant Cell*, **16**, 3059–68.

121 Fyfe, P.K., Jones, M.R. and Heathcote, P. (2002) Insights into the evolution of the antenna domains of Type-I and Type-II photosynthetic reaction centres through homology modeling. *FEBS Letters*, **530**, 117–23.

14
The Structure of Purple Bacterial Antenna Complexes

Richard J. Cogdell, Alastair T. Gardiner, Mads Gabrielsen, June Southall, Aleksander W. Roszak, Neil W. Isaacs, Ritsuko Fujii, and Hideki Hashimoto

14.1
Introduction

Although a very large amount of solar energy reaches the surface of the earth each day, it represents a rather diffuse source of energy. In order to concentrate this solar energy before transducing it into useable chemical energy, photosynthetic organisms have evolved light-harvesting systems. This chapter describes the structure of the light-harvesting complexes from purple non-sulfur photosynthetic bacteria. These bacteria are anaerobic photosynthetic organisms, and, as a consequence, occupy ecological niches below chlorophyll containing oxygenic photosynthetic organisms. This means that the solar energy that reaches them has been pre-filtered by chlorophylls. In other words, only blue-green and red-light, wavelengths beyond 750 nm, are available to the purple bacteria [1]. Their photosynthesis is, therefore, based on pigments which can harvest blue-green or far-red light. They use carotenoids to absorbed blue-green light and bacteriochlorophylls to absorb the far-red light. These light-absorbing pigments are non-covalently attached to specific, integral membrane proteins to form their light-harvesting complexes. This chapter presents an overview of the current understanding of the structure of the purple bacterial antenna complexes.

The long wavelength (Q_y) absorption band of bacteriochlorophyll *a* (Bchl *a*) in 7 : 2 v/v solution of acetone and methanol is at 770 nm. When Bchl *a* is non-covalently bound into the antenna complexes, this long wavelength absorption band is red-shifted. The extent of the red-shift depends upon the association of the antenna complex and Bchl. This is illustrated in Figure 14.1 which shows the near-IR absorption spectrum of membranes from the purple photosynthetic bacterium *Rhodopseudomonas* (Rps.) *acidophila* and the absorption spectra of the two types of antenna complex (LH1 and LH2) that this species contains. The absorption spectrum of the whole membranes is a combination of the absorption spectra of LH2 and LH1. The LH1 complex has a single strong absorption band centered at ~880 nm. As described later, LH1 forms a well-defined complex with the reaction centers (RCs); together, they constitute the core of the purple bacterial

Photosynthetic Protein Complexes: A Structural Approach. Edited by P. Fromme

ISBN: 978-3-527-31730-1

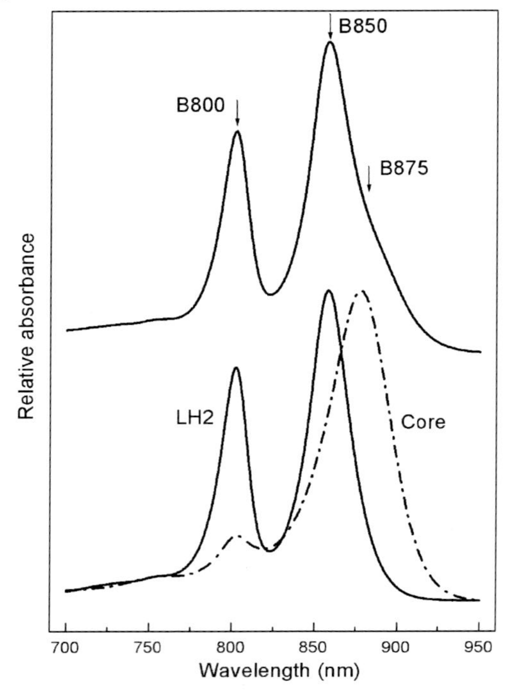

Figure 14.1 The absorption spectra of the light-harvesting apparatus from *Rps. acidophila* strain 10050. Top panel, membranes; bottom panel, isolated complexes.

photosynthetic unit [2]. All known species of purple bacteria have LH1 complexes. Most species, such as *Rps. acidophila*, also have additional peripheral antenna complexes such as LH2. The LH2 complex from *Rps. acidophila* has two strong near-IR absorption bands, one at ~800 nm and one at ~850 nm. These two absorption bands reflect the presence of two distinct groups of Bchl *a* molecules in LH2, as will be shown in Section 14.3 when the detailed structure of LH2 is described. The different absorption maxima of the groups of Bchl *a* in LH1 and LH2 are functionally very important. Since blue-light has higher energy than red-light, these different absorption bands set up an energy gradient within the photosynthetic unit. This energy gradient naturally acts as a funnel and directs energy from the periphery of the photosynthetic unit into the LH1 complexes and, therefore, onto the RCs [3] (see Chapter 12).

14.2
Knowledge of the Structural Arrangements before the Determination of the Crystal Structure

The purple bacterial antenna complexes are constructed on a modular principle [4]. The fundamental module appears to be a dimer of two low molecular weight, very hydrophobic apoproteins called α and β. These two apoproteins typically have about 50–60 amino acids and, therefore, are able to cross the photosynthetic membrane once as a single membrane spanning α-helix. The dimer of these apoproteins non-covalently bind 2–3 Bchl α's and 1–2 carotenoid molecules. These dimer units aggregate to form the intact antenna complexes. Before 3D structures of these antenna complexes were available, detailed information about their structure was obtained by a careful comparison of a large database of the primary sequences of the antenna apoproteins in combination with detailed spectroscopic analysis [2, 5, 6]. The apoproteins all have a tri-partite structure. Their N-termini are hydrophilic, a central region of about 20–24 amino acids is very hydrophobic, and their C-termini are hydrophilic. In the center of the hydrophobic region (the membrane spanning region), there is a conserved histidine (αHis31) residue that has been shown to be ligated to the central magnesium atom (in the middle of the bacteriochlorin ring of the Bchl *a* molecules that are responsible for the 850 nm absorption band in the LH2 complexes and the 880 nm absorption band in the LH1 complexes). Other conserved amino acid residues such as αTyr44 and αTyr45 were suggested to form hydrogen bonds with the bacteriochlorin rings of the 850 nm absorbing Bchl *a* molecules; the presence or absence of these amino acids correlates with where the Q_y absorption band of these Bchl *a* molecules is located [7–9]. Detailed spectroscopic studies had indicated that the B800 Bchl *a* molecules have a largely monomeric character, while the B850 and B880 Bchl *a* molecules represented excitonically coupled aggregates [6, 10]. This information set the scene prior to the determination of the crystal structures of the antenna complexes and was helpful in the interpretation of those structures.

14.3
The X-Ray Crystal Structure of the LH2 Complex from *Rps. acidophila* strain 10050

In 1995, the X-ray crystal structure of the LH2 complex from *Rps. acidophila* strain 10050 was first resolved at 2.5Å [11]. This has subsequently been improved to 2Å [12]. Figure 14.2 shows the overall structure of LH2. It is a circular nonamer composed of nine pairs of α,β-dimers. As predicted, before the crystal structure was determined, each apoprotein contains a single transmembrane spanning α-helix. Looking down on the top of the complex, in a direction that would be equivalent to looking down on the surface of the photosynthetic membrane, the outer wall of the complex is formed by the nine α-helices of the β-apoproteins. The α-helices from the nine α-apoproteins form the inner wall of the complex. All of the pigments are arranged between these rings of α-helices. The structure is capped on either side by the N- and C-termini folding over and interacting with each other. Within the transmembrane region of the complex, the α- and β-apoprotein helices do not interact with each other. The N-termini of both apoproteins are found at the cytoplasmic surface of the membrane and both the C-termini are located at the periplasmic surface.

Beginning at the cytoplasmic side, and proceeding downward into the structure, the first group of Bchl *a* molecules encountered are nine monomeric B800

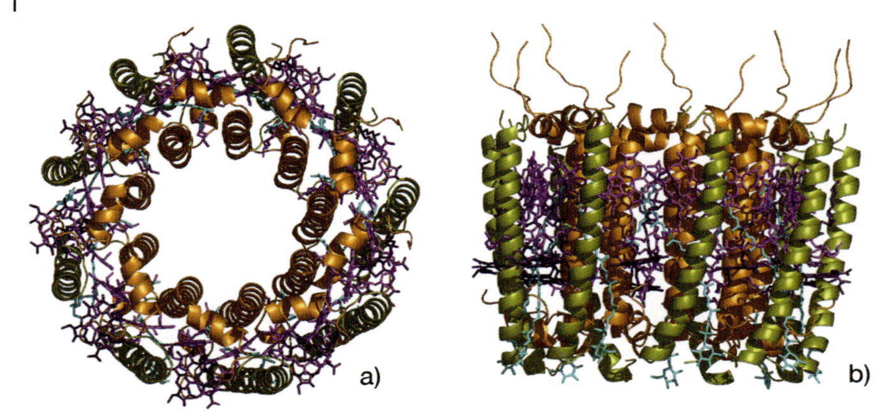

Figure 14.2 A top **(a)** and side **(b)** view of the overall structure of LH2 from *Rps. acidophila* strain 10050. The α-apoproteins are shown in brown, the β-apoproteins in green. The Bchls are purple, the carotenoids are turquoise.

molecules (Figure 14.3). Each α,β-dimer binds one B800 Bchl *a*. Their bacteriochlorin rings lie flat, parallel to the plane of the membrane and at right angles to the direction of the transmembrane α-helices. The central magnesium atom of these B800 Bchl *a* molecules is ligated to an extension of N-terminal methionine residue of the α-apoproteins. This extension appears to be a carboxyl group. This is a relatively unusual modification of a protein, but it is not unique ([12] and references therein). The B800 binding pocket is relatively hydrophilic. Each B800 Bchl *a* molecules has hydrogen bonds with β−Arg 20, β−His 12, and α−Gln 3. These B800 Bchl *a* molecules are separated center to center by ~21Å. The assignment of this group of Bchl *a* molecules to the B800 band was based on the fact that they are monomeric; and because it is possible to reversibly, and non-destructively, remove them−whereupon, the B800 absorption band disappears [13, 14].

Toward the middle of the complex, there is a second group of Bchl *a* molecules (Figure 14.3a and b). There are 18 Bchl *a* molecules in this group, 2 for each α,β-dimer. The bacteriochlorin rings of this group lie parallel to the transmembrane α-helices and, therefore, at right angles to the membrane plane. The central magnesium atoms of these Bchl *a* molecules are ligated to the apoproteins by the conserved histidine residues highlighted in Section 14.2. One of these histidine residues comes from the α-apoprotein (His31), while the other comes from the β-apoprotein (His30). The bacteriochlorin rings of these Bchl *a* molecules are in close proximity to each other, being separated center to center by 9.5Å within an α,β-dimer, and by 8.8Å to the nearest Bchl *a* molecule in the next dimer around the ring. Of course, spacing between the edges of the bacteriochlorin rings is much closer. This ring of strongly coupled Bchl *a* molecules is assigned to the 850 nm absorption band [3, 15].

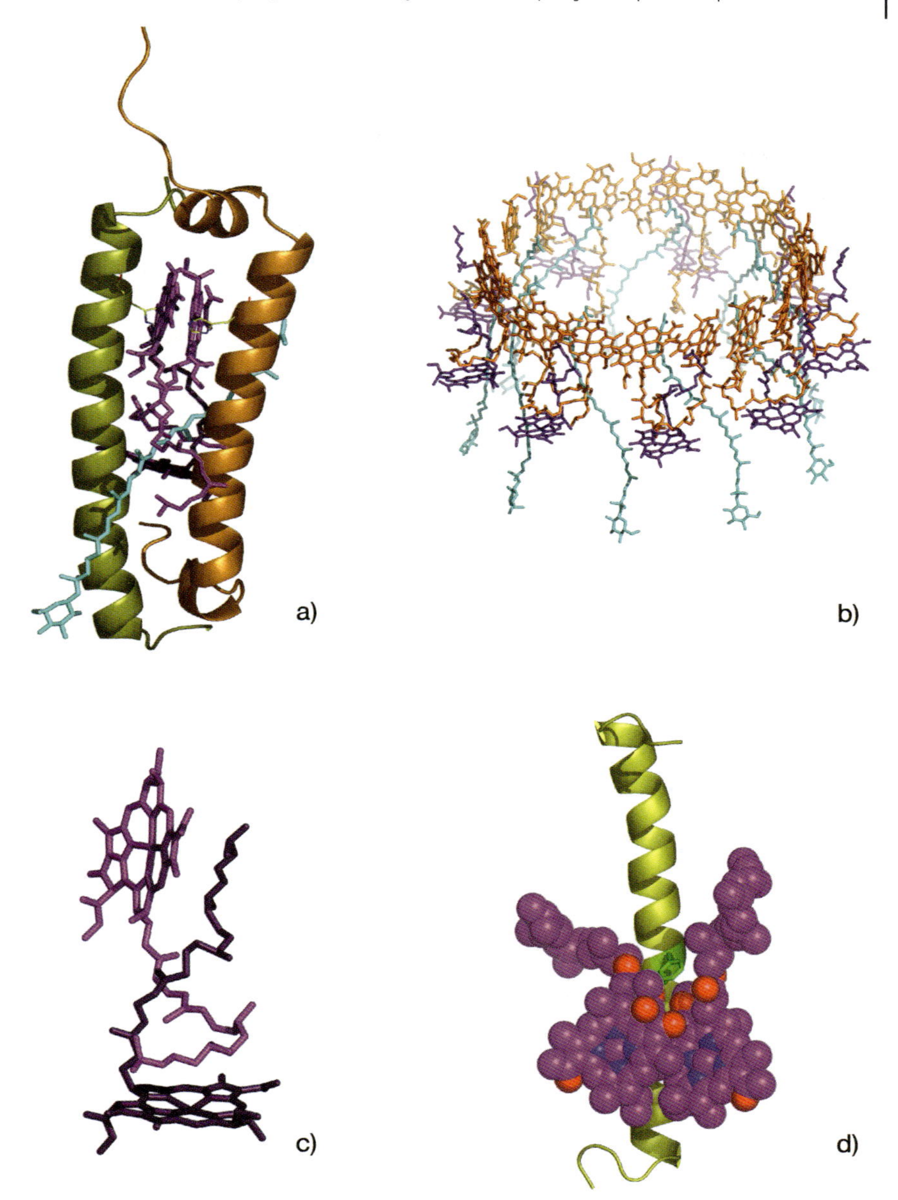

Figure 14.3 Detailed views of the arrangements of the pigments in the LH2 complex from *Rps. acidophila* strain 10050. **(a)** A single α/β-protomer, α-apoprotein coloured brown and the β-apoprotein coloured green. The Bchls are purple and the carotenoid is turquoise; **(b)** The arrangement of all the pigments in LH2 with the apoproteins removed for clarity. The B850 Bchls are brown, the B800 Bchls purple, and the carotenoids turquoise; **(c)** A view of how the phytol-chains of one B850 Bchl and the B800 Bchl intertwine; **(d)** A view of how the ester oxygen atoms of the two B850 Bchls wrap around β-Phe29.

Bchl *a* molecules contain both a bacteriochlorin ring and a phytol chain. The phytol chains are important structural elements [13]. This is illustrated in Figure 14.3c and d. The phytol chains from the B800 Bchl *a* molecules interact strongly with phytol chains from the β-bound B850 Bchl *a* molecules. This interaction is similar to two long fingers clasping each other. The B800 Bchl *a* molecules are peripherally bound; the strong interaction between these phytol chains is important for the correct positioning of the B800 molecules in the complex. There is another interesting interaction of the phytol chains with the protein. The oxygen from the ester-linkage between the phytol group of the B850 Bchl *a*'s and the ester on ring 5 of the bacteriochlorins of these Bchl *a*'s forms a ring around the aromatic side-chain of β-Phe 22. This is one of the few examples of amino acid side-chains that protrude into the central volume of the complex where the pigments are located. The interactions with these aromatic amino-acids appear to be the way that the protein is able to control the orientation of the bacteriochlorin rings. Once the protein has the Bchl *a* molecules by the tail, the orientation of the head group is also fixed. The orientation of the head group is important because the Q_x and Q_y transition dipole moments lie in a set direction within the molecular frame of the bacteriochlorin ring. The relative orientation of the Q_y transition dipole-moments between the Bchl *a* molecules is important because this is one of the parameters which affects the efficiency of energy-transfer. When these transition dipole-moments are parallel to each other, energy transfer is most favorable; when they are orthogonal to each other, energy transfer is least favourable. Interestingly, even though the bacteriochlorin rings of the B800 and B850 molecules are orthogonal to each other because of the direction of the Q_y transition dipole moment within the bacteriochlorin ring [13], the relative orientations of their respective Q_y transitions are rather parallel to each other, that is, favorable for B800 to B850 energy-transfer.

As well as binding three Bchl *a* molecules per dimer, each α,β-pair binds one carotenoid molecule, in this case, a rhodopin-glucoside [16]. Rhodopin-glucoside contains 11 conjugated double-bonds and a glucoside head group. The glucoside head group is located in a hydrophilic pocket on the cytoplasmic side of the complex. It appears that the glucoside group can adopt more than one conformation in this binding pocket because the electron density for this group is partially disordered. The conjugated portion of this carotenoid passes in close contact with the edge of the B800 bacteriochlorin ring (3.4Å). It then proceeds further through the complex passing into the next α,β-dimer and runs over the face of the α-bound B850 bacteriochlorin ring at a distance of 3.7Å. This carotenoid, therefore, adopts an elongated all-*trans* configuration, but is also slightly twisted along its long axis to form about half a turn of a helix. The carotenoid is an essential structural component of the LH2 complex. In the absence of this carotenoid the LH2 complex cannot be stably assembled [17]. Rhodopin-glucoside effectively acts to secure adjacent α,β-dimers together.

Soon after the crystal structure of LH2 from *Rps. acidophila* was described, the structure of another species, *Rhodospirillum* (*Rs.*) *molischianum*, was also determined [18]. Remarkably this complex turned out to be an octamer, not a nonamer.

The overall structure of LH2 from *Rs. molischianum* is very similar to that from *Rps. acidophila*. A major difference involves the B800 Bchl Molecules. In the case of LH2 from *Rs. molischianum*, the bacteriochlorin rings of the B800 Bchl *a*'s are rotated by 90° and tip into the plane of the membrane by about 20° relative to the *Rps. acidophila* structure. The central magnesium atoms of these B800 Bchl *a* molecules are ligated to the protein via α-Asp 6. It is interesting to note that, even though these two structures have been described for more than 10 years, and studied intensively during that period, a clear understanding is not generally available of what controls the ring size of LH2 complexes, or whether or not changes in ring size are functionally important. Very recently there has been a molecular dynamics study designed to try to understand what features in the structures of individual α,β-dimers affect the oligomerization state of LH2 [19]. The main conclusions of this study were that the angles between adjacent α,β-dimers (and therefore ring size) are mainly controlled by surface contacts within their transmembrane domains. It will be interesting to see whether these findings are confirmed by experimental results of mutagenesis studies in the future.

Some strains of *Rps. acidophila*, such as strain 7050, are able to synthesize additional types of LH2 when grown at low-light intensities [20]. These low-light complexes are the products of different α,β-apoprotein structural genes and have altered spectroscopic properties. In some of these complexes, the B850 band is blue-shifted to 820 nm. Both *Rps. acidophila* strains 10050 and 7050 contain multiple α,β-apoprotein gene pairs in their genomes [21]. Under high-light conditions, regarding strain 10050, only one gene-pair appears to be expressed; only one type of α and one type of β are found in LH2 under these conditions [16]. In strain 7050, under low-light conditions, the apoprotein composition of the B800–820 complex is not uniform. This may explain why the resolution of the crystals of this complex is lower than that of LH2 from strain 10050. The origin of this type of spectral shift had been investigated, prior to determination of the structure of LH2 from *Rps. acidophila*, by making a set of site-directed mutants of the homologous LH2 complex from *Rhodobacter sphaeroides* (Fowler and co-workers [7, 8]). This study [8] showed that when tyrosines αTyr44 and αTyr45 (corresponding to αTyr44 and αTrp45 in *Rps.* acidophila), were changed to non-hydrogen bonding amino acids, the 850 nm absorption band shifted to 820 nm. Moreover, this shift was correlated with a loss in hydrogen bonding to the B850 Bchl *a*'s, as monitored by resonance Raman spectroscopy. The crystal structure of the low light B800–820 complex, from strain 7050, has been determined [22]. In most regions, the structure of this complex can be exactly overlaid with that of LH2. Interestingly, the main changes are seen in the structure of the B850 bacteriochlorin rings. In LH2 residues, α-Trp 45 and α-Tyr 44 are hydrogen bonded to the carbonyl groups of the bacteriochlorin rings. As a result of these hydrogen bonds, the carbonyl groups are held in positions where they are able to extend the degree of conjugation of the bacteriochlorin rings. In the structure from strain 7050, these two residues are replaced by α-Leu 45 and α-Phe 44. These two residues cannot form hydrogen bonds and, as a consequence, the carbonyl groups are able to rotate, relative to the plane of the bacteriochlorin rings. Thereby, this rotation reduces the degree of conjugation and causes the blue-shift of the absorption band [23].

14.4 The Structure of RC–LH1 Core Complexes

Over the past few years there, there have been several detailed investigations into the structure of the RC–LH1 core complexes from a range of different species of purple bacteria. This is a unified view of their structure. Please note, however, that, because there is no high resolution structure of a purple bacterial core complex, this view may change in the future. It appears that there are two types of core complexes [3]. Some species, such as *Rps. viridis* [24], or *Rps. palustris* [25], have monomeric core complexes that consist of a central RC (see Chapter 12) surrounded by an LH1 complex. Other species, such as *Rb. sphaeroides*, have dimeric core complexes [26]. In this case, two RCs are each surrounded by their own LH1 complex and the pair of complexes somehow interact tightly to give the intact dimeric structure. Both of these types of core complex have been visualized in intact photosynthetic membranes using AFM and electron microscopy [26–31]. This is illustrated in Figure 14.4.

How were the current modes of the structure of the core complexes developed? In 1995, the first projection map of the LH1 complex from *Rs. rubrum* was described. Readers interested in information from studies carried out before this are referred to the following reviews [33, 34]. Karrasch et al. [32] made 2D crystals of reconstituted LH1 complexes from *Rs. rubrum* and used electron crystallography to determine a projection map of the complex at a resolution of 8.5Å. This projection map revealed a circular LH1 structure, composed of 16 α,β-dimers, together with 32 molecules of strongly interacting Bchl *a* that are responsible for the Q_y absorption band at 880 nm (and that are structurally equivalent to the B850 Bchl *a*'s seen in LH2). At this resolution, the structures of the LH1 α- and β-apoproteins are very similar to those from LH2; therefore, it was possible to use the structural information from an α,β-dimer, from LH2, to produce a somewhat high-resolution model of LH1. Furthermore, as pointed out by Karrasch et al. [32], with a 16-mer, the hole in the middle is now large enough to house the RC. Figure 14.5a illustrates such a high-resolution model incorporating an RC into the middle of the ring. This central location of the RC is strongly supported by a large body of EM data. For many years, this model was generally accepted, until information about a protein called PufX raised doubts about the validity of this simple model.

PufX is a protein that was initially found in *Rb. sphaeroides* and *Rb. capsulatus* [35, 36]. Mutants in which PufX was deleted failed to grow photosynthetically [37–42]. Photosynthetic growth could be re-established in these mutants if LH1 was also deleted [38]. The idea allows that PufX was required for growth because it allowed the exit of reduced ubiquinol through the LH1 ring. Electron transport in the RC produces reduced ubiquinol. This molecule has to deliver its reducing

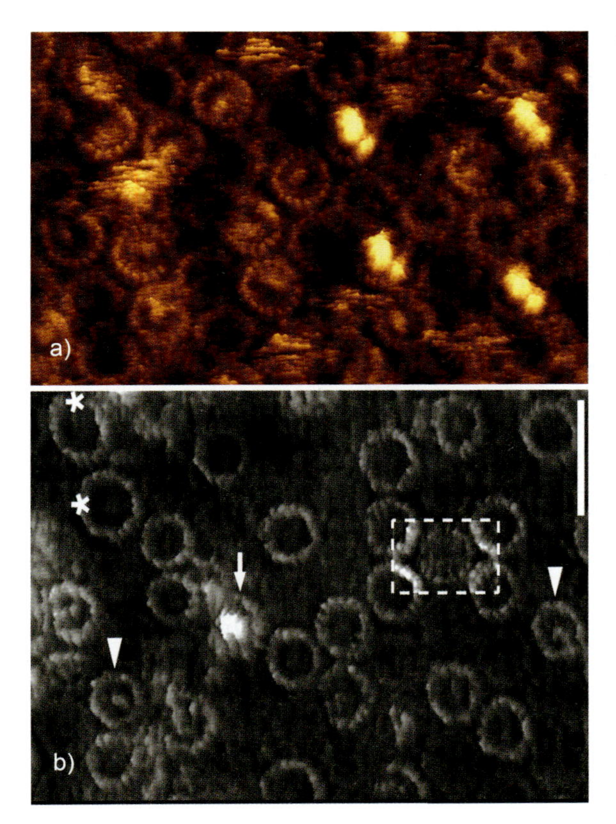

Figure 14.4 AFM images of RC–LH1 complexes. **(a)** *Rps. viridis* complexes imaged from native membranes [30]; **(b)** Image of the cytoplasmic side from reconstituted *Rb. sphaeroides* RC–LH1–Pufx 2D-crystals [28]. Spacebar is 25nm.

equivalents to the cytochrome bc_1 complex so that cyclic electron transport can proceed. The hypothesis, therefore, was that PufX would be located within LH1 ring and somehow provide a way out for ubiquinol [26, 43]. Interestingly, $PufX^-$ mutants of *Rb. sphaeroides* no longer produce dimeric core complexes [26]. It is not yet clear how the presence of PufX is involved in dimer formation. There are two models that have been proposed to explain this, based on what appears to be very similar raw data. These two models, however, are structurally very different and it remains to be seen which, if either, is correct.

In 2003, the crystal structure of the RC-LH1 complex from *Rps. palustris* was described at a resolution of 4.8Å [25]. It is important to make a few comments about what can, and what cannot, be deduced from such a low resolution crystal structure. At this resolution, the general structural features can be visualized, but the assignment of amino acid side chains is not possible. A model of this RC-LH1

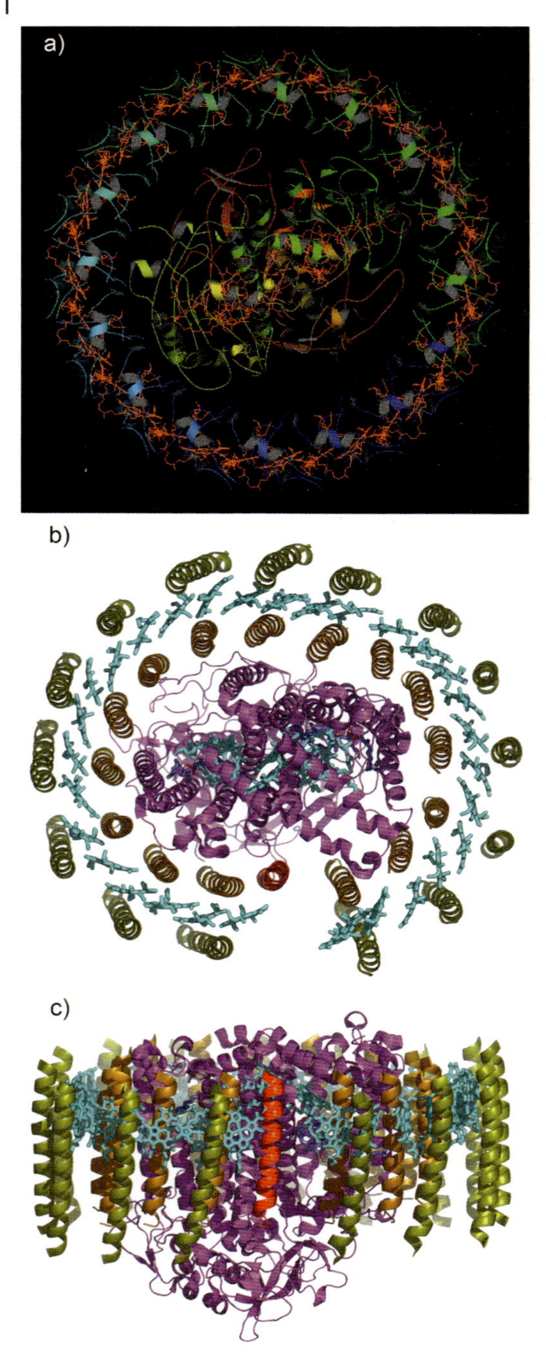

Figure 14.5 Structural models of the RC–LH1 complex. **(a)** A model based on the Karrash and coworkers [32] 2-D projection map of the reconstituted LH1 from *R. rubrum*. The RC has been placed in the center of the symmetrical LH1 ring that contains 16 α/β-apoprotein pairs. This model was kindly provided by Dr. Miroslav Papiz; **(b)** A section through the crystal structure of the RC-LH1 complex from *Rps. palustris*. The α-apoproteins are brown, the β-apoproteins green, the RC purple, and Protein W red; **(c)** A side view of the complete RC-LH1 complex from *Rps. palustris*, colour coding as in **(b)**.

complex is shown in Figure 14.5b and c. The top-view shows a section through the structure at the level of the B880 Bchl *a*'s. The side-view shows the organization of the transmembrane α-helices. The following details can be revealed from the structure: LH1 complex is elliptical; it consists of 15 α,β-dimers that surround the RC; and, where the sixteenth α,β-dimer would have been in the model of the LH1 complex from *Rs. rubrum,* there is a single transmembrane helix, which has been called protein-W. Additionally, it appears that W is analogous to PufX, and it structurally interrupts the LH1 ring. Interestingly, W is located exactly opposite the Q_B binding site in the RC from which the reduced ubiquinol molecule must leave in order to complete cyclic electron transport. Unfortunately, at this resolution, no more details are available. Does W form a gate? Can this gate be opened and shut? These detailed questions will only be answered when a high-resolution structure has been determined.

The idea that the RC-LH1 core complex is elliptical has been strongly supported by AFM studies on intact photosynthetic membranes from *Rps. viridis* [30]. This species only contains monomeric RC-LH1 core complexes. It does not have any LH2 complexes. Moreover, its core complexes are found in para-crystalline arrays within the membranes. The AFM pictures of these membranes show that the core complexes are elliptical. It should be emphasized that the core complexes in these membranes are fully native and have not been treated by detergents.

The presence of a gap in the LH1 ring in core complexes from *Rps. palustris* has been confirmed by detailed single molecule studies [44]. The fluorescence emission spectra of single core complexes from *Rps. palustris* and from a $PufX^-$ mutant from *Rb. sphaeroides* were compared. The core complexes from *Rps. palustris* showed a narrow, intense fluorescence emission line at the red-edge of their emission spectrum. This feature was absent from the emission spectra of the core complexes from the $PufX^-$ mutant. The details of the quantum mechanical explanation of this phenomenon are beyond the scope of this chapter; however, interested readers can consult [3] for a detailed discussion of this point. This narrow spectral feature can only be accounted for if one assumes the presence of a gap in the B880 ring of Bchl *a* molecules. Further detailed analysis of the single molecule data has revealed that this data can only be well accounted for by also assuming that the LH1 ring is elliptical. This single molecule data, therefore, provides strong evidence in support of the 4.8Å resolution crystal structure of the RC–LH1 core complex of *Rps. palustris.*

14.5
Energy Transfer

Purple bacterial antenna complexes have proved to be ideal pigment–protein complexes to study photosynthetic energy-transfer reactions (for a review, see [45]). This is not only because of the availability of high-resolution structures, but also because the groups of light-absorbing pigments are spectrally well resolved. It is possible, therefore, to selectively excite individual groups of pigments and then to be able to resolve the subsequent inter-pigment energy-transfer steps. Information on these different energy-transfer processes has followed the progress in the development of ultra fast laser technology. The energy-transfer reactions take place on the femtosecond to picosecond time-scale. Now, with the advent of laser systems with excitation pulses as short as a few femtoseconds, it is possible to time-resolve the individual energy-transfer events that occur between absorption of a photon by LH2 and the arrival of that energy at the RC. Figure 14.6 shows a model for the photosynthetic unit of a typical purple bacterium with the times of all of the Bchl *a* to Bchl *a* energy transfer steps indicated. When a Bchl *a* molecule absorbs a photon, the excited singlet state can last for 1–2 ns before the energy is lost and the Bchl *a* molecule returns to the ground state. The energy-transfer reactions must be faster than the relaxation to the ground state to be efficient. When a B800 molecule in LH2 is excited by a 200 femtoseconds excitation pulse it takes 0.9 picoseconds for that absorbed energy to be transferred to the B850 molecules. If the LH2 complexes are cooled to 4 K the rate of this energy transfer step only slows down to about 2.5 picoseconds. This illustrates that energy transfer within LH2 is essentially temperature independent. The Bchl molecules are arranged by the

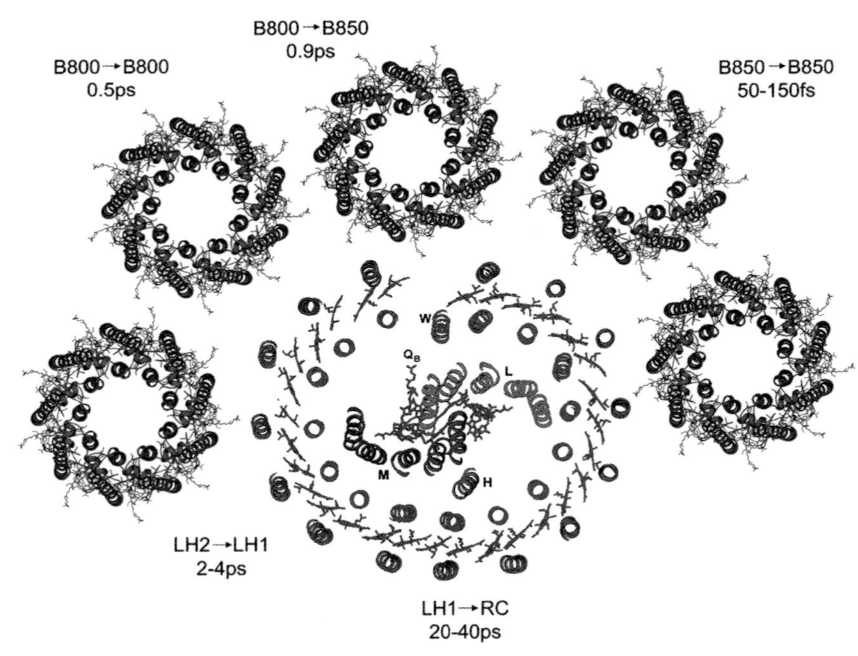

Figure 14.6 A schematic view of a generic purple photosynthetic bacterial photosynthetic unit. This picture shows five LH2 complexes, based on the structure from *Rps. acidophila* strain 10050, and one RC–LH1 complex, based on the *Rps. palustris* structure. The times of all of the Bchl to Bchl energy transfer steps are indicated. The data shown here is a composite of the currently available structural and time resolved energy transfer information (see [3] for full details).

apoproteins in such a way that the energy transfer process is independent of any protein motions and, therefore, the system is rather similar to a solid state device. Once the excitation energy reaches the B850 manifold it is initially delocalized over most, if not all, the B850 Bchls. The extent of this delocalization quickly reduces due to inhomogeneities in the system. In the photosynthetic membrane, if an RC–LH1 complex is close enough to the excited LH2 molecule, then the excitation energy will be transferred from LH2 to LH1. This step takes 2–5 picoseconds. Once the excitation energy arrives in the B875 (B890) ring it takes about 50 picoseconds for this excitated state to be transferred to the special pair of Bchl molecules in the RC. This final energy transfer step is the slowest because of the relatively large distance between the LH1 Bchls and the special pair in the RC. This distance is essential to prevent any possibility of electron transfer between LH1 and the RC. If even one Bchl molecule in LH1 is oxidized then the excited state lifetime of LH1 is dramatically reduced and LH1 will cease to function as a light harvester [46]. Since the time taken to transfer the absorbed light-energy from LH2 all the way to the RC is much faster than 1–2 nanoseconds, the purple bacterial photosynthetic unit operates at high quantum yields of greater than 95 [34].

Acknowledgments

Richard J. Cogdell thanks the Biotechnology and Biological Sciences Research Council for financial support. Hideki Hashimoto was funded by grants from the Japanese Ministry of Education, Culture, Sports, Science, and Technology (grants-17204026 and 17654083) and the Strategic International Cooperative program of Japan Science and Technology Corporation.

References

1 Pfennig, N. (1967) Photosynthetic bacteria. *Annual Review of Microbiology*, **21**, 285–324.

2 Zuber, H. and Cogdell, R.J. (1995) Structure and organization of purple bacterial antenna complexes, in *Anoxygenic Photosynthetic Bacteria* (eds R. E. Blankenship, M.T. Madigan and C.E. Bauer), Kluwer Academic Press, Dordrecht, pp. 315–48.

3 Cogdell, R.J., Gall, A. and Kohler, J. (2006) The architecture and function of the light-harvesting apparatus of purple bacteria: from single molecules to in vivo membranes. *Quarterly Reviews of Biophysics*, **39**, 227–324.

4 Law, C.J., Roszak, A.W., Southall, J., Gardiner, A.T., Isaacs, N.W. and Cogdell, R.J. (2004) The structure and function of bacterial light-harvesting complexes. *Molecular Membrane Biology*, **21**, 183–91.

5 van Grondelle, R., Dekker, J.P., Gillbro, T. and Sundström, V. (1994) Energy transfer and trapping in photosynthesis. *Biochimica et Biophysica Acta – Bioenergetics*, **1187**, 1–65.

6 Kramer, H.J.M., Van Grondelle, R., Hunter, C.N., Westerhuis, W.H.J. and Amesz, J. (1984) Pigment organization of the B800–850 antenna complex of Rhodopseudomonas sphaeroides.

Biochimica et Biophysica Acta – Bioenergetics, **765**, 156–65.

7 Fowler, G.J., Sockalingum, G.D., Robert, B. and Hunter, C.N. (1994) Blue shifts in bacteriochlorophyll absorbance correlate with changed hydrogen bonding patterns in light-harvesting 2 mutants of Rhodobacter sphaeroides with alterations at alpha-Tyr-44 and alpha-Tyr-45. *The Biochemical Journal*, **299** (*Pt 3*), 695–700.

8 Fowler, G.J., Visschers, R.W., Grief, G.G., van Grondelle, R. and Hunter, C. N. (1992) Genetically modified photosynthetic antenna complexes with blueshifted absorbance bands. *Nature*, **355**, 848–50.

9 Robert, B. and Lutz, M. (1985) Structures of antenna complexes of several rhodospirillales from their resonance raman-spectra. *Biochimica et Biophysica Acta*, **807**, 10–23.

10 Cogdell, R.J. and Scheer, H. (1985) Circular-dichroism of light-harvesting complexes from purple photosynthetic bacteria. *Photochemistry and Photobiology*, **42**, 669–78.

11 McDermott, G., Prince, S.M., Freer, A.A., Hawthornthwaite-Lawless, A.M., Papiz, M.Z., Cogdell, R.J. and Isaacs, N. W. (1995) Crystal structure of an integral membrane light-harvesting complex from photosynthetic bacteria. *Nature*, **374**, 517–21.

12 Papiz, M.Z., Prince, S.M., Howard, T., Cogdell, R.J. and Isaacs, N.W. (2003) The structure and thermal motion of the B800–850 LH2 complex from *Rps. acidophila* at 2.0 Å resolution and 100K: new structural features and functionally relevant motions. *Journal of Molecular Biolology*, **326**, 1523–38.

13 Freer, A., Prince, S., Sauer, K., Papiz, M., Hawthornthwaite-Lawless, A., McDermott, G., Cogdell, R. and Isaacs, N.W. (1996) Pigment-pigment interactions and energy transfer in the antenna complex of the photosynthetic bacterium Rhodopseudomonas acidophila. *Structure*, **4**, 449–62.

14 Fraser, N.J., Dominy, P.J., Ucker, B., Simonin, I., Scheer, H. and Cogdell, R.J. (1999) Selective release, removal, and reconstitution of bacteriochlorophyll a molecules into the B800 sites of LH2 complexes from Rhodopseudomonas acidophila 10050. *Biochemistry*, **38**, 9684–92.

15 Sauer, K., Cogdell, R.J., Prince, S.M., Freer, A., Isaacs, N.W. and Scheer, H. (1996) Structure-based calculations of the optical spectra of the LH2 bacteriochlorophyll-protein complex from Rhodopseudomonas acidophila. *Photochemistry and Photobiology*, **64**, 564–76.

16 Gardiner, A.T., Cogdell, R.J. and Takaichi, S. (1993) The effect of growth-conditions on the light-harvesting apparatus in Rhodopseudomonas-Acidophila. *Photosynthesis Research*, **38**, 159–67.

17 Lang, H.P. and Hunter, C.N. (1994) The relationship between carotenoid biosynthesis and the assembly of the light-harvesting LH2 complex in Rhodobacter sphaeroides. *Biochemical Journal*, **298**, 197–205.

18 Koepke, J., Hu, X., Muenke, C., Schulten, K. and Michel, H. (1996) The crystal structure of the light-harvesting complex II (B800–850) from Rhodospirillum molischianum. *Structure*, **4**, 581–97.

19 Janosi, L., Keer, H., Kosztin, I. and Ritz, T. (2006) Influence of subunit structure on the oligomerization state of light-harvesting complexes: a free energy calculation study. *Chemical Physics*, **323**, 117–28.

20 Heinemeyer, E.A. and Schmidt, K. (1983) Changes in carotenoid biosynthesis caused by variations of growth-conditions in cultures of Rhodopseudomonas-acidophila strain-7050. *Archives of Microbiology*, **134**, 217–21.

21 Gardiner, A.T., MacKenzie, R.C., Barrett, S.J., Kaiser, K. and Cogdell, R.J. (1996) The purple photosynthetic bacterium Rhodopseudomonas acidophila contains multiple puc peripheral antenna complex (LH2) genes: cloning and initial characterisation of four beta/alpha pairs. *Photosynthesis Research*, **49**, 223–35.

22 McLuskey, K., Prince, S.M., Cogdell, R.J. and Isaacs, N.W. (2001) The crystallographic structure of the B800–820 LH3 light- harvesting complex from the purple bacteria *Rhodopseudomonas acidophila* strain 7050. *Biochemistry*, **40**, 8783–9.

23 Cogdell, R.J., Howard, T.D., Isaacs, N.W., McLuskey, K. and Gardiner, A.T. (2002) Structural factors which control the position of the Q(y) absorption band of bacteriochlorophyll a in purple bacterial antenna complexes. *Photosynthesis Research*, **74**, 135–41.

24 Jay, F., Lambillotte, M., Stark, W. and Muehlethaler, K. (1984) The preparation and characterization of native photoreceptor units from the thylakoids of Rhodopseudomonas viridis. *EMBO Journal*, **3**, 773–6.

25 Roszak, A.W., Howard, T.D., Southall, J., Gardiner, A.T., Law, C.J., Isaacs, N.W. and Cogdell, R.J. (2003) Crystal structure of the RC-LH1 core complex from Rhodopseudomonas palustris. *Science*, **302**, 1969–72.

26 Siebert, C.A., Qian, P., Fotiadis, D., Engel, A., Hunter, C.N. and Bullough, P. A. (2004) Molecular architecture of photosynthetic membranes in *Rhodobacter sphaeroides*: the role of PufX. *EMBO Journal*, **23**, 690–700.

27 Bahatyrova, S., Frese, R.N., Siebert, C.A., Olsen, J.D., van der Werf, K.O., van Grondelle, R., Niederman, R.A., Bullough, P.A., Otto, C. and Hunter, C. N. (2004) The native architecture of a photosynthetic membrane. *Nature*, **430**, 1058–62.

28 Fotiadis, D., Qian, P., Philippsen, A., Bullough, P.A., Engel, A. and Hunter, C. N. (2004) Structural analysis of the reaction center light-harvesting complex I photosynthetic core complex of *Rhodospirillum rubrum* using atomic force microscopy. *Journal of Biological Chemistry*, **279**, 2063–8.

29 Scheuring, S., Rigaud, J.L. and Sturgis, J.N. (2004) Variable LH2 stoichiometry and core clustering in native membranes of Rhodospirillum photometricum. *EMBO Journal*, **23**, 4127–33.

30 Scheuring, S., Seguin, J., Marco, S., Levy, D., Robert, B. and Rigaud, J.L. (2003) Nanodissection and high-resolution imaging of the Rhodopseudomonas viridis photosynthetic core complex in native membranes by AFM. *Proceedings of the National Academy of Science of the United States of America*, **100**, 1690–3.

31 Stahlberg, H., Dubochet, J., Vogel, H. and Ghosh, R. (1998) Are the light-harvesting I complexes from *Rhodospirillum rubrum* arranged around the reaction center in a square geometry? *Journal of Molecular Biology*, **282**, 819–31.

32 Karrasch, S., Bullough, P.A. and Ghosh, R. (1995) The 8.5 Å. projection map of the light-harvesting complex I from *Rhodospirillum rubrum* reveals a ring composed of 16 subunits. *EMBO Journal*, **14**, 631–8.

33 Papiz, M.Z., Prince, S.M., HawthornthwaiteLawless, A.M., McDermott, G., Freer, A.A., Isaacs, N.W. and Cogdell, R.J. (1996) A model for the photosynthetic apparatus of purple bacteria. *Trends in Plant Science*, **1**, 198–206.

34 Hu, X.C., Ritz, T., Damjanovic, A., Autenrieth, F. and Schulten, K. (2002) Photosynthetic apparatus of purple bacteria. *Quarterly Reviews of Biophysics*, **35**, 1–62.

35 Donohue, T.J., Kiley, P.J. and Kaplan, S. (1988) The *puf* operon region of Rhodobacter sphaeroides. *Photosynthesis Research*, **19**, 39–61.

36 Klug, G. and Cohen, S.N. (1988) Pleiotropic effects of localized *Rhodobacter capsulatus puf* operon deletions on production of light-absorbing pigment-protein complexes. *Journal of Bacteriology*, **170**, 5814–21.

37 Barz, W.P., Francia, F., Venturoli, G., Melandri, B.A., Vermeglio, A. and Oesterhelt, D. (1995) Role of the PufX protein in photosynthetic growth of *Rhodobacter sphaeroides*. 1. PufX is required for efficient light-driven electron transfer and photophosphorylation under anaerobic conditions. *Biochemistry*, **34**, 15235–47.

38 Barz, W.P., Vermeglio, A., Francia, F., Venturoli, G., Melandri, B.A. and Oesterhelt, D. (1995) Role of the PufX protein in photosynthetic growth of *Rhodobacter sphaeroides*. 2. PufX is required for efficient ubiquinone/ubiquinol exchange between the reaction center QB site and the cytochrome bc1 complex. *Biochemistry*, **34**, 15248–58.

39 Farchaus, J.W. and Oesterhelt, D. (1989) A *Rhodobacter sphaeroides pufL, M* and *X* deletion mutant and its complementation

in *trans* with a 5.3 kb puf operon shuttle fragment. *EMBO Journal*, **8**, 47–54.
40 Lilburn, T.G. and Beatty, J.T. (1992) Suppressor mutants of the photosynthetically incompetent *pufX* deletion mutant *Rhodobacter capsulatus* DRC6 (pTL2). *FEMS Microbiological Letters*, **100**, 155–60.
41 Lilburn, T.G., Haith, C.E., Prince, R.C. and Beatty, J.T. (1992) Pleiotropic effects of *pufX* gene deletion on the structure and function of the photosynthetic apparatus of Rhodobacter capsulatus. *Biochimica et Biophysica Acta – Bioenergetics*, **1100**, 160–70.
42 Recchia, P.A., Davis, C.M., Lilburn, T.G., Beatty, J.T., Parkes-Loach, P.S., Hunter, C.N. and Loach, P.A. (1998) Isolation of the PufX protein from *Rhodobacter capsulatus* and *Rhodobacter sphaeroides*: evidence for its interaction with the α-polypeptide of the core light-harvesting complex. *Biochemistry*, **37**, 11055–63.
43 Cogdell, R.J., Fyfe, P.K., Barrett, S.J., Prince, S.M., Freer, A.A., Isaacs, N.W., McGlynn, P. and Hunter, C.N. (1996) The purple bacterial photosynthetic unit. *Photosynthesis Research*, **48**, 55–63.
44 Richter, M.F., Baier, J., Prem, T., Oellerich, S., Francia, F., Venturoli, G., Oesterhelt, D., Southall, J., Cogdell, R.J. and Kohler, J. (2007) Symmetry matters for the electronic structure of core complexes from Rhodopseudomonas palustris and Rhodobacter sphaeroides PufX. *Proceedings of the National Academy of Sciences of the United States of America*, **104**, 6661–5.
45 Sundström, V., Pullerits, T. and van Grondelle, R. (1999) Photosynthetic light-harvesting: reconciling dynamics and structure of purple bacterial LH2 reveals function of photosynthetic unit. *Journal of Physical Chemistry B*, **103**, 2327–46.
46 Law, C.J. and Cogdell, R.J. (1998) The effect of chemical oxidation on the fluorescence of the LH1 (B880) complex from the purple bacterium Rhodobium marinum. *FEBS Letters*, **432**, 27–30.

15
Ferredoxin and Flavodoxin Mediated Electron Transfer in Photosystem I

Raimund Fromme

15.1
Introduction

In photosynthetic organisms, light is captured by a variety of antenna systems (see Chapter 1) and finally brought to the reaction centers, where charge separation takes place. Once this occurs, the electrons move downhill along the reaction coordinate to find the final electron acceptor and convert light energy into chemical energy. In the overall reaction of linear electron transport, the reduction of $NADP^+$ to NADPH is the final step. In this process, ferredoxin has the role of accepting the electron from the membrane-bound reaction center, Photosystem I, and bringing it into the mobile aqueous phase to reduce the ferredoxin-$NADP^+$ reductase (FNR). The schematic position of the ferredoxin between the Photosystem I and ferredoxin reductase is represented in Chapter 1 Figure 1.1.

In this chapter, docking of ferredoxin to Photosystem I and accepting of the electron is described to the depth of current knowledge. The next step in the process of linear electron transport is the release of ferredoxin from Photosystem I and docking to ferredoxin reductase, and will also be discussed. Under conditions of iron stress, in several organisms ferredoxin can be replaced by flavodoxin–therefore, the current structures and models for this mechanism are briefly explained. The state of cyclic electron transport, along with the consequences this process presents for ferredoxin and the binding to Photosystem I is the final topic of this chapter. Here, we mainly focus on the structure of the proteins involved in this important part of photosynthesis. Their kinetic properties are described in excellent reviews elsewhere like Setif [1] in Golbeck, ed. (2006).

Photosynthetic Protein Complexes: A Structural Approach. Edited by P. Fromme

ISBN: 978-3-527-31730-1

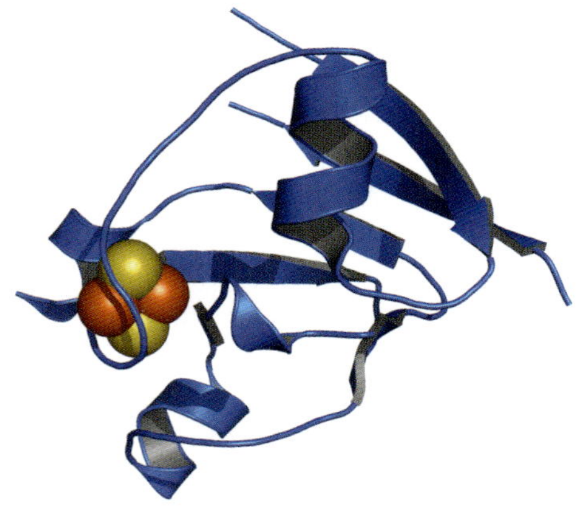

Figure 15.1 Crystal structure of ferredoxin from *Mastigocladus laminosus* (PDB accession code 1RFK) at 1.25 Å. The secondary structure of the backbone is shown in blue. The structure represented is according to the model of Fish and coworkers [2].

15.2 The Structure of Ferredoxin in Plants, Red Algae and Cyanobacteria

Ferredoxins are soluble proteins characterized by the iron-sulfur cluster present in their interior. In this chapter, only those containing the [2Fe-2S] clusters, or plant type, ferredoxins are described. [3Fe-4S] and [4Fe-4S] cluster binding ferredoxins, having completely different functions, are present in mammalian cells, and will not be discussed here.

As a typical cyanobacterial ferredoxin (Fd), the 3-D structure of *Mastigocladus laminosus* is depicted in Figure 15.1. This thermophilic cyanobacterium has an optimal ferredoxin activity at 65 °C. *Mastigocladus laminosus* ferredoxin has 98 residues that sum to a molecular weight of 8500 Da. This Fd has four alpha helices and seven beta strands as characteristic of its secondary structure. The [2Fe-2S] cluster is contained in a solvent-exposed loop region and is ligated by four highly conserved cysteines. The midpoint redox potential is at –420 mV – one of strongest in nature – and necessary for using the provided electrons from Photosystem I in order to reduce $NADP^+$ to NADPH.

15.3 Ferredoxin in Different Organisms

Ferredoxin is a well-characterized protein with a wide number of high resolution structures in different organisms. Table 15.1 summarizes the most important structures of plant type ferredoxin and flavodoxin (as the replacing protein), as

Table 15.1 Fe-2S ferredoxin structures in the PDB Data Bank.

Species	PDB Code	Resolution, in Å	Method
Cyanobacteria			
Synechocystis PCC6803	1OFF	1.8	X-ray
	1DOX		NMR
	1DOY		
Binary and ternary complexes with	2PVG	2.4	X-ray
thioredoxin and thioredoxin reductase	2PVO	3.4	
Spirulina platensis	4FXC	2.5	X-ray
Aphanothece sacrum	1FXI	2.2	X-ray
Anabaena PCC7120	1FXA	2.5	X-ray
Anabaena PCC7119	1CZP	1.17	X-ray
Mastigocladus laminosus [2]	1RFK	1.25	X-ray
Thermosynechococcus elongatus			NMR
Green algae			
Chlorella fusca	1AWD	1.4	X-ray
Higher plants			
Parsley	1PFD		NMR
Spinach (*Spinacia oleracea*)	1A70	1.7	X-ray
Zea mays, complex with ferredoxin reductase [3]	1GAW	2.59	X-ray
Equisetum arvense	1FRR	1.8	X-ray
Protozoa			
Plasmodium falciparum	1IUE	1.7	X-ray
Archaea and Halobacteria			
Halobacterium salinarium	1E10		NMR
	1E0Z		
Haloarcula marismortui	1DOI	1.9	X-ray
Flavodoxin			
Desulfovibrio vulgaris	1J8Q	1.35	X-ray
Anabaena [8]	1OBO	1.25	X-ray
Species	PDB Code	Resolution, in Å	Method
Photosystem I			
Cyanobacteria			
Thermosynechococcus elongatus [13]	1JB0	2.5	X-ray
Higher Plants			
Pea (*Pisum sativum*) [15]	1QZV	4.44	X-ray
Pea (*Pisum sativum*) [14]	2O01	3.4	X-ray

well as cocrystals of these with other proteins. Photosystem I structures are added for reference.

15.4
Electron Transfer between Ferredoxin and Ferredoxin NADP$^+$ Reductase

Ferredoxin NADP$^+$ reductases (FNR's) are enzymes that catalyze electron transfer between ferredoxin (flavodoxin) and NADP(H). These are proteins of approximately 300 residues, with a resulting molecular weight of about 35 kDa. FNRs are monomeric, with a non-covalently bound FAD (flavin adenine mononucleotide) as prosthetic group. They are known to be present in mitochondria, plastids, bacteria and cyanobacteria – as well as in the apioplasts of intracellular plastids. In forward electron transfer, FNR is part of the following reaction:

$$NADP^+ + 2Fd(Fe^{3+}) + H^+ \rightarrow 2Fd(Fe^{3+}) + NADPH$$

Several structures of FNR's are available from higher plants (maize, 1FNR, spinach, 2FNR and pea, 3FNR). A further two structures have been obtained from the cyanobacteria *Anabaena* and *Azobacter vinilandii*. There are also structures from *E. coli* and *Rhodobacter capsulatus*.

The topology is reminiscent of a sandwich formed from 5 beta sheets and surrounded by 5 alpha helices. The main structural differences are in the loop composition and the protein surface.

In the following, we will examine the structure of Fd–FNR complex from zea mays (PDB accession code 1gaq), depicted in Figure 15.2. In the illustration, the

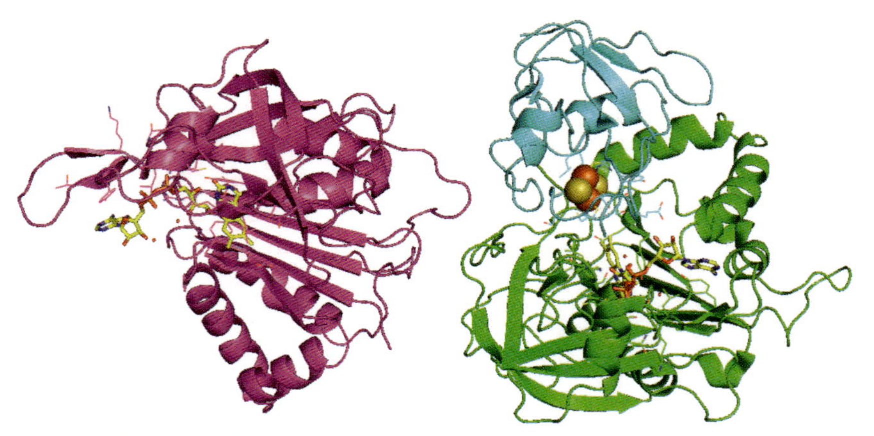

Figure 15.2 Structure of FNR (purple) and FNR-Ferredoxin complex (green and cyan). Both the FNR and ferredoxin proteins are from *Zea mays* (PDB accession code 1GAW, see Kurisu and coworkers [3]).

structure consists of FNR (in pink) and the FNR-Fd complex in green and cyan. In this structure, there is a ratio of two FNR to one Fd is in the asymmetric unit. This is interesting because it allows us to see what happens to the non-binding and binding FNR during the binding process of Fd. For unbiased comparison, the free FNR (PDB accession code 1gaw) is described in the same original paper from Kurisu and coworkers [3].

Interaction between Fd and FNR occurs via salt bridges, originating at the Fd residues Glu 29, Arg40, Asp 60, Asp65 and Asp 66. These residues are highly conserved over a wide range of species. Furthermore, site-directed mutagenesis studies point out the residues Asp26 and Glu30 as also contributing to salt-bridge formation. In addition, molecular modeling studies, combined with information gained from NMR experiments with the cyanobacterium *Synechocystis* PCC6803 have led to two models with new potential salt-bridge makers – Gly32 and Glu92. Another study, using surface calculations, also suggested Glu11 and Glu92 as new potential salt-bridge residues [4, 5].

15.5 Flavodoxin

An alternative enzyme to ferredoxin in accepting an electron from Photosystem I is flavodoxin, which also participates in a broad range of other electron transfer reactions. The prosthetic group in flavodoxin is a flavin mononucleotide (FMN). Interestingly, the use of flavodoxin is always preferred in iron limited environments [6, 7] – however, replacement of ferredoxin with flavodoxin has only been recognized in a small number of published papers. In contrast to FNR structures with ferredoxin, cocrystallization of flavodoxin with FNR has not been successful. Therefore, the best knowledge is about the structures of only flavodoxin and FNR alone. A homology model has been built for the purpose of visualization. The highest resolution structures of flavodoxin are at 1.2 to 1.35 Å, from *Anabaena* [8] and *Desulfuvibrio* [9]. Depicted in Figure 15.3 is the flavodoxin from *Anabaena* (PDB accession code 1OBO). In this cartoon representation, the 5 alpha helices are like poles in a picket fence in relationship to the 5 beta sheets in each monomer. Two monomers form the asymmetric unit. The newest kinetic studies, performed with *Anabaena* for flavodoxin mediated electron transfer to Photosystem I, are from Goñi and coworkers, 2008 [10].

15.6 Docking to Photosystem I

On the stromal side of Photosystem I are three subunits thought to be involved in docking ferredoxin to PSI – PsaC, PsaD and PsaE. We will briefly discuss these subunits and their participation in the interaction.

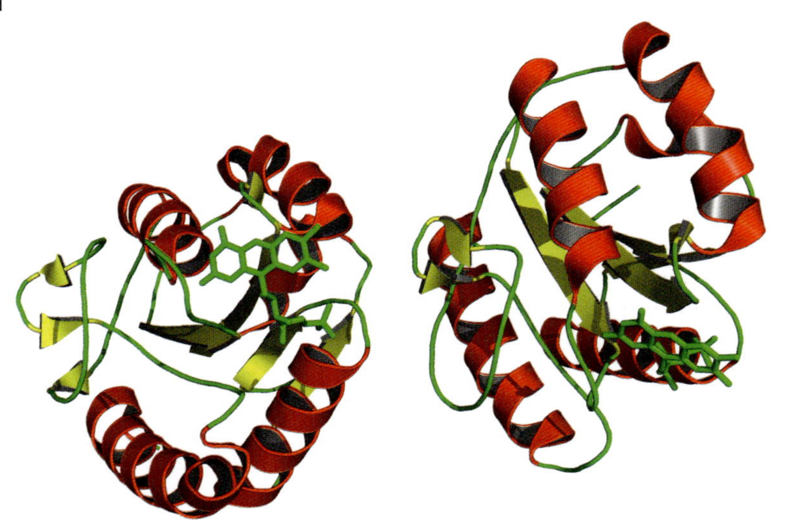

Figure 15.3 Flavodoxin from *Anabaena* (PDB accession code 1OBO see Lostao and coworkers[8]). The alpha helices are drawn in red, the beta strands in yellow and the loops and FMN (flavin mononucleotide) are in green.

15.6.1 Subunit PsaC

The terminal FeS clusters, F_A and F_B, of Photosystem I are contained by the 8.9 kDa Protein PsaC. The nature of each of the clusters is defined by distinct lines in the EPR spectra characteristic of [4Fe-4S]-clusters. From the protein sequence the binding region of the [4Fe-4S]-clusters is conserved in the motif CXXCXXCXXXCP. The structure was very well-predicted (Dunn and collegues and Golbeck) during homology modeling of the bacterial reaction centers [11, 12]–and was later confirmed in acquisition of the structures of cyanobacterial (Jordan [13]) and plant PSI (Ben-Shem [14] and Amunts [15]).

15.6.2 Subunit PsaD

Subunit D is located at the stromal hump, close to the "connecting domain". Several studies have confirmed that it is also essential for electron transfer from PSI to ferredoxin [16–20]. PsaD consists, in the main part, of a large, antiparallel, four-stranded β-sheet. The fourth strand is connected to the only α-helix of PsaD by a short loop. This helix forms interactions both with PsaC and with PsaA. A short antiparallel β-sheet after the helix is followed by a very prominent and remarkable feature of PsaD: its stromal clamp, consisting of the sequence region between D95 and D123. This part of PsaD wraps around PsaC, forming several contacts between PsaD and both PsaC and PsaE. The stromal clamp confirms the important function of PsaD as a critical stabilization factor of the electron acceptor sites in PSI, and its important role in keeping PsaC in its correct orientation [21, 22].

The C-terminal part of PsaD is in close vicinity to PsaB and PsaL. The close vicinity of PsaD and PsaL was already suggested from crosslinking and mutagenesis results [23, 24]. The C-terminal region of subunit PsaD is exposed on the stromal surface of Photosystem I. This fact was already suggested by previous work on protease accessibility and NHS-biotin labeling, as well as investigations of mutant Photosystem I, which lacks the 24 residues from its carboxyl terminus [23, 25].

In addition to its stabilizing function, PsaD is actively involved in the docking of ferredoxin. The direct interaction of these two proteins has been shown by chemical cross linking [26–28]. Lelong *and coworkers* [28] showed that the cross-linked complex is fully competent in the transfer of electrons from Photosystem I to ferredoxin. The position of ferredoxin in these cross-linked complexes was identified by electron microscopy [29]. The same docking site was also found for flavodoxin [30]. The negatively charged ferredoxin may be guided towards it binding site by the basic patch provided mainly by PsaD and PsaC. It is also very remarkable that overexpressed PsaD can still bind ferredoxin [31]. Cocrystals between PSI and ferredoxin have been reported, and may serve as a basis for a structure of the PSI–ferredoxin complex [32]. This is confirmed in the first cocrystal structure at 3.8 A resolution from Yu and coworkers, 2008 [33]. The arrangement of Photosystem I and ferredoxin is depicted in Figure 15.4. Ferredoxin is in Figure 15.4 depicted in magenta and has in this position contact to PsaD (yellow) and just one residue in a distance below 5 Å to PsaC (in salmon). In the cocrystal structure subunit PsaE has no close contact to ferredoxin. The distance between F_B, the terminal [4Fe-4S] cluster of Photosystem I, and ferredoxin is within 16.5 Å of either the docking or the transfer position. Therefore, a final move from the ferredoxin before the electron transfer is possible but seemed directed by PsaD and stromal loops of PsaA.

15.6.3 Subunit PsaE

The structure of subunit PsaE (8 kDa) in solution was determined by ^{1}H and ^{15}N NMR spectroscopy [34]. It shows a compact structure of five anti-parallel stranded ß-sheets. The core structure of PsaE, consisting of five α-strands is essentially the same in solution and in PsaE attached to the PSI complex. The main difference between the free and bound PsaE are the conformations of the loops and the C- and N-terminus. The flexible loop connecting the β-sheets β3 (C) and β4 (D) was not well resolved in the NMR structure and therefore seems to be flexible and involved in the interaction with the PsaA/PsaB core. This loop has a different conformation and is twisted when PsaE binds to the core complex. The twist of this loop has already been reported at 4 Å [35], and is fully confirmed in the

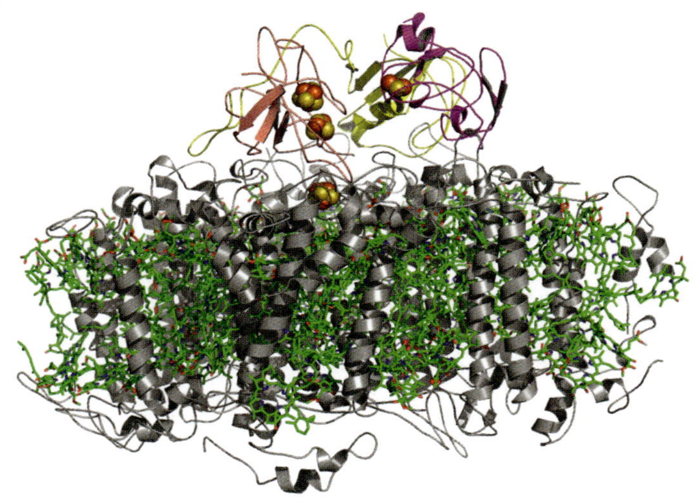

Figure 15.4 Structure of Photosystem I from *Thermosynechococcus elongatus* with bound ferredoxin. Ferredoxin is drawn in pink and PsaD in yellow. The orientation of the view is perpendicular to the membrane, on the top side is the stromal hump with the docked ferredoxin [33].

structural model at 2.5 Å resolution [13]. This loop is involved in interactions with PsaA, PsaB and PsaC, suggesting a change of the loop conformation during assembly of the Photosystem I complex. More recent studies show that PsaE can assemble into the PSI complex without help of assembly factors, and that it is driven by electrostatic interactions [36].

Different functions have been reported for PsaE. It may be directly involved in the anchoring of ferredoxin [37, 38], plays a role in cyclic electron transport [39], and can be cross-linked in barley with the ferredoxin-$NADP^+$-oxidoreductase (FNR) via its N-terminal extension [40].

The first evidence indicating that the location of subunit PsaE might be at the periphery of the stromal hump came from electron microscopy of a mutant lacking the gene of PsaE in cyanobacteria [41]. This was subsequently confirmed by the 4 Å and 2.5 Å structural models of Photosystem I [13, 25].

The interactions of PsaE with PsaC and PsaD are relatively weak, which explains the finding that the geometry of the stromal structure formed by PsaC and PsaD is not dramatically changed in the absence of PsaE [18]. However, the C-terminal region of PsaD, which forms a clamp surrounding PsaC, is in direct contact with PsaE (loop ß2/ß3), confirming previous crosslinking studies [42].

Subunit PsaE was predicted to be directly involved in anchoring the ferredoxin, [36, 38, 43] and, as mentioned above, also plays a role in cyclic electron transport [44] and can be cross linked in barley with FNR via its N-terminal extension [40]. Site-directed mutagenesis on subunit E in *Synechocystis* shows that the eight

C-terminal amino acids are necessary for the precise anchorage of subunit C into Photosystem I. Furthermore, several mutants on PsaE have a dramatic effect on the binding of ferredoxin (for a detailed discussion on the function of subunit E see [19, 20]. The fact that PsaE is involved in the docking of ferredoxin and flavodoxin [30, 45] has been questioned by the finding that PsaE deletion mutants are still able to grow photoautotrophically. This contradiction has been, nevertheless, solved by the discovery that PsaE deletion mutants increased the level of ferredoxin in the cells by orders of magnitude to compensate defects caused by the lack of PsaE [46]. However, the new crystal structure of Photosystem I and ferredoxin did not indicate any close contacts found between subunit E and ferredoxin [33], but shows that the ferredoxin docking site is formed by subunits PsaC, PsaD and PsaA.

Acknowledgments

I like to thank many colleagues from Arizona State University for the pleasant atmosphere and fruitful discussions. I am grateful for reading and remarks given by Yana Bukhman-deRuyter to improve the quality of this manuscript. Last but not least I want thank Petra Fromme for the urgent opportunity to write this chapter.

Funding for our research by NIH R01 GM71619–71601 and NSF MCB 04127142 is gratefully acknowledged.

References

1 Setif, P. (2006) *Photosystem I: The Light-Driven Plastocyanin: Ferredoxin Oxidoreductase* (ed. H. Golbeck) Springer, Dordrecht, pp. 439–54.

2 Fish, A., Lebendiker, M., Nechushtai, R. and Livnah, O. (2005) Structural basis for the thermostability of ferredoxin from the cyanobacterium *Mastigocladus laminosus*. *Journal of Molecular Biology*, **350** (*3*), 599–608.

3 Kurisu, G., Kusunoki, M., Katoh, E., Yamazaki, T., Teshima, K., Onda, Y., Kimata-Ariga, Y. and Hase, T. (2001) Structure of the electron transfer complexbetween ferredoxin and ferredoxin-NADP(+) reductase. *Nature Structural Biology*, **8**, 117–21.

4 Palma, P.N., Lagoutte, B., Krippahl, L. *et al.* (2005) Synechocystis ferredoxin / ferredoxin-NADP(+)-reductase / NADP(+) complex: structural model obtained by NMR-restrained docking. *FEBS Letters*, **579** (*21*), 4585–90.

5 Lelong, C., Setif, P., Bottin, H. *et al.* (1995) H-1 and N-15 NMR sequential assignment, secondary structure and tertiary fold of [2Fe-2S] ferredoxin grom Synechocystis sp. PCC-6803. *Biochemistry*, **34** (*44*), 144623–73.

6 Laroche, J., Boyd, P.W., McKay, R.M. and Geider, R.J. (1996) Flavodoxin as an *in situ* marker for iron stress in phytoplankton. *Nature*, **382**, 802–5.

7 Erdner, D.L. and Anderson, D.M. (1999) Ferredoxin and flavodoxin as biochemical indicators of iron limitation during open-ocean iron enrichment. *Limnology and Oceanography*, **44** (*7*), 1609–15.

8 Lostao, A., Daoudi, F., Irun, M.P., Ramon, A., Fernandez-Cabrera, C., Romero, A. and Sancho, J. (2003) How FMN binds to Anabaena apoflavodoxin: a hydrophobic encounter at an open binding site. *The*

Journal of Biological Chemistry, **278**, 24053–61.

9 Artali, R., Bombieri, G., Meneghetti, F., Gilardi, G., Sadeghi, S.J., Cavazzini, D. and Rossi, G.L. (2002) Comparison of the refined crystal structures of wild-type (1.34 A) flavodoxin from Desulfovibrio vulgaris and the S35C mutant (1.44 A) at 100 K. *Acta Crystallographica Section D*, **58**, 1787–92.

10 Goñi, G., Serrano, A., Frago, S., Hervás, M., Peregrina, J.R., De la Rosa, M.A., Gómez-Moreno, C., Navarro, J.A. and Medina, M. (2008) Flavodoxin-Mediated Electron Transfer from Photosystem I to Ferredoxin-NADP+ Reductase in Anabaena: Role of Flavodoxin Hydrophobic Residues in Protein-Protein Interactions. *Biochemistry*, **47**, 201–17.

11 Dunn, P.P., Packman, L.C., Pappin, D. and Gray, J.C. (1988) N-terminal amino acid sequence analysis of the subunits of pea photosystem I. *FEBS Letters*, **228**, 157–61.

12 Golbeck, J.H. (1993) The structure of Photosystem I. *Current Opinion in Structural Biology*, **3**, 508–14.

13 Jordan, P., Fromme, P., Klukas, O., Witt, H.T., Saenger, W. and Krauß, N. (2001) Three dimensional structure of cyanobacterial photosystem i at 2.5 Å resolution. *Nature*, **411**, 909–17.

14 Ben-Shem, A., Frolow, F. and Nelson, N. (2003) Crystal structure of plant photosystem I. *Nature*, **426**, 630–5.

15 Amunts, A., Drory, O. and Nelson, N. (2007) The structure of a plant photosystem I supercomplex at 3.4 angstrom resolution. *Nature*, **447** (*7140*), 58–63.

16 Chitnis, V.P., Jungs, Y.S., Albee, L., Golbeck, J.H. and Chitnis, P.R. (1996) Mutational analysis of photosystem I polypeptides. Role of PsaD and the lysyl 106 residue in the reductase activity of the photosystem I. *The Journal of Biological Chemistry*, **271**, 11772–80.

17 Chitnis, V.P., Ke, A. and Chitnis, P.R. (1997) The PsaD subunit of photosystem I. Mutations in the basic domain reduce the level of PsaD in the membranes. *Plant Physiology*, **115**, 1699–705.

18 Barth, P., Lagoutte, B. and Setif, P. (1998) Ferredoxin reduction by photosystem I from Synechocystis sp. PCC 6803: toward an understanding of the respective roles of subunits PsaD and PsaE in ferredoxin binding. *Biochemistry*, **37**, 16233–41.

19 Setif, P. (2001) Ferredoxin and flavodoxin reduction by photosystem I. *Biochimica et Biophysica Acta*, **1507**, 161–79.

20 Setif, P., Fischer, N., Lagoutte, B., Bottin, H. and Rochaix, J.D. (2002) The ferredoxin docking site of photosystem I. *Biochimica et Biophysica Acta*, **1555**, 204–9.

21 Li, N., Warren, P.V., Golbeck, J.H., Frank, G., Zuber, H. and Bryant, D.A. (1991) Polypeptide composition of the Photosystem I complex and the Photosystem I core protein from Synechococcus sp. PCC 6301. *Biochimica et Biophysica Acta*, **1059**, 215–25.

22 Lagoutte, B., Hanley, J. and Bottin, H. (2001) Multiple functions for the C terminus of the PsaD subunit in the cyanobacterial photosystem I complex. *Plant Physiology*, **126**, 307–16.

23 Xu, Q., Jung, Y.S., Chitnis, V.P., Guikema, J.A., Golbeck, J.H. and Chitnis, P.R. (1994) Mutational analysis of photosystem I polypeptides in Synechocystis sp. PCC 6803. Subunit requirements for reduction of $NADP^+$ mediated by ferredoxin and flavodoxin. *The Journal of Biological Chemistry*, **269**, 21512–18.

24 Janson, S., Andersen, B. and Scheller, H.V. (1996) Nearest-neighbor analysis of higher-plant photosystem I holocomplex. *Plant Physiology*, **112**, 409–20.

25 Chitinis, P.R., Xu, Q., Chitnis, V.P. and Nechustai, R. (1995) Function and organization of Photosystem I polypeptides. *Photosynthesis Research*, **44**, 23–40.

26 Zanetti, G. and Merati, G. (1987) Interaction between photosystem I and ferredoxin. Identification by chemical cross-linking of the polypeptide which binds ferredoxin. *European Journal of Biochemistry*, **169**, 143–6.

27 Zilber, M. and Malkin, R. (1988) Ferredoxin cross-links to a 22 kDa subunit of Photosystem I. *Plant Physiology*, **88**, 810–14.

28 Lelong, C., Setif, P., Lagoutte, B. and Bottin, H. (1994) Identification of the

amino acids involved in the functional interaction between photosystem I and ferredoxin from Synechocystis sp. PCC 6803 by chemical cross-linking. *The Journal of Biological Chemistry*, **269**, 10034–9.

29 Lelong, C., Boekema, E.J., Kruip, J., Bottin, H., Rögner, M. and Setif, P. (1996) Characterization of a redox active cross-linked complex between cyanobacterial photosystem I and soluble ferredoxin. *The EMBO Journal*, **15**, 2160–8.

30 Mühlenhoff, U., Kruip, J., Bryant, D.A., Rögner, M., Setif, P. and Boekema, E. (1996) Characterization of a redox-active cross-linked complex between cyanobacterial photosystem I and its physiological acceptor flavodoxin. *The EMBO Journal*, **15**, 488–97.

31 Pandini, V., Aliverti, A. and Zanetti, G. (1999) Interaction of the soluble recombinant PsaD subunit of spinach photosystem I with ferredoxin I. *Biochemistry*, **38**, 10707–13.

32 Fromme, P., Bottin, H., Krauss, N. and Setif, P. (2002) Crystallization and EPR characterization of a functional complex of Photosystem I with its natural electron acceptor ferredoxin. *Biophysical Journal*, **83**, 1760–73.

33 Yu, H., Fromme, R., Grotjohann, I., Wang, M., Setif, P., Bottin, H. and Fromme P. X-Ray Structure analysis from co-crystals of Photosystem I and its natural electron acceptor ferredoxin. (In preparation).

34 Falzone, C.J., Kao, Y.H., Zhao, J., Bryant, D.A. and Lecomte, J.T. (1994) Three-dimensional solution structure of PsaE from the cyanobacterium Synechococcus sp. strain PCC 7002, a photosystem I protein that shows structural homology with SH3 domains. *Biochemistry*, **33**, 6052–62.

35 Klukas, O., Schubert, W.D., Jordan, P., Krauss, N., Fromme, P., Witt, H.T. and Saenger, W. (1999) Photosystem I, an improved model of the stromal subunits PsaC, PsaD, and PsaE. *The Journal of Biological Chemistry*, **274**, 7351–60.

36 Lushy, A., Verchovsky, L. and Nechushtai, R. (2002) The stable assembly of newly synthesized PsaE into the photosystem I complex occurring via the exchange mechanism is facilitated by electrostatic interactions. *Biochemistry*, **41**, 11192–9.

37 Sonoike, K., Hatanaka, H. and Katoh, S. (1993) Small subunits of Photosystem I reaction center complexes from Synechococcus elongatus. II. The psaE gene product has a role to promote interaction between the terminal electron acceptor and ferredoxin. *Biochimica et Biophysica Acta*, **1141**, 52–7.

38 Weber, N. and Strotmann, H. (1993) On the function of subunit PsaE in chloroplast Photosystem I. *Biochimica et Biophysica Acta*, **1143**, 204–10.

39 Rousseau, F., Setif, P. and Lagoutte, B. (1993) Evidence for the involvement of PSI-E subunit in the reduction of ferredoxin by photosystem I. *The EMBO Journal*, **12**, 1755–65.

40 Zhao, J., Snyder, W.B., Mühlenhoff, U., Rhiel, E., Warren, P.V., Golbeck, J.H. and Bryant, D.A. (1993) Cloning and characterization of the psaE gene of the cyanobacterium Synechococcus sp. PCC 7002: characterization of a psaE mutant and overproduction of the protein in Escherichia coli. *Molecular Microbiology*, **9**, 183–94.

41 Andersen, B., Scheller, H.V. and Moller, B.L. (1992) The PSI-E subunit of photosystem I binds ferredoxin: NADP$^+$ oxidoreductase. *FEBS Letters*, **311**, 169–73.

42 Kruip, J., Chitnis, P.R., Lagoutte, B., Rögner, M. and Boekema, E.J. (1997) Structural organization of the major subunits in cyanobacterial photosystem 1. Localization of subunits PsaC, -D, -E, -F, and -J. *The Journal of Biological Chemistry*, **272**, 17061–9.

43 Mühlenhoff, U., Zhao, J. and Bryant, D.A. (1996) Interaction between photosystem I and flavodoxin from the cyanobacterium Synechococcus sp. PCC 7002 as revealed by chemical cross-linking. *European Journal of Biochemistry*, **235**, 324–31.

44 Strotmann, H. and Weber, N. (1993) On the function of PsaE in chloroplast Photosystem I. *Biochimica et Biophysica Acta*, **1143**, 204–10.

45 Yu, L., Zhao, J., Mühlenhoff, U., Bryant, D.A. and Golbeck, J.H. (1993) PsaE is required for in vivo cyclic electron flow around Photosystem I in the cyanobacterium Synechococcus sp. PCC 7002. *Plant Physiology*, **103**, 171–80.

46 Meimberg, K., Lagoutte, B., Bottin, H. and Mühlenhoff, U. (1998) The PsaE subunit is required for complex formation between photosystem I and flavodoxin from the cyanobacterium Synechocystis sp. PCC 6803. *Biochemistry*, **37**, 9759–67.

Index

Photosynthetic Protein Complexes: A Structural Approach. Edited by P. Fromme

ISBN: 978-3-527-31730-1

d

e

q

r